AF616095

OXFORD
POLYTECHNIC
LIBRARY

00 337144 01

MODERN POWER STATION PRACTICE

IN 8 VOLUMES

VOLUME 5

CHEMISTRY AND METALLURGY

EDITORIAL PANEL

CONTRIBUTORS TO VOLUME 5

J. Anderson, A.R.I.C., A.B.I.M., M.Inst.F.
L. A. Didlick, A.R.I.C., M.Inst.F.
M. G. Gemmill, B.Sc., A.I.M.
W. R. Godfrey, A.R.I.C.
L. A. Huntington, A.M.C.T., A.R.I.C., M.Inst.F.
A. J. W. Jackson.
C. Kennedy, A.M.C.T., A.R.I.C., M.Inst.F., A.F.Inst.Pet.
N. S. Musgrave, A.R.I.C.
D. H. Turner, A.Met., A.I.M., A.M.Inst.W.
R. S. Varley, B.Sc. (Lond.), A.R.I.C.

MODERN POWER STATION PRACTICE

SECOND REVISED AND ENLARGED EDITION

VOLUME 5

CHEMISTRY AND METALLURGY

Published for and on behalf of the
CENTRAL ELECTRICITY GENERATING BOARD
BY

PERGAMON PRESS
OXFORD · NEW YORK · TORONTO
SYDNEY · BRAUNSCHWEIG

Pergamon Press Ltd., Headington Hill Hall, Oxford
Pergamon Press Inc., Maxwell House, Fairview Park, Elmsford, New York 10523
Pergamon of Canada Ltd., 207 Queen's Quay West, Toronto 1
Pergamon Press (Aust.) Pty. Ltd., 19a Boundary Street,
Rushcutters Bay, N.S.W. 2011, Australia
Vieweg & Sohn GmbH, Burgplatz 1, Braunschweig

First published in five volumes 1963/64
Second revised and enlarged edition 1971
Library of Congress Catalog Card No. 75-86200

Printed in Hungary

08 015568 5
08 016436 6 (set of 8 volumes)

CONTENTS

List of Illustrations xv

Chapter 1. Fuel and Oil 1

1.1. Introduction to Power Station Chemistry 1
- 1.1.1. Historical 1
- 1.1.2. The Work of the Power Station Chemist 1

1.2. Coal 3
- 1.2.1. Introduction 3
- 1.2.2. British Coal Resources 3
- 1.2.3. British Coal Production and Utilisation in C.E.G.B. Power Stations 3

1.3. Classification of Coal 5
- 1.3.1. Dr. Seyler's Classification 5
- 1.3.2. Dr. Seyler's Fuel Charts 7
- 1.3.3. The N.C.B. Coal Rank Code Number Classification 7
- 1.3.4. Coal Rank Code Numbers in Power Station Practice 10

1.4. Classification of Coal by Size 16

1.5. Preparation of Coal 17
- 1.5.1. Introduction 17
- 1.5.2. Coal Preparation for Power Station Use 17
- 1.5.3. Coal Cleaning 18
- 1.5.4. Wet Processes 19
- 1.5.5. Dry Cleaning 24
- 1.5.6. Partial Preparation of Coal for Modern Power Station Use 24

1.6. Coal—the Price Structures 26
- 1.6.1. Industrial Coal Price Structure 26
- 1.6.2. Carbonisation Coal Price Structure 28
- 1.6.3. The Slurry Price Schedule 28

1.7. Transport of Coal to Power Stations 28
- 1.7.1. Rail 28
- 1.7.2. Coastwise Shipment 29
- 1.7.3. Road 29
- 1.7.4. Other Methods 29

1.8. Coal Storage 29

1.9. Coal Sampling 30
- 1.9.1. Purpose of Sampling 30
- 1.9.2. Theory of Sampling 31
  - 1.9.2.1. Unbiased Increments 32
  - 1.9.2.2. Weight of Increments 32
  - 1.9.2.3. Number of Increments 32
  - 1.9.2.4. Consistency 33
  - 1.9.2.5. Sampling Behaviour 33
- 1.9.3. Replicate Sampling 33
- 1.9.4. Sampling Processes 34

1.9.4.1. Automatic Samplers 35
1.9.4.2. Hand Sampling 35
1.9.5. Sampling for Moisture 39
1.9.5.1. Moisture Sample taken Separately 40
1.9.5.2. Moisture Sample collected with General Sample 40
1.9.5.3. Moisture Sample Preparation 40
1.9.6. Storage of Samples 40

1.10. Sample Preparation 41
1.10.1. Introduction 41
1.10.2. Errors of Sample Preparation 41
1.10.3. Equipment for Sample Preparation 42
1.10.3.1. Crushers, Grinders, etc. 42
1.10.3.2. Sample Dividers 43
1.10.3.3. Sample Mixing 43
1.10.3.4. Final Stages of Sample Preparation 43

1.11. Analysis and Testing of Coal and Coke 47
1.11.1. Introduction 47
1.11.2. Proximate Analysis 48
1.11.3. Moisture Content 48
1.11.3.1. Free Moisture 49
1.11.3.2. Inherent Moisture 49
1.11.3.3. Air-dry Moisture 50
1.11.4. Ash 50
1.11.5. Volatile Matter 51
1.11.6. Calorific Value 51
1.11.6.1. Introduction 51
1.11.6.2. Determination of Calorific Value 52
1.11.6.3. Adiabatic Bomb Calorimeter 56
1.11.7. Sulphur 58
1.11.8. Chlorine 59
1.11.9. The Ultimate Analysis of Coal 59
1.11.9.1. Carbon and Hydrogen 60
1.11.9.2. Nitrogen 60
1.11.10. Phosphorus 61
1.11.11. Arsenic 61
1.11.12. Swelling Index 61
1.11.13. Gray–King Coking Test 62
1.11.14. Carbon Present as Carbonates 62
1.11.15. Reporting of Results 63
1.11.16. Instrumental Analysis 66
1.11.16.1. Mineral Matter by Gamma-rays 66
1.11.16.2. Moisture 68

1.12. Other Tests on Coal 69
1.12.1. Bulk Density 69
1.12.2. Grindability 69
1.12.3. Handleability 71

1.13. Size Analysis 72
1.13.1. Size Analysis of Coal 72
1.13.2. Sampling and Size Analysis of Pulverised Fuel 73

1.14. Ash from Coal 74
1.14.1. Boiler Fouling 74
1.14.2. Analysis of Gas-side Deposits 76
1.14.3. Analysis of Ash 78
1.14.4. Ash Fusion Temperature 78

1.15. Gas-side Cleaning of Boilers 79
1.15.1. On-load Cleaning 79

1.15.2. Off-load Cleaning 80

1.16. Commercial Utilisation of Ash 80
1.16.1. Characteristics of P.F.A. 81
1.16.2. Commercial Utilisation of P.F.A. 81

1.17. Oil Fuels 83
1.17.1. Introduction 83
1.17.2. World Oil Supplies 83
1.17.3. Crude Oil Refining 84
1.17.4. Refining in the U.K. 86
1.17.5. Utilisation of Petroleum in the U.K. 87

1.18. British Standard Classification of Oil Fuels 88
1.18.1. Introduction 88
1.18.2. Engine Fuels 88
1.18.3. Burner Fuels 88

1.19. Burner Fuels 89
1.19.1. Typical Characteristics 89

1.20. Use of Oil Fuel by the C.E.G.B. 90
1.20.1. Lighting-up Uses 90
1.20.2. Boilers converted to Burn Heavy Fuel Oil 90
1.20.3. Other Fuel Oil Conversions 91
1.20.4. Fuel for Gas Turbines 92
1.20.5. Future use of Oil Fuel by the C.E.G.B. 93

1.21. Oil Firing—Delivery and Sampling 95
1.21.1. Delivery 95
1.21.2. Measurement 96
1.21.3. Sampling 96
1.21.4. Sampling from Pipelines 96
1.21.5. Sampling from Ships 97
1.21.6. Sampling from Road and Rail Tank Cars 97
1.21.7. Sampling from Storage Tanks 99
1.21.8. Sample Containers 99

1.22. Testing 99
1.22.1. Safety Precautions 99
1.22.2. Specific Gravity 99
1.22.3. Viscosity 99
1.22.4. Cloud and Pour Point 101
1.22.5. Calorific Value 101
1.22.6. Sulphur 101
1.22.7. Sodium 101
1.22.8. Vanadium 102
1.22.9. Ash Content 102
1.22.10. Sediment 102

1.23. Storage and Handling 103
1.23.1. Policy 103
1.23.2. Practice 103

1.24. Oil Firing—Combustion and Associated Problems 105
1.24.1. Combustion 105
1.24.2. High-temperature Fouling and Corrosion 106
1.24.3. Low-temperature Fouling and Corrosion 106
1.24.4. Factors Affecting the Acid Dew Point 107
1.24.5. Control of Low-temperature Corrosion 108
1.24.6. Smut Emission 112
1.24.7. Gas-side Cleaning of Oil-fired Boilers 113
1.24.8. Ash Disposal from Oil-fired Boilers 113

1.25. Testing of Fuel Oils 113
1.25.1. Viscosity 115
1.25.2. Specific Gravity 115
1.25.3. Flash Point 115
1.25.4. Water Content 117
1.25.5. Sediment 117
1.25.6. Ash—Sodium and Vanadium 117
1.25.7. Cloud and Pour Points 117
1.25.8. Gross C.V. and Ultimate Analysis 119
1.25.9. Sulphur 119

1.26. Testing of Flue Gases 119
1.26.1. Introduction 119
1.26.2. Dew point Measurement and Rate of Acid Build-up 119
1.26.3. Corrosion Probe Tests 121
1.26.4. Acid Deposition Rate 121
1.26.5. High-temperature Deposition Measurement 126
1.26.6. Dust Burden Measurement 126
1.26.7. Gas Analysis—Carbon Dioxide and Oxygen 128

1.27. Air Pollution 129
1.27.1. Introduction 129
1.27.2. The Nature of the Pollutants 129
1.27.2.1. Solid Emissions 129
1.27.2.2. Gaseous Emissions 129
1.27.3. Measurement of Pollution 130
1.27.3.1. Introduction 130
1.27.3.2. C.E.R.L. Dust Pollution Gauge 131
1.27.3.3. D.S.I.R. Volumetric Sulphur Dioxide Apparatus and Smoke Filter 132
1.27.3.4. C.E.R.L. Sulphur Dioxide Recorder 132
1.27.3.5. Ringelmann Smoke Chart 134
1.27.3.6. Siting of Pollution Gauges 134
1.27.4. Results 134

1.28. Lubricating Oils 136
1.28.1. Turbine Oil 136
1.28.2. Other Lubricating Oils 137
1.28.3. Additives and Inhibitors 139
1.28.3.1. Introduction 139
1.28.3.2. Gear Oil Additives 139
1.28.3.3. Industrial Oil Improvers 139
1.28.3.4. Oxidation and Rust Inhibitors 139
1.28.3.5. Detergent Additives 140
1.28.3.6. Viscosity Index Improvers 140
1.28.3.7. Pour Point Depressants 140
1.28.3.8. Emulsifying Agents 140
1.28.3.9. Anti-foam Agents 141
1.28.3.10. Demulsifiers 141
1.28.4. Sampling 141
1.28.5. Tests and Their Significance 141
1.28.6. Purification of Lubricating Oil 145
1.28.7. Operating Troubles in Turbine Oil Systems 148
1.28.8. Fire-resistant Oils 149

1.29. Greases 151
1.29.1. Introduction 151
1.29.2. Lime-base Greases 151
1.29.3. Soda-base Greases 152
1.29.4. Aluminium-base Greases 152
1.29.5. Lithium Greases 152

1.29.6. Complex Greases 152
1.29.7. Fluid Greases 152
1.29.8. Block Greases 153
1.29.9. New Greases 153
1.29.10. Graphite and Molybdenum Disulphide Greases 153
1.29.11. Greases for Nuclear Applications 153
1.29.12. Problems 154
1.29.13. Properties of Greases 154
1.29.14. Consistency 154
1.29.15. Drop Point 155
1.29.16. Resistance to Shear 155
1.29.17. Resistance to Oxidation 156
1.29.18. Resistance to Water 156

1.30. Insulating Oil 156
1.30.1. Introduction 156
1.30.2. Characteristics 156
1.30.3. Sampling 157
1.30.4. Testing 161
1.30.5. Purification 162
1.30.6. Fire-resistant Synthetic Insulating Oils 163

1.31. Buchholz Gas Analysis 163
1.31.1. Introduction 163
1.31.2. Analysis 163
1.31.3. Chemical Reactions 165
1.31.4. Test Apparatus 165

Exercises 167

Chapter 2. Corrosion. Feed and Boiler Water 169

2.1. Introduction 169
2.1.1. Chemistry of Reactions in Solution 169
2.1.1.1. A Simple Model of the Atom 169
2.1.1.2. Ions in Solution 173
2.1.1.3. Solubility and Mechanism of Precipitation Reactions 177

2.2. The Corrosion of Metals 180
2.2.1. Introduction 180
2.2.2. An Electrochemical Cell 181
2.2.3. A Simple Corrosion Cell 182
2.2.4. Electromotive Force and Polarity of Bimetallic Couples 184
2.2.5. Galvanic Corrosion 186
2.2.6. The Effect of Polarisation on Currents Flowing in Corrosion Cells 186
2.2.7. The Corrosion of Single Metals 189
2.2.8. Differential Aeration 191
2.2.9. A Review of Factors Affecting Corrosion Rates 191

2.3. Feed and Boiler Water 192
2.3.1. Introduction 192
2.3.2. Historical Survey 192
2.3.3. Boiler Water 194
2.3.3.1. Scale Formation 194
2.3.3.1.1. Phosphate treatment 195
2.3.3.2. Alkalinity 198
2.3.3.3. Dissolved Oxygen and its Removal 200
2.3.3.4. Silica 203
2.3.3.5. Chloride and Condenser Leakage 206
2.3.3.6. Protection of Boilers Off Load 207

2.3.4. Feed Water 208
2.3.4.1. Metal Oxides 208
2.3.4.2. Dissolved Oxygen 210
2.3.4.3. Carbon Dioxide 210
2.3.4.4. Treatment of Feed Water 211
2.3.4.4.1. pH Value 211
2.3.4.4.2. Ammonia 211
2.3.4.4.3. Hydrazine 211
2.3.4.4.4. Cyclohexylamine and morpholine 213
2.3.4.4.5. Filming amines 214
2.3.5. Once-through Boiler Plant 214

2.4. On-load Corrosion of Boilers 217
2.4.1. Occurrence 217
2.4.2. Protective Magnetite and the Effect of Caustic Soda 219
2.4.3. Effect of Heat Transfer and Related Phenomena 220
2.4.4. Non-protective Magnetite 222
2.4.5. Corrosion Processes in the Boiler 224
2.4.6. Prevention of On-load Corrosion 226

2.5. Generation Operation Memoranda Nos. 67 and 72 228
2.5.1. Introduction 228
2.5.2. Generation Operation Memorandum No. 67 228
2.5.3. Generation Operation Memorandum No. 72 230

2.6. Sampling and Analysis of Feed Water, Boiler Water and Steam 232
2.6.1. Introduction 232
2.6.2. Sampling and Analysis 233
2.6.3. Boiler Water Sampling 233
2.6.4. Steam Sampling 235
2.6.5. Feed Water Sampling 235

2.7. Instrumentation for Water Quality Control 236
2.7.1. Introduction 236
2.7.2. Measurement of Electrical Conductivity 236
2.7.3. Measurement of pH Value 238
2.7.4. Measurement of Dissolved Oxygen 240

Suggestions for Further Reading 245

Exercises 246

Chapter 3. Water Treatment Plant: Cooling Water Systems 249

3.1. Introduction 249
3.2. Sources and Quality of Raw Water 251
3.2.1. Chemical Composition 253

3.3. Boiler Feed Water Make-up 254
3.3.1. Town Supply Water Treatment 256
3.3.2. River Water Treatment 258

3.4. Water Softening 259
3.4.1. Lime/Soda Softening 259
3.4.2. Types of Lime and Lime/Soda Softeners 261
3.4.3. Calculation of Softening Charges and Control of Water Softening 266
3.4.4. Base-exchange or Zeolite Softening 269
3.4.5. Lime/base-exchange Softener 273
3.4.6. Blend Softener 274
3.4.7. Weakly Acidic Cation Exchange/base-exchange Softener 275
3.4.8. Other Softening Processes 275

3.5. Evaporators 276
3.5.1. Scale Prevention 278
3.5.2. Evaporator Blowdown 279
3.5.3. Evaporator Cleaning 279
3.5.4. Evaporator Distillate Quality 279

3.6. Demineralisation 280
3.6.1. Synthetic Ion-exchange Resins 281
3.6.2. Regeneration 286
3.6.3. Demineralisation Processes 290
3.6.4. Practical Considerations 309
3.6.5. Recent Developments 315

3.7. Choice of Make-up Water Treatment Plant 317

3.8. Cooling Water Systems 321
3.8.1. Scale Deposition on Heat Exchange Surfaces 321
3.8.2. Control of Organic Growth 328
3.8.3. Maintaining Condenser Cleanliness 330
3.8.4. Control of Marine Growth 331
3.8.5. Corrosion in Cooling Water Systems 333
3.8.6. Dry Cooling Towers 343
3.8.7. Protection of Cooling Tower Timber 345

References 348

Exercises 348

Chapter 4. Plant Cleaning and Inspection 349

4.1. Introduction 349

4.2. Pre-Commissioning Cleaning of Steam/Water Circuits 349
4.2.1. Reasons for Pre-commissioning Cleaning 350
4.2.2. Development of Cleaning Processes 352
4.2.3. Chemistry of the Cleaning Processes 356
4.2.3.1. Action of Acids 356
4.2.3.2. Inhibitors 358
4.2.3.3. Effect of Fluorides 359
4.2.3.4. Effect of Passivation 362
4.2.4. Administration and Planning 363
4.2.5. Criteria of Satisfactory Results 364
4.2.6. Inspection before Cleaning 365
4.2.7. Alkali Boil-out 365
4.2.8. Acid Cleaning 368
4.2.8.1. Phosphoric Acid 369
4.2.8.2. Sulphuric Acid 369
4.2.8.3. Hydrochloric Acid 369
4.2.8.4. Organic Acids 369
4.2.8.5. Application of Acid Cleaning 371
4.2.9. Draining Acid and Flushing 372
4.2.10. Passivation 374
4.2.11. Final Inspection 375
4.2.12. Control of the Cleaning Process 377
4.2.13. Disposal of Effluents 380
4.2.14. Purging of Superheater and Reheater 380
4.2.15. Safety 385
4.2.16. Works Cleaning and Protection of Plant 386
4.2.16.1. Protective Coatings 387
4.2.16.2. Closure of Cleaned Components Vapour-phase Inhibitors 389
4.2.16.3. Application of Protective Methods 390

4.2.17. Site Erection under Clean Conditions 391

4.3. Pre-commissioning Preparation of Turbine Oil Systems 392
4.3.1. Introduction 392
4.3.2. Purposes 393
4.3.3. Works Cleaning and Protection 393
4.3.4. Flushing 393

4.4. Plant Cleaning during Service Life 394
4.4.1. Boilers and Economisers 394
4.4.1.1. Incidence and Nature of Fouling 394
4.4.1.2. Deciding when to Clean the Boiler 395
4.4.1.3. Methods of Cleaning 396
4.4.2. Turbine Oil Systems 397
4.4.3. Feed Heaters 398
4.4.4. Condensers 398
4.4.4.1. Steam-side Cleaning 399
4.4.4.2. Water-side Cleaning 399
4.4.5. Evaporators 399
4.4.6. Turbine Blading 400

4.5. Inspection of Plant 402
4.5.1. Introduction 402
4.5.2. Inspection at Planned Outage or Overhaul 403
4.5.3. Inspection Techniques 406
4.5.3.1. Visual Examination: Sampling and Examination of Deposits 406
4.5.3.2. Instrumental Methods 408
4.5.3.2.1. Condenser tube examination 408
4.5.3.2.2. Metal sorters 409
4.5.3.2.3. Metal flaw detectors 410
4.5.3.2.4. Boiler tube inspection 410

4.6. Recent Developments 414
4.6.1. Protection of Austenitic Parts 414
4.6.2. Boiler Plant Purging with Nitrogen 415
4.6.3. Safety 415
4.6.4. Sub-surface Cavitation 415
4.6.5. Passivation 416

Appendix. Record of the Cleaning of the Boiler System of a 350 MW Unit 416

References 427

Exercises 427

Chapter 5. Metallurgy and Welding 429

5.1. Introduction 429

5.2. Ferrous Metallurgy 429
5.2.1. Production of Iron and Steel 430
5.2.2. Solidification Processes 432
5.2.3. Crystallographic Structure and the Allotropy of Iron 434
5.2.4. Constitution of Steel 436
5.2.5. Heat Treatment of Steel 438
5.2.5.1. Effects of Cooling Rate 440
5.2.5.2. Effects of Alloying 444
5.2.6. Cast Irons 447
5.2.7. Deformation and Fracture 448

5.3. Criteria for the Assessment of Materials 450
5.3.1. Mechanical Properties at Ambient Temperature 450
5.3.2. Physical Properties 453

5.3.3. Temperature-dependent Mechanical Properties 454
5.3.4. Weldability 457
5.3.5. Corrosion Resistance 457

5.4. Fabrication 458
5.4.1. Material Production and Forming 459
5.4.2. Welding 460
5.4.2.1. Gas Welding 460
5.4.2.2. Arc Welding 460
5.4.2.3. Electric Resistance Welding 464
5.4.2.4. Electroslag Welding 465
5.4.2.5. New Welding Processes 465
5.4.2.6. Automatic Welding 467
5.4.3. Weldability and Defects in Welds 468
5.4.3.1. Porosity 469
5.4.3.2. Cracking in Weld Metal 469
5.4.3.3. Cracking in the Heat-affected Zone 470
5.4.3.4. Low-hydrogen Welding Processes 471
5.4.3.5. Weldability Tests 471
5.4.3.6. Stainless Steels 473
5.3.4.7. Post-weld Heat Treatment 476
5.4.4. Desirable Weldment Properties 477
5.4.5. Welding Applications in Construction of C.E.G.B. Plant 477
5.4.5.1. Boiler Drums 477
5.4.5.2. Boiler Tubes 478
5.4.5.3. Superheater and Reheater Tubes 478
5.4.5.4. Headers and Manifolds 480
5.4.5.5. Steam Pipes and Valves 480
5.4.5.6. Turbine Casings and Steam Chests 480
5.4.5.7. Turbine Rotors 480
5.4.5.8. Feed Heaters 480

5.5. Non-destructive Examination 482
5.5.1. Visual Methods 482
5 5.2. Magnetic Crack Detection 483
5.5.3. Penetrant Inspection 483
5.5.4. Radiography 484
5.5.5. Ultrasonic Flaw Detection 486
5.5.6. Electrical/Magnetic Methods 490

5.6. Metallurgical Aspects of Power Plant Failure 491
5.6.1. Boiler Plant 492
5.6.1.1. Boiler Tubes 492
5.6.1.2. Boiler Drums 493
5.6.1.3. Superheater Tubes 493
5.6.2. Steam Pipework and Valves 496
5.6.3. Turbo-generator Plant 498
5.6.3.1. Turbine Cylinder Casings 498
5.6.3.2. Turbine Rotors 500
5.6.3.3. High-temperature Bolting 500
5.6.3.4. Turbine Blading 500
5.6.3.5. Alternator Rotor End Rings 500

5.7. Concluding Remarks 501

Glossary of Welding Terms 503

References 504

Recommended Reading 505

Exercises 505

Contents of Volumes 1–8 507

LIST OF ILLUSTRATIONS

Fig. 1.2.3.	The coalfields of Great Britain	2/3
Fig. 1.3.2.	The Seylers classification of coal main features—DMMF basis	6
Fig. 1.3.3a.	N.C.B. coal classification system	8
Fig. 1.3.3b.	Relationship of caking power to volatile matter	9
Fig. 1.5.4a.	Coal preparation plant—flow diagram	20
Fig. 1.5.4b.	Baum jig washer	22
Fig. 1.5.4c.	Baum jig washer—flow diagram	23
Fig. 1.5.6.	Loss of combustibles in power station coals—washed and untreated	25
Fig. 1.9.4a.	Bretby automatic coal sampler	36
Fig. 1.9.4b.	Birtley automatic coal sampler	37
Fig. 1.9.4c.	Geco automatic coal sampler	37
Fig. 1.9.4d.	Seaborne automatic coal sampler	38
Fig. 1.9.4e.	Pollock automatic coal sampler	38
Fig. 1.9.4f.	Hand sampling from wagons	39
Fig. 1.10.3a.	Laboratory coal grinder (Raymond mill)	42
Fig. 1.10.3b.	Rotary mechanical sample dividers	44
Fig. 1.10.3c.	Cone mechanical sample divider	45
Fig. 1.10.3d.	"Flip-flap" mechanical sample divider	45
Fig. 1.10.3e.	Riffle hand sample divider	46
Fig. 1.10.3f.	Double cone sample mixer	46
Fig. 1.11.6.2.	"Static" calorimeter	53
Fig. 1.11.6.3.	Adiabatic calorimeter	57
Fig. 1.11.12.	Crucible swelling index—profiles	62
Fig. 1.11.16.1.	Mineral matter in coal by gamma-ray analysis	67
Fig. 1.12.2.	Hardgrove grindability testing machine	70
Fig. 1.14.4.	Ash fusion test—cone profiles	79
Fig. 1.17.3.	Refinery flow diagram	85
Fig. 1.21.4.	Pipeline sampling	97
Fig. 1.21.5.	Oil sampling containers	98
Fig. 1.22.3.	Viscosity temperature diagram—fuel oils	100
Fig. 1.24.4.	Typical Relationship, dew point/excess oxygen	107
Fig. 1.24.5.	Additive feeder	110
Fig. 1.25.1.	Redwood viscometer	114
Fig. 1.25.3.	Pensky-Martens flash point apparatus	116

FIG. 1.25.4.	Dean and Stark apparatus	118
FIG. 1.26.2A.	Dew point meter (flue gas testing)	120
FIG. 1.26.2B.	Relationship, rate of acid build-up temperature	121
FIG. 1.26.3A.	Relationships, corrosion/temperature	122
FIG. 1.26.3B.	Relationships, corrosion/time	123
FIG. 1.26.3C.	B.C.U.R.A. corrosion probe	124
FIG. 1.26.4.	C.E.R.L. low-temperature deposition probe	125
FIG. 1.26.5.	C.E.R.L. high-temperature deposition probe	125
FIG. 1.26.6.	Flue gas dust burden sampling	127
FIG. 1.26.7.	Apparatus for analysis of SO_2 and SO_3	128
FIG. 1.27.3.2.	C.E.R.L. dust pollution gauge	132
FIG. 1.27.3.3.	Apparatus for smoke and SO_2 in air	133
FIG. 1.27.3.5.	Ringelmann chart	135
FIG. 1.28.5A.	Oxidation of turbine oil	143
FIG. 1.28.6A.	De Laval centrifuge	146
FIG. 1.28.6B.	Edge type filter	147
FIG. 1.28.6C.	Filter press	148
FIG. 1.29.15.	Grease drop point test	155
FIG. 1.30.3A.	Thief dipper for sampling oil in bulk	158
FIG. 1.30.3B.	Pressure thief	159
FIG. 1.30.3C.	Glass sampling tube for insulating oils	160
FIG. 1.30.5.	Metafilter for purification of insulating oils	162
FIG. 1.31.1.	Buchholz protection relay	164
FIG. 1.31.4.	Buchholz gas analysis tube	166
FIG. 2.1.1.1.	Some atomic and ionic groupings	171
FIG. 2.1.1.2A.	Formation of sodium chloride	174
FIG. 2.1.1.2B.	A molecule of water	174
FIG. 2.1.1.3A.	Solubility curves of sodium salts	177
FIG. 2.1.1.3B.	Solubility curves for scale-forming salts	178
FIG. 2.1.1.3C.	Solubility curve for trisodium phosphate	178
FIG. 2.2.2.	Daniell cell	182
FIG. 2.2.3.	A simple corrosion cell	183
FIG. 2.2.6A.	Corrosion of steel in hydrochloric acid at a break in mill-scale	187
FIG. 2.2.6B.	Corrosion rate/polarisation relationships	188
FIG. 2.2.6C.	Cathodic protection of iron/brass couple	189
FIG. 2.2.7.	Corrosion by salt solution on abraded steel	190
FIG. 2.2.8.	Crevice corrosion taking place between riveted plates	191
FIG. 2.3.3.2A.	Attack on mild steel by acid and alkali at 310°C	197
FIG. 2.3.3.2B.	Co-ordinated phosphate–pH curves for medium- and high-pressure boilers	199
FIG. 2.3.3.3A.	Formation of an oxygen pit	201
FIG. 2.3.3.3B.	Effect of time and temperature on reactions between hydrazine and dissolved oxygen	202
FIG. 2.3.3.4A.	Permissible levels of silica in boiler water	204

FIG. 2.3.3.4B. Distribution ratios (P) of silica, sodium hydroxide and sodium chloride, between steam and water (Styrikovich) 205
FIG. 2.3.4.1A. Copper concentration in feed water during start-up 209
FIG. 2.3.4.1B. Iron concentration in feed water during start-up 209
FIG. 2.3.4.3. pH and conductivity of CO_2 in water at 20°C 210
FIG. 2.3.4.4.2. pH and conductivity of ammonia in water at 20°C 212
FIG. 2.3.5. Nuclear drum and once-through boilers 215
FIG. 2.4.1. Examples of corrosion 218
FIG. 2.4.3. Diagrammatic representation of a concentrated film of sodium hydroxide 221
FIG. 2.4.4A. Formation of adherent magnetite on mild steel in various environments 223
FIG. 2.4.4B. Countercurrent mechanism of protective magnetite growth 223
FIG. 2.6.5. General arrangement of sampling probe (high pressure) 234
FIG. 2.7.2. Principle of measurement of electrical conductivity of solutions 237
FIG. 2.7.3A. Simple potentiometer for measurement of electrode potential 238
FIG. 2.7.3B. Simple thermionic valve circuit for measurement of electrode potential 239
FIG. 2.7.3C. Circuit for pH measurement with glass and calomel electrodes 240
FIG. 2.7.4A. Electrochemical dissolved oxygen analyser—schematic (Cambridge) 241
FIG. 2.7.4B. Electrochemical dissolved oxygen analyser (Wallace and Tiernan) 244
FIG. 3.3. Boiler feed water preparation processes 255
FIG. 3.3.1. Treatment for removal of organic matter 257
FIG. 3.4.2A. Conventional lime/soda softener 262
FIG. 3.4.2B. Sludge blanket lime/soda softener 263
FIG. 3.4.2C. Catalyst lime softener 265
FIG. 3.4.4. Base-exchange softener 272
FIG. 3.4.5. Lime/base-exchange softener 273
FIG. 3.4.6. Blend softener 274
FIG. 3.4.7. Starvation/base-exchange softener 275
FIG. 3.5A. Evaporator and associated plant flow diagram 277
FIG. 3.5B. Single effect P and B high-efficiency bled steam evaporator 278/9
FIG. 3.6.1. Mixed bed ion exchange unit-operation and regeneration 284
FIG. 3.6.2A. Permutit mixed bed unit (Reproduced by kind permission of the Permutit Co. Ltd.) 288
FIG. 3.6.2B. Freezing point of aqueous caustic soda solutions (Reproduced by kind permission of the Permutit Co. Ltd.) 291
FIG. 3.6.2C. Freezing point of aqueous sulphuric acid (Reproduced by kind permission of the Permutit Co. Ltd.) 292
FIG. 3.6.2D. Regenerant dilution and injection equipment 292
FIG. 3.6.3A. Demineralisation plants 293
FIG. 3.6.3B. Diagrammatic arrangement of "Deminrolit" two-bed mixed bed plant (Reproduced by kind permission of the Permutit Co. Ltd.) 294/5
FIG. 3.6.3C. Exchange capacity vs. concentration of sodium ions in solution (sodium/hydrogen cycle) 294

FIG. 3.6.3D. Changes in ion concentration within the resin bed 296
FIG. 3.6.3E. Composition of solution from a fully regenerated cation unit in the sodium/hydrogen cycle when treating sodium chloride solution 297
FIG. 3.6.3F. Composition of spent regenerant during regeneration cycle 298
FIG. 3.6.3G. Composition of water leaving an operational cation unit 298
FIG. 3.6.3H. Factors affecting cation leakage (influent alkalinity 50% of total cations) (Reproduced by kind permission of the Permutit Co. Ltd.) 299
FIG. 3.6.3J. Demineralisation plant at West Thurrock 301
FIG. 3.6.3K. Control and instrument panel—West Thurrock demineralisation plant 302
FIG. 3.6.3L. Change in conductivity during passage through plant 303
FIG. 3.6.3M. Make-up system, Ironbridge "B" 305
FIG. 3.6.3N. Feed system condensate polishing 306
FIG. 3.6.3P. Powdex process cartridge unit sectional view (Reproduced by kind permission of the Graver Water Conditioning Co.) 308
FIG. 3.7A, 3.7B Make-up water treatment plant 318
FIG. 3.7C. Comparison of costs for various make-up systems 320
FIG. 3.8.1. Carbonate and bicarbonate alkalinity of a water containing initially only bicarbonate alkalinity, after bringing into equilibrium with the carbon dioxide content of the atmosphere 326
FIG. 3.8.2. V-notch chlorinator flow diagram 328
FIG. 3.8.3. The Taprogge system of on-load condenser tube cleaning 331
FIG. 3.8.5A. Galvanic cell 335
FIG. 3.8.5B, C. Cathodic protection anodes 335, 336
FIG. 3.8.5D. Protection afforded by a single anode 337
FIG. 3.8.5E. Cathodic protection—standard half-cell 338
FIG. 3.8.5F. Silver/silver chloride half-cell and copper/copper sulphate half-cell 339
FIG. 3.8.5G. The effect of cathodically protecting a condenser water box 340
FIG. 3.8.6A, B. Dry cooling tower system at Rugeley 344
FIG. 3.8.7A. Collapse of cooling tower timber after soft rot attack 346
FIG. 3.8.7B. Soft rot attack on timber 346
FIG. 3.8.7C. Appearance of timber after soft rot attack 347
FIG. 4.2.1. Boiler tube corrosion at a weld protrusion 351
FIG. 4.2.2. Cleaning circuit for boiler, superheater, reheater and feed system 355
FIG. 4.2.3.3. Dissolution of mill-scale under dynamic conditions 361
FIG. 4.2.6A. Debris from a feed heater flash box drain 366
FIG. 4.2.6B. Boiler drum condition before cleaning 367
FIG. 4.2.8.3. Stainless steel pin damaged by hydrochloric acid 370
FIG. 4.2.8.5A. Contractors' pumps and temporary pipework 373
FIG. 4.2.8.5B. Temporary drum door and connection 374
FIG. 4.2.11A. Drum-end "penthouse" enclosure 376
FIG. 4.2.11B. (a), (b) Boiler tube camera photographs, before and after acid-cleaning 376
FIG. 4.2.11C. Radiograph showing "debris" in superheater bend 377

FIG. 4.2.14A. Steel target plates after steam blowing 382
FIG. 4.2.14B. Aluminium target plate after steam blowing 384
FIG. 4.2.16.2. End sealing of large bore unflanged pipe 389
FIG. 4.4.6. Deposit on turbine l.p. rotor 401
FIG. 4.5.2A. Photograph by telephoto lens of furnace wall tubes 405
FIG. 4.5.2B. Picture by closed circuit television of chimney interior 407
FIG. 4.5.3.2.4A. Boiler tube "still" camera 411
FIG. 4.5.3.2.4B. Defects shown in boiler tubes by $1\frac{3}{4}$-in. tube camera 413
FIG. A.1. Circulation during hydrochloric acid cleaning 419
FIG. A.2. Circulation during citric acid cleaning 421
FIG. A.3. Circulation during reheater flushing 423
FIG. A.4. Circulation during superheater flushing 424
FIG. A.5. Water velocities during flushing 425
FIG. A.6. Flow path for purging the complete circuit 426
FIG. 5.2.2A. Solidification of metals—crystal growth (From *Metallurgy for Engineers*, Rollason, Edward Arnold Ltd.) 431
FIG. 5.2.2B. Structure, contraction effects and segregation in ingots (From *Metallurgy for Engineers*, Rollason, Edward Arnold Ltd.) 433
FIG. 5.2.3A. Crystal structure of alpha-iron (body-centred). 434
FIG. 5.2.3B. Crystal structure of gamma-iron (face-centred) 435
FIG. 5.2.4A. The iron/iron carbide equilibrium diagram (Reproduced by kind permission of the American Welding Society.) 437
FIG. 5.2.4B. Pearlite ($\times 1000$) 438
FIG. 5.2.5. "Steel" portion of iron/iron carbide diagram (Reproduced by kind permission of Sandvik U.K. Ltd.) 439
FIG. 5.2.5.1A. Typical isothermal transformation diagram of a steel (TTT diagram) 441
FIG. 5.2.5.1B. Bainite ($\times 500$) 442
FIG. 5.2.5.1C. Martensite ($\times 500$) 443
FIG. 5.2.5.1D. Spheroidised pearlite ($\times 500$) 444
FIG. 5.3.1A. Stress–elongation curves for mild steel (From *Metallurgy for Engineers*, Rollason, Edward Arnold Ltd.) 452
FIG. 5.3.1B. Typical S–N curves 453
FIG. 5.3.3A. Typical "Charpy" V-notch values for a semi-killed low carbon steel (Reproduced from J. G. Tweedale, *Mechanical Properties of Metals*, Allen & Unwin) 454
FIG. 5.3.3B. Typical family of creep curves 455
FIG. 5.3.5. Corrosion of steel in steam, related to chromium content 458
FIG. 5.4.2. Welding processes—power and power intensity (Reproduced by kind permission of the Welding Institute) 461
FIG. 5.4.2.1. Basic gas welding equipment (Reproduced by kind permission of the American Welding Society) 461
FIG. 5.4.2.2A. Schematic representation of shielded metal arc welding (Reproduced by kind permission of the American Welding Society) 462

FIG. 5.4.2.2B Schematic diagram of inert-gas metal-arc nonconsumable (TIG) process (Reproduced by kind permission of the American Welding Society) 463
FIG. 5.4.2.2C. Schematic diagram of inert-gas metal-arc consumable (MIG) process (Reproduced by kind permission of the American Welding Society) 463
FIG. 5.4.2.3. Flash welding of tube (Reproduced by kind permission of the American Welding Society) 464
FIG. 5.4.2.4. Fundamentals of electroslag welding (From *The Metallurgy of Welding*, D. Seferian, Chapman & Hall Ltd.) 466
FIG. 5.4.3. Influence of the temperature cycle in welding on the parent metal grain size (curve B—oxy-acetylene welding; curve A—arc welding) (From *The Metallurgy of Welding*, D. Seferian, Chapman & Hall Ltd.) 468
FIG. 5.4.3.5. Controlled thermal severity test for weldability—assembly (Reproduced by kind permission of the Welding Institute) 472
FIG. 5.4.3.6. Constitutional diagram for stainless steel weld metal (Schaeffler) 475
FIG. 5.4.5.2. (a) Membrane panel of boiler tubes. (b) Showing profile of welds between tubes, A, B and C 479
FIG. 5.4.5.8A. Heater tube/tube plate—external plain weld 481
FIG. 5.4.5.8B. Heater tube/tube plate—external spigot weld 481
FIG. 5.4.5.8C. Heater tube/tube plate—internal stub weld 481
FIG. 5.5.4A. Basic arrangement for radiography (Reproduced by kind permission of the Welding Institute) 484
FIG. 5.5.4B. Radiographic detection of defects in 1-in. welded plate 486
FIG. 5.5.5A. Use of ultrasonic waves in N.D.T. (Reproduced by kind permission of the Welding Institute) 488
FIG. 5.5.5B. Principles of echo reflection method (Reproduced by kind permission of the Welding Institute) 489
FIG. 5.5.6. "Probolog" traces of aluminium–brass 490
FIG. 5.6.1.2. Boiler drum reinforcement at manhole. (a) Original method of manhole reinforcement. (b) Modified (forged) section with integral stiffening 494
FIG. 5.6.1.3A. Rupture of cold bent 18 Cr, 12 Ni, 1 Nb austenitic superheater tube 495
FIG. 5.6.1.3B. Reheater tube attachment weld, showing crack propagating into tube (×5) 496
FIG. 5.6.2A. Corrosion fatigue cracks in steam receiver (×135) 497
FIG. 5.6.2B. Designs of steam pipe/flange welds: (a) unsatisfactory, (b) satisfactory 499
FIG. 5.6.3.5. Rotor end bell showing stress corrosion cracking 502

CHAPTER 1

FUEL AND OIL

1.1. INTRODUCTION TO POWER STATION CHEMISTRY

1.1.1. Historical

During the pioneer years 1880 to 1920, technical development in the generation and supply of electricity was comparatively slow and chemical services to power stations were almost non-existent.

As plant design and efficiency advanced, it was soon recognised that applied scientific knowledge was essential for the safe, economic and efficient day-to-day running of power plant. Once the design of the plant had been established, and the plant installed, some of the main economies in generation were those entailing the scientific control of fuels, combustion, water treatment and corrosion.

In the second decade of the twentieth century, in order to meet technological advances in the expanding electricity supply industry, the chemist appeared for the first time in the power station to work alongside his electrical and mechanical engineering colleagues. Among the early pioneers in power station chemistry were G. W. Hewson, W. S. Coates, A. B. Owles and J. Dunn of the London, Newcastle, Manchester and Leicester undertakings respectively, and their laboratories were the first to be established in Britain to provide a chemical service for the generation and transmission of electricity. During the last forty years, particularly the last ten years or so, chemical technology has advanced rapidly and, as an essential service to power stations, has been joined by other sciences such as metallurgy and biology. Of these, metallurgy has become of major importance and Chapter 5 provides an introduction to this subject.

1.1.2. The Work of the Power Station Chemist

In the early development of a chemical service, the chemist was primarily concerned with the sampling and analysis of coal and coke for evaluating the quality/price relationship and for station efficiency purposes, the testing of lubricating and insulating oils, and the quality control of raw, treated, feed and boiler waters, including the technical control of the processes of water softening, evaporation and chlorination. Corrosion control, plant investigations and plant inspection became complementary responsibilities.

Today, the range of fuels has widened to include distillate and residual fuel oils, and the sampling and reduction problems associated with the handling of the huge tonnages of coal and oil involved, have intensified considerably. Lubricating and insulating oils now include numerous additives to prolong operational life and require specialised apparatus and testing techniques to maintain their special properties. Analytical techniques generally in fuel, oil and water analysis have advanced in accuracy and speed. The sensitivity and accuracy of modern analytical methods are generally some five to ten times better than were obtained twenty years ago and, for example, it is now not unusual to measure impurities in boiler feed water at a level of 10 μg/l (0·010 ppm) with a precision of 1 μg/l (0·001 ppm).

As a single example of the way in which analyses have been speeded up by use of modern techniques, the estimation of carbon and hydrogen in fuels, which once took some 12 h to do, can now be completed automatically in approximately 12 min.

New analytical methods using infrared, X-ray diffraction and gas chromatographic techniques are now appearing in some laboratories, together with instrumentation for automatic chemical analyses. Infrared gas analysis techniques are also used for locating air leaks into parts of the turbine and condenser which are at sub-atmospheric pressure.

During the last fifteen years or so, it has become necessary to revise previous standards of feed-water and boiler water quality, in order to minimise internal boiler corrosion which can result in extremely costly outages of plant. Simultaneously, steam purity requirements have become progressively more stringent. For these reasons principally, feed and boiler water control now requires, first the complete elimination of all kinds of casual impurity (for example, condenser leakage and dissolved gases) and second the injection of very small quantities of chemicals to maintain closely controlled conditions. Chapter 2 is concerned with these aspects.

Similarly, the standards of purity for make-up water have become more exacting. The earlier processes of water softening and evaporation have given way to, or are augmented by, ion-exchange methods of purification to produce make-up of extreme purity. These processes are considered in Chapter 3.

There have also been remarkable developments in the preparation for service of the internal surfaces of new plant—mainly as a prerequisite for the avoidance of boiler corrosion in later service. These developments include elaborate chemical and physical cleaning processes, together with instrumental methods of internal inspection, and they are discussed in Chapter 4.

In nuclear stations, even though the technology of conventional fuels does not arise, the chemist's work is no less significant than in conventional stations. All other aspects of his work apply equally to the nuclear stations and are extended by some specialised functions, of which radiochemistry and reactor gas analysis are examples. Radiochemistry involves the identification and measurement of small quantities of radioactive isotopes, in terms of their radioactivity, whilst both novel and familiar chemical reactions occurring in the reactor gas, have to be examined. This particular work is not included in this volume, but is discussed in Volume 8.

The station chemist of today, then, provides a skilled scientific service in order to improve availability, efficiency and the economics of generation by controlling the raw materials and minimising chemical damage to the very high cost capital plant. He applies specialised knowledge and experience to the many problems arising in daily operation inside the station, and to the increasing problems now arising outside the station in the fields of atmospheric pollution, marine fouling and river pollution. To ensure further continuous technical progress, he also has access to the scientific resources and specialists of Regional Research and Development Departments and the research facilities of the Central Electricity Research Laboratories at Leatherhead, the Engineering Laboratories at Marchwood and the Nuclear Laboratories at Berkeley.

1.2. COAL

1.2.1. Introduction

Coal may be defined as that part of the earth's crust which has formed as a result of the accumulation of decayed plant remains millions of years ago, and its subsequent consolidation over the years by a complex series of chemical and physical changes. Oil shales are excluded from this definition.

Today coal is known to be a complex mixture of the degradation products of these plant remains and their associated mineral matter, and its properties vary considerably according to location, geological, bacterial, heat and pressure influences and chemical change. As these parameters all vary in their intensities, the final coal substance consequently also varies, to produce a large range of coals of differing characteristics.

1.2.2. British Coal Resources

Britain has huge reserves of accessible coal estimated to be of the order of 170,000 million tons. The coal substance is of high quality, broadly classified by international standards as "hard". There is no brown coal in the U.K. and only insignificant deposits of lignite. Although deposits of peat are available in Scotland, this fuel has not proved economical to use on a large scale.

1.2.3. British Coal Production and Utilisation in C.E.G.B. Power Stations

Whereas some fifty years ago up to 287 million tons were mined annually—vast quantities were exported—production has fallen to around 185 million tons and exports to 5 million tons per annum. Inland consumption stands at about 187 million tons. Clearly the N.C.B. are facing intense competition both in the home and export markets and current estimates are that a tonnage of the order of 155 million may be sufficient for 1970—in which year C.E.G.B. would not expect to require more than 70 million tons for power generation.

The Board's coal consumption has been rising and its requirement now exceeds 60 million tons per annum. Table 1 shows the pattern of coal production in the year 1964/5 together with the coal used by C.E.G.B.

A map showing the coalfields of the U.K. is included as Figure 1.2.3.

Table 2 shows the tonnages of the various qualities of coals purchased by the Board in 1964.

TABLE 1

N.C.B. COAL PRODUCTION AND C.E.G.B. PURCHASES IN YEAR 1964/5

N.C.B. Division	N.C.B. production		C.E.G.B. purchases		C.E.G.B. as % of N.C.B.
	million tons	%	million tons	%	
Scottish	15·49	8·43	1·23	1·91	7·93
Northern {Northumberland & Durham	31·23	17·01	10·39	16·19	33·27
Yorkshire	43·49	23·68	14·35	22·35	33·00
North-Western	12·36	6·73	5·17	8·05	41·80
East Midlands	46·62	25·38	20·25	31·55	43·44
West Midlands	14·16	7·71	6·64	10·35	46·93
South-Western	18·84	10·26	5·63	8·77	29·87
Kent	1·47	0·80	0·53	0·83	36·03
	183·66		64·19*		34·95

* In addition C.E.G.B. purchased 0·6 million tons of miscellaneous coals.

TABLE 2

C.E.G.B. COAL PURCHASES IN 1964 (THOUSANDS OF TONS)

Description	Deep-mined N.C.B.	Opencast N.C.B.	Small mines	Total
Small graded	625	30	10	665
Washed smalls	8292	15	109	8416
Washed fines	254	2	—	254
Run-of-mine	189	24	116	329
Untreated smalls	31,862	2492	1819	36,173
Blended smalls	13,301	—	80	13,381
Dry cleaned smalls	727	—	—	727
Dry fines	2411	16	—	2427
Slurry	1020	—	—	1020
	58,681	2579	2134	63,394*

* In addition, the Board purchased 567,000 tons of miscellaneous other coals.

Table 3 shows the average proximate analyses and calorific values of the main qualities given above in Table 2.

A broad indication of the way in which some of the properties of coal vary from coalfield to coalfield is given in Table 4 where the average proximate analyses and calorific values of coal purchased by C.E.G.B. in 1964 are shown. Particular note should be taken of the total moisture content which varies mainly according to the amount of moisture naturally present in the coal substance. The calorific values recorded are also related to the property of the coal substance, as well as to the proportions of ash and moisture present.

TABLE 3

AVERAGE PROPERTIES OF COALS DELIVERED TO C.E.G.B. IN 1964

Description	Moisture %	Ash %	Volatile %	Calorific value Btu/lb
Washed smalls	14·0	9·8	28·2	11,110
Washed fines	12·4	13·3	17·6	11,150
Run-of-mine / Untreated smalls	10·2	18·5	26·6	10,320
Blended smalls	11·6	17·6	27·8	10,180
Dry fines	9·4	22·3	20·6	9990
Slurry	21·9	17·1	22·4	8800
All coal:	11·2	17·0	26·7	10,375

TABLE 4

AVERAGE PROPERTIES OF C.E.G.B. COALS ACCORDING TO SOURCE (1964)

N.C.B. Division	Moisture %	Ash %	Volatile %	Calorific value Btu/lb
Scottish	14·8	10·4	28·2	10,680
Northern { Northumber-land &	10·4	17·8	27·9	10,395
Durham	6·2	17·2	27·6	11,485
Yorkshire	9·2	19·4	27·4	10,410
North-Western	9·4	16·5	28·3	10,840
East Midlands	14·2	16·3	28·1	9845
West Midlands	13·5	16·6	28·5	9820
South-Western	7·7	18·1	14·3	11,320
Kent	5·5	18·4	18·2	11,680

1.3. CLASSIFICATION OF COAL

1.3.1. Dr. Seyler's Classification

In an attempt to predict the properties of coals from analytical data, Seyler, using dry mineral matter-free percentages of carbon and hydrogen as the main parameters, plotted the positions of all known coals, in a graphical form.

Seyler gave each class of coal a distinctive name based on its rank (the rank of the coal being a function of its carbon content calculated on the pure coal basis) and named five main classes as in Table 5.

TABLE 5

SEYLER'S COAL CLASSIFICATION

No.	Class	% carbon in coal substance
1	Lignitous	75–84
2	Bituminous	84–91
3	Carbonaceous and semi-anthracite	91–92·5
4	Anthracite	Over 92·5

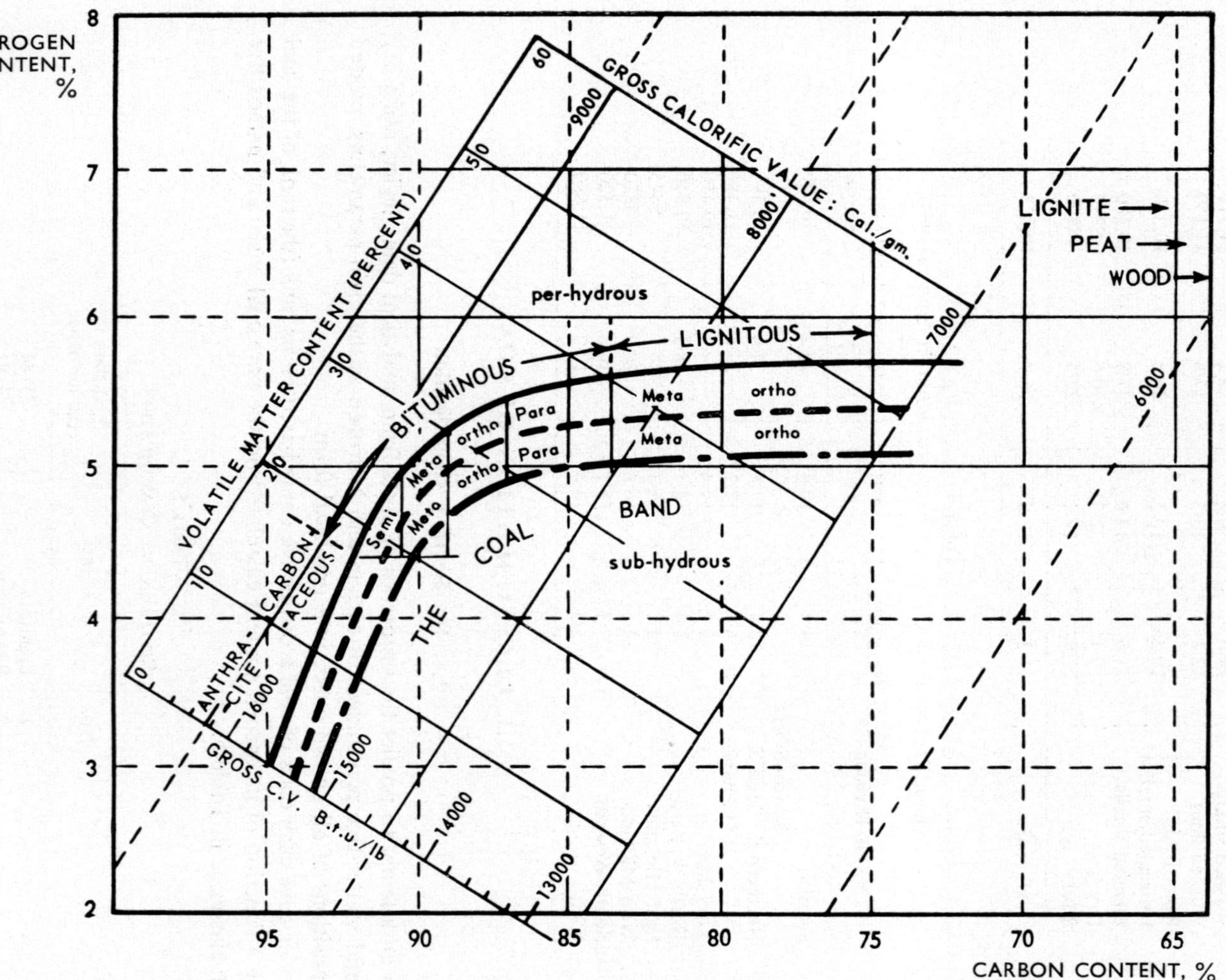

FIG. 1.3.2. The Seyler classification of coal main features—DMMF basis

1.3.2. Dr. Seyler's Fuel Charts

Seyler's fuel chart, Figure 1.3.2. (which is an enlarged section of a more general chart), may be used to calculate any two unknowns of the quantities carbon, hydrogen, volatile matter and calorific value when the other two are known. For example, if the calorific value and volatile matter are known, the carbon and hydrogen contents can be read off the charts at the intersection of the appropriate calorific value line (isocal) and volatile matter line (isovol) provided the coal falls within the coal band. For dull coals or coals outside the band, a correction is made to the volatile matter figure which is reduced by 0·5% for every 10% of durain (the dark grey, hard fraction of coal which contains much of the inorganic matter), estimated to be present. The intersection of the corrected isovol with the isocal is then found. Seyler's chart can also be used for checking boiler efficiency, but this is outside the scope of this lesson and further reference may be made to: H. M. Spiers, *Technical Data on Fuel*, page 290 (6th edition), where the subject is discussed in detail.

1.3.3. The N.C.B. Coal Rank Code Number Classification

During the Second World War, the Fuel Efficiency Committee of the Ministry of Fuel and Power investigated methods for the classification of commercial coals, and suggested a simple scheme utilising only two properties of coal; volatile matter expressed on the dry mineral-matter-free basis (D.M.M.F.) and the caking power of the clean coal. For this purpose a clean coal is defined as a coal with 10% or less ash content. The N.C.B. adopted this system for survey and marketing purposes, and for convenience substituted code numbers to describe the various classes of coal, instead of the former names designated by older methods of classification. One of the many advantages of the N.C.B. classification is that it recognises the wide variation of coal types possible in coals of similar volatile matter content. The classification is given graphically in Figure 1.3.3A.

The Coal Rank Code System was devised some years ago as a simple means of classifying coals. It was first published in Survey Paper No. 58 by the Fuel Research Division of the Department of Scientific and Industrial Research in 1946; since then it has been modified in 1952, 1956 and again in 1964. Because the system was first used in an investigation requiring mechanical analysis of the data, numbers instead of names were used to designate the different classes; apart from this particular application the numerical system provides a useful form of shorthand description, especially suitable for use in tables of analytical data.

The classification is based upon the volatile contents (expressed on a dry mineral matter-free basis) and the caking properties of the *clean* coals. Using the criterion of volatile matter alone a first division into the following groups is obtained:

	Volatile matter	Code number
Anthracites	Up to and including 9·0%	100
Low-volatile steam coals	9·1–19·5%	200
Medium-volatile coals	19·6–32·0%	301*
High-volatile coals	Over 32·0%	(see below)

* Certain coals have been affected by the heat from nearby igneous intrusions, with the result that their caking properties are generally subnormal compared with those of other coals of similar volatile content. These affected coals are distinguished by the code numbers 302 and 303, and by adding H after the code number when they fall in the 100 and 200 classes. They occur mainly in Scotland but some are also found in Durham.

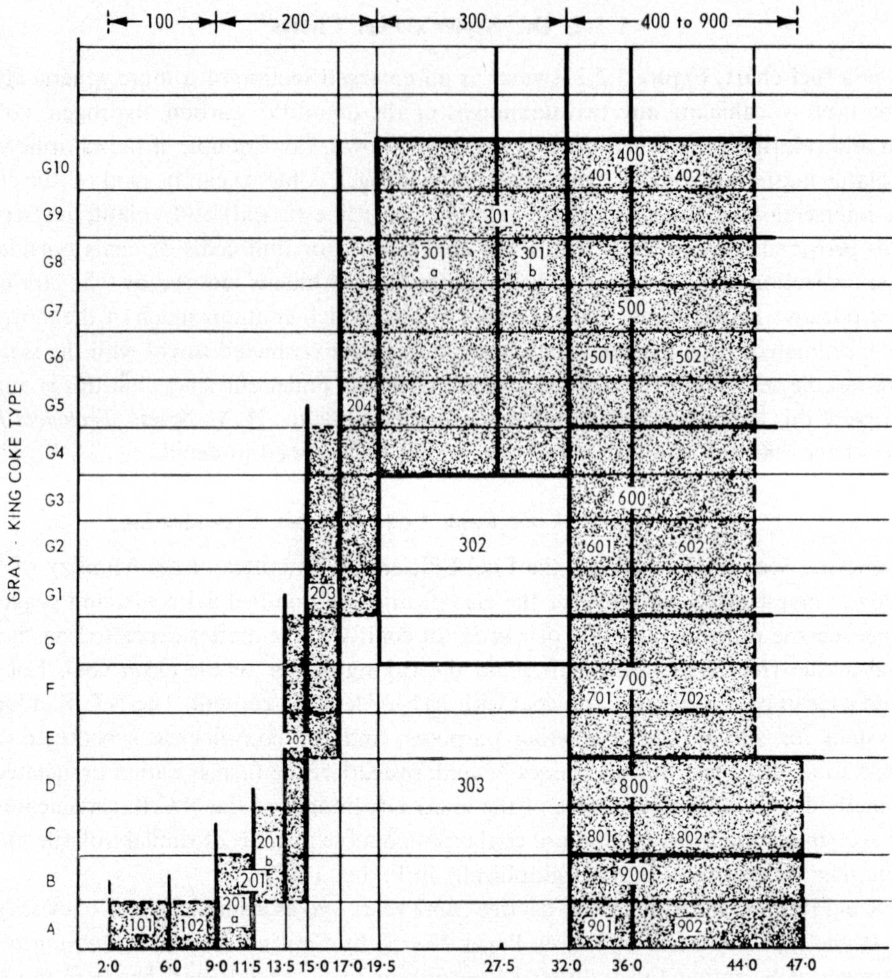

Fig. 1.3.3A. N.C.B. coal classification system

VOLATILE MATTER ON DRY MINERAL-MATTER-FREE BASIS
(per cent)

---- Defines a general limit as found in practice, althought not a boundary for classification purposes.

━━ Defines a classification boundary.

▬ Defines the limit for most British coals.

NOTES

1. Coals that have been affected by igneous intrusions ('heat-altered' coals) occur mainly in classes 100, 200 and 300, and when recognized should be distinguished by adding the suffix H to the coal rank code, e.g. 102H, 201bH.

2. Coals that have been oxidised by weathering may occur in any class, and when recognized should be distinguished by adding the suffix W to the coal rank code, e.g. 801W.

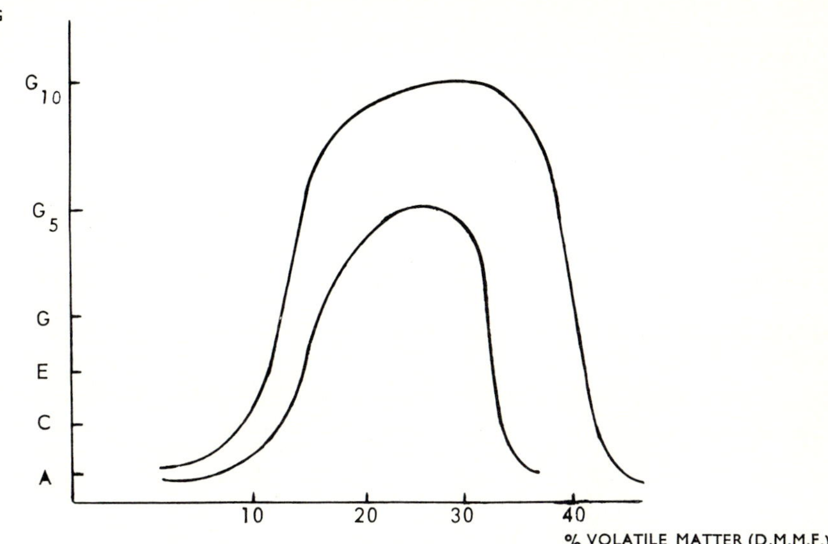

FIG. 1.3.3B. Relationship of caking power to volatile matter

In the first three groups—i.e. in coals of volatile matter up to 32%—there is a close relationship between volatile matter content and caking properties. Consequently, the effect of sub-dividing into progressive ranges of volatile matter content is also to produce classes with progressive ranges of caking power.

In the fourth group—i.e. in coals with more than 32% of volatile matter—there is a wide range of caking properties at any given volatile matter content, and subdivision has been made on the basis of caking power, as indicated by the type of coke produced in the Gray–King assay. Six ranges of caking properties are recognised:

	Gray–King coke type	Code number
Very strongly caking	G_9, G_{10} and higher	400
Strongly caking	G_5, G_6, G_7 and G_8	500
Medium caking	G_1, G_2, G_3 and G_4	600
Weakly caking	E, F and G	700
Very weakly caking	C and D	800
Non-caking	A and B	900

Each of the classes 400–900 can be further subdivided according to volatile matter content: a 1 in the third figure of the code number indicates that the volatile matter content of a coal lies between 32·1 and 36·0%, and a 2 that it is over 36%.

When Gray–King coke types are plotted against dry mineral-matter-free volatile contents for British deep-mined coals, the main bulk lie within the band illustrated above (see Fig. 1.3.3B).

The frequency of occurrence of the various ranks, as defined by the N.C.B. Coal Rank Code Numbers, is shown in Table 6 with the tonnages and proportions used by C.E.G.B. in 1964.

TABLE 6

DISTRIBUTION OF BRITISH COAL INTO N.C.B. RANK CODE NUMBERS

Coal Rank Code No.	100	200	300	400	500	600	700	800	900	All coal ranks
N.C.B. saleable output: 1963/64:										
N.C.B. Division	% of divisional output by Coal Rank Code									million tons
Scotland	1	2	3	—	2	11	19	36	26	16·5
N. and D.	—	1	17	20	30	12	17	3	—	32·4
Yorkshire	—	—	—	4	29	25	29	13	—	43·1
N. Western	—	—	5	3	25	30	25	12	—	12·7
E. Midlands	—	—	—	—	7	16	16	34	27	46·7
W. Midlands	—	—	—	1	7	12	19	44	17	14·7
S. Western	17	44	35	1	2	—	1	—	—	19·3
S. Eastern	—	44	56	—	—	—	—	—	—	1·6
Great Britain	2	5	8	5	16	16	19	19	10	187·0
C.E.G.B. purchases: 1964: By Coal Rank Code: *										
%	½	5	2½	2	14	14	26	17½	18½	100
million tons	0·2	3·0	1·4	1·1	7·8	7·7	14·3	9·8	10·3	55·6

* This analysis excludes 8·3 million tons of coal of which the rank was unknown.

It will be of interest to observe that C.E.G.B. obtained about the national mined proportion of Rank Numbers: 200, 500, 600 and 800; very high proportions of Ranks 700 and 900 were purchased. The smaller proportions of Ranks 300 and 400 are understandable because these are the prime coking coals of the country. Rank 100 (anthracite) is of course mined for its value in heating plants and appliances.

1.3.4. Coal Rank Code Numbers in Power Station Practice

Table 7 sets out for the main ranks of British coals, the average properties which are of importance.

A brief description is given below of the characteristic behaviour of small coals of the main ranks when used for steam raising purposes on chain grate stokers and as pulverised fuel. In indicating by these notes the general characteristics of coal, no account has been taken of the effect of the quantity or nature of ash, e.g. fusion temperature or hardness of mineral matter. The size range and distribution, and free moisture content have also been ignored. Although all these factors are relevant to combustion, it is the behaviour resulting from the nature of the clean coal substance which it is sought to emphasise here.

N.C.B. Rank Number	Remarks
100	On chain grates is very difficult to ignite, and slow in burning. Burns without flame, and produces a heavy "carry-over" of high carbon grit. Grates run hot as almost all the heat release is on the grate. Can be fired as P.F. in special combustion chambers, but a high mill power is required in order to secure a sufficiently fine product. Carbon in ash is high and is likely to be approaching 20% even with special plant. (The only existing plant burning this coal shows about 40% carbon in fly-ash.)
201	On chain grates the main difficulty is due to "carry-over" of high carbon grit. The burn off is good but rather slow, and a low ash content gives rise to overheated grates. Small graded sizes are expensive, but can be used very efficiently. As P.F., combustion is rather easier than for Rank 100, but carbon in ash is still high (between 7 and 17%) and mill power above average.
202/3/4	On chain grates the first two groups are extremely satisfactory provided that fines in smalls are not excessive. Caking power is just sufficient to bind small particles satisfactorily, without forming large coke masses which prevent a good burn off. The rapidly increasing caking property causes some 204 coals to give a high carbon in ash loss due to swollen coke formation. As P.F. these coals are more satisfactory, being softer but still requiring fine milling. Carbon in dust is usually 5–12%.
301	On chain grates these are moderately satisfactory. The carbon loss due to swollen cokes is balanced by reduced grit loss and the higher proportion of heat released in the form of volatile constituents. These soft coals are quite satisfactory as P.F., and the carbon in fly-ash continues to decrease as the rank decreases. Carbon in P.F. fly-ash is usually 3–10%.
400	On chain grates these coals are very unsatisfactory. The strong caking characteristic leads to the formation of large coke masses which will not burn out. This feature is particularly marked in the 401 group, and results in poor efficiency and boiler output. For P.F. these coals are very satisfactory. The coals are soft and result in a lower carbon in dust.
5/6/700	The two latter groups are excellent on chain grate stokers. The caking property is sufficient to hold fines in the fuel bed. Less primary air is required, and the coal can be fired at very high loadings per square foot of grate. The strong coking nature of the 500 rank could lead to a high carbon in ash at high grate loadings. These coals are all extremely satisfactory for P.F. and yield a low carbon in refuse.
800/900	These low rank coals on chain grate stokers can give rise to grit "carry-over", but secondary air can be used to retard this tendency. Grits carried over are very low in carbon, and these free burning coals give a very low carbon in ash. Fired as P.F., these coals are extremely satisfactory, the only drawback being the low inherent calorific value. Carbon in refuse is low.

TABLE

AVERAGE PROPERTIES OF A

Certain basic characteristics of British coals are given below, together with notional examples of washed smalls representing the various types of coal. The moisture and ash content of commercial grades vary with the rank of coal, seam characteristics, mining conditions, and the degree of preparation.

The moisture content of large and graded coals is usually a little higher than that shown for air-dried coal; for untreated smalls the moisture content will be about 2% higher (except where this is affected by water infusion during mining).

The moisture content of washed smalls is commonly 6 to 8% above the moisture content for air-dried coal.

	Anthracite		Dry steam	Coking steam		Medium volatile coking coals	
National Coal Board–Rank Code Number	101	102	201	202	204	301a	301b
Moisture content of air-dried coal	2	1	1	1	1	1	1
Moisture content at 96% R.H. and 30°C	4	2	1	1	1	1	1
Analysis of dry-mineral-free coal							
Volatile matter	5·0	7·5	11·5	14·0	18·0	23·0	30·0
Calorific value (Btu/lb)	15,400	15,600	15,700	15,750	15,800	15,800	15,650
Carbon	94·4	93·0	92·4	92·0	91·6	90·4	89·3
Hydrogen	2·9	3·7	4·0	4·2	4·5	4·9	5·0
Nitrogen	1·1	1·3	1·4	1·4	1·4	1·5	1·5
Sulphur	0·7	0·7	0·9	0·7	0·7	0·6	0·8
Oxygen	0·9	1·3	1·3	1·7	1·8	2·6	3·4
Caking properties							
B.S. Swelling number	0	0	1	3	9	8	8
Gray–King coke type (600°C)	A	A	B	E	G3	G6	G7
Maximum dilation (Audibert–Arnu)	—	—	—	—	25	140	256
Principal coalfields of origin	South Wales, Scotland		South Wales	South Wales, Kent		South Wales, Durham, Kent	

7

Representative Range of British Coals

Values of ash content commonly found commercially are:

Large and graded coal	3 to 6%
Washed smalls	5 to 10%
Untreated small	12 to 18%

The analysis of a given commercial product of known coal rank code number can be estimated very approximately by conversion of the appropriate dry-mineral-free analysis to the basis of the actual moisture and ash contents of the sample, as follows:

If X is the value of the parameter on the dry-mineral-free basis, then the value of that parameter on a basis of $M\%$ moisture and $A\%$ ash is:

$$\text{Approximately } x \times \frac{100-(M+1{\cdot}15A)}{100}$$

High volatile coking coals					General purposes coals				Heat altered coals		
Very strongly caking	Strongly caking		Medium caking		Weakly caking		Very weakly caking	Non-caking	Low volatile	Medium volatile	
401	501	502	601	602	701	702	802	902	101H to 204H	302	303
2	2	3	4	4	5	5	8	10	2	1	2
2	3	4	6	5	6	7	11	13	3	2	3
35·0	35·0	37·5	35·0	37·5	35·0	38·5	39·0	40·0	15·0	25·0	28·0
15,500	15,350	15,250	15,000	15,000	14,800	14,750	14,500	14,100	15,600	15,600	15,450
87·5	86·8	85·2	85·0	84·5	84·5	83·5	82·0	81·0	90·6	88·8	87·8
5·3	5·3	5·4	5·5	5·5	5·2	5·4	5·3	5·1	4·3	5·0	5·1
1·9	1·7	1·8	1·8	1·8	1·8	1·7	1·7	1·6	2·0	2·0	1·7
1·0	1·0	0·9	0·8	0·9	0·8	1·0	1·0	0·7	0·9	0·8	1·0
4·3	5·2	6·7	7·1	7·3	7·7	8·4	10·0	11·6	2·1	3·4	4·4
9	8	$7\frac{1}{2}$	6	6	$3\frac{1}{2}$	$3\frac{1}{2}$	$1\frac{1}{2}$	1	$\frac{1}{2}$	$8\frac{1}{2}$	6
G10	G7	G6	G2	G3	F	F	C–D	B	A	G2	D/E
280	133	88	20	21	−12	−19	−20	—	—	—	−28
Durham, Yorks., Northumberland	Yorkshire, Durham, Northumberland, Lancs., Notts., and North Derbyshire		Yorkshire, Notts. and North Derbyshire, Lancs., Northumberland, Scotland		Yorkshire, Notts. and North Derbyshire, Northumberland, Lancs., North Staffordshire, Scotland		Notts. and North Derbyshire, Yorkshire, Cannock Chase, Warwicks., Lancs., Scotland	Notts. and North Derbyshire, Leics., Scotland, Warwicks., South Derbyshire	Durham, Scotland	Durham, Scotland	

TABLE 7

	Anthracite		Dry steam	Coking steam		Medium volatile coking coals	
Analysis of notional washed smalls							
(as fired)							
Moisture	8	7	7	7	7	7	7
Ash	8	8	8	8	8	8	8
Carbon	78·2	77·9	77·4	77·1	76·8	75·8	74·8
Hydrogen	2·4	3·1	3·4	3·5	3·8	4·1	4·2
Nitrogen	0·9	1·1	1·2	1·2	1·2	1·3	1·3
Sulphur	1·0	1·0	1·0	1·0	1·0	1·2	1·2
Oxygen	1·5	1·9	2·0	2·2	2·2	2·6	3·5
Calorific value (Btu/lb)							
Gross	12,750	13,050	13,150	13,200	13,250	13,250	13,100
Net	12,440	12,680	12,750	12,790	12,820	12,790	12,630
Theoretical air requirements per kg of fuel:							
kg air	9·84	10·05	10·10	10·09	10·16	10·15	10·03
m^3 air at 0°C and 760 mm	7·61	7·77	7·81	7·80	7·86	7·85	7·76
Waste gases per kg of fuel:							
m^3 wet waste gas at 0°C and 760 mm	7·84	8·03	8·09	8·08	8·16	8·17	8·09
CO_2 content of dry waste gases:							
%	19·5	19·1	18·9	18·9	18·7	18·5	18·5
Composition of moist waste gases:							
CO_2%	18·6	18·1	17·9	17·8	17·6	17·3	17·3
H_2O%	4·5	5·2	5·6	5·7	6·1	6·5	6·7
Oxides of N and S %	0·3	0·3	0·3	0·3	0·3	0·3	0·3
N_2%	76·6	76·4	76·2	76·2	76·0	75·9	75·7
Dew point of waste gases:							
°C	31	34	35	35·5	37	38	38·5

(Contd.)

High volatile coking coals					General purposes coals				Heat altered coals		
Very strongly caking	Strongly caking		Medium caking		Weakly caking		Very weakly caking	Non-caking	Low volatile	Medium volatile	
9	9	10	11	11	13	13	16	18	8	7	8
8	8	8	8	8	8	8	8	8	8	8	8
71·6	71·0	68·8	67·8	67·4	65·7	65·0	61·3	59·0	75·0	74·4	72·7
4·3	4·3	4·4	4·2	4·4	4·0	4·2	4·0	3·7	3·6	4·2	4·2
1·6	1·4	1·5	1·4	1·4	1·4	1·3	1·3	1·2	1·7	1·7	1·4
1·7	1·7	1·7	1·7	1·7	1·7	1·7	1·7	1·7	1·2	1·2	1·2
3·8	4·6	5·6	5·9	6·1	6·2	6·8	7·7	8·4	2·5	3·5	4·5
12,700	12,550	12,300	11,950	11,950	11,500	11,500	10,850	10,250	12,900	13,050	12,800
12,200	12,050	11,780	11,440	11,420	10,980	10,960	10,300	9710	12,470	12,580	12,320
9·74	9·61	9·37	9·16	9·17	8·84	8·79	8·26	7·85	9·93	10·03	9·76
7·53	7·44	7·24	7·08	7·08	6·84	6·80	6·39	6·07	7·68	7·75	7·55
7·89	7·80	7·63	7·47	7·49	7·23	7·21	6·83	6·52	7·98	8·08	7·89
18·3	18·3	18·3	18·4	18·3	18·4	18·4	18·4	18·6	18·7	18·4	18·5
16·9	17·0	16·8	16·9	16·8	17·0	16·8	16·8	16·9	17·5	17·2	17·2
7·3	7·4	7·8	7·8	8·0	8·0	8·3	8·9	9·1	6·1	6·7	7·0
0·5	0·4	0·5	0·5	0·5	0·5	0·5	0·5	0·5	0·4	0·4	0·4
75·3	75·2	74·9	74·8	74·7	74·5	74·4	73·8	73·5	76·0	75·7	75·4
40	40	41	41	42	42	42·5	44	44	37	38·5	39

1.4. CLASSIFICATION OF COAL BY SIZE

Various attempts have been made in the U.K. (and in other parts of the world) to produce standard size limits with standard commercial descriptions for coals. In practice these have been defeated by the need of the individual collieries to supply their own particular customers with the size ranges and qualities they require, in a way that ensures the disposal of the whole of the coal they produce.

Nevertheless it may be useful (Table 8) to give some indication of the typical size limits against commercial descriptions prevalent in the United Kingdom.

TABLE 8

CLASSIFICATION OF COAL SIZES

	Description	Round hole screen sizes (in) Upper	Round hole screen sizes (in) Lower	Remarks
	Large	—	2	Except where large graded products are also produced
Graded coals	Large cobbles	6	3	
	Cobbles	4	2	
	Trebles	3	2	Sometimes called "Large nuts".
	Doubles	2	1	Sometimes called "Nuts" or "Small nuts"
	Singles	1	$\frac{1}{2}$	Sometimes called "Beans".
	Peas	$\frac{1}{2}$	$\frac{1}{4}$	
	Grains	$\frac{1}{4}$	$\frac{1}{8}$	
Coals with no bottom size limit	Run-of-mine	No limit		Unscreened.
	Through	Say 3 or 4		Sometimes called "Run-of-mine" but the "large" fraction has been screened off.
	Smalls	Top size depends on the need to extract large and graded coals to satisfy other markets. It also depends on the need to make a "smalls" product handleable by re-introducing or leaving in the singles (or doubles as well). Screen to sizes can be 2, $1\frac{1}{2}$, 1, $\frac{3}{4}$, $\frac{1}{2}$, $\frac{3}{8}$ in. Smalls produced by screening at sizes between $\frac{1}{2}''$ and $\frac{1}{4}''$ are usually known as duffs.		
	Fines and slurry	Any top size under $\frac{3}{8}$ in. down to $\frac{1}{2}$ mm.		

As will be described later when dealing with the Price Structures for coal, the price of coal varies with the size; in addition graded coals are more expensive than coals which are of the same top size but have no bottom size limit. The cheapest coals for power station use are therefore smalls, duffs, fines and slurry. Graded coals, because of their high cost, are only purchased when it is essential to improve the size—consist of the small coal in general use, for example, when a small stoker-fired plant has grit emission problems, or in order to ensure that stocks are suitable for handling in times of wet weather or in thawing snow.

1.5. PREPARATION OF COAL

1.5.1. Introduction

Raw coal as mined underground is a mixture varying in size from lumps of 3 ft in length down to dust and it includes "dirt" or shale unavoidably removed from the roof and floor of the seam or running as a band in the coal seam itself—this in addition to the smaller amount of mineral matter distributed generally throughout the coal itself. This type of shale varies in combustible content from 30 to 60% (it must not be confused with the oil-bearing type rock shales).

Modern mining relies increasingly on mechanised coal-getting and power loading; the proportion so won in 1964/5 was 75%, double that of 1960. Concentration upon mechanical methods of coal extraction in order to raise maximum quantities at minimum cost has led to a deterioration in the size—consist of the run-of-mine (raw) coal and also of its ash content. The N.C.B. are now attending to the problem of maintaining output without continued loss of "large" coal and are endeavouring to avoid the inclusion of material from roof and floor. Attempts are also being made to find machines suitable for cutting out a main dirt band from the coal seam before the main cutter/loader comes through.

At collieries which have to provide a wide range of products, the run-of-mine (raw) coal is passed over a screen to separate large coal and then over further screens to separate the sizes required to make graded products. These potentially valuable sizes are then prepared for their markets. The remaining coal, which is the raw smalls, can either be sold direct for use in power stations or can be subjected to a cleaning process to make it fit for industrial steam raising use or it may be required for carbonisation purposes, if of the correct rank and sulphur content. At some collieries the requirements may be as simple as a screen to separate large from small coal (two products only being required) and it is conceivable that in the future, with C.E.G.B. taking a greater proportion of the total coal mined, the whole output of some collieries may be crushed to provide small coal for power generation. It is interesting to note that about 60% of all coal produced by N.C.B. in 1964/5 was "smalls". Of the small coal they produced, 43% was classified as treated (cleaned) and 57% untreated. On a total coal tonnage of 183·7 million this gives:

46 million tons treated smalls and 63 million tons untreated of which the C.E.G.B. used:

8·3 million tons treated smalls and 50·8 million tons untreated smalls.

1.5.2. Coal Preparation for Power Station Use

It is possible to operate stoker-fired plant on comparatively clean untreated small coals but it is difficult to find sufficient for the purpose; consequently it has been necessary to order a certain amount of washed small coal (and indeed small graded sizes). In the year 1964/5, 25% of the Board's coal was burnt on stoker-fired plant. Washed smalls and graded coals purchased by C.E.G.B. in that year together amounted to only 14% of the total taken.

As mentioned in 1.5.1, there has been a steady upward trend in ash content of untreated (raw) smalls obtainable from the N.C.B. and although it would be possible for C.E.G.B. to design new coal-fired plants to operate on the dirtier smalls it is preferable for the N.C.B. to maintain a satisfactory standard for coal to be made available for use in power stations. The advantage of using untreated small coal is that it bears the cheapest possible production costs and can therefore be sold at a price which just covers these. The disadvantages are that the transport of the inert material in them has to be paid for, the combustion process is affected resulting in increased carbon loss and there are increased costs and problems of ash extraction and disposal. Furthermore, for a given grit arrestor or electrostatic precipitator efficiency, the amount of dust emitted to atmosphere increases directly with the ash content of the coal. Another point is that the wider the range of quality of the coals used in any plant, the more difficult it will be to approach the optimum operating boiler efficiency. In this connection for P.F. plants, variation in coal size and usually in rank are of less importance than variation in ash content.

Although taking into account the overall economics of coal winning, preparation, transport, utilisation, ash disposal and "clean air" considerations, it is not possible to justify the selection of a single figure as the maximum acceptable ash content for all coals (because the economics vary from colliery to colliery and power station to power station), it has been decided that ash contents for smalls provided for power station use should not exceed 20% *as a general rule* and 25% in exceptional cases.

In order to comply with C.E.G.B. requirements for the quality standards of small coal the N.C.B., apart from their attempts to raise cleaner coal at the collieries, are embarking upon schemes of partial preparation which will achieve the desired objectives by restricting the ash content of coal available to the power stations. They should not, of course, overclean the coal because of the waste of heat incurred and because this would automatically raise the cost of electricity generation from coal.

1.5.3. **Coal Cleaning**

Pure coal has a specific gravity of about 1.3, shale about 2.4, and pyrites—the mineral which is responsible for high sulphur contents—4·9. Therefore, theoretically it would appear a relatively simple matter to separate them by flotation in liquid media of appropriate specific gravity. However, individual pieces of bituminous coal have a specific gravity range of 1·25 to 1·45, and the range for anthracite is 1·4 to 1·7. This in itself is not a great complication and the real difficulty arises in connection with the proportion of "middlings" in a coal.

Middlings is the term used to describe those pieces which are neither coal nor shale, and have specific gravities intermediate between that of the coal and the shale. There will clearly be little absolutely pure coal and pure shale present, but coals which can be economically and efficiently improved in quality by flotation processes will approximate to a mixture of coal and shale with only a small proportion of middlings present.

A middlings lump can either be composed of a piece of shale adhering to a piece of coal, or of a piece of inferior coal—one which has a high inherent mineral content. The latter is the true middlings; the former can be made to yield more free coal in a rewash if it is first separated and then nipped in a crusher in order to give opportunity for the coal to break away from the shale.

1.5.4. **Wet Processes**

In practice high-density liquids are not used, but they are simulated by a mixture of finely divided solids kept suspended in water—like "liquid mud". Such liquids are not suitable for floating small coal particles since they would get irretrievably mixed with the medium, and in practice no pieces of coal of less than $\frac{1}{2}$ in. in size (usually 1 in.) can be processed in this way.

Another older method of simulating a high density liquid is to float the coal in a mixture of sand and water. The sand and water are held together by rapid upward recirculation. In this method all coal below 1/16 in. in size (approximately the sand particle size) must first be removed. This type of washer is known as the Chance Cone and it was developed in the days when coal was mined dry and could be screened at 1/16 in. or $\frac{1}{2}$ in.—the large +4 in. being first removed for cleaning by hand selection:

Small coal (below 2 in. and below 1 in. according to need) is most effectively cleaned in jigwashers, which exploit the different rates of fall in water of coal and shale particles. These two main types of washer frequently operate side by side, the one dealing with the larger pieces and the other with the small.

When smalls are washed, material below 30-mesh ($\frac{1}{2}$ mm) remains in the wash water and would accumulate. It can no longer be discharged to rivers and all must be recovered. This can be done after flocculation and settling, by vacuum filtration or filter press to yield a slurry. The oldest slurry separation process was to settle in ponds or lagoons (rather like P.F. fly ash disposal) and then drain off the water to recover the slurry by forming mounds or banks. Some slurry can be added and mixed back to the washed smalls product, but the amount may be limited by the percentage ash that can be tolerated and by the adverse effect that too large a percentage of slurry may have on handleability. Some coal fires can be effectively cleaned by the froth flotation process and this gives opportunity to dispose of more in a washed small coal. The process is expensive but valuable in connection with small coals to be used for metallurgical coke production, where the amount of fines present is not important, but the ash content is. Figure 1.5.4A shows the flow sheet of a typical coal preparation plant.

(a) The Dense Medium Process

Dense medium plants are generally used for cleaning coal in the size range 8 in. to 1 in. by using suspensions of finely divided magnetite, barytes, even the shale itself or sand in water. The dense medium must be carefully maintained at appropriate specific gravity levels. By this means separation into three products, coal, middlings and shale can be efficiently carried out. When raw coal is fed into a dense medium of specific gravity 1·4, clean coal separates and floats to the surface for easy removal by rakes or paddles. Middlings and shale sink, and are continuously removed from the bath by means of an elevator, to a second bath of dense medium say of specific gravity 1·8, in which the shale sinks and the middlings are recovered. These can then be crushed and rewashed in a jig washer along with raw coal of 1 in. to dust size. Well-known makes are the Barvoys and Simon Carves.

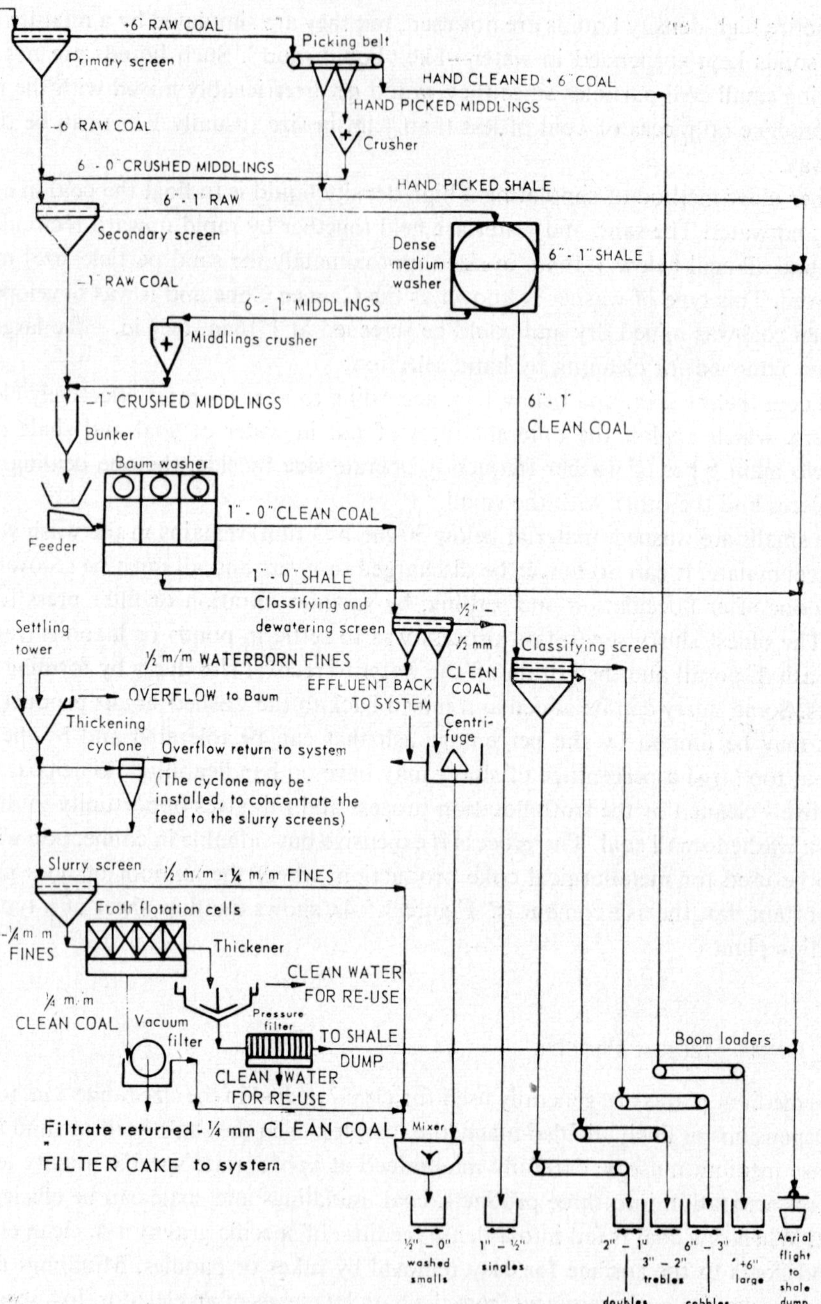

FIG. 1.5.4A. Coal preparation plant—flow diagram

(b) The Jig-washing Process

Cleaner, less dense coal is separated from the heavier dirt when water is moved rapidly up and down through a bed of raw coal on a perforated plate. The pulsation may be produced by a mechanically driven piston or pneumatically as in a Baum washer (see Fig. 1.5.4B).

The jig washer can handle coal of any close size range up to 6 in. top size, but most installations are designed for coals below 2 in. top size. Jig washers are frequently referred to as Baum or Acco boxes. The movement of the water causes dirt to settle out on the plate while clean coal forms an upper layer which is removed by a stream of water.

Jig washers can yield three products if required—cleaned coal, middlings and discard, and are generally installed where there is a marked difference in specific gravity of the coal and dirt particles with few middlings. Smalls which cannot be washed efficiently can often be used without further treatment in power station boilers.

Centrifugal driers are increasingly being used to de-water washed smalls from jig washers before loading for transport. Smalls are slow draining, taking as much as 48–72 h to reach equilibrium in free draining wooden rail wagons, and longer in steel wagons. If any were required for new power stations, the "merry-go-round" trains working transit times of 4 h or less would not permit their being received in a handleable condition. Further, the avoided cost of transporting water (and the enhanced boiler efficiency with low moisture fuel contents) can be set against the cost of providing and running the centrifuges. Figure 1.5.4C shows the flow diagram of a Baum jig wash box.

(c) Cyclone Washers

Cyclone washers or hydroclones as they are sometimes called, are something of an innovation and as yet are not in great use in the U.K. It is claimed that for the treatment of smalls below $\frac{3}{8}$ in. no single washery system is known which can equal the accuracy of separation of the cyclone washers. They can handle coal up to $1\frac{1}{4}$ in. in size or the very smallest sizes.

Although N.C.B. may make some use of this type of washer where fully washed coals are required, they will as far as possible avoid wetting the small size fractions of the great bulk of coal required for use in power stations because of the cost and difficulty of removing the water again and the effect upon handleability of the processed coal.

(d) Froth Flotation

With increase in mechanisation in the coal industry, the production of fines of less than 1 mm in size has increased considerably. This fraction of coal may require special treatment in cleaning. For this purpose the froth flotation plant has been developed from a method used to reclaim metal ores. A foaming agent, usually creosote containing cresylic acid, is added to the fine coal in water. A mixing impeller which also draws in air to the system, forms a stream of bubbles which adhere to the coal particles but not to the shale. The clean

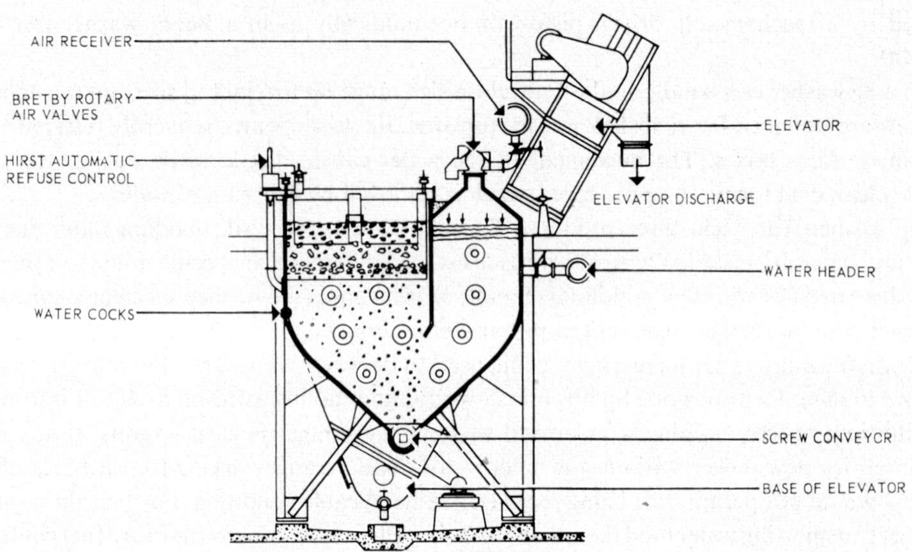

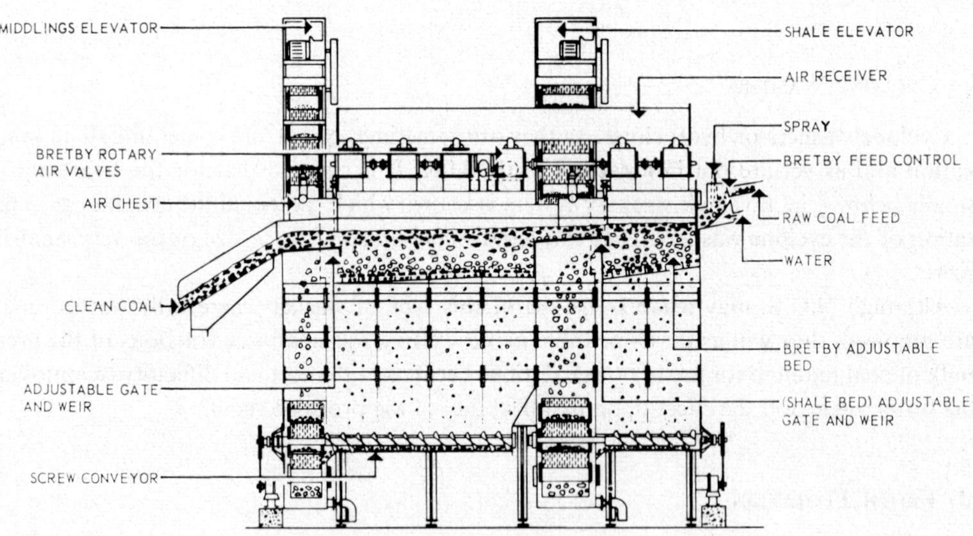

FIG. 1.5.4B. Baum jig washer

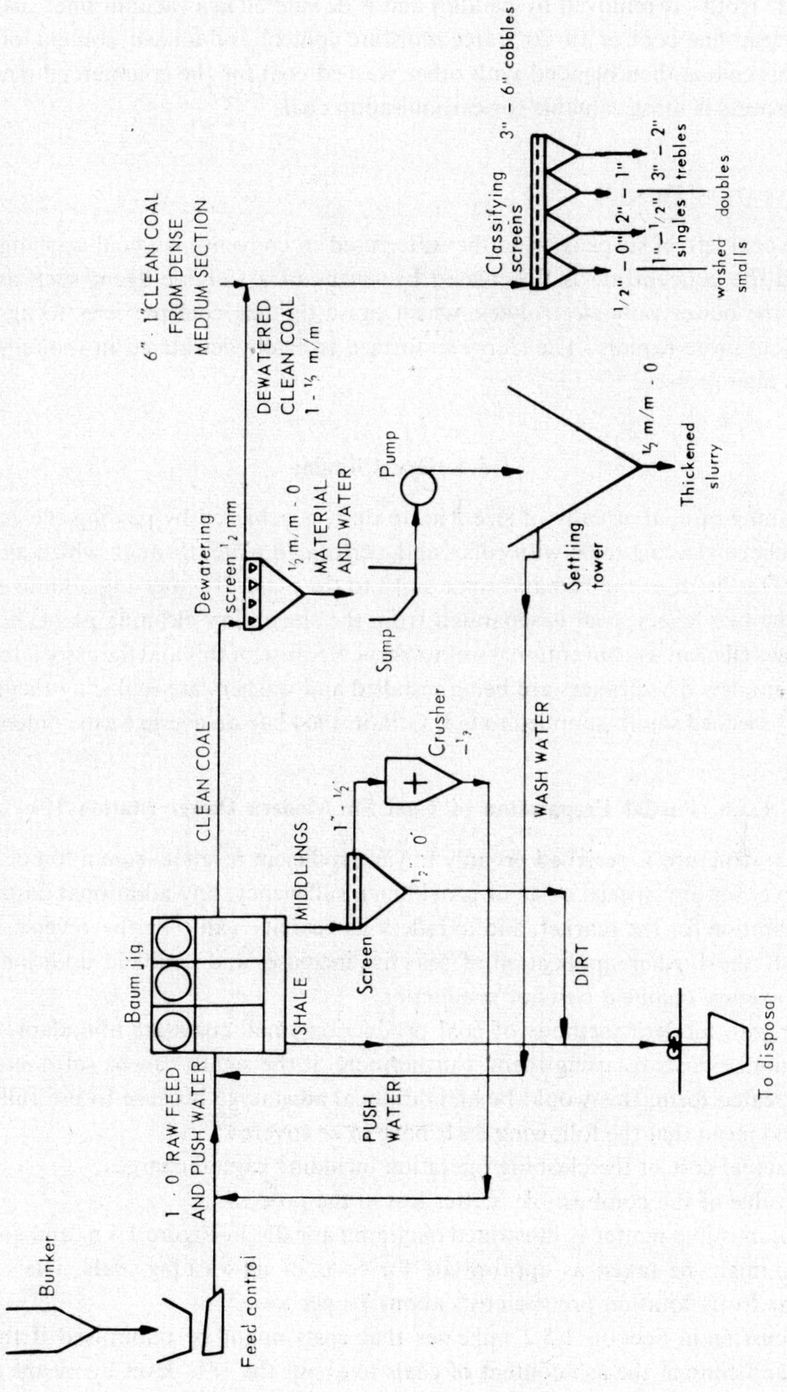

Fig. 1.5.4c. Baum jig washer—flow diagram

coal then floats to the surface as a "froth" whilst the dirt sinks to the bottom of the flotation cell.

The coal "froth" is removed by paddles and is dewatered in a vacuum filter plant to give a cake of clean fine coal of 18–20% free moisture content and an ash content of between 5–13%. This coal is then blended with other washed coal for the commercial market. The flotation process is most valuable for carbonisation coals.

(e) FLOCCULATION PROCESS

The fine coal left in suspension in the water used in conventional coal washing plants is settled and the concentrate is flocculated by means of a suitable agent such as alkaline starch, or the newer poly-electrolytes, which cause the fine coal particles to agglomerate and settle out more rapidly. The slurry so formed is finally dewatered in vacuum filters or sometimes filter presses.

1.5.5. **Dry Cleaning**

Dry cleaning of coal, usually of size 2 in. to dust, is achieved by passing the coal over a slightly inclined shaking table with riffles and perforated plate, through which air is blown upwards. This fluidises the bed and shale sinks to the plate. By inserting a knife edge plate between the two layers, coal is separated from the shale. Dry cleaning plant, however, is not nearly as efficient as conventional wet washers. Because of this and the associated hazards to health, no new dry-cleaners are being installed and washers are replacing them. 467,000 tons of dry cleaned smalls supplied to C.E.G.B. in 1964 had an average ash content of 18%.

1.5.6. **Partial Preparation of Coal for Modern Power Station Use**

The price structure is designed broadly to yield sufficient revenue from a ton of each coal product to cover any special costs of providing a sufficiency, any additional costs incurred in its preparation for the market, and to reflect its scarcity value or the reverse. This, together with the further application of selective increases and coalfield additions, largely meets the average coalfield costs of production.

Because with modern methods of coal production small coals are abundant, C.E.G.B. would minimise costs by using them. Furthermore, if the smalls can be satisfactorily used in the untreated form, this would be an additional advantage because to use fully cleaned coals would mean that the following costs have to be covered:

(a) the actual cost of the cleaning operation including capital charges;

(b) the value of the combustible matter lost in the process.

Loss of combustible matter is illustrated diagrammatically in Figure 1.5.6, and some 3/- to 4/- per ton might be taken as appropriate for costs of jig-washing coals. The washing of fines by the froth flotation process costs about 7/- per ton.

The discussion in Section 1.5.2 indicates that costs might be minimised if the N.C.B. can limit and control the ash content of coals to about the 18% level by means of partial preparation and blending. The preparation can involve either:

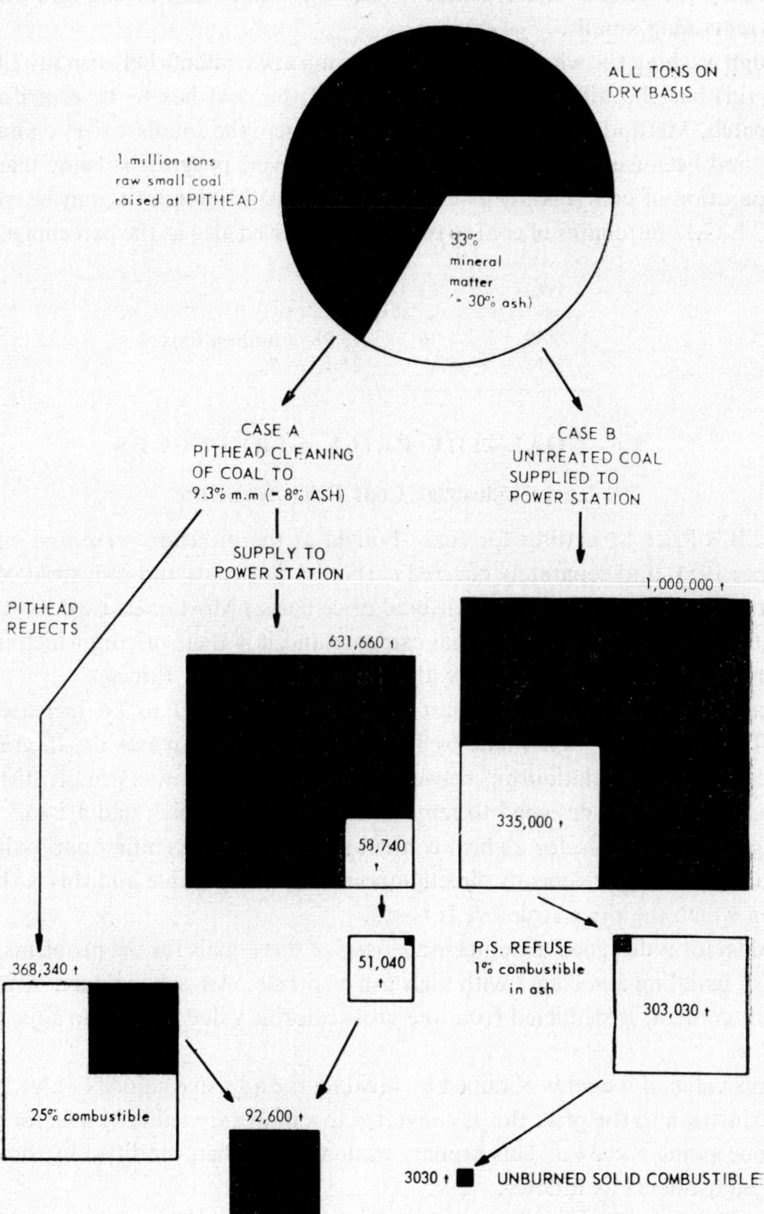

Case A is equivalent to operating a power station with 22·1% combustible in refuse.

FIG. 1.5.6. Loss of combustibles in power station coals—washed and untreated

(i) Splitting the stream of raw small coal in order to wash an appropriate proportion of the whole, and returning it to the remainder after drying.
(ii) Washing the dirtiest size fraction of the raw small coal; drying and blending into the remaining smalls.
(iii) Rough washing the whole of the smalls (giving ash contents between say 14 and 20%).

Method (iii) has the difficulty that the whole of the coal has to be centrifugally dried before despatch. Method (ii) would not be suitable where the smallest sizes contain the bulk of the dirt, and because of the drying problem. However, progress is being made with the partial preparation of coal (mainly in Yorkshire and East Midlands) as may be seen from the following C.E.G.B. purchases of coal so treated—expressed also as the percentage of all coal:

1962	1963	1964	
0·59	5·36	13·38	million tons
1·1	8.8	21·1	%

1.6. COAL—THE PRICE STRUCTURES

1.6.1. Industrial Coal Price Structure

The N.C.B.'s Price Structures for coal—bought at the pithead—were first employed on 31 December 1951, and separately covered carbonisation coals and industrial coals. (Coals for domestic use are sold on a zone-delivered price basis.) Most coals used in the electricity supply industry fall within the industrial category and it is their pricing which is discussed. (Slurries are purchased in accordance with a Schedule of Slurry Prices.)

During and immediately after the last war, coal prices had to be increased by large amounts. These increases were made by imposing flat-rate increases on all grades of coal with the result that the relationship between prices and quality was grossly distorted. The Coal Price Structure was designed to remove the anomalies which had arisen.

Coal in general is bought for its heat content and, therefore as a first approximation, the value of coal to the buyer depends directly upon the calorific value and this is the principal factor upon which the price structure is based.

A second factor is designed to compensate users of dirty coals for the problems of combustion and ash handling associated with high ash contents. An ash penalty, rising with increasing ash content, is deducted from the gross calorific value to yield an adjusted calorific value.

The points value of a coal is obtained by dividing the adjusted calorific value by 100. As a first approximation to the price this is converted to a monetary value by a factor; the present worth of one point is 8·34 *d*. This primary evaluation is then modified by the application of various adjustments as follows:

(i) *Size and mode of preparation*
—washed smalls are marked up and untreated smalls are marked down, the amount depending upon size; the large sizes are higher priced than the smaller sizes.

(ii) *Sulphur content*
—a monetary addition for low sulphur or deduction for high, varying according to content.

(iii) *Commercial adjustment*

—to allow for special characteristics which have commercial significance, for example low ash fusion temperature; fines content.

(iv) *Selective increases*

—starting in May 1954 these have been employed from time to time to regulate the supply and demand of certain classes of coal and to deal with increasing production costs in certain coalfields.

(v) *Price grouping*

—calculated prices are grouped by rounding off into "boxes" or steps of 1/- per ton differences.

(vi) *Coalfield adjustments*

—additions originally designed to maintain approximately the sales to the various markets from each main producing district. Without these the use of the Price Structures would have produced chaos. The adjustments which are flat rate additions ranging from zero to 13*s.* per ton, partly reflected differences in coalfield average production costs. By 1962 these adjustments ranged from zero to 23/8*d.* per ton and since 1966 from zero to 53/8*d.* per ton. The prices of coals from some coalfields are now so distorted that, when costed on a heat basis, washed coals may be cheaper than untreated.

The application of this method of pricing is illustrated by the examples below. Full details of the scales and adjustments are given in *Coal: the Price Structure*, a handbook published by the Confederation of British Industry (formerly F.B.I.).

EXAMPLE:

	A (Untreated smalls)	B (Washed smalls)
Coal type and analysis	$1\frac{1}{2}$ in. U.T.S.	$\frac{3}{4}$ in. W.S.
C.V., Btu/1b	10,500	11,800
Ash, %	20	8
Sulphur, %	2·2	1·4
Industrial coal evaluation		
Ash penalty	minus 875	minus 200
Adjusted C.V.	9625	11,600
Industrial points	96·25	116·00
Primary evaluation at 8·34d.	803*d.*	967*d.*
	= 66/11*d.*	= 80/7*d.*
Size and preparation adjustment	minus 1/3*d.*	plus 3*d.*
	= 65/8*d.*	= 80/10*d.*
Sulphur adjustment	minus 11*d.*	Nil
	= 64/9*d.*	= 80/10*d.*
Selective and commercial adjustment	assumed Nil	
Grouped price	65/2*d.*	81/2*d.*
Coalfield addition say	8/2*d.*	8/2*d.*
Pithead price:	73/4*d.*	89/4*d.*
Cost of 1 million Btu at pithead:	37·4*d.*	40·5*d.*

1.6.2. Carbonisation Coal Price Structure

The carbonisation price structure is applied to most coals of N.C.B. Rank Code Numbers: 300, 400, 500 and 600. These coals are held to be of intrinsically greater value than the industrial steam-raising coals. Their technical evaluation, therefore, results in higher pithead prices than would be obtained by the use of the industrial formula (the same coalfield additions are applied to both).

The technical evaluation takes into account the yield of gas obtainable and the strength, quality and quantity of the coke that can be produced. Coals of Rank 602 would give the same price by the use of either price formula, while coals of Rank 301 command the very highest prices.

1.6.3. The Slurry Price Schedule

This schedule gives the price per ton of slurry loaded free into road vehicle. The price varies in accordance with the calorific value group, and with the coalfield of origin to take some account of coalfield additions.

1.7. TRANSPORT OF COAL TO POWER STATIONS

1.7.1. Rail

The railways are now the most economic means of carrying and unloading the very large quantities of coal required by modern power stations, as shown in Table 9.

TABLE 9

DISTRIBUTION OF COAL CARRIED TO POWER STATIONS BY METHOD OF TRANSPORT

	1967 mn tons	%
Rail	39·0	61·8
Sea	8·9	14·1
Road	10·2	16·2
Inland waterway	2·2	3·5
Conveyor belt	2·8	4·4

It is expected that in the foreseeable future at least 40 million tons per annum will be transported in permanently-coupled trains for rapid discharge at power stations. These trains will each carry up to 1000 tons of coal which can be discharged in well under 1 h as the train moves ($\frac{1}{2}$ mile per hour) without uncoupling the wagons or locomotive. The rail wagons are built of steel, of hopper shaped design, with large bottom doors which are automatically opened and closed. The carrying capacity is either 26 or 32 tons, depending on whether or not the supplying colliery can load them with a height extension — known as a "canopy".

Other types of steel hopper wagons carrying 21 and $24\frac{1}{2}$ tons of coal are also in use. For power stations which use tipplers, the wagon carrying capacities are 16 and 21 tons. These wagons are of steel construction—wooden wagons are fast disappearing.

1.7.2. Coastwise Shipment

This method is used for coastal and estuarial stations remote from coalfields. The quantity of coal carried by sea can be expected to decrease for two reasons: (a) Most of the new stations which are being built remote from the coalfields are likely to be oil-fired or nuclear; (b) the exceptionally high transport costs of seaborne coal which involve movement by train from pit to port and loading into ship for final delivery.

The bulk of the coal carried by sea is handled in vessels carrying some 4300 tons, but the Board have recently chartered six 6500-ton vessels for trading to new Thames estuarial power stations. Smaller ships which carry only 2500 tons are required for certain stations, notably those which have to pass the Thames bridges. For some stations even smaller ships are necessary, for example, those operating in the Bristol Channel. The Board owns a fleet of ships and charters the remainder required.

1.7.3. Road

The carriage of coal by road is only employed where the producing colliery has no rail connections, for transport from small mines, for moving slurries and similar fuels not satisfactorily dischargeable from rail wagons, and in emergencies.

1.7.4. Other Methods

Conveyor, inland waterway and pipeline are other methods of delivering coal. However, if for any reason production at the supplying pit stops, alternative delivery arrangements would have to be made.

1.8. COAL STORAGE

It is necessary for power stations to have a reserve store of coal on site at all times to enable them to continue to generate during interruptions in either the production or the transport organizations. The Board has decided that power stations should always be prepared to operate for at least four weeks, from their own stocks. The heaviest generation is expected to occur in winter time and stocks will normally be lowest in March in any year. Consequently, the basic reserve for stations should be equivalent to four estimated average winter weeks' consumption.

Assuming the summer period to be from April to September inclusive, approximately 48% of coal production takes place in this period. For the reason stated in the preceding paragraph, power stations have to be equipped to handle coal in and out of stock at a rate sufficient to satisfy the station's needs. Extra expense would be incurred if the coal industry had also to stock and destock coal because power stations use less coal in summer than winter. Therefore, the power stations having the facilities, must use them to take coal from the collieries at an even rate throughout the year, or in other words, to take coal as it becomes available. There is an added advantage in so doing — the load on the transport systems is evenly spread and consequently less costly to operate.

The amount of seasonal coal stocking and destocking is increasing; there are two reasons for this. Firstly, the pattern of power generation is tending to peak more in the winter; secondly, there is the tendency for coal to become a peak-load fuel. The introduction of oil as a fuel for power generation has been a contributor to this and as more nuclear power becomes available—this may at some time in the future take the whole of the base load—so will the disparity between the quantity of coal consumed in the summer and winter period increase. Some five years ago it was necessary only to stock and destock some $2\frac{1}{2}$ million tons each year; we now put in and take out about 4 million tons.

Small coals are preserved from spontaneous combustion in stock by attention to compaction of the layers and sides of the heaps. Compaction is necessary to prevent the entrance of air in quantity sufficient to cause and sustain combustion, yet insufficient to carry away heat and cool the pile. When graded coals are stocked, it is necessary to do so in small heaps without attempts at compaction so that air flows freely through the piles, cooling them and thus preventing a dangerous temperature. Small coals are best dealt with in large heaps in order to minimise the risks. Nevertheless, it is expedient at many power stations to lay down a stock equivalent to four winter weeks' supply to be left undisturbed for as long as possible, keeping a separate stock heap for seasonal use. The separation of the basic reserve is also useful when stocktaking since as long as it remains untouched, the errors in stock assessment are reduced. However, the reserve stock should consist of specially selected coal so as to ensure that it will be satisfactory for generation when an emergency arises. It may also be necessary to create a smaller additional stockpile of specially selected good quality coals to act as a "sweetener" for the seasonal stock when using it under adverse weather conditions.

1.9. COAL SAMPLING

1.9.1. Purpose of Sampling

The Board requires to know for its power stations the quality:

(a) of coal paid for when delivered;

(b) of coal consumed during any given period.

In 1964/5 the Generating Board paid some £220 million at the pithead—and nearly £40 million for transport—for over 63 million tons of coal; it is important, therefore, to monitor the quality of deliveries to ensure that value for money is obtained. In addition, a knowledge o fthe quality of coals produced enables fuel supplies officers to plan the supply of coals to individual stations in accordance with their technical needs in the most economical way.

Secondly, the performance of plant is frequently measured using coal test results so that any falling away from standards may be detected and steps taken to restore efficiency.

Since it is not possible to check every piece of coal in a consignment, laboratory tests are carried out on a sample which is taken, crushed and reduced to represent the bulk of coal from which it was drawn. These processes are scientifically specified and must be rigidly observed in practice, for the final test sample to be dependably representative of the consignment. If a sample does not represent the bulk accurately no reliable test results can be obtained. No laboratory expertise can make up for the failure in sampling.

1.9.2. Theory of Sampling

The ways in which coals should be sampled to provide the information required, are set out in B.S. 1017: Part 1:1960; the sampling of coke is dealt with in B.S. 1017: Part 2:1960. In addition sampling for screen analysis of coal and coke is discussed in B.S. 1293 and B.S. 2074, 1965. (See Section 1.12 for a discussion of miscellaneous tests, size analysis of raw coal and P.F., grindability, handleability and bulk density.) These Standards must be consulted for full practical details, but a consideration of the theory is necessary for their proper understanding.

All discussion of coal sampling theory for general analysis is based on the fact that the ash content is the most variable characteristic, and that a sample satisfying accuracy requirements in respect of ash will give better than that accuracy in other characteristics. Sampling for moisture in coal is dealt with separately only because of the different handling needed to minimise drying losses.

Experience shows that the moisture content is the biggest variable in coke and accordingly this occupies the place taken by ash in coal sampling theory. Because of the consequential similarity, and the lesser importance of coke to the Board, coke is not considered further in this chapter.

Sampling methods can be classified within three types, which can be called arbitrary, selective and representative.

Arbitrary sampling describes the situation in which a labourer is instructed to obtain a sample for test; the labourer naturally goes to a conveniently accessible part of the consignment and collects the required amount. This method is often claimed to be random, but with a material of variable quality, it can produce samples with characteristics quite different from the bulk. This is particularly the case with coal because the large pieces tend to segregate at the edges of coal streams or heaps.

Selective sampling is used if the worst of a consignment can be taken and tested on the assumption that if the sample is acceptable, the consignment must be satisfactory. There are few circumstances where this method is useful and it needs experience for it to be successful. Test results on coal sampled this way would not be of any great value.

Representative sampling is the most logical method: a consignment is divided into a number of roughly equal parts—grabfuls, wagons, minutes of discharge from a belt or hopper, etc., and an increment (part sample) is taken, without discrimination, from each part; the increments are then combined to form a gross sample. Ideally, a second sample should give test results not significantly different from the first.

When sampling coal by the methods of B.S. 1017, a number of unbiased increments, each of a prescribed minimum weight, are combined to yield a representative sample, the test results from which can be relied upon not to show errors greater than a previously selected level.

1.9.2.1. Unbiased Increments

An unbiased increment is one in which all particles presented for sampling have an equal and random chance of inclusion, there being no selective rejection of either large or small sizes, which would lead to results of analysis being persistently higher or persistently lower than the correct value. Bias in sampling can be detected only by reference to a standard method of sampling which is known to be free from bias. There are no certain ways of eliminating bias, only the application of some general rules and commonsense.

1.9.2.2. Weight of Increments

For any given weight of increment, its accuracy will depend upon the size of the largest pieces: this follows from a theory of size–weight ratio first propounded by E. G. Bailey in 1909 (see B.S. 403:1930, General Principles of Sampling, by Grummell and Dunningham). Clearly, the chance inclusion or exclusion of a piece of rock 1 in. across will make for greater error in a 1-lb increment than in a 2-lb increment. In practice the general rule is:

$$\text{top-size in inches} \times 2 = \text{minimum increment weight in pounds.}$$

It is important that increments should be of approximately equal weight. Serious loss of accuracy arises if increments are less than the stated minimum weight: on the other hand there is no appreciable gain in accuracy if the minimum weight is exceeded. Nevertheless, designers of automatic samplers generally arrange for increments to be overweight in an attempt to avoid biased sampling. This yields embarrassingly large gross samples which are usually dealt with by coarse crushing, mixing, extracting a convenient sub-sample and returning the excess to the coal handling system.

1.9.2.3. Number of Increments

Increments should be evenly distributed over the bulk of coal being sampled. The number of increments necessary for a sample to be representative of a consignment depends upon three factors:

(i) the accuracy aimed at;
(ii) the uniform or variable quality of the coal, and
(iii) whether this consistency is known or is to be ascertained.

Accuracy is usually expressed in the form $\pm 1\%$ with a probability of 95%. This means that if a coal with a true ash content of 18% is sampled 100 times, 95 of the results will be within the range 17–19%, the remainder tending to be evenly divided above and below these limits.

The purpose of sampling will set the accuracy. For test purposes, a "high" accuracy of $\pm\frac{1}{2}\%$ requiring hundreds of increments may be indicated but for commercial purposes a "normal" accuracy of $\pm 1\%$ is usually chosen, so that only scores of increments are needed. At the other end of the scale, if we wished to know the ash content within $\pm 5\%$, then less than five increments would suffice.

It should be noted that, in general, to double the accuracy requires the taking of four times the number of increments.

1.9.2.4. Consistency

This has been found to be a function of ash content: the higher the ash content, the less uniform the coal and consequently the greater the number of increments needed to yield the same accuracy of sampling. For the same accuracy, the number of increments is approximately related to the square of the ash content.

1.9.2.5. Sampling Behaviour

The sampling of known coals in routine supply, in accordance with the above rules, can be very simple. If a consignment is one of a series whose sampling behaviour has been determined, the minimum number of increments can be taken with confidence. If, however, an isolated consignment, or a first delivery has to be sampled, then procedure must be based on the presumed ash content or, if that is not known, it must be that used for the most difficult coal known at the site in order that the accuracy does not fall below that required. With the sampling of subsequent deliveries, as discussed in the following paragraph, experience will show what accuracy is achieved by the sampling procedure and if the sampling effort should be intensified or relaxed.

These principles apply whether increments are taken by hand or mechanically at predetermined intervals (see Sections 1.9.4.1 and 1.9.4.2).

1.9.3. **Replicate Sampling**

The accuracy of sampling attained with any given coal is assessed by comparing the test results from replicate samples. The same total number of increments are taken, but they are not bulked together in one gross sample; instead the increments are systematically allocated in rotation to two or more containers for sub-samples which are prepared and analysed separately.

To assess the accuracy of sampling of a single consignment at least six sub-samples are required. The range of the ash contents of these sub-samples is found and compared with "limiting values of range" for the different standards of accuracy which are given in B.S. 1017. These limiting values are set by statistical concepts. As an example, a consignment of coal sampled in replicate had six sub-samples which gave on analysis the following ash contents:

$$15{\cdot}3,\ 17{\cdot}1,\ 16{\cdot}5,\ 17{\cdot}2,\ 15{\cdot}8,\ 16{\cdot}4:\ \text{mean} = 16{\cdot}4\%$$

The difference between the lowest and highest in this series is 1·9. In B.S. 1017 the lower and upper limiting values of range for the high standard of accuracy (i.e. $\pm 0{\cdot}5\%$) are given as 0·7 and 2·7 respectively. Therefore, it may be assumed that the sample has been collected to at least this accuracy so that the true value of the consignment lies within the limits $16{\cdot}4 \pm 0{\cdot}5$, i.e. between 15·9 and 16·9.

If a coal is markedly heterogeneous, the range of ash contents may well be greater than the upper limit. In this case the desired accuracy has not been achieved. On the other hand a more homogeneous coal will give a narrower range of ash contents which may fall below the lower limit, in which case it may be assumed that a greater accuracy than required, has been achieved.

If more than six sub-samples are taken, then for a given coal, the range of ash contents can be expected to be wider since there is more chance of including sub-samples of unusually high or low ash content. Sampling theory takes this into account and appropriately wider "limiting values of range" are used in the assessment.

When dealing with a coal in regular supply, a different method of assessing accuracy is used. Ten successive consignments are sampled in duplicate and the mean difference in ash contents between duplicates is compared with a theoretical value. This theoretical value calculated from statistical theory depends on the accuracy required and whether it is required of an individual sample or of the average of a number of samples. For instance when sampling to the normal standard of accuracy ($\pm 1\%$) the mean difference between duplicates of ten samples should be less than 0·8 for the accuracy of each individual consignment to be $\pm 1\%$, but need only be less than 2·5 for the average ash of ten samples to be accurate to $\pm 1\%$. Further tables indicate how the sampling procedure ought to be modified for any further consignments of the same coal, should the accuracy be too low or too high for the purpose of the user (high accuracy is only achieved at high cost).

1.9.4. **Sampling Processes**

To take an unbiased sample, the whole of the coal to be sampled must be accessible to the sampler. In practice this means that the sample should be taken while the coal is moving, for example, in a falling stream or on a conveyor. This presents some difficulties since coal particles in motion are liable to segregate, larger pieces tending to roll to the sides. For test purposes unbiased increments are sometimes taken by stopping a conveyor belt and brushing off all particles from a given length of belt. This is inconvenient and it is not always possible to restart inclined belts if they are fully laden. It is, therefore, impracticable as a routine procedure.

Automatic samplers are generally preferred since they can take a full cross-section from a stream of coal on a high capacity conveyor; a belt carrying 800 tons/hour discharges at the rate of 500 lb/sec. Hand-sampling may be necessary in some circumstances but instructions to the sampler have to be precise if human errors are to be avoided, and special consideration has to be given to the precautions needed to ensure the safety of the sampler.

1.9.4.1. Automatic Samplers

Five principal makes of automatic sampler are used in power stations:

(a) Bretby (Fig. 1.9.4A)
(b) Birtley (Fig. 1.9.4B)
(c) Geco (Fig. 1.9.4C)
(d) Seaborne (Fig. 1.9.4D)
(e) Pollock (Fig. 1.9.4E)

Their characteristics have been reviewed by R. G. S. Johnson (C.E.G.B., South Western Region, R. and D. Note S44, Automatic Coal Samplers).

The Bretby, Birtley and Seaborne samplers collect an increment in a "bucket" which passes through a stream of coal falling from a belt, and discharges the increment at the end of its travel through the stream. The Geco sampler operates similarly but the increment consists of the coal falling into a slot which traverses the coal stream; as it is collected, it is diverted into a chute. The Pollock sampler sweeps a transverse band from the coal on a moving belt, by means of a blade pivoted above the belt.

Whilst the basic principles are good for all of them, in practice none is free from fault, because of bias introduced when particles are deflected in a non-random way by the lips of the bucket or slot.

Johnson takes the Bretby as his first choice of these five makes, and finds that the Seaborne and Pollock samplers tested gave excessively biased samples.

In view of the faults that have been found in existing types of coal samplers, a C.E.G.B. Standard (No. 50031, Sept. 1968) has been drawn up, specifying equipment capable of dealing with high rates of coal flow yet providing small representative duplicate samples accordance with the requirements of B.S. 1017, Part 1. At present its application is limited to sampling bunker coal for the assessment of station thermal efficiency; because consignments tend to mix wagon discharge hoppers leading to loss of identity, it is not possible to apply the equipment to the sampling for commercial monitoring of coal delivered to power stations in permanently coupled trains. Other methods of achieving this end are being considered.

1.9.4.2. Hand Sampling

(a) *From conveyors and chutes.* When hand sampling from a stream of coal, the increments can be drawn from only part of the stream at a time and care has to be exercised to make the sample representative of all parts of the coal stream.

Instructions to samplers should be as complete and simple as possible. Increments should be collected at specified times and positions to reduce the element of personal bias introduced even by experienced samplers. The sampler should be able to reach with a ladle or scoop the whole cross-section of the stream in safety and without strain.

If sampling at a point of discharge is not possible, increments must be taken from the belt and they should represent the whole thickness and both sides of the coal stream. Sampling from a moving belt demands skill and good judgement.

Coal in bucket conveyors must not be sampled by scoop; the whole contents of a bucket should be taken as an increment, preferably at the point of discharge. To keep down the total weight of gross sample, increments may be split and a part discarded, before adding to the bulk.

(b) *Sampling from road and rail wagons.* It is not possible to use any of the above methods during the filling or emptying of wagons; samples may have to be taken when wagons are tipped or by collecting increments by augers or from holes dug into the coal. Such samples are always suspect since in neither case is there free access to all parts of the coal. Where two products have been loaded separately into one wagon for example, untreated smalls and wet fines, it is virtually impossible to take a representative sample. Even for a fairly homogeneous coal, segregation caused by loading in conical heaps from chutes or

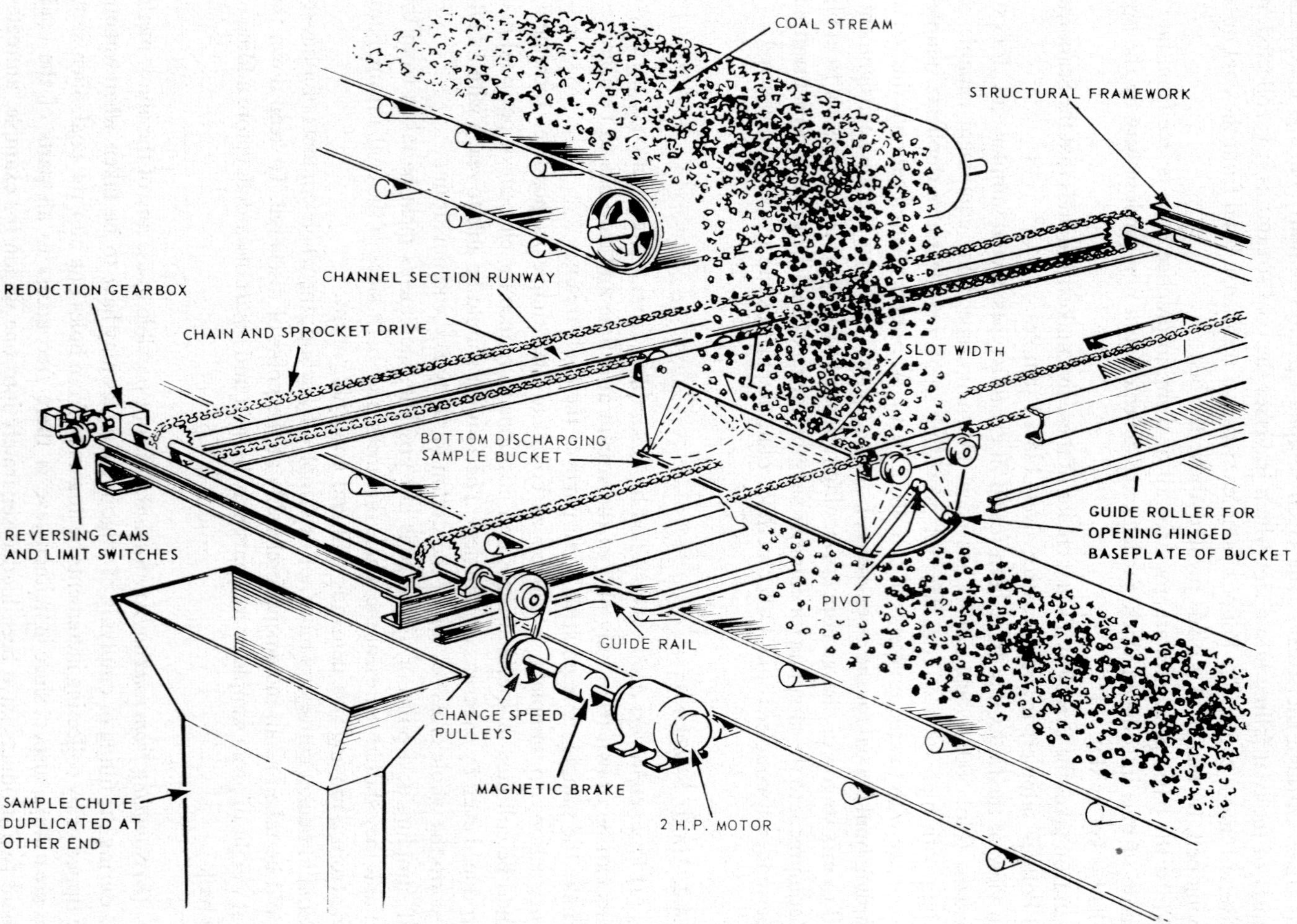

Fig. 1.9.4a. Bretby automatic coal sampler

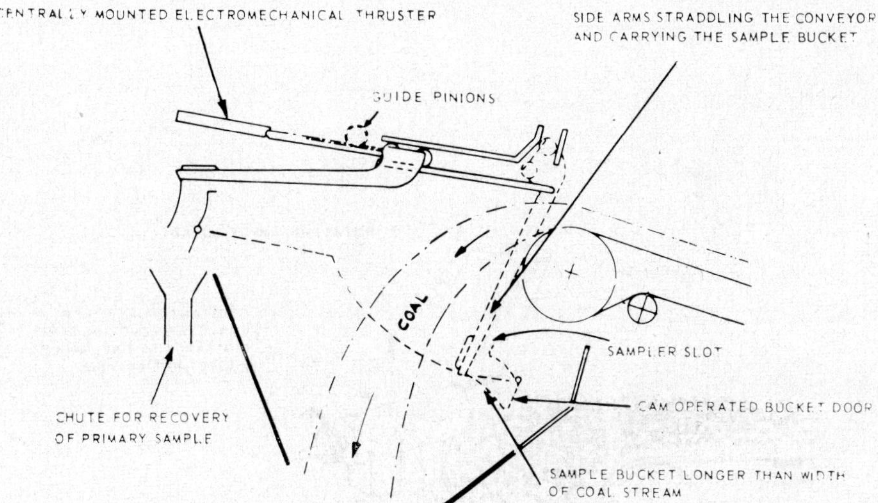

Fig. 1.9.4B. Birtley automatic coal sampler

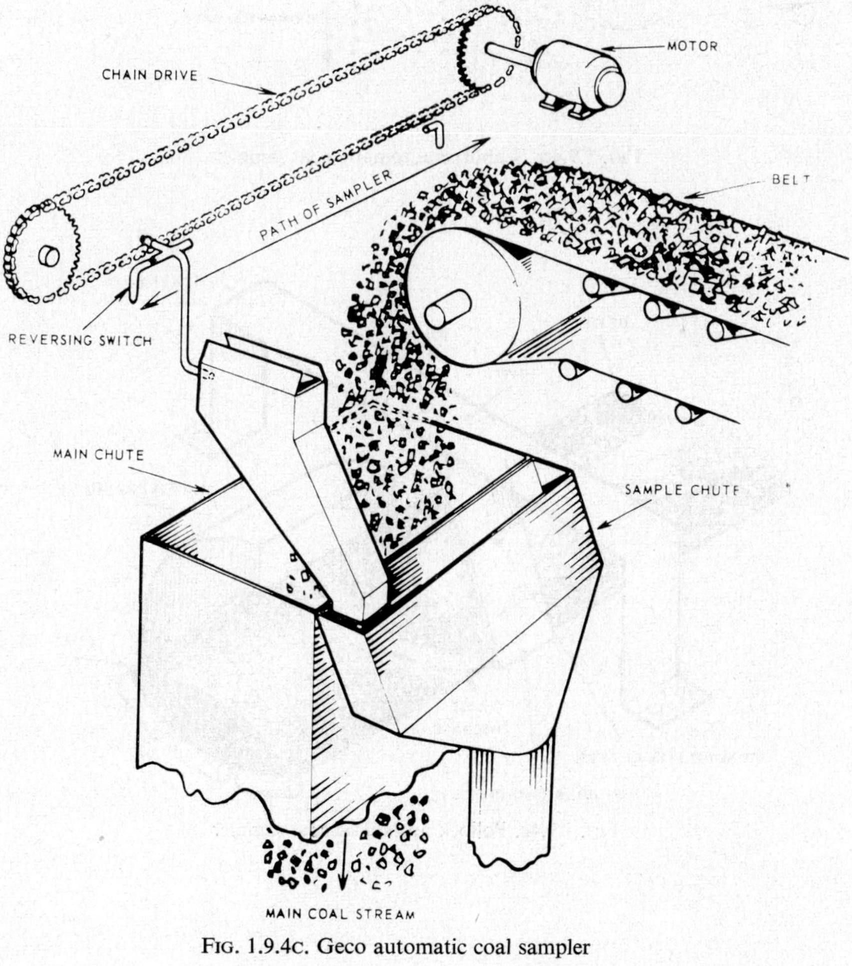

Fig. 1.9.4C. Geco automatic coal sampler

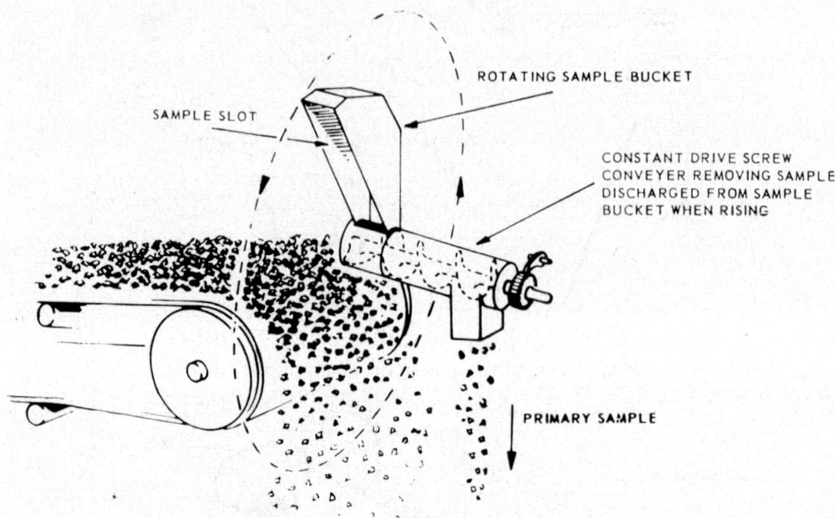

FIG. 1.9.4D. Seaborne automatic coal sampler

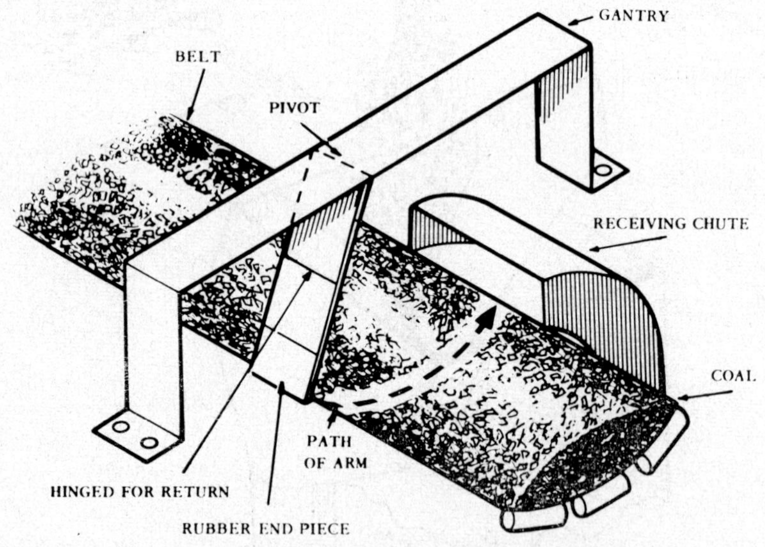

FIG. 1.9.4E. Pollock automatic coal sampler

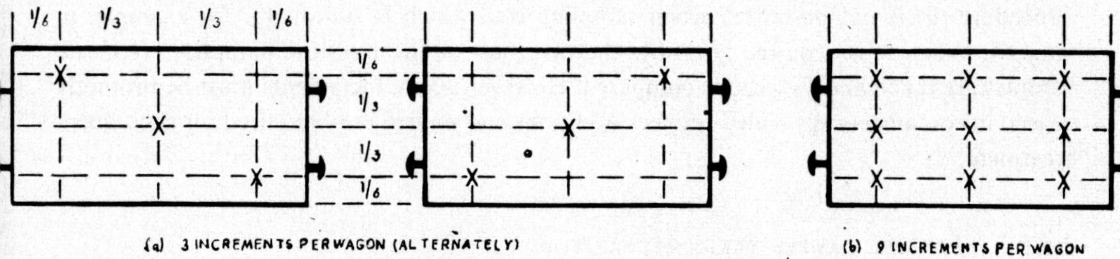

FIG. 1.9.4F. Hand sampling from wagons.

by shunting of wagons, can lead to bias. In this case the bias, checked periodically against other methods of sampling, may vary only slightly and results may be used for detecting changes in quality.

(i) Sampling during tipping: This is one large increment taken as the coal falls from the wagon. Not all parts can be sampled and the point of entry of the container should be varied from wagon to wagon.

(ii) Sampling from wagon tops: Augers and probes have been used to draw a "core" sample and where they are available they are preferred, since they penetrate all layers in the wagon. It is more usual, however, to dig holes not less than 12 in. deep and to take scoopfuls from the bottom. In all cases the pattern of sampling points is such that the centres, ends and sides of wagons contribute their fair share to the gross sample (see Fig. 1.9.4F).

(c) *Sampling from heaps, stock-piles or in ships.* In these cases there is greater possibility of segregation and accessibility of coal is more difficult than in any other sampling situation. Clearly this is a function of size of the heap; if the heap is small enough and space permits, a fair sample can be taken while it is turned over or by dividing the area and sampling as if it were a series of wagon tops. Samples taken from a large heap, say 30 ft high, are never satisfactory even when sampled with an auger and test results for such heaps always have a doubtful validity.

1.9.5. **Sampling for Moisture**

The principles of sampling for moisture are the same as those of sampling for general analysis—the sample must be unbiased and made up of increments spaced evenly over the consignment.

In practice, however, there are two important differences. Firstly, fewer increments are needed for the same accuracy because moisture is more evenly distributed than ash. Secondly, special handling is necessary to avoid loss of moisture. Unless samples are put immediately into a sealed container, they almost invariably become drier with the passage of time. When sampling for moisture, regard should be paid to the condition of the coal, particularly whether the moisture distribution is stable or not. For example, it would be futile to try to measure a rapidly-changing variable, such as moisture, in coal direct from a washer.

Samples for total moisture may be taken:

(i) separately from the gross sample for general analysis, or

(ii) with the gross sample and subsequently abstracted from it.

Procedure (ii) is not permitted when sampling coal which is stationary, for example, in wagons. When it is required to know the precision of the moisture sampling, replicate samples are taken and the results compared. However taken, increments must be promptly placed in containers kept sealed between additions and so held until required for subsequent treatment.

1.9.5.1. Moisture Sample taken Separately

Ten increments are taken at regular intervals in the manner appropriate to the circumstances as if a sample for general analysis were being taken. Each increment should weigh 1 lb for coals up to 1 in., 2 lb for coals up to 2 in. and 3 lb for coals up to 3 in.

1.9.5.2. Moisture Sample Collected with General Sample

The total weight of sample must not be less than the minimum for the general sample plus that for the moisture sample. Until the moisture sample has been abstracted, the sample during and after collecting must be contained so as to avoid drying off.

If the coal is visibly dry and of high rank, the coal may be crushed to $\frac{1}{2}$ in. and ten 1 lb increments immediately taken from the heap of coal. Where the coal is neither visibly dry nor excessively wet, a sub-sample is drawn by ten increments from the gross sample as if taking a separate moisture sample (see above).

If the coal is very wet, two-stage moisture determination is necessary. The sample and containers are weighed, air-dried and the loss in weight noted. The moisture left in the air-dried sample is then determined as for visibly dry coal. In such a case it is always preferable to take the moisture sample separately from the analysis sample (see previous section).

1.9.5.3. Moisture Sample Preparation

Unless visibly dry the sample is air-dried, the loss of moisture being measured, and then crushed to either $\frac{1}{2}$ in. or 6-mesh B.S. (0·11 in.). Reduction to 2 lb or $\frac{1}{2}$ lb, respectively, follows (see section on sample preparation). The air-dried sample is in equilibrium with the air and may be submitted to test to determine the inherent moisture. Visibly dry coals may be treated as air-dried coals from the outset.

1.9.6. **Storage of Samples**

If not treated immediately, sample containers should be properly labelled, stored in a cool place and protected from the effects of heat and weather. Precautions should also be taken to guard against contamination with dust or other samples of coal or ashes.

1.10. SAMPLE PREPARATION

1.10.1. Introduction

The minimum weight of gross sample taken by the procedures described, may be as little as 12 lb or as much as 6 cwt depending on the accuracy required, top size and ash content. Automatic samplers usually take more than the minimum number of increments, each weighing several times the prescribed weight and so the total sample may weigh as much as 30–40 cwt. It is the object of sample preparation to reduce this bulk to manageable proportions.

Just as a gross sample must be representative of the bulk of coal from which it was drawn, so the "analysis" sample must represent the gross sample from which it was produced. As described in some detail in succeeding paragraphs, this can be effected by crushing in stages to small size, provided that careful attention is given to mixing the crushed coal before the bulk is reduced in weight for the next stage. The reductions in weight are made by use of mechanical sample dividers or riffles which ensure proper representation.

The "analysis" sample will consist of only $\frac{1}{2}$ lb of coal contained in a bottle, with a particle size not greater than 72-mesh (0·0083 in aperture) and at this very small size, provided that the coal in the bottle is carefully mixed, each portion taken from it for analytical purposes will possess the properties of the consignment from which it was derived.

1.10.2. Errors of Sample Preparation

It is important that contamination of the sample with extraneous material is avoided, that none of the sample is lost, and that bias is not introduced during sample division. These errors are minimised by careful handling and the use of suitable equipment.

Error is introduced if the weight of sample retained after splitting is less than a specified limit. For any given accuracy, this minimum weight depends upon the size of the coal (see Section 1.9.2.2, Bailey's size–weight ratio theory) and upon the ash content of the coal. This is demonstrated in the following extract from B.S. 1017, Table 10: preparation of samples to normal accuracy (±1%), showing the minimum weight to which a sample may be reduced, according to ash and top size.

TABLE 10

MINIMUM WEIGHT OF SAMPLE AFTER REDUCTION (NORMAL ACCURACY)

Top size of coal	Minimum weight of sample after sub-division, when ash is:				
	5%	5–10%	10–15%	15–20%	20%
$\frac{1}{2}$ in.	35 lb	90 lb	150 lb	200 lb	250 lb
$\frac{1}{4}$ in.	4 lb	12 lb	20 lb	25 lb	30 lb
$\frac{1}{8}$ in.	$\frac{1}{2}$ lb	$1\frac{1}{2}$ lb	$2\frac{1}{2}$ lb	$3\frac{1}{4}$ lb	4 lb
72 B.S. sieve	$\frac{1}{2}$ lb	$\frac{1}{2}$ lb	$\frac{1}{2}$ lb	$\frac{1}{2}$ lb	$\frac{1}{2}$ lb

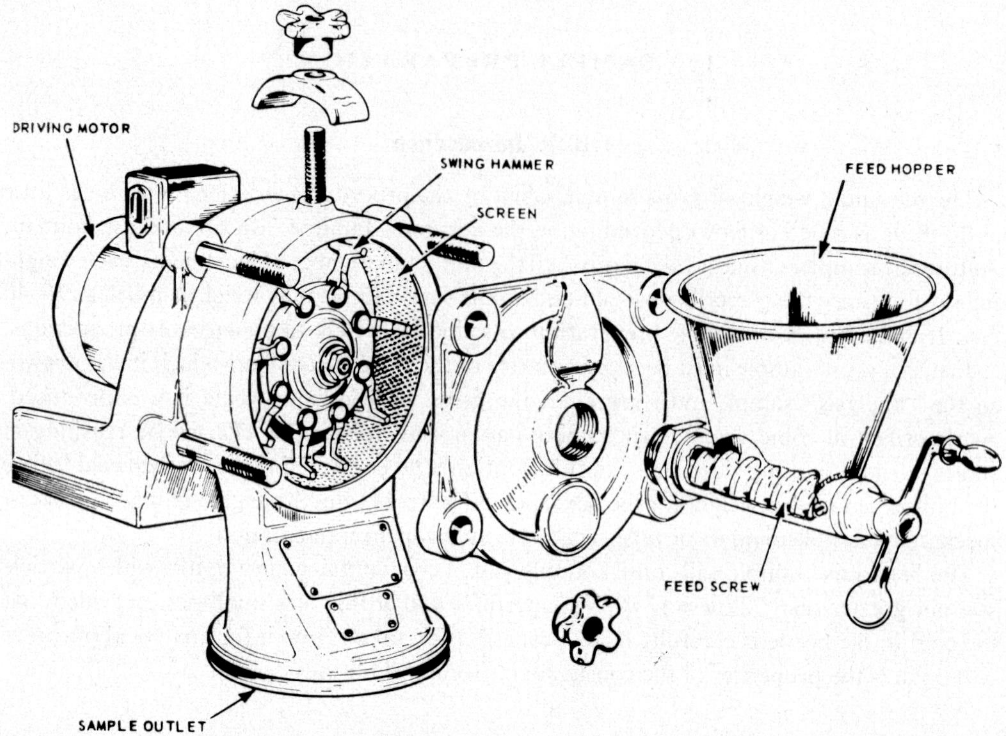

FIG. 1.10.3A. Laboratory coal grinder (Raymond mill)

1.10.3. **Equipment for Sample Preparation**

Sample preparation rooms should be easily cleaned and draught-free. The risks of contamination and loss of sample are thereby reduced.

Samples which are visibly wet must be air-dried before preparation and large driers with flues (occasionally fans) which will not heat the coal excessively, should be installed.

1.10.3.1. CRUSHERS, GRINDERS, ETC.

For particle size reduction, power driven machinery should be used. The only exception are the breaking, with a hand-punner, of large lumps so that they can be fed to a mechanical crusher, and the grinding in a mortar of the small parts of the final sample retained on the test sieve of 72-mesh.

Machinery should be readily accessible for cleaning to avoid contamination and residues left in machines should be removed and added back to the sample.

Two other aspects need attention, particularly with medium- and high-speed machines. Firstly, breather bags should be fitted to trap fine dust which would otherwise be blown away. Secondly, moving parts should not have to be set to fine clearances to obtain the required size of coal; not only does this lead to expensive maintenance and contamination but, more important, the temperature of the coal is raised so that oxidation changes the coal substance. Figure 1.10.3A illustrates a Raymond laboratory grinder.

1.10.3.2. Sample Dividers

Whenever possible, mechanical methods of sample division are preferred to manual methods. Mechanical dividers collect a large number of increments from a stream of coal separating them into "retained" and "discard" portions (see Figs. 1.10.3B, C and D). Adjustments to the settings provide a means of retaining any required fraction; in some machines this can be as low as 1:24th. If the capacity of the receiver is less than the total gross weight, sample division is carried out in a series of steps. A portion of the gross sample is treated at a time and the necessary fractions of each part are combined.

The only manual method now recognised is the riffle; this operates on a basis of halving (see Fig. 1.10.3E). If less than half is required, repassing the sample is necessary; if more than half then the "discard" from the sample is repassed. By a combination of passes and re-uniting of fractions, samples may be reduced to any desired weight. The slot width of the riffle should be about four times the topsize of the coal. Ratios very much greater than 5 or less than 3, lead to bias.

It is important to note that although a sample can be split into two or more parts for analysis, this is not duplicate sampling. Such splitting and testing checks only those operations carried out after the sample is divided. Replicate sampling requires the taking of separate gross samples which are never combined, and lead to separate test results.

1.10.3.3. Sample Mixing

The errors of division can be considerably reduced if the sample is thoroughly mixed beforehand. Hand-mixing frequently leads to segregation and mechanical mixers, e.g. double-cone types (see Fig. 1.10.3F) are preferred.

Replicate riffling and re-uniting the halves is an acceptable alternative mixing method.

1.10.3.4. Final Stages of Sample Preparation

The $\frac{1}{2}$ lb "analysis" sample is intended just to pass a 72-mesh sieve.

If there is more than 1% oversize, it should be sieved out, ground by hand with mortar and pestle and replaced. The exception is when only a routine "ash" determination is required, where a 36-mesh sieve will suffice.

For coal samples it is important that the sample is not overground so that nearly all passes say 120-mesh, because such treatment tends to overheat and oxidise the coal substance. (This is not true for cokes which are more resistant to oxidation. The testing of cokes is in fact made easier and more accurate by superfine grinding.)

Samples should be thoroughly mixed before placing in a sealed tin or bottle bearing a properly detailed label, which must not be easily detachable.

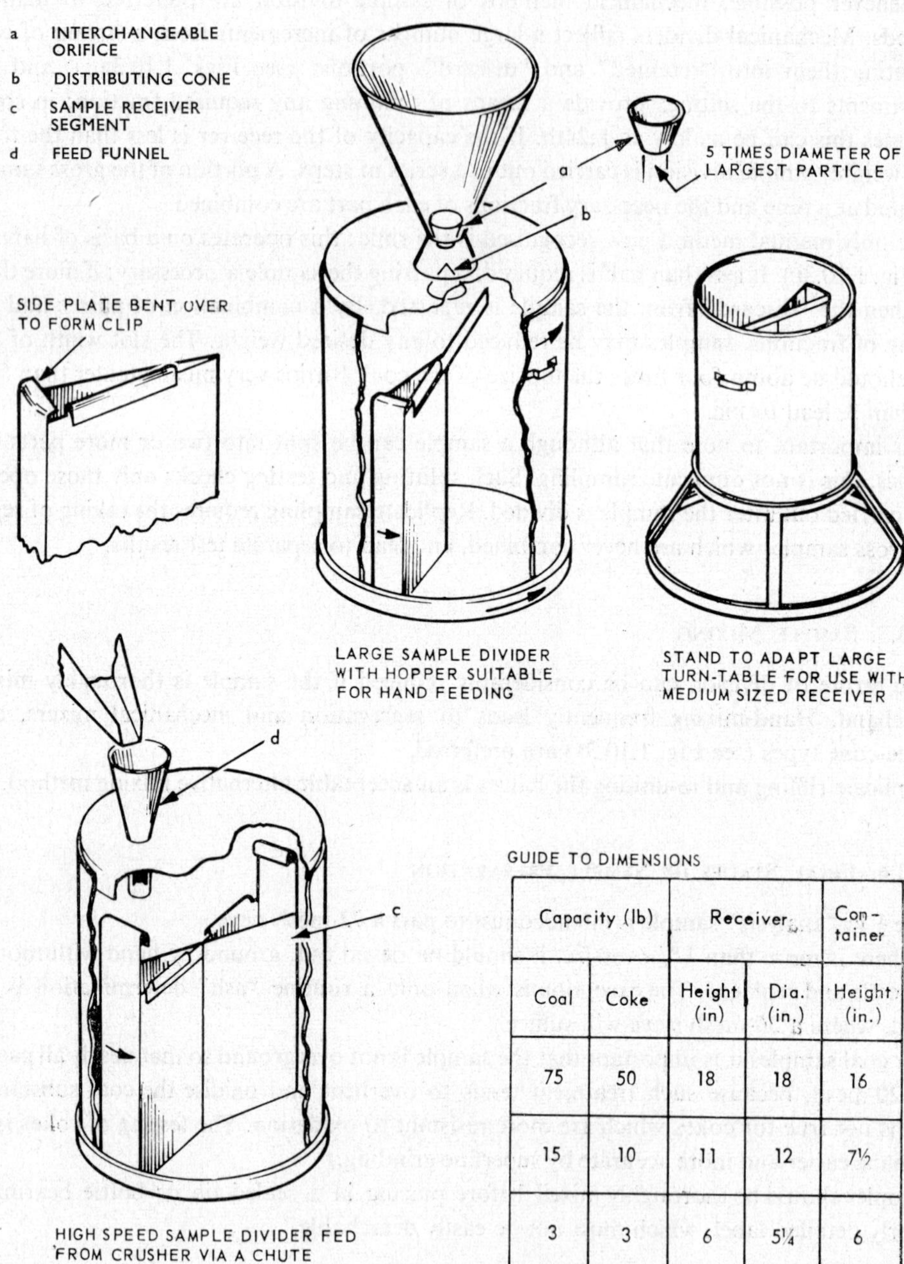

GUIDE TO DIMENSIONS

Capacity (lb)		Receiver		Container
Coal	Coke	Height (in)	Dia. (in.)	Height (in.)
75	50	18	18	16
15	10	11	12	7½
3	3	6	5¼	6

FIG. 1.10.3B. Rotary mechanical sample dividers

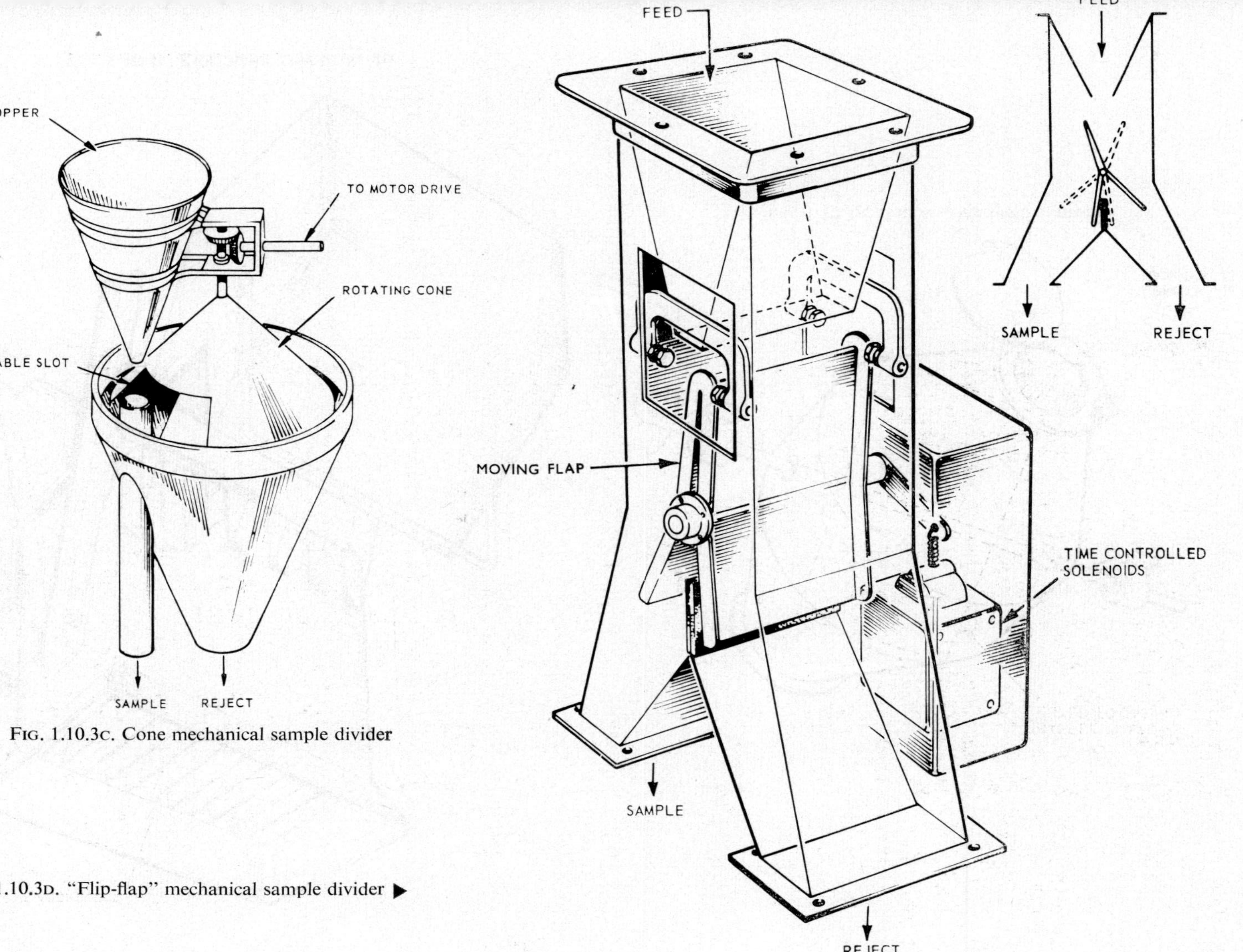

FIG. 1.10.3C. Cone mechanical sample divider

FIG. 1.10.3D. "Flip-flap" mechanical sample divider ▶

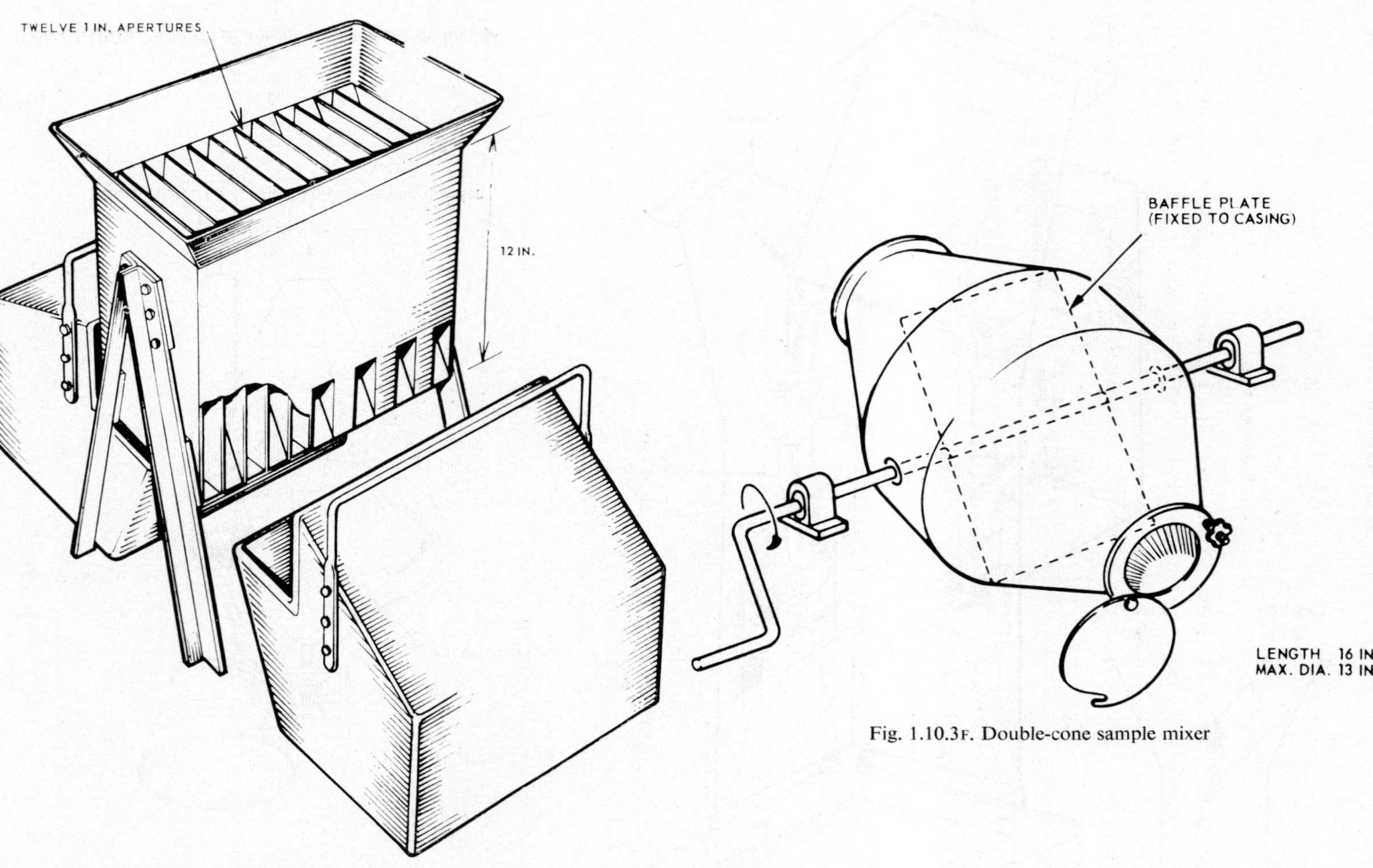

FIG. 1.10.3E. Riffle hand sample divider

Fig. 1.10.3F. Double-cone sample mixer

1.11. ANALYSIS AND TESTING OF COAL AND COKE

1.11.1. Introduction

B.S. 1016: Parts 1–15, sets out detailed methods of analysis and test for coal and coke. The results obtained when coal or coke is tested under conditions which differ from those described, cannot be considered reliable, since the expected accuracy will not be attained. Where the tests are arbitrary, for example, determination of volatile matter, they will not necessarily be comparable with the figures reported by other workers.

The Standard is published in separate parts, 15 dealing with methods of analysis and test, and the 16th with tolerances, accuracy and reporting generally:

Part 1 Total Moisture in Coal
2 Total Moisture in Coke
3 Proximate Analysis of Coal
4 Proximate Analysis of Coke
5 Gross Calorific Value of Coal and Coke
6 Ultimate Analysis of Coal
7 Ultimate Analysis of Coke
8 Chlorine in Coal and Coke
9 Phosphorus in Coal and Coke
10 Arsenic in Coal and Coke
11 Forms of Sulphur in Coal
12 Caking and Swelling Properties of Coal
13 Tests Special to Coke
14 Analysis of Coal Ash and Coke Ash
15 Fusibility of Coal Ash and Coke Ash
16 Reporting of Results

To keep pace with developing techniques and new methods, the Standards are revised from time to time and the latest amendments should be consulted. For this reason, it will be more useful here to consider testing and the tests in a general way, omitting the detail to be found in the Standards. A final section deals with instruments being developed for control purposes.

In addition to the B.S. tests, details have been published in technical journals, of ways to determine other coal constituents, for example, fluorine or germanium, but these tests are carried out only in exceptional circumstances. Of the tests listed, not all are carried out on all coal samples. For routine purposes, testing of coal delivered and consumed is limited to:

Moisture content
Ash content
Calorific value
Sulphur content

However, each power station laboratory will be making routine determinations of such other characteristics of coal as are important to them in connection with their plant and needs.

Samples of delivered coal are also taken and screened to assess the size consist (see Section 1.13).

The frequency with which any test is carried out depends upon four factors:

(i) the use made of the result to calculate the price, or the control that is to be exercised on the plant;
(ii) the cost of carrying out the test;
(iii) the variability of recent results;
(iv) the need to acquire information not of immediate use, about any particular supply of coal.

For special tests—boiler acceptance and performance tests—the same laboratory tests are carried out but the higher accuracy required is ensured by taking replicate samples to the best standard as discussed above and by reporting the mean result of repeated laboratory tests.

1.11.2. Proximate Analysis

This is defined as the analysis of coal expressed in terms of moisture in the analysis sample, volatile matter, ash and fixed carbon.

Ash is the residue remaining when coal is incinerated to constant weight under standard conditions.

Volatile matter is the loss in weight (other than that due to moisture) that occurs when coal is carbonised to a coke under standard conditions. The test is an arbitrary one: if conditions are changed the results will be different. Volatile matter is not, therefore, an entity contained in the coal which can be extracted and measured like water, and in consequence it is not strictly correct to speak of volatile matter "content".

Fixed carbon is a calculated figure:

Fixed carbon = 100 − (moisture + volatile matter + ash), all expressed as per cent on the same basis. It purports to measure the coke residue (except ash) from the volatile matter determination; like volatile matter it is not a substance contained in the coal. Since the two figures are the expression of one determination, chemists are tending to omit reporting fixed carbon, regarding it as a relic of the days when coke and gas quantities were important to the gas industry. It is likely that the next revision of B.S. 1016 will drop the term entirely.

1.11.3. Moisture Content

Moisture exists in all coals in three forms:

(a) *Free moisture*—"surface" moisture, also present on apparently dry coals, which is dried off when coal is exposed to the air without heating.

(b) *Inherent moisture*—the moisture retained in the pores of the coal substance when free moisture has evaporated.

(c) *Water of hydration*—this is water of constitution chemically bound to the shales. Such water is not driven off by heating to 105–110°C, as are free moisture and inherent moisture. As a part of the minerals present in coals, it is discussed further with ash content.

Air-dry moisture is a term used to describe that part of the total moisture retained in the 72-mesh analysis sample, after it has been exposed to the laboratory atmosphere and has attained approximate equilibrium with it. This is necessary so that there shall be no change in condition of the sample, either during the weighing of a portion for test or whilst the sample is standing between tests. It is not possible to handle the sample in the bone-dry condition because finely-ground coal will pick up moisture rapidly; on the other hand the presence of free moisture would lead to evaporation losses not only during weighing but during preparation of the sample. In amount, the air-dry moisture is similar to the inherent moisture.

Total moisture is the sum of free and inherent moisture. It is an inert constituent of coal and it reduces the calorific value; it costs as much as coal to transport and its latent heat of evaporation when the coal is burned, contributes to flue-gas losses.

1.11.3.1. Free Moisture

All coals contain some free moisture. Its origin may be underground mine water, water infused into coal seams and sprays used for dust suppression, washing processes and exposure to rain or snow.

At most C.E.G.B. stations, free moisture is determined as the first stage in arriving at the total moisture, and it is the proportion which is evaporated on bringing a sample into equilibrium with the ambient air. Free moisture is determined, as percentage, from the loss in weight of a 10–30 lb sample after exposure to freely-circulating air at not more than 15°C above ambient temperature, for 16 to 24 h. The sample is spread out into trays about 1 in. deep, and if very wet the drying time may be extended beyond 24 h.

The difficulty in sampling coals with excessive free moisture, for example, coal from the drainage deck of a washer, has been noted. Drainage of free moisture from smalls with a high proportion of fine coal (passing 30-mesh B.S. sieve; 0·023 in.) is very slow and it is these coals which are difficult to handle (see Section 1.9). Although the determination of free moisture as a stage in the determination of total moisture is desirable in all except the visibly dry coals, for visibly wet coals it is essential.

1.11.3.2. Inherent Moisture

This is related to the rank of the coal, being least in Ranks 200 to 300 (1–2%), rising to 3% for Rank 101, and to a maximum (10–12%) for Rank 900.

Where total moisture is determined on $\frac{1}{2}$ in. coal, the water in a 2-lb sample is distilled off with toluene (b.pt. 110°C) and the amount of condensed water collected in the receiver during 6–8 h is noted. An alternative method suitable for high-rank coals only (Rank 100–400) is to measure the loss in weight when a 2-lb sample is heated in an oven to 105°–110°C for 5–6 h in a slow air current.

If the sample is crushed to 6-mesh size, two methods suitable for any hard coal may be used; distillation of 100 g (4 oz approx.) in toluene as above (time about 1–$1\frac{1}{2}$ h) or measuring the loss of weight of 10 g of sample when heated to 105–110°C for $1\frac{1}{2}$–3 h in an atmosphere of nitrogen. For high rank coals, oven-drying in air suffices.

1.11.3.3. Air-dry Moisture

To determine the moisture content of the laboratory sample for general analysis (the purpose of this determination will be referred to in Section 1.11.15, Reporting of Results) three methods are available: drying 1 gram in a nitrogen oven for $1\frac{1}{2}$–3 h as above; drying in a vacuum oven in a similar manner, and lastly a direct weighing of the moisture collected by an absorbent, from dry nitrogen gas passed over the coal in a heated tube.

It should be noted that when indirect methods based on loss of weight are used, the finely-ground coal in a bone-dry condition will pick up moisture from the laboratory atmosphere, and speedy cooling and weighing are essential.

When coal is heated in air to above 100°C but below ignition temperature, other changes occur besides the loss of moisture:

(a) a loss of weight due to the evolution of occluded gases and the slow decomposition of the coal;

(b) an increase in weight due to the formation of solid peroxides, when the coal is heated in air. The use of nitrogen to exclude oxygen prevents this.

Both these changes are more marked the higher the temperature and the lower the rank.

1.11.4. Ash

Ash (defined in Section 1.11.2) is of three types:

(i) *Inherent ash*—that ash content which cannot be reduced by any method of cleaning. It may be thought of as the mineral constituents of the vegetable matter from which the coal was derived plus the silt on which it grew. Inherent ash is usually constant for a given coal seam and can range from $\frac{1}{2}$% to nearly 20%.

(ii) *Associated ash*—present in the coal seam as bands, lenticles and partings. One form of middlings consists of such mineral matter which has not been split off the coal lumps during mining (see section on Coal Preparation).

(iii) *Adventitious ash*—not present in the seam, but introduced from floor and roof during cutting, as a result of deliberate policy or because geological weaknesses dictate its removal to ensure a safe working roof. Adventitious ash may be fire-clay or carbonaceous shales from the clay deposited in shallow water when the coal was laid down.

The incineration of coal is accompanied by visible changes, burning of the volatile gases evolved and the slower combustion of the coke residue. There are, however, other less obvious changes taking place—water of hydration is driven off, carbonates (like limestone) break down to oxides and gaseous carbon dioxide, pyrites (iron sulphide) burns to iron oxide and oxides of sulphur, some sulphates may decompose to oxides and sulphur trioxide. The picture is further complicated by the possibility of the oxides from one sample picking up and "fixing" the sulphur oxides emitted from an adjacent sample.

For these reasons the determination of ash has, to a lesser degree, the empirical nature of the determination of volatile matter, and has to be carried out under standard conditions if reproducible and comparable results are to be obtained.

Two methods are described. In the first a single muffle furnace is used; a known weight of coal (1–2 g) is heated in air to 500°C in 30 min; the temperature is then raised to 815°C in a further 60–90 min. The sample is heated to constant weight. In the second, two muffle furnaces are required. The sample is heated in one kept at 500°C for 30 min; the dish is then transferred to the other kept at 815°C and is heated to constant weight. The second method is preferable in laboratories which are required to test a large number of samples. In each case the furnace must be well ventilated to minimise fixation of oxides of sulphur.

1.11.5. **Volatile Matter**

Volatile matter is used in most systems of coal classification, which is not surprising since it reflects the coal type and its behaviour in a combustion appliance. An early distinction was drawn between anthracites, gas/coking coals and free-burning steam-raising coals; the connection with volatile matter is obvious. The present N.C.B. Coal Rank Code Number links volatile matter with coking behaviour in a way which is helpful to the user.

It has been emphasised above that the test conditions are standardised to yield reproducible and comparable results. This is of importance if price in any way depends on volatile matter, as it does when the carbonisation coal price formula is invoked.

The determination is made by heating 1 g of coal in a crucible to 900°C for 1 min, out of contact with air. The crucible has a lid to exclude air and stands in a wire frame. The volatile matter is calculated from the loss of weight and a deduction is made for the loss due to moisture.

1.11.6. **Calorific Value**

1.11.6.1. Introduction

The calorific value is the basic standard of value of any fuel; it is a measure of its heating power and is the primary factor in coal pricing.

Calorific value is the number of heat units liberated per unit weight of fuel when completely burned in oxygen. In Britain and the U.S. it is measured in British Thermal Units per pound (Btu/lb); the metric equivalent may be calories per gram (cal/g) or kilocalories per kilogram (kcal/kg)—these are numerically equivalent. The relationship between these two systems is simple:

$$\text{cal/g} \times 1{\cdot}8 = \text{Btu/lb}.$$

It is worth noting here that the calorific value of liquid fuels is occasionally stated in Btu per gallon; for gaseous fuels the unit is usually Btu per cubic foot.

It is necessary to distinguish between four calorific values, viz.:

1. Gross calorific value at constant volume.
2. Net calorific value at constant volume.
3. Gross calorific value at constant pressure.
4. Net calorific value at constant pressure.

The laboratory bomb calorimeter determination is the first of these—gross at constant volume; the others may be calculated from it if the composition of the fuel is known. The word "gross" signifies that the latent heat of evaporation of the water present in the fuel plus that formed during combustion, is recovered by condensing the products; "net" signifies that the latent heat is lost, the water being discharged as vapour. For a "typical" Midlands coal (low rank, having 10% moisture and 4% hydrogen) with gross calorific value 10,500 Btu/lb, the net calorific value will be 10,050 Btu/lb.

It will be realised that this latent heat of evaporation is not recoverable under the boiler operating conditions and this has led to the practice of continental manufacturers reporting boiler efficiencies based on the lower or net calorific value. These efficiencies are some 4% higher than the figures based on the higher or gross calorific value, which is normal for British and American installations, and this should be borne in mind when comparing reported performances.

Combustion of fuel in a bomb differs from that in a boiler; the first is at constant volume, the other at nearly constant pressure. When the products of combustion are allowed to expand at constant pressure, work is done and the gross calorific value at constant pressure is higher than the bomb-determined calorific value, by the heat equivalent of this work. The correction on this account is about 12 Btu/lb for coal and, being less than the uncertainty of sampling and analysis, is normally neglected.

There are a number of formulae whereby the calorific value may be calculated to a fairly close approximation; it is necessary to know the ultimate analysis (q.v.) but as this is rarely available, it is simpler and more reliable to carry out a direct determination of calorific value. Formulae also exist for predicting calorific value from the proximate analysis; the results may be in error by as much as $\pm 6\%$ and should be used only as a rough guide when a determination cannot be made.

The calorific value of the pure coal substance (that is, dry and free from all mineral matter) is virtually constant for the coal from any given seam in one locality. Once its value has been established (see section on Reporting), it may be used with the total moisture and ash content, to calculate the calorific value of any consignment from the same seam.

This approach has some advantage for the routine monitoring of coal deliveries. The time and money employed in carrying out a number of determinations of moisture and ash contents and calculating the calorific value, provides much more useful information about trends and average quality in the long run, than the same effort devoted to fewer direct determinations of calorific value in a bomb calorimeter.

In the short term, and for checking, there is no alternative to the standard procedure of bomb calorimetry.

1.11.6.2. Determination of Calorific Value

The method of determining the calorific value of coal or coke is prescribed in full in B.S. 1016, Part 5, but since it is of particular importance to the C.E.G.B., the method will be described in some detail here. The essential parts of the commonly-used "static" bomb calorimeter are shown diagrammatically in Figure 1.11.6.2 and the principles of the method are as follows.

A small quantity of the sample is burnt in compressed oxygen inside a stainless steel cylinder, or "bomb", which is immersed in water contained in a cylindrical calorimeter

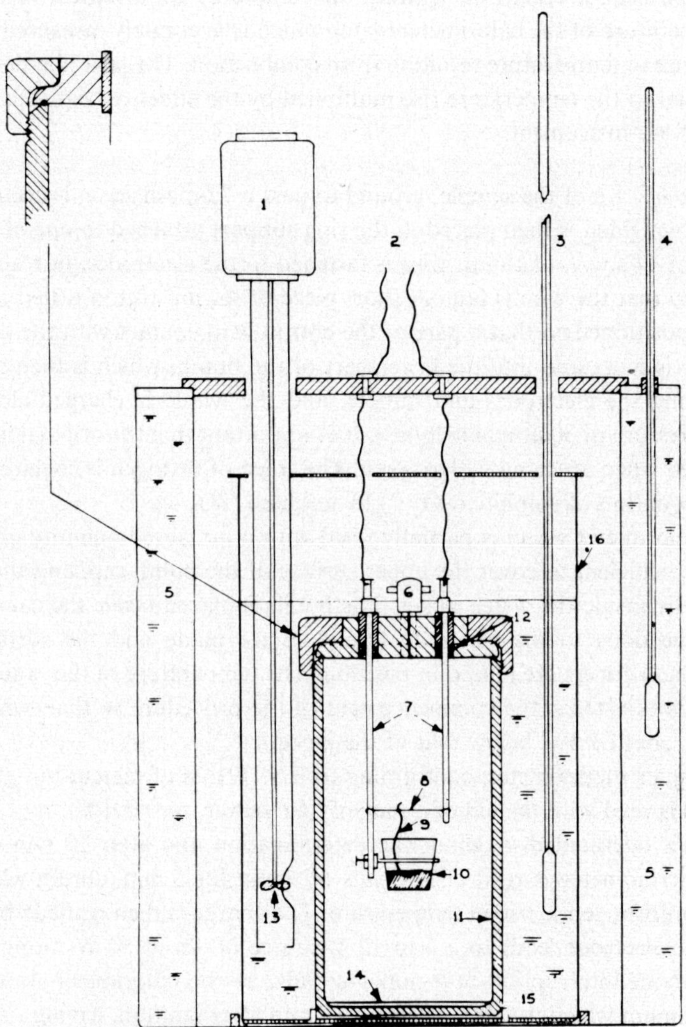

FIG. 1.11.6.2. "Static" calorimeter

vessel. This vessel with its contents is situated inside a larger vessel of which the hollow cylindrical walls are filled with water to form the "jacket", and which has the purpose of minimising heat transfer between the calorimeter vessel and the environment of the whole equipment. The heat liberated on igniting the sample by an electrical fuse, is calculated from the temperature of the calorimeter water which is accurately measured before, during and after the rise in temperature resulting from combustion. The heat liberated is, of course, essentially equal to the temperature rise multiplied by the effective heat capacity, or "water equivalent" of the instrument.

(a) *Procedure*. 1 g of the sample, ground to pass a 72-mesh sieve, is accurately weighed into the crucible which is then placed in the ring support attached to one of the electrodes, and a length of 44 s.w.g. platinum wire is fastened to the electrodes, just above the top of the crucible, so that the wire is taut. A short piece of sewing cotton is tied to the centre of the wire and positioned so that a part of the cotton is in contact with the sample. 1 ml of distilled water is measured into the lower part of the bomb, which is then assembled with the cap carrying the electrodes and sample, and the whole is charged slowly with pure oxygen to a pressure of 30 atm (440 lb/in^2). It is important that the original air in the bomb is not displaced when charging with oxygen. The trace of nitrogen is required as a catalept to promote oxidation of sulphite (SO_3^{--}) to sulphate (SO_4^{--}).

Next, the calorimeter vessel is partially filled with a measured quantity of water (preferably distilled), sufficient to cover the upper surface of the bomb cap, and the vessel is then located centrally inside the water jacket. The bomb is placed inside the calorimeter vessel, electrical connections to the electrode terminals are made and the stirrer, calorimeter thermometer, and covers are placed in position. The temperature of the water in the calorimeter vessel is adjusted at the commencement of the procedure so that, when assembly is complete, it is about 1·5°C below that of the jacket.

The calorimeter thermometer, conforming to B.S. 791, is of mercury-in-glass, graduated in 0·01°C and is read with the aid of a magnifying viewer, to 0·001°C.

The stirrer is operated throughout the determination and after 10 min of stirring, the calorimeter thermometer is read at intervals of 1 min for 5 min, during which time there should be a uniform gentle rise in temperature. The charge is then ignited* by momentarily connecting the electrode leads to a 6 to 12 V source of about 50 W rating. Readings are continued without interruption at 1-min intervals, as the calorimeter water temperature rises to a maximum which usually occurs 5 to 7 min after ignition, having risen by approximately 2·5°C. Readings are ended after a further five or six observations, during which a uniform fall in temperature should occur.

The bomb is then removed from the calorimeter, pressure is released slowly, and the bomb is dismantled. It can then be inspected for any sooty deposits on the internal parts and in the absence of such deposits, combustion is regarded as having been satisfactory. The inner surfaces of the bomb and the internal fittings are carefully washed with distilled water into a beaker, for the subsequent determination of the corrections for nitric and sulphuric acids, as mentioned below.

* As a safety precaution the firing key should be operated from a position some distance away from the calorimeter.

(b) *Corrections.* A number of corrections must be applied in arriving at the calorific value, as follows:

(i) *Temperature corrections*

A "cooling correction" is calculated as shown in B.S. 1016 : Part 5, to allow for heat transfer between the calorimeter vessel and its environment during the main rise in temperature, and a further correction is made for errors in the thermometer. Calorimeter thermometers should be sent for test by the National Physical Laboratory regularly—at 3–5 year intervals usually—and the thermometer correction is obtained from the test certificate which, at the points of calibration by the N.P.L., is accurate to $\pm 0{\cdot}002°C$.

(ii) *Fixed corrections*

These are necessitated by heat released by the fuse wire and combustion of the cotton, and are kept to a constant amount, about 20 cal, for every determination.

(iii) *Nitric acid and sulphuric acid corrections*

Some of the nitrogen in the sample and air enclosed in the bomb, is oxidised to nitric acid with the evolution of heat (about 10 cal), whereas this reaction does not occur perceptibly in boiler furnaces. Additionally, sulphur in the sample is oxidised to sulphuric acid in the bomb, liberating more heat than it would in a boiler, where practically all of the sulphur is converted to sulphur dioxide. The correction here is usually between 15 and 60 cal/g of coal, depending on its sulphur content, and is either calculated (together with the nitric acid) from analysis of the bomb washings, or from the sulphur content of the sample, where this has been separately determined.

(c) *Mean effective heat capacity of the instrument.* In order to complete the calculation of the calorific value, it is necessary to know what weight of water and metal (in terms of an equivalent weight of water), has absorbed the heat evolved during combustion in the bomb. This is obtained periodically as the mean of five determinations carried out as has been described, but substituting 1·2 g of dry benzoic acid in place of the sample. The benzoic acid, which is a crystalline organic substance containing only carbon, hydrogen and oxygen, is specially purified for this purpose and is obtainable as a "thermochemical standard" having a calorific value certified by the N.P.L., which value is close to 6319 cal/g, or 11,374 Btu/lb.

(d) *Calculation of the result.* To summarise the calculation, we have

$$CV = \frac{(Tc \times H) - C_f - C_s - C_n}{g}$$

where CV is the calorific value in calories per gram,

Tc the temperature rise in degrees C of calorimeter water, corrected for cooling and thermometer errors,

H the mean effective heat capacity, in calories per degree C,

C_f the fixed correction, in calories,

C_s the correction for sulphuric acid, in calories,

C_n the correction for nitric acid, in calories,

g the weight of sample in grams.

When determining the effective heat capacity, the certified calorific value of benzoic acid is taken as a standard, the temperature rise and corrections are calculated and a measured weight of benzoic acid is used; the effective heat capacity H can thus be evaluated from the formula

$$H = \frac{(CV \times g) + C_f + C_n}{Tc}.$$

Having determined the mean effective heat capacity, the calorific value of coal or other fuel can be calculated and the final result, cal/g, is converted to Btu/lb on multiplying by the factor 1·8. The difference between duplicates determined in one laboratory should not exceed 50 Btu/lb and the difference between the means of duplicate determinations from different laboratories should not exceed 120 Btu/lb; an experienced operator usually expects to obtain duplicate results within 20 Btu/lb.

1.11.6.3. Adiabatic Bomb Calorimeter

In the determination of calorific value using the static calorimeter, the magnitude and accuracy of the cooling correction depend on the temperature difference between the calorimeter water and the jacket, and it is not always convenient to control this so as to produce a small cooling correction, particularly when several determinations are to be made in one day. Further, the calculation of the cooling correction takes time, as does the need to read temperatures carefully at 1 min intervals during 15 to 20 min. For these reasons the "adiabatic" bomb calorimeter, which eliminates the cooling correction, was developed commercially 12 to 15 years ago and has displaced the static type in modern stations.

The innovations of the adiabatic calorimeter are in the provision of the means to change the jacket temperature rapidly, particularly to increase it, and this is done by installing electric heaters in the jacket through which the water is circulated. The uppermost cover of the calorimeter is also made in hollow form and the jacket water is passed through this in turn. Sensitive thermistors are placed, one in the calorimeter water and one in the jacket, and these control the jacket heaters via a bridge circuit and relay, so that as the calorimeter water temperature rises, the jacket water is heated rapidly to the same temperature. In this way there is a negligible heat transfer between the calorimeter and the jacket, and it is only necessary to read the temperature accurately just before firing (when initial equilibrium has been reached), and again at a convenient time after the final equilibrium temperature has been attained. Calculation of the result is the same as has just been described, modified only by deleting the cooling correction.

Although at present the accuracy of an adiabatic calorimeter is not significantly greater than that of a properly controlled static instrument, it takes less time and is potentially simpler to operate. For these reasons, although they are costlier and more complex instruments, adiabatic calorimeters have been widely adopted. A typical commercial instrument is shown in Figure 1.11.6.3.

FIG. 1.11.6.3. Adiabatic calorimeter

1.11.7. Sulphur

The determination of sulphur is part of the ultimate analysis of coal (q.v.), but is discussed separately because of its special importance in pricing, as well as technically. Sulphur in coal is found in three combinations:

(i) Sulphate sulphur (negligible).
(ii) Organic sulphur (0·8% average).
(iii) Pyritic sulphur (0·8% average).

Organic sulphur tends to be relatively constant but pyritic sulphur varies widely.

Sulphate sulphur exists as small quantities of ferrous sulphate ($FeSO_4 \cdot 7H_2O$) from the oxidation of iron pyrites (FeS) and as gypsum ($CaSO_4 \cdot 2H_2O$). They are found as thin plates in partings of the coal where the solution has evaporated.

Organic sulphur is combined with the carbon and nitrogen to form the coal substance. In consequence it is not removable by cleaning.

Pyrites (also known as marcasite) is ferrous sulphide (FeS). It occurs as massive nodules and lenticles, as "brassy" bands, and as finely dispersed particles. The first form is easily removed by the various coal cleaning methods. Pyrites more closely associated with the coal is removable only with considerable loss of saleable carbonaceous material, or "vend".

Methods are available for assessing the amount of each of the forms of sulphur but usually they are not important for power station purposes. Total sulphur may be determined in the bomb-washings from the determination of calorific value; this is a simple procedure and is usually adopted since the result is required for the calorific value calculation. If total sulphur has to be determined, and the number of samples is not large, the "Eschka" or bomb method is used; large numbers of determinations are more conveniently carried out by the high-temperature (tube) method.

In the Eschka method, 1 g coal is burned by heating in a crucible with a mixture of magnesium oxide and sodium carbonate. This alkaline mixture retains all the sulphur as soluble sulphates which are determined by the standard chemical method of precipitation as barium sulphate, which is finally dried and weighed.

The high-temperature (tube) method requires costly equipment which can be made largely automatic and needs little special training to operate. The justification for its use depends on the number of sulphur determinations to be carried out, or on results being required more rapidly than can be provided by the other methods. The coal sample (0·5 g mixed with alumina) in a small refractory "boat", is pushed into a tube kept at 1350°C through which passes a rapid current of oxygen; the sulphur oxides formed are absorbed as the oxygen from the tube is bubbled through hydrogen peroxide, which converts them wholly to sulphuric acid. The acid is measured by titration with standard alkali solution, a correction being applied for chlorine carried forward by the oxygen. (This is also a method for the estimation of chlorine (q.v.).)

Mention should also be made that sulphur can be determined by the "oxygen flask" technique. In this method, 1/30 g of coal, wrapped in paper, is burned in a glass flask filled with oxygen at atmospheric pressure. The flask contains hydrogen peroxide solution to dissolve and oxidise the sulphur gases to sulphuric acid. This is estimated by titration with standard barium perchlorate solution with a special indicator. Research at Marchwood Engineering Laboratories has shown that the accuracy of the method is almost as good as the classical methods and because of its simplicity, it may usefully be applied to the routine determination of sulphur in coal (Ratcliffe and Cunningham, *Fuel* **47**, 89–92 (1968)).

1.11.8. **Chlorine**

Chlorine has been regarded as a significant constituent of coal since about 1939 when a correlation between chlorine content of the coal and the degree of gas-side boiler fouling resulting was observed. This aspect is discussed, with ash and sulphur contents, in a later section.

Chlorine occurs almost entirely as sodium chloride and potassium chloride. These salts derive (like the brine and salt deposits of the North-East and the Midlands) from the shallow seas which covered the coal measures in a later geological period. Although the chlorides can in part be dissolved out by repeated water washing, some will always remain in the coal. In practice the washing processes used in coal preparation do not significantly reduce the chloride content, partly because of insufficient residence time, but also because the washery water is largely recycled and tends to reach a state of equilibrium, where the wash water adds nearly as much chloride as it removes. It is not often possible to obtain sufficient clean water to rinse the washed coal or to dilute the wash water, and in any case surplus washery water must be expensively cleaned to a rigid standard before disposal.

The methods available for estimation resemble those described for sulphur, namely the Eschka and high-temperature methods, and for routine analyses the oxygen flask method has much to recommend it (Ratcliffe and Young, *Fuel* **47**, 185–192 (1968)). The finish in all cases is by a volumetric titration.

1.11.9. **The Ultimate Analysis of Coal**

Strictly, the ultimate analysis of coal is the determination of the elements of which its molecules are compounded and as such would include elements like chlorine, phosphorus, etc.; for fuel technology purposes the term is defined as the analysis of coal in terms of its carbon, hydrogen, nitrogen and sulphur contents. Oxygen is usually taken as being 100 minus the sum of the former constituents, expressed as percentages. As such it includes the errors of estimation of the other elements and a direct method is being developed.

The ultimate analysis is of great scientific interest and has important practical application in combustion calculations and heat balances. As already stated, the C.V. can be calculated to a fairly close approximation from the ultimate analysis. Until recently, elemental analysis by the classical Liebig method required high skill and experience and it was time-consuming. Accordingly, for heat balance purposes, the ultimate analysis was frequently calculated from analyses which are on record for the particular seam or coal rank. More recently the high-temperature tube method, cutting down the combustion time from 2 h to 9 min and introducing other time-saving modifications, has been developed. In spite of these improvements the determination of carbon and hydrogen is not a simple operation and there is an intensive search for a rapid method based on instruments which will always yield reliable results—so far none proposed has been found wholly acceptable.

1.11.9.1. Carbon and Hydrogen

Two standard methods are available, combustion at 800°C (classical Liebig) and at 1350°C (Sheffield high temperature). In both methods the calculated result is "total carbon", that is, it includes carbon from coal as well as from any carbonates present and a correction is needed at a later stage. In each the hydrogen is converted to water which is absorbed and weighed together with the water present, as moisture in the coal and as water of hydration, in the associated mineral matter and shale. In this case corrections have to be applied for these interfering factors.

(a) In the low-temperature method, 0·3 g of coal is burned in a slow stream of oxygen in a heated tube 50 in long, capable of close temperature control in zones. The tube is packed so that the products of combustion pass through a plug of copper oxide (18 in. long) kept at 800°C and then successively over a lead chromate plug (4 in. long) and then a 1-in. roll of silver gauze, both kept at 600°C. The copper oxide ensures that all the carbon is converted to carbon dioxide and all the hydrogen to water. Oxides of sulphur are removed by the lead chromate and chlorine by the silver gauze.

The products of combustion carried by the stream of oxygen pass through an absorption train consisting of tubes which are packed with an absorbing substance. Water is first removed by magnesium perchlorate and measured by the increase in weight. Traces of nitrogen dioxide which would lead to erroneous results for carbon are then removed by manganese dioxide. Finally, carbon dioxide is absorbed by soda-asbestos; again the increase in weight of the absorption tube measures the carbon dioxide evolved.

(b) The high-temperature method differs from the Liebig method in having a shorter unpacked tube (only 26 in. long), using a rapid stream of oxygen, and operating at 1350°C. At this temperature both sulphur and chlorine are trapped by a roll of silver gauze; nitrogen dioxide is not formed under these conditions and no special precautions are therefore necessary. The assessments of water and carbon dioxide produced can be made as in the low temperature method, or can be attempted by automatic instruments if speed is a necessary factor.

1.11.9.2. Nitrogen

Nitrogen in coal is an inert substance which takes no part in combustion processes. It may be estimated by the Kjeldahl method of heating the coal with concentrated sulphuric acid to destroy the organic material; the presence of a catalyst (selenium, mercuric sulphate, vanadium pentoxide) ensures that all the nitrogen is rapidly converted to ammonium sulphate. When cool, excess sodium hydroxide is added to free the ammonia, which may then be steam-distilled and collected for estimation in the condensate by titration with standard acid solution.

The estimation may be carried out, with appropriate differences in detail, on 1 g (macro-method) or 0·1 g (semi-micro method) of coal. In practice because of the shorter times of digestion ($\frac{1}{2}$ h as against 2 h) and the smaller scale of the apparatus, nearly all laboratories use the semi-micro method.

1.11.10. Phosphorus

Phosphorus occurs in British coals and ranges between 0·005% and 0·15%. High phosphorus coals can be very troublesome in stoker-fired plants, especially in economisers where a slow but continuous build-up of hard deposits occurs. On gilled tubes a heavy draught loss occurs which is sufficient to reduce the outputs of boilers. The deposits are the result of chemical attack and digestion of fly ash particles and are not removed by water washing, while chipping accessible parts is slow and laborious. The deposits, which always contain a high proportion of associated boron compounds, can be dissolved by alternately spraying with inhibited hydrochloric acid and hot caustic soda. This method has its difficulties and dangers and is therefore rarely employed; in addition the cast iron gills suffer corrosion.

Fortunately the high phosphorus coals are comparatively rare except in South Wales, although severe phosphatic fouling occurred at Dunston power station when using certain Northumberland coals. The phosphorus compounds are only volatilised from deep fuel beds with high temperatures, and consequently the element causes no trouble with P.F. firing.

For measurement, a known weight of coal ash is digested with a mixture of nitric and hydrofluoric acids to extract the phosphorus, which is precipitated under controlled conditions as ammonium phosphomolybdate. This is dissolved in an excess of standard sodium hydroxide solution and the amount of alkali remaining is titrated with standard nitric acid.

Alternatively the phosphorus can be estimated colorimetrically; the optical density of "molybdenum-blue" developed by reduction under standard conditions is measured with a suitable spectrophotometer and compared with a calibration curve.

The British Coke Research Association has reported on the use of the oxygen–flask technique for the determination of phosphorus in coke (Kirk and Wilkinson, *Fuel* 1964, **43**, 105). Recent work combining oxygen–flask combustion, with colour development by a method of Severn and James (BCURA Doc. No. ANAL/183, April 1968) and the spectrophotometric finish has shown a tolerance which is practically the same as is currently specified in the existing B. S. Standard.

1.11.11. Arsenic

Arsenic is present only in trace quantities, usually less than 10 ppm in British coals. Its determination may be important, however, particularly where there is the possibility of effluents from ash dumps, which may contain arsenic, draining into natural water sources.

It is determined in the laboratory by mixing a known weight of the coal analysis sample with magnesium oxide and potassium permanganate, and heating in oxygen to remove organic matter. The residue is extracted with acid and the arsenic is converted to the gas, "arsine", AsH_3. The gas is passed through a prepared paper impregnated with mercuric bromide, to produce a coloured stain for comparison with quantitative standards.

1.11.12. Swelling Index

The swelling property of a coal indicates its behaviour as a fuel for stoker-fired boilers. High-swelling coals lead to uneven and patchy fire beds and water conditioning is essential for good combustion. This property also leads to difficulty in obtaining sufficient air flow

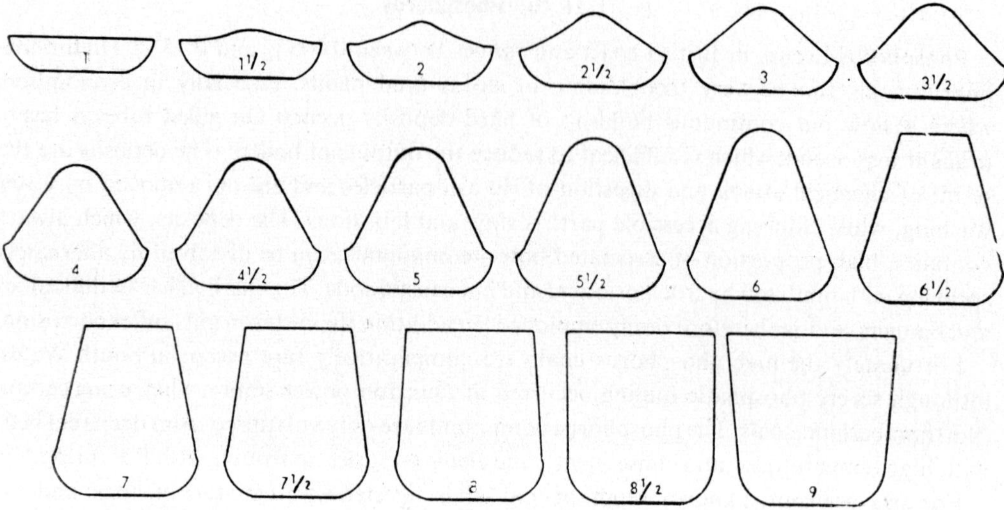

FIG. 1.11.12. Crucible swelling index—profiles (British Standards Institution)

through the fuel bed, while the burn-off of the massive cokes produced, is slow and usually far from complete.

In this test, 1 g of air-dried 72-mesh coal is heated under standard conditions in a closed crucible so that a temperature of 800°C is attained after $1\frac{1}{2}$ min, and 820°C within $2\frac{1}{2}$ min. After $2\frac{1}{2}$ min heating or when the flame of the burning volatile matter has died out, whichever is the longer period, the crucible is cooled and the coherent coke button compared with a series of standard profiles numbered 1 to 9 (Fig. 1.11.12). If the residual coke is not coherent it is designated: Swelling number, Zero.

1.11.13. **Gray–King Coking Test**

Devised originally for the gas industry, the test together with the volatile matter is the basis for the N.C.B. Coal Rank Code Number. Both the Gray–King and Swelling Index tests involve caking phenomena—particle adherence, softening, pyrolytic swelling and shrinkage.

The Gray–King type is identified by the letters A to G (this last is subdivided into 11 subtypes). It is measured by heating 20 g of coal sample in a tube out of contact with air. Starting at 325°C the temperature is raised during 55 min to 600°C and maintained for a further 15 min. The coke formed is classified by comparing it with photographs of a standard series of coke types. If the coke is very swollen (higher than G_2) the test is repeated with the coal mixed with a suitable quantity of inert electrode carbon. The minimum amount of electrode carbon needed to prevent swelling above the original volume of coal and electrode carbon mixture, is a further definition for these highly caking coals.

1.11.14. **Carbon Present as Carbonates**

The carbon dioxide content of the mineral carbonates of a coal is required

(a) To correct the determined % carbon when calculating to a dry-mineral-matter-free basis, and for calculating the combustible carbon.

(b) When calculating the volatile matter of low volatile coals and anthracites of high carbonate content, to a dry-ash-free basis.

(c) When calculating the volatile matter of a coal to the dry-mineral-matter-free basis.

The analysis is carried out in the laboratory by dissolving a known weight of the analysis sample in hydrochloric or phosphoric acids and absorbing the resultant carbon dioxide evolved. The increase in weight of the absorbent is directly due to the carbon dioxide liberated.

1.11.15. **Reporting of Results**

It will be appreciated that the results obtained from the tests discussed above, will be on the basis of the air-dried condition of the laboratory sample. It will be most frequently necessary to convert them to the condition of the coal as it was sampled, i.e. with its full moisture content.

It is also possible to convert them to the "dry" basis and to the "dry-ash-free" basis (D.A.F.), in order to be able to compare results under standard conditions. For example it may be required either to check whether a result for a particular coal is in line with the expected value, or to study the consistency of a single product (from one colliery) over a period of time. The "dry" basis can be used to study variations in ash; the D.A.F. basis to study variations in volatile matter and calorific value.

Unfortunately the D.A.F. calorific value can never be constant for any coal because the value changes with its ash content. However, the change can be quite regular and, therefore, serve as a guide to the usefulness of a result or as to whether inconsistent results can be expected from a colliery. The following is an indication of how the D.A.F. calorific value changes with ash content for a single colliery product and if the results are graphed, a norm for some coals can be established:

Dry ash %	Dry calorific value calculated from tests Btu/lb	D.A.F. calorific value calculated Btu/lb
9·6	13,160	14,560
13·8	12,490	14,490
16·6	12,010	14,400
18·4	11,740	14,390
19·1	11,610	14,370
22·4	11,060	14,260
26·3	10,450	14,180

The reason for the progressive change in D.A.F. calorific values is that ash in coal and the mineral matter originally present are not the same thing. The ash weighs less than the mineral matter from which it was derived, because of chemical interactions occurring during the combustion of the coal (described in Section 1.11.3). The loss of water of hydration of the mineral matter is a predominant feature accounting for a loss of nearly 10% so that as a broad approximation the weight of ash obtained from coal is about nine-tenths of the weight of the original mineral matter present.

If mineral matter can be used in the conversion calculations in place of ash, it would appear that the basis for comparisons would be improved. The best formulae for obtaining an approximation to the original mineral matter content, require the determination of more constituents of the coal than are made in most routine checks, and unless the coal is low in sulphur, chlorine and carbonate content, the use of the more simple formulae give little or no advantage over the D.A.F. method. Formulae for calculating mineral matter are included below:

Symbols:

M = % of total moisture as sampled.
M_1 = % of moisture in the analysis sample.
A = % of ash in the analysis sample.
MM = % mineral matter in the analysis sample.
Cl = % chlorine in the analysis sample.
CO_2 = % carbon dioxide (from carbonate) in the analysis sample.
S_p = % pyritic sulphur in the analysis sample.
S_a = % ash sulphur in the analysis sample.
S_s = % sulphate sulphur in the analysis sample.
S_t = % total sulphur in the analysis sample.
K = a constant.

(a) "*As sampled*" *basis.* Multiply the % of the constituent estimated in the analysis sample by

$$\frac{100-M}{100-M_1}$$

(b) "*Dry*" *basis.* Multiply the % of the constituent estimated in the analysis sample bs

$$\frac{100}{100-M_1}$$

(c) "*Dry, ash-free*" *basis.* Multiply the % of the constituent estimated in the analysiy sample by

$$\frac{100}{100-(M_1+A)}$$

(d) "*Dry mineral-matter-free*" *basis.* For calculation to the pure coal substance basis, the mineral matter content may be obtained from the following:

(i) $MM = 1{\cdot}13\,A + 0{\cdot}5\,S_p + 0{\cdot}8\,CO_2 - 2{\cdot}8\,S_a + 2{\cdot}8\,S_s + 0{\cdot}5\,Cl$
(ii) $MM = 1{\cdot}10\,A + 0{\cdot}53\,S_t + 0{\cdot}74\,CO_2 - 0{\cdot}32$
(iii) $MM = 1{\cdot}1\,A + K$

Whenever possible, method (i) should be used. Method (iii) is crude but effective for coals low in CO_2, Cl and S. For instance, for many South Wales coals it works well when K is taken as 0·2, 0·3 or 0·4.

The dry mineral-matter-free basis is calculated by multiplying the constituent estimated in the analysis sample by

$$\frac{100}{100-(M_1+MM)}$$

Note that when the "constituent" being calculated is volatile matter, it is essential to make a deduction for the volatile constituents (such as CO_2 in carbonates) of the mineral matter, as shown in B. S. 1016, part 16.

Table 11 shows a coal analysis reported on various bases.

TABLE 11

A COAL ANALYSIS REPORTED ON VARIOUS BASES

Property	As received basis	Air-dried basis	Dry basis	Dry ash-free basis	Dry mineral-matter-free basis
Proximate Analysis					
Moisture	13·0*	9·1	—		
Volatile matter	32·8	34·3	37·7	41·6	40·3
Fixed carbon	46·1	48·1	52·9	58·4	59·7
Ash	8·1	8·5	9·4	—	—
Mineral matter	9·8	10·2	11·2	—	—
Calorific value (Btu/lb)	11,300	11,810	12,990	14,330	14,580
Sulphur (total)	1·69	1·77	1·95	—	—
Sulphate sulphur	0·08	0·08	0·09	—	—
Pyritic sulphur	0·77	0·80	0·88	—	—
Organic sulphur	0·84	0·89	0·98	—	1·10
Sulphur in ash (as sulphur in coal)	0·25	0·26	0·29	—	—
Carbon dioxide	0·47	0·49	0·54	—	—
Chlorine	0·58	0·61	0·67	—	0·19†
Phosphorus	0·035	0·037	0·041	—	—
Ultimate Analysis					
Moisture	13·0	9·1	—	—	—
Ash	8·1	8·5	9·4	—	—
Carbon	63·4	66·3	72·9	80·5	82·0
Hydrogen	4·3	4·5	5·0	5·5	5·5
Nitrogen	1·3	1·4	1·5	1·7	1·7
Sulphur (combustible)	1·4	1·5	1·7	1·8	1·1‡
Difference (oxygen, etc.)	8·5	8·7	9·5	10·5	9·7

Results are expressed as percentages unless otherwise indicated.
* Free moisture 4·3, inherent moisture 8·7.
† Assumed organic chlorine.
‡ Organic sulphur.

1.11.16. **Instrumental Analysis**

The classical methods of sampling and analysing coal have long been open to the objection that although they were precise, they were costly in time and equipment. Attempts have been made since 1945 to relate the average specific gravity of the particles in the mixture of coal, to its ash content; a method for rapid moisture assessment depended on measuring the amount of water needed to bring the coal from its moisture content as sampled, to a saturation point considered to be a constant for the coal.

These early post-war methods produced poor correlation, partly because of defects in the test methods but principally because they were still not fast enough to permit many samples to be taken and tested. The results therefore showed a large scatter about the true result in the same way as classical methods would, if the sampling was based on a small number of increments.

During the last five years the introduction of computer control of boilers has intensified the search for faster and more reliable instruments whose results could be used as control input data.

It has been noted above that the calorific value of the pure coal substance for any given seam is almost constant. Instrumental tests are intended to take advantage of this by frequently measuring the moisture and mineral matter, using those results together with an independently assessed dry-mineral-matter-free calorific value to compute the calorific value of a coal as it is being fired. Although this goal has not been reached, enough progress has been made to encourage the hope that it will be achieved. Already there are a number of N.C.B. coal washers where the quality of output is monitored by instruments which then control the operation so that the washed coal has a consistent and standard ash content.

The problem of insufficient increments, noted above, is solved by using the speed of response of the instrument. Individual test results show scatter, as would the ash contents of individual increments found by classical methods of analysis. Although the instantaneous test results indicated by the instrument may each be subject to significant error, provided the errors are truly random, it is feasible to eliminate them by averaging the results from a large number of tests; the instrument signal could be fed into an integrator whose output could be either a smoothed or "running" average, or a cumulative average over a period. Figure 1.11.16.1 shows an actual record from a Cendrex instrument and a calculated example of how a running average and a cumulative average would appear.

It must be noted that the question of bias in taking the sample for presentation to the instrument, still demands great care if it is to be kept to acceptable levels. The classical methods previously described are standard "reference" methods by which instruments are calibrated.

1.11.16.1. Mineral Matter by Gamma-rays

The most successful instruments so far developed depend on taking an increment from a coal stream and either (a) packing it in a standard way (Simcar), or (b) crushing and drying it to a standard condition (Cendrex); it is then exposed to gamma-rays from a radioactive substance or an X-ray tube, for measurement of the mineral matter content.

The Simcar industrial model takes a 30-lb increment of 1-in. coal (larger sizes need crush-

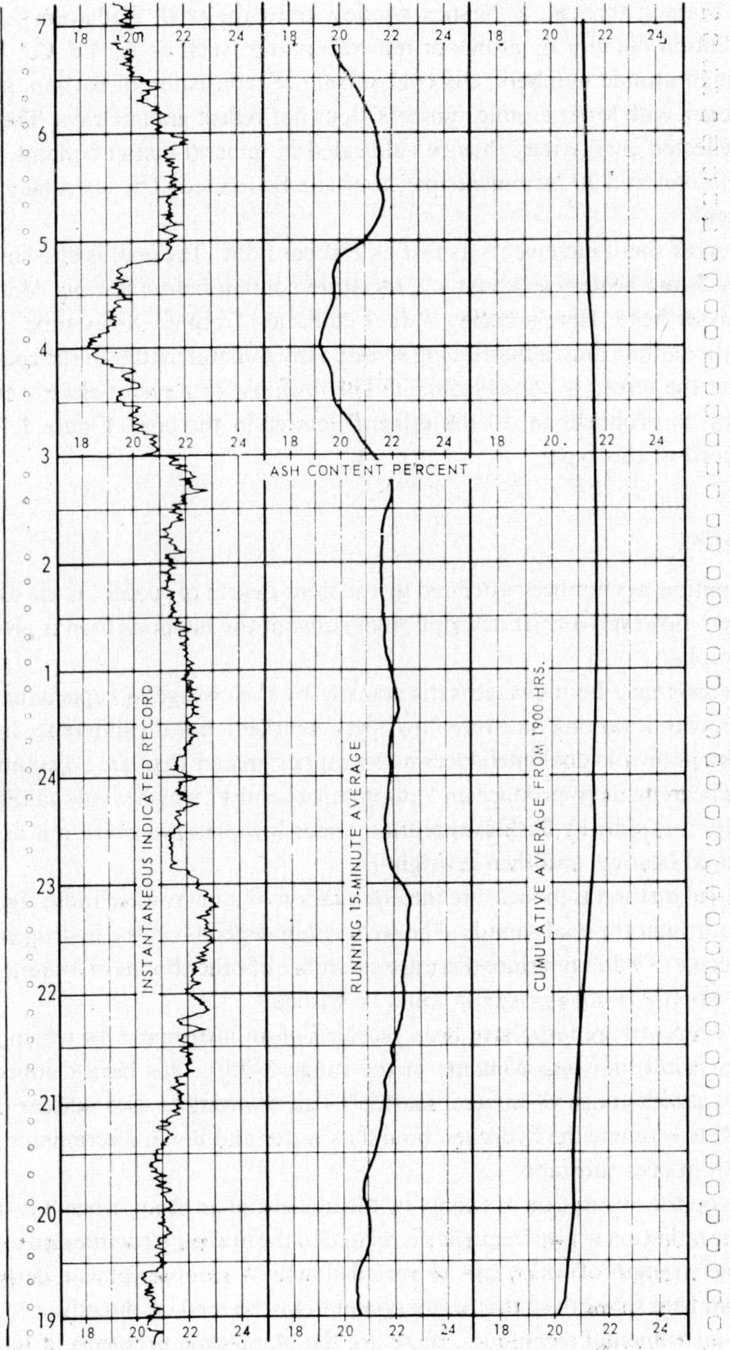

FIG. 1.11.16.1 Mineral matter in coal by gamma-ray analysis

ing), contained in a steel box which is then flooded with water to eliminate moisture as a variable factor. Weak gamma-rays from a radio-active source (for example, thulium or americium) are beamed through a Perspex window onto the coal. Radiation is reflected back to a scintillation counter by atoms of mineral matter, such as Al, Fe, Ca, Si, which have relatively high atomic numbers; the coal substance, consisting of carbon, hydrogen, oxygen and nitrogen with lower atomic numbers, does not reflect gamma-rays. The detector measuring the reflected energy may thus be calibrated in mineral matter content. The process takes two minutes and 30 increments per hour can be treated. The instrument range is 2–30% ash content.

The Cendrex takes small increments as fast as 200 per hour. The coal is crushed to 1–10 mm and dried by 3-min heating to about 1% moisture content before flowing continuously through the analyser head. Here a beam of weak radiation from an X-ray tube is allowed to pass through the sample onto a fluorescing screen. Any mineral matter in the coal diffuses the X-rays so that the screen is not as brilliant. The response of a photo-electric cell to the varying brightness is proportional to the mineral matter in the coal. Figure 1.11.16.1 is a continuous record of this type.

1.11.16.2. MOISTURE

Moisture estimation has not been attended by the same degree of success as the estimation of mineral matter; however, a brief description of some of the methods tried is given in the following paragraphs.

Moisture in cereals may be indicated satisfactorily by the change in capacitance of two plates when grain with a varying moisture flows between them, but the difference in particle behaviour between grain and coal introduces uncertainties greater than are acceptable.

Attempts have been made to produce an "automatic chemist" which would take a known weight of coal, dry it rapidly by flash-evaporation under low pressure whilst it is exposed to high energy infrared heating, and then reweigh it.

A more successful method is to measure the attenuation of a microwave radio signal when it is transmitted through the coal sample. The weakening depends on the hydrogen–oxygen bonds in the molecule of water but since there are a number of other bonds of a similar nature in the coal, the response is not as clear as could be wished.

However, more recently, reports have been received of an instrument for which an accuracy of $\pm 1\%$ for total moisture contents in the range 3–25%, has been claimed for all ranks of coal and a wide range of mineral matter.* This instrument uses nuclear magnetic resonance (NMR) to estimate the hydrogen bound as water and it can discriminate between this and hydrogen in coal substance.

NMR depends on the absorption of energy by the nucleus of an atom, when it is subjected to electromagnetic radiation whose frequency is related to the nuclear structure and to a strong magnetic field, the strength of which can be varied slowly. A recorder plots a derivative of the absorption and field strength so that water content may be read off directly.

As with other instrumental techniques, there are the plant-scale problems of taking and handling a representative coal sample.

* Ladner and Wheatley, *B.C.U.R.A. Information Circular*, No. 288, June 1965.

1.12. OTHER TESTS ON COAL

1.12.1. Bulk Density

The measurement of the bulk density of coal in small heaps and in wagons is a fairly simple matter. It is much more difficult when the coal is a large stockpile of compacted coal.

Bulk density of coal in a wagon is the ratio of net weight to the volume calculated from the wagon dimensions and the levelled coal surface.

The same principles apply when using a special bin or box. It must be noted, however, that when dealing with smaller volume, the effect of voids between the walls and coal lumps, introduces error, the smaller the box the larger the error.

It is not possible to measure the bulk density of coal in stockpiles by these methods, since the sample when extracted has a larger volume than the coal *in situ*. Accordingly the technique is reversed. A coal sample is dug out and weighed; the volume it occupied is assessed by lining the hole with polythene sheeting and measuring the volume of sand or water needed to fill it.

A further difficulty has to be considered. It is argued that the surface results are not typical of the density at lower levels. To avoid this, it has been proposed that an auger be used to extract a coal sample which can be weighed; the volume it occupied is calculated from the cylinder that is represented by the inserted length of the auger.

In the last two or three years attempts have been made to adapt the gamma-ray back-scatter method to this determination. A tube is driven into the pile and an emitter and detector are lowered to various depths at which readings on an indicator are taken. The method is not only subject to instrument error as in the ash determination, but there is the practical difficulty of driving a tube into compacted coal, without bending it—a bend can prevent the instrument from reaching the required depth. Gamma-ray back-scatter instruments for density measurement are claimed to have an accuracy of $\pm 5\%$ when used with care, and $\pm 3\%$ after experience has been gained with particular coals packed by particular methods.

1.12.2. Grindability

75% of coal consumed in our boilers is pulverised for firing. It is desirable to know how the properties of the coal and mineral matter affect the grinding process, how easily the fuel can be reduced to the required fineness and what power is needed to do this. A variety of tests have been proposed; each has its merits and defects and an international group is now studying them.

The best known tester for grindability is the Hardgrove machine which is a miniature ring-ball type pulveriser. The grindability index from this machine was originally based on the increase in surface area produced on the coal particles, but because its measurement was tedious, a simplified procedure was introduced.

A 50-g sample, passing a No. 16 U.S. sieve and retained on a No. 30 U.S. sieve (0·0469 in. and 0·0232 in. respectively) is ground for 60 revolutions with 64 lb loading on the ball-ring. The amount of coal now passing No. 200 U.S. sieve (0·0029 in.) is weighed (W). Although this can be used directly as a measure of the ease with which the coal was ground, in order

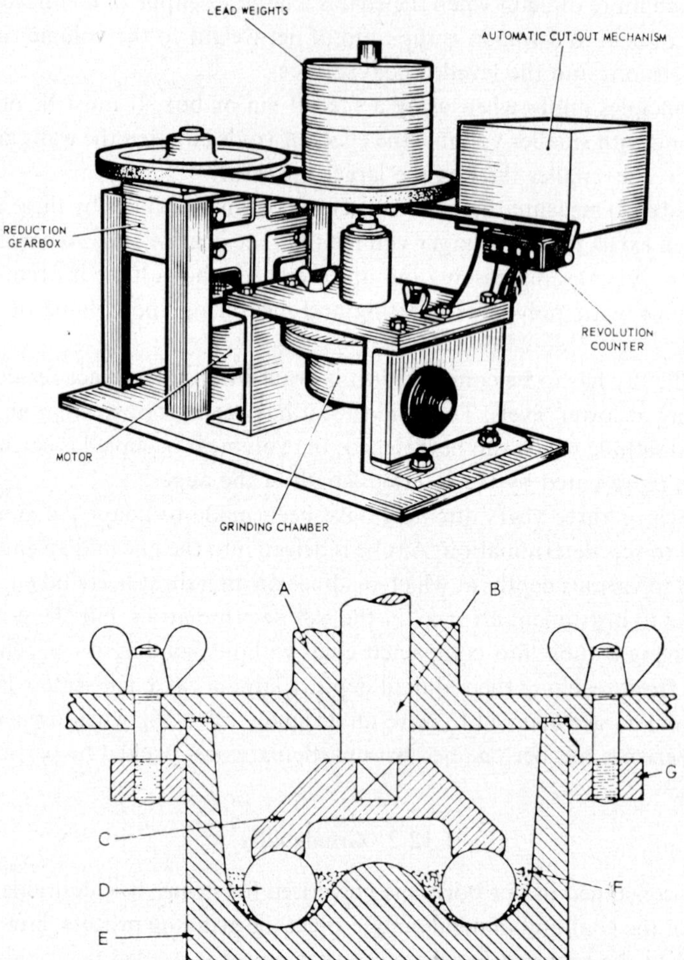

HARDGROVE MACHINE-SECTION OF GRINDING CHAMBER

A	Frame	E	Lower grinding
B	Driving shaft	F	Approximate level of coal sample
C	Top grinding element	G	Clamping ring
D	Steel balls		

FIG. 1.12.2. Hardgrove grindability testing machine

to retain comparability with the original Index based on surface area, it is transformed by the formula: Hardgrove Index = $13+6{\cdot}93$ W.

The Hardgrove Index so calculated can range from over 100 for easily ground coals (Ranks 200 and 300) down to rather less than 50 for coals which are more difficult (Ranks 900 and 100). This broad relationship also holds for other properties associated with rank, such as volatile matter, carbon, etc.

This particular method of test has been criticised because (a) during preparation for the test, some 50% of the sample is sieved out leaving a test portion which may not be properly representative, (b) only part of the sample is taken to the final ground state, and (c) the fines are not removed as they are produced. (See Fitton, Hughes and Hurley, *J. Inst. F.* 1956, p. 54, and Callcott, *J. Inst. F.* 1956, p. 207.)

1.12.3. **Handleability**

One of the problems which has required attention during recent years, has been that of ensuring the continuous feed of coal to boilers, particularly in winter weather when supply interruptions have the most embarrassing consequences. Most difficulty is caused by fine wet coals containing slurries, and coals reclaimed from compacted stocks. Tests to compare the handleability of different coals are useful as a means of sorting those which could be expected to handle well after storage, from those which should be used directly or mixed with other grades.

The Durham handleability testing machine developed by the N.C.B. resembles most others. (A detailed description of its construction, use and test results is given by Hall and Cutress, *J. Inst. F.* 1960, p. 63.) This type consists of a vibrating stainless steel conical hopper fitted with a slide on the 4½-in. outlet. 12 kg of coal (in the undried condition) are loaded, the slide door opened and the time to empty all the coal is noted. The mean of three tests is reported. Hall and Cutress report that the rate of flow for a given coal decreases as its free moisture increases, reaching a minimum between 12% and 17%. Thereafter flow rates increase as the coal tends towards liquid consistency. The time of flow is increased by a rise in the percentage of superfines (through 30-mesh sieve) present, but for a given size consist the free moisture at the slowest flow rate varies with coal rank: 17% for Ranks 300–400 and 12% for Rank 900. They also note that handleability is improved by oiling coal with small amounts (less than 1%) of gas oil. The method is most effective with coals of high rank.

With the increasing dependence of power stations on their coal stocks in the Christmas and New Year colliery holidays, an interest has been taken in anti-freeze agents and methods of thawing for coal. The N.C.B. and British Rail are also interested because of their need to continue to sell and carry coal in times of extremely cold weather. Because of its low cost calcium chloride has been selected for the greatest number of experiments. It can be added to the washed coals (which are the most vulnerable) at the washery, either in solution or as the solid; in the latter case it dissolves slowly in excess water draining through the coal. Dryish untreated coals can be protected by scattering the chemical on the surfaces of stockpiles, and on rail wagons (with heavier doses near the sides.) The chemical can be used in solution to loosen frozen coal left in rail wagons after tipping. The North-Western Region have reported no adverse effects in boiler fouling when calcium chloride is used, but corrosion of rail wagons has been observed. More work will be done to study the effect on austenitic steel superheaters and reheaters.

The thawing of coal in rail wagons has been attempted by steam lances, gas flames and infrared heaters, but the most effective method may be found in spraying the outside of the wagons with hot water (90°C). The latter method is to be carefully investigated and a refrigeration chamber has been constructed at High Marnham.

Prevention rather than cure is aimed at, and in this connection high rank coals seem to react well to oiling with gas oil. Although no ill effects of gas oil on conveyor belts was reported by the South-Western Region, it is expected that the mineral oil would accelerate ageing and impair the life of the belts.

1.13. SIZE ANALYSIS

1.13.1. Size Analysis of Coal

The size analysis of a coal is important as a factor in its pricing and because it affects its handleability, combustion, and subsequent grit emission. The growing N.C.B. practice of feeding fines and slurries back into the smalls, together with the effects of increased mechanised mining, means that more monitoring of this aspect of coal quality is called for.

Attention is drawn to the discussion of sizing in the section on Coal Preparation (Section 1.5) and to the definitions of the various kinds of coal (Section 1.4). In this connection it should be noted that these are in terms of round hole screens but that laboratory tests are frequently carried out using square mesh sieves. Tables relating these two systems are included in the B.S. publications listed; as an approximation a 1-in. round hole is equivalent to $\frac{7}{8}$-in. square hole.

The following B.S. publications deal with the subject of sampling and sieving:

B.S. 1293 and 2074: 1965 Methods for the size analysis of Coal and Coke—deals with sampling and testing (other than P.F.).

B.S. 1796: 1952: Methods for the Use of Test Sieves.

B.S. 410: 1962: Specification for Test Sieves (Supplement No. 1: 1965: PD 5490).

This Standard specifies a range of sieves with intervals based on inch fractions. A revised edition has been drafted for publication and is based on S.I. metric units in the Renard series of Preferred Numbers; the apertures are halved at every third sieve in the series and there is always a standard sieve very close to any desired size.

The sampling theory is based on similar considerations as are outlined in the section dealing with sampling for analysis. A gross sample is taken by combining a number of increments each of minimum weight. The number of increments for the sampling of isolated consignments of coal is set at 40 by B.S. 1293 and 2074 since it has been shown that errors due to sieving and sample division tend to offset the increase in precision that might be expected from increasing this number. The minimum weight of an increment depends on the nominal top size of the coal.

The sampling, preparation and sieving of samples of coal will inevitably give rise to breakage, particularly in the larger sizes. There is, therefore, an inherent bias in size analysis results and the terms "precision" and "precise" are used and not the terms "accuracy" and "accurate". The level of precision depends on the percentage of the size fraction concerned. If it is 5% the error may be ±1·8%; if it is 25% the error may be ±3·2%.

The sampling method must, of course, guard against breakage of lumps of coal. Mechanical screening of the sample may be used in control testing and other comparative tests, but in cases of dispute or where a reference method is required hand sieving in accordance with the British Standard is essential.

1.13.2. **Sampling and Size Analysis of Pulverised Fuel**

Until the British Standards Institution produces a standard dealing with the sampling of pulverised fuel, C.E.G.B. practice is based on a manual of operation issued in July 1956 (Memorandum/GO/0323/ET). "Pulverised Fuel Sampling Equipment, Mark II and Mark III".

The size analysis of powders may be carried out by a number of methods, elutriation in air and sedimentation rate in water being among the best known.

For power station purposes the need is for a simple size indication to check mill performance and for combustion control, which depends greatly on particle size. Too fine a powder is wasteful of mill power, too coarse a powder does not burn completely in the combustion chamber and leads to an excessive proportion of unburned carbon in the precipitator dust.

TABLE 12

ANALYSES OF COAL ASH AND RELATED FUSION CHARACTERISTICS

Sample Number	1	2	3	4	5	6
	%	%	%	%	%	%
Silica, SiO_2	30·0	33·7	38·6	45·3	45·0	49·4
Alumina, Al_2O_3	20·2	22·4	21·8	25·5	29·7	29·2
Ferric oxide, Fe_2O_3	20·7	16·7	9·1	7·5	6·7	3·7
Calcium oxide, CaO	16·3	12·1	16·2	11·6	3·4	2·8
Magnesium oxide, MgO	5·6	6·4	6·9	3·1	6·4	6·5
Sodium oxide, Na_2O	0·9	1·5	0·8	0·9	2·3	2·6
Potassium oxide, K_2O	0·8	1·4	2·1	3·9	2·6	1·5
Titanium oxide, TiO_2	0·7	0·7	1·4	0·8	1·0	1·2
Manganese oxide, Mn_3O_4	0·4	0·4	0·2	0·1	0·1	0·0
Sulphate, SO_3	4·1	4·0	2·6	1·1	2·1	2·1
Phosphate, P_2O_5	0·3	0·8	0·3	0·2	0·7	0·9
Silica ratio	41·1	48·9	54·4	66·9	72·9	79·1
Fusion characteristics:						
	°C	°C	°C	°C	°C	°C
Deformation	1110	1140	1210	1200	1200	1150
Hemisphere	1140	1190	1250	1280	1360	1400
Flow	>1550	>1550	>1550	>1550	>1550	>1550

For most practical purposes in a P.F. station, a simple sieving test is adequate, in which the fractions (as percentage) retained on a 100-mesh and passing 200-mesh, are measured. Accurate sampling procedure is an essential pre-requisite to useful results.

1.14. ASH FROM COAL

The colour of coal ash varies from almost pure white to a dark chocolate colour, depending mainly on the iron content. Its fusibility varies widely according to its chemical composition. Ash containing much iron from a highly pyritic coal, forms the readily fusible ferrous silicate which is a common source of clinker in furnaces (see Dixon, Skipsey and Watts, *J. Inst. F.* 1964, p. 485, "Distribution and Composition of Inorganic Matter in British Coals"; and *J. Inst. F.* 1965, p. 455, for a useful survey).

The simplest coal ash consists of the refractory clay kaolin, composed only of alumina and silica, to which may be added as the composition increases in complexity, lime and oxides of iron in varying proportions. With increase of lime and iron contents, the fusion point of ash falls significantly. In Table 12, the chemical analysis of laboratory prepared samples of coal ash is shown in relation to the corresponding fusion temperatures. The method of determining ash fusion temperature, and the significance of "silica ratio", are given in Section 1.14.4, which follows.

1.14.1. **Boiler Fouling**

Trends in the development of boiler plant have been towards larger unit capacities and more advanced steam cycles. Coal quality has become of increasing significance, particularly with reference to the amount and properties of the mineral constituents. Efficient operation is required over long periods with a minimum of outage for boiler cleaning and maintenance, and sustained boiler availability is essential with the current installations of large boiler-turbine units. Unit boilers at base-load power stations should be capable of almost 100% availability, since their replacement in case of loss of availability, entails the use of less efficient plant with increased cost of generation.

A major loss of availability over the last twenty-five years has been due to fouling and corrosion of boiler heat-exchange surfaces by deposits. These are often hard, strongly adherent and not easily removed by on-load cleaning techniques such as sequence soot

TABLE 13

CLASSIFICATION OF CHLORINE, SULPHUR AND PHOSPHORUS IN COAL

Description	Ash %	Cl %	S %	P %
Very high	>15·1	>0·61	>4·01	>0·071
High	10·1–15·0	0·31–0·60	2·51–4·00	0·031–0·070
Moderately high	7·6–10·0	–	2·01–2·50	–
Moderate	5·1–7·5	0·16–0·30	1·51–2·00	0·011–0·030
Moderately low	–	–	1·01–1·50	–
Low	2·6–5·0	0·06–0·15	0·51–1·00	0·0051–0·010
Very low	<2·6	<0·06	<0·51	<0·0051

TABLE 14(a)

LIMITING AMOUNTS OF CHLORINE, ASH AND SULPHUR IN VARIOUS CATEGORIES OF COALS

Category	Chlorine %	Ash %	Sulphur %
1	<0·20	<8·0	<1·2
2	0·2–0·39	8–13·9	1·2–1·8
3	0·4–0·60	14–19·9	1·9–2·5
4	>0·60	20–25·0	>2·5
5	–	>25·0	–

TABLE 14(b)

ASSESSMENT OF CHANGES OF SUPERHEATER CORROSION AT HIGH TEMPERATURE

	Categories, defined above		
	Chlorine	Ash	Sulphur
Free from corrosion	1	1, 2, 3, 4, 5	1, 2, 3
Free from corrosion	2	3, 4, 5	1, 2
Probably free from corrosion	1	4, 5	4
	2	4, 5	3
	2	2	1
	3	5	1, 2
Possibly free from corrosion (slightly better than an even chance, of not corroding)	1	4	4
	2	2	2
	2	3	3
	2	5	4
	3	4	1, 2
	3	5	3
Risk of corrosion, increasing progressively in severity	1	1, 2, 3	4
	2	1	1, 2, 3, 4
	2	2	3, 4
	2	3	4
	2	4	4
	3	1	1, 2, 3, 4
	3	2	1, 2, 3, 4
	3	3	1, 2, 3, 4
	3	4	3, 4
	3	5	4
	4	1, 2, 3, 4, 5	1, 2, 3, 4

blowing, and air or water lancing. They affect all surfaces from the tubes in the combustion chamber to the coolest metal surfaces of the air heater, where they may be so corrosive that frequent air heater maintenance is necessary. As deposits develop, gas passages may be reduced until the load can no longer be maintained. The conditions are further aggravated by the loss of thermal efficiency due to reduction in heat transfer.

In 1944 this loss in availability, attributed directly to external deposits, was of the order of 15% and the Fuel Research Organisation was requested to join the Boiler Availability Committee (formed in 1939) to assist its work on deposit formation and corrosion by investigating the properties of the fuels which cause boiler fouling.

The Boiler Availability Committee Papers also report their investigations to improve availability of plant, by improved methods of gas-side cleaning. Work by Dr. H.E. Crossley drew attention to the possibility of minimising boiler fouling by a planned distribution of coals to power stations to enable the correct type of coal to be burned by boilers most suited to burn them.

Noting that deposits were sulphatic or phosphatic and that fouling was correlated with chlorine content of the fuel burned, Crossley classified coals with high, medium and low levels of chlorine, sulphur and phosphorus. Using this as a guide, a marked alleviation of fouling troubles followed. Table 13 (page 74) shows his classification as extended by N.C.B.

Experience showed that fouling with coals high in these constituents was much less severe in pulverised fuel boilers than in stoker-fired boilers, and this was attributed to the distillation processes occurring in a glowing firebed. (Here it should be noted that chlorine is not troublesome of itself. Coals which are high in chlorine are salt bearing and it is the alkali metals, mainly sodium, combined with the chlorine, which are distilled. Chlorine is therefore used as an indicator.)

More recent work has revealed that two further considerations need to be taken into account, (a) fouling of a p.f. boiler with fuel of a fixed chlorine content increases as the ash content decreases, and (b) severe corrosion of superheater tubes can arise when the steel temperature is above 1100°F (593°C) (corresponding to a steam temperature of 1050°F (566°C)), even if the sulphur and chlorine contents are classed as medium. The interactions of chlorine, ash and sulphur are very complex and have been discussed by Crossley in the 1962 Melchett Lecture (*J. Inst. F.* 1963, p. 228). Tables 14(a) and 14(b) has been taken from this paper.

1.14.2. **Analysis of Gas-side Deposits**

The analysis of gas-side deposits from furnaces, superheaters, air heaters and economisers is of importance to identify the fouling elements in the coal supplied and to assess the problem of deposit removal. Some boiler deposits occur as a series of strata of various colours, hardness and thickness, which can be carefully separated and analysed. Bonded deposits on superheater tubes are white near the tube surface and have a coloured layer, usually of fly ash, on the outside. Other deposits may appear as scales with an outer layer of dust, when again, identifiable separate layers can be removed and analysed separately. For the purpose of analysis it is convenient to divide deposits into two classes, fly ash material and secondary products. The fly ash type of deposit under the microscope is seen as fused coal ash and is in fact similar in chemical composition to coal ash. Flue grit also is of this type and fly ash deposits generally can be analysed by using the standard methods for coal ash. Secondary product deposits appear as white, rust coloured, or pink masses and usually contain the sulphates, and sometimes the phosphates, of sodium, potassium, iron and aluminium, together with free silica. There may also be varying amounts of many other elements, and they frequently contain free sulphuric acid. The analysis of these deposits may be carried out by classical chemical methods of analysis or by the modified methods described in "The Analysis of External Deposits from Boilers" by H.E. Crossley, A.H. Edwards and D. Flint, Technical Paper No. 2, The Boiler Availability Committee.

Typical analyses of superheater, economiser and air-heater deposits are given in Tables 15, 16 and 17.

TABLE 15

COMPOSITION OF SUPERHEATER DEPOSITS (CROSSLEY)

Constituent %	Phosphatic type			Alkali-matrix type		Re-fusion type	
	White inner layer	Brown friable core	Brown hard outer layer	White inner layer	Brown hard outer layer	White inner layer	Blackish hard outer layer
SiO_2	49·9	43·0	42·8	10·0	18·6	17·3	35·5
Fe_2O_3	7·8	6·2	5·7	14·5	15·6	28·3	23·4
Al_2O_3	6·2	22·4	18·1	2·0	17·5	11·1	21·7
$(NaK)_2O$	4·4	0·5	20·5	24·9	12·8	8·6	3·5
P_2O_5	10·6	3·5	8·2	2·6	2·5	0·2	0·3
SO_3	6·5	1·8	1·8	42·5	27·3	30·9	6·0

TABLE 16

COMPOSITION OF ECONOMIZER DEPOSITS (CROSSLEY)

Constituent %	Sulphatic type		Phosphatic type	
	Hard scale	Dust	Hard scale	Dust
SiO_2	7·2	20·4	23·6	35·1
Fe_2O_3	11·6	18·7	3·0	5·6
Al_2O_3	0·6	16·3	8·5	18·8
$(NaK)_2O$	17·7	8·7	4·4	4·2
P_2O_5	2·3	1·2	25·0	6·9
SO_3	44·0	23·4	12·6	9·0

TABLE 17

COMPOSITION OF AIR-HEATER DEPOSITS (CROSSLEY)

Constituent %	Coal of high phosphorus, low sulphur content, dry dust deposit	Coal of low phosphorus high sulphur content, sticky dust deposit
SiO_2	47·1	6·7
Fe_2O_3	4·7	25·2
Al_2O_3	5·0	3·8
$(NaK)_2O$	4·8	2·2
P_2O_5	3·1	0·4
SO_3	13·8	43·8

1.14.3. **Analysis of Ash**

The recommended methods for the analysis of the major constituents common to all ashes are those first published by B.C.U.R.A in 1958 based on the rapid methods devised by Archer, Flint and Jordan, *Fuel, London* 1958, **37,** 421, and now adopted in B.S. 1016, Part 14, 1963. Much saving of time over the classical methods of analysis has been accomplished by the elimination of lengthy chemical separations and replacing them with selective complexometric and physicochemical methods using flame photometers and spectrophotometers. To meet the growing need for increasing the speed and accuracy of ash analyses, spectrographic methods are used for the determination of some of the elements.

TABLE 18
CHARACTERISTIC COMPOSITION OF BRITISH COAL AND COKE ASHES

Constituent	Expressed as	Range %	Typical analysis (E. Midlands coal)
Silica	SiO_2	15–55	49·5
Alumina	Al_2O_3	10–40	26·3
Ferric oxide	Fe_2O_3	1–40	8·5
Calcium oxide	CaO	1–25	3·4
Magnesium oxide	MgO	0·5–5	2·6
Sodium oxide	Na_2O	0–8	0·8
Potassium oxide	K_2O	0–5	3·8
Titanium oxide	TiO_2	0–3	1·1
Manganese oxide	Mn_3O_4	0–1	0·2
Sulphate	SO_3	0–12	1·1
Phosphate	P_2O_5	0–3	0·3

Table 18 gives the usual range of chemical composition of British coal and coke ash analyses. (See also Dixon, Edwards, Flint and James, *Fuel*, 1964, **43,** 331–348.)

1.14.4. **Ash Fusion Temperature**

The fusion characteristics of ash are determined as a guide to possible deterioration of combustion chamber performance by slagging, and to the propensity of a coal to produce clinker. These characteristics are considered when selecting fuels for the various designs of boiler. For most boilers comparatively high fusion temperatures are obviously desirable, but for slag-tap and cyclone types of furnace, a readily fusible ash is necessary, and it is desirable that the molten slag should flow readily. Reid and Cohen in the U.S. found (1940) that slag viscosity was related to the "silica ratio" defined as

$$\frac{100 \times SiO_2}{SiO_2 + Fe_2O_3 + CaO + MgO}$$

where the components are expressed as per cent of each oxide (see Table 12). B.C.U.R.A. have carried out extensive work on this subject for the C.E.G.B. (*The Chemical Composition and the Viscometric Properties of Slags*, 1963.)

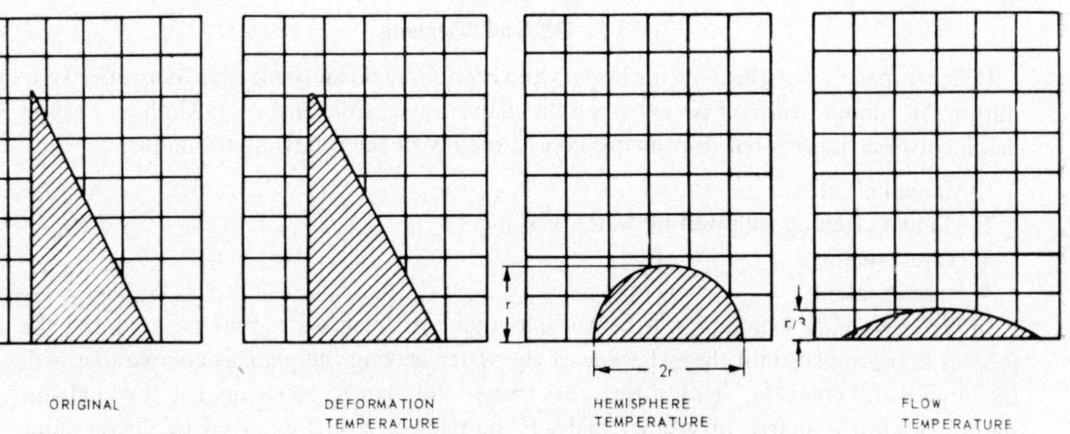

FIG. 1.14.4. Ash fusion test—cone profiles

The determination of ash fusion temperature is carried out in the laboratory according to B.S. 1016, Part 15, 1960, using ash obtained under standard conditions, ensuring complete combustion of the prepared coal sample. In brief, the method consists of the controlled raising of temperature of a specimen of the ash (moulded to a standard pyramid shape) first to the point of deformation of the apex, then through the temperature at which the sample has formed a hemisphere, to the final flow temperature (see Fig. 1.14.4). This is carried out in a small furnace wherein a reducing atmosphere, consisting of carbon dioxide 50%, and hydrogen 50%, is maintained, and the temperature is measured by means of a suitable thermocouple or an optical pyrometer.

In an alternative method, the ash is moulded into a 3-mm cube or cylinder and melted in a Leitz "heating" microscope with a controlled atmosphere. The profile of the ash specimen can be examined visually or photographed as a permanent record of its flow characteristics.

Owing to the very different conditions in which ash fusion temperature is measured, compared with combustion conditions in the boiler, a close correlation between measured fusion temperature and experience in operation, is not usually obtained. Nevertheless, this measurement provides a useful guide to the likely slagging propensity of many coals, particularly in p.f. boilers.

1.15. GAS-SIDE CLEANING OF BOILERS

1.15.1. On-load Cleaning

On-load gas-side cleaning is normally carried out by means of sequence soot-blowing or water-lancing, the principle being that of mechanical and thermal shock removal of deposits by directing high-pressure steam or water on to the metal surfaces to be kept free of deposits. This is usually quite effective in controlling the secondary deposits, but once fused fly-ash deposits are formed, on-load cleaning becomes less effective and boiler performance falls.

1.15.2. **Off-load Cleaning**

Boiler furnaces, superheaters, air heaters and economisers can be cleaned more effectively during the annual overhaul period when the plant is accessible and cool. Methods vary in detail between stations but in principle consist mainly of the following techniques:

1. Manual cleaning.
2. Manual cleaning followed by water washing.
3. Water soaking.
4. Steam soaking.

In water-washing techniques, a convenient source of fresh water may be used and the process is continued until the pH value of the water leaving the plant is comparable with the clean water entering, or until the areas treated are seen to be clean. The final effluent should be clear and free of visible solids. If the deposit is not removed by direct water washing, water soaking may be adopted. This consists of soaking the deposits with slowly running jets of water for a period of about 3 days. Manual cleaning, however, may be necessary following water soaking, if the deposit is particularly troublesome. The addition of a wetting agent in the washing process is often useful and ensures deeper penetration of the water through the deposit to the metal surface. Soda ash is sometimes used in the initial stages of water washing to assist in neutralising acidic deposits, and to liberate carbon dioxide in order to break up the deposit. All water-washing processes should be followed immediately by a brief "drying out" period to dry out the boiler and minimise superficial rusting.

When water washing fails, steam soaking may be attempted. The principle of the method is to fill the boiler with cold water and inject low-pressure steam into the furnace so that it penetrates the deposits and condenses on the metal surfaces of the tubes, thus causing loss of adhesion of the deposit. This process may take 2 or 3 days and in difficult cases may have to be followed by further manual cleaning.

In all of these techniques, care must be taken to protect other metal surfaces, ducting and refractories from the deleterious effects of the wash water and effluents.

1.16. COMMERCIAL UTILISATION OF ASH

Ash from the Board's power stations can be classified into two main groups.

(a) Clinker Ash

This is obtained from the grates of stoker-fired boilers. It has for long been a saleable commodity, being used in the building and civil engineering industries for the manufacture of clinker blocks, in road construction, etc. All new coal-fired power stations are fired by pulverising the coal and so the production of clinker ash will decline as the load factor drops on the older, stoker-fired stations.

(b) Pulverised Fuel Ash

The ash from the combustion of pulverised fuel can be subdivided into two groups. The larger proportion (75% to 90%) is a fine powder which is carried from the furnace by the flue gases and is collected mainly by electrostatic precipitators. This ash is now marketed under the trade name P.F.A. The remaining proportion of pulverised fuel ash consists of an agglomerate collected at the base of the boiler and is known as furnace bottom ash. This ash continues to gain acceptance in those industries previously relying on clinker ash as a raw material.

Production of P.F.A. in 1964/65 amounted to 9·2 million tons.

1.16.1. Characteristics of P.F.A.

P.F.A. consists principally of minute glass spheres, a few of which are hollow, together with some crystalline matter and a varying amount of carbon. Inorganic analysis shows the following radicals to be present (Table 19).

TABLE 19

ANALYSIS OF P.F.A.

		Maximum %	Minimum %	Typical %
Silica	(as SiO_2)	58	38	48
Alumina	(as Al_2O_3)	40	20	26
Iron oxides	(as Fe_2O_3)	16	6	10
Calcium	(as CaO)	10	2	4
Magnesium	(as MgO)	3·5	1	2
Sulphate	(as SO_3)	2·5	0·5	1·2
Alkalies	(as Na_2O, K_2O)	5·5	2·0	4·5

Carbon is present in P.F.A. in amounts which vary with the efficiency of combustion. Older power stations may have 10% or higher figures, whereas many modern base load stations have below 3%.

A small proportion of P.F.A. is soluble in water, the solution being generally alkaline in reaction (sometimes after a transient acidity), and containing principally calcium and sulphate ions.

P.F.A. varies in specific gravity within the range 1·8 to 2·4 and the bulk dry density averages about 57 lb/ft³. Specific surfaces of 2200 to 3500 cm²/g are usual, although finer grades (4000 to 6000 cm²/g) may be obtained from those stations fitted with cyclones followed in series by electrostatic precipitators.

1.16.2. Commercial Utilisation of P.F.A.

P.F.A. from the majority of power stations has two characteristics of vital importance to its marketing:

(a) It is pozzolanic, that is, it is similar to the volcanic ashes or pozzolanas used by the Romans, in combination with calcined limestone, as a cementing material.

(b) When "conditioned" with water and compacted it is self-hardening. The self-hardening property is associated with the presence of soluble salts in the ash.

In recent years the possibilities of using this potentially valuable raw material have begun to be systematically investigated. Latest figures show that commercial utilisation of P.F.A. is now running at well over 3 million tons per annum—more than four times what it was in 1960/61. Thus it can be seen that it is becoming an accepted material in the building and civil engineering industries. Its uses are in the manufacture of building blocks and bricks, the production of a lightweight aggregate by sintering, as a constituent of concrete, and large tonnages are being used as load-bearing fill on many motorway and trunk road schemes and on building sites. The production of a lightweight aggregate by sintering P.F.A. will very substantially increase in the next few years, when it will be of the order of at least 1 million tons per annum. A more recent development, with good possibilities of exploitation, is the construction of low cost roads by the technique of stabilisation with 10% of cement. It is also being used as a raw material for the manufacture of a pozzolanic cement.

Some idea of the economy which P.F.A. can obtain is the saving of between £20,000 and £30,000 in the cost of construction of the recently impounded Stithians Dam in Cornwall (where a 25% cement replacement was specified) in spite of the dam's remoteness from a P.F.A. producing station. Economy is not the sole reason for such use of P.F.A. Increased workability, lower heat of hydration and greater resistance to aggressive ground waters are among the other benefits of such use. The publication in 1965 of British Standard 3892: Pulverised Fuel Ash for Use in Concrete, is expected to boost the sale of P.F.A. for this purpose.

Table 20 shows the tonnages of P.F.A. utilised under the main headings in the year ended 31 March 1965, and Table 21 gives the total quantities utilised in the same period.

TABLE 20

MAIN USES P.F.A.

(a) *Total usage*	tons
Bricks	34,800
Lightweight aggregate	60,000
Concrete and mortar	60,600
Building blocks	871,200
Fill: Roads and embankments	822,500
Fill: Building sites	514,000
Other uses	21,500
Total:	2,384,600

(b) *C.E.G.B. own works included above*	tons
Concrete	6,400
Fill: Roads and embankments	13,700
Fill: Building sites	310,300
Total:	330,400

TABLE 21

TOTAL COMMERCIAL UTILISATION
(YEAR ENDED 31 MARCH)

P.F.A.*		Furnace bottom ash	
1963/4	1964/5	1963/4	1964/5
1,555,500	1,799,300	341,500	585,300

* Includes some furnace bottom ash recovered from lagoons and dumps.

1.17. OIL FUELS

1.17.1. Introduction

The mechanism of the formation of crude oil under the earth's crust is not fully understood. Young crude oil is, however, estimated to be over 10,000,000 years old and the oldest oil some 440,000,000 years old. Over this wide range of time the earth was constantly changing, and in the process, much of it was being repeatedly covered by sea in which existed millions of minute sea animals. These microorganisms depended on marine plant life for food and the latter in turn depended on the sun for a continuous supply of energy. On death, the minute animal remains mixed with decaying vegetable matter and the fine silts washed by rivers into the sea to form the basic raw material for the formation of oil.

1.17.2. World Oil Supplies

Crude petroleum has been discovered and production developed in six main areas in the world. The rate of discovery of petroleum deposits is related to the amount of exploration carried out and the total of proved reserves in the world has been mounting rapidly. Table 22 shows the main areas from which crude petroleum is drawn, together with the proved reserves as known at the end of 1964.

TABLE 22

	Million long tons	% of world total
North America	5900	12·8
South America	3600	7·7
Middle East	28,400	61·1
Africa	2500	5·4
Western Europe	400	0·8
U.S.S.R. (Caspian-Volga-Ural)	4100	8·9
Far East	1600	3·3
	46,500	100·0

The amount of crude petroleum produced in the year 1964 is given in Table 23 together with the current capacity of refining plant installed (in terms of tons per annum):

TABLE 23

	1964 production million long tons	1964 refining capacity million long tons per annum
North America	479·5	556
South America	217·6	166
Middle East	374·2	81
Africa	81·4	32
Western Europe	20·6	357
U.S.S.R.	245·3	225
Far East	27·4	123
Other countries	4	13
World total:	1450	1553
U.K. (included above)	0·13	65

The main source of crude petroleum used in British refineries is the Middle East. However, quantities are also taken from Venezuela, and Libyan oilfields are supplying an increasing proportion. Crude oils from Nigeria have started to flow into the U.K.

The crude petroleum deposits of the world vary considerably in their composition. They are rarely used direct in the unprocessed state as a fuel because they contain volatile light hydrocarbons and cannot be handled without fire and explosion risks. A further reason is that it is advantageous to extract from them all the products of intrinsically high value. (One exception is the crude oil from Taraken (Borneo) which has a sufficiently high flash point and suitable pumpability to allow it to be used directly as a fuel.)

Crude oils are commonly classified according to their distillation residue as:

(a) Asphaltic base (contain little wax).
(b) Paraffin base (contain waxes, but little asphaltic matter).
(c) Mixed base (contain some waxes, some asphaltic matter).

Apart from differences in chemical structure of their components, crude oils vary widely according to the proportions of high and low molecular weight fractions they contain. Those with a high percentage of light molecules have a low specific gravity and vice versa. The gravity of the crude indicates the yields of light fractions and residue which will be obtained on distillation.

Heavy crudes (sp. gr. 0·97–1·01)—from the wide range found in Venezuela—give up to 90% fuel oil, while the light crudes (sp. gr. 0·78–0·83) of the Sahara can give as low as 25%. The bulk of crude from the Middle East is intermediate (average sp. gr. about 0·86) and yields 40/50% fuel oil.

1.17.3. Crude Oil Refining

Oil refineries prepare a large number of different products for consumer use by means of a combination of fractional distillation and processes such as catalytic cracking and reforming. The proportions of the different products can to a great extent be varied according

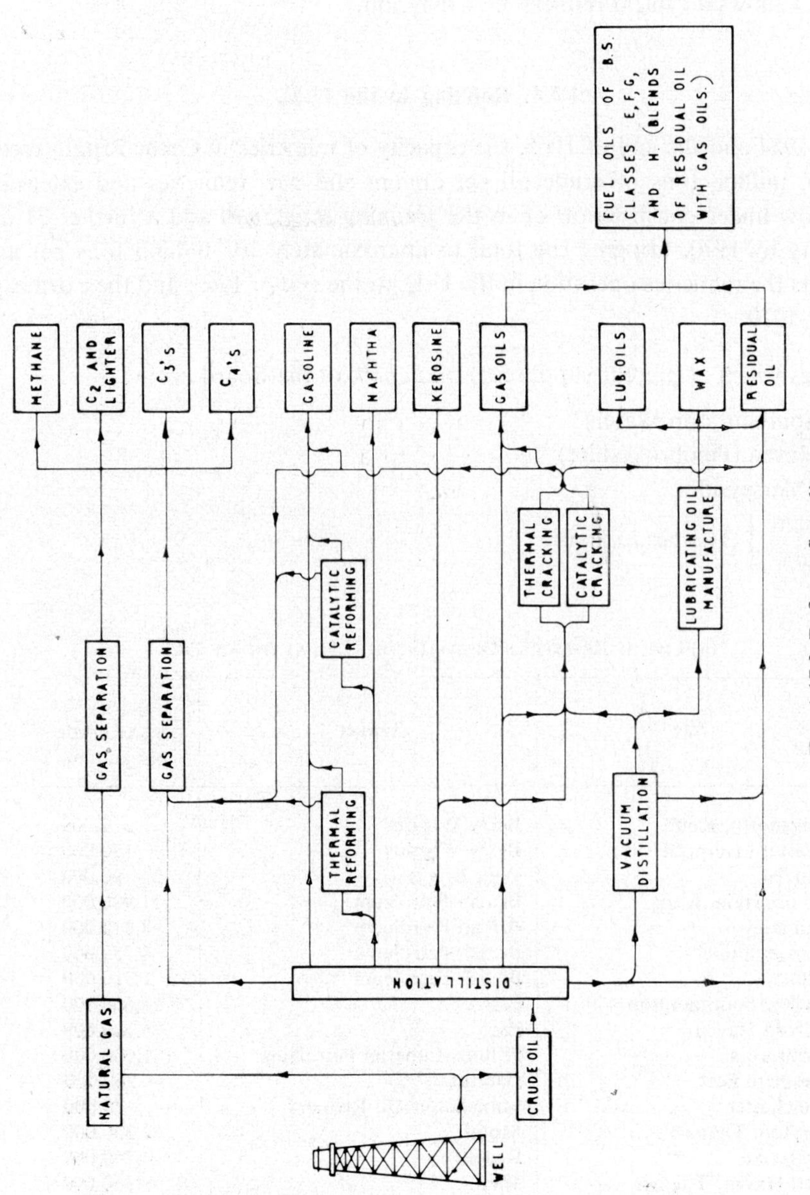

FIG. 1.17.3. Refinery flow diagram

to the source of crude oil used by the refinery, because the crude petroleum deposits found in the various oilfields vary in composition and general properties to a significant degree. The proportions of the various products can also be varied by the choice of refinery processes used. The choice of crudes and processes will depend upon market requirements. Figure 1.17.3 shows a typical refinery flow diagram.

1.17.4. **Refining in the U.K.**

Between 1954 and the end of 1965, the capacity of refineries in Great Britain rose from 29·6 to 74·6 million tons of crude oil per annum and new refineries and extensions to refineries now under construction or in the planning stage, will add a further 31 million tons capacity by 1970, bringing the total to approximately 105 million tons per annum. Table 24 lists the refineries operating in the U.K. at the end of 1965, and the extra capacity planned for 1970.

The refineries which at present supply fuel oil *in bulk* to the Board are:

Fawley (Southampton Water)
Milford Haven (Pembrokeshire)
Stanlow (Merseyside)
Isle of Grain } Thames Estuary
Shellhaven }

TABLE 24

(a) LIST OF REFINERIES OPERATING IN U.K. AT END OF 1965

Site	Owner	Capacity—tons of crude per annum
Kingsnorth, Kent	Berry Wiggins	310,000
Weaste, Liverpool	Berry Wiggins	190,000
Dundee	Wm. Briggs	90,000
Isle of Grain, Kent	British Petroleum	11,900,000
Llandarcy	British Petroleum	8,310,000
Grangemouth	British Petroleum	4,750,000
Belfast	British Petroleum	1,310,000
Fawley, Southampton	Esso	11,600,000
Milford Haven	Esso	6,340,000
North Tees	Phillips-Imperial Petroleum	1,000,000
Ellesmere Port	Lobitos	360,000
Manchester	Manchester Oil Refinery	170,000
Coryton, Thames	Mobil	2,350,000
Pembroke	Regent	4,750,000
Shell Haven, Thames	Shell	9,350,000
Stanlow	Shell	9,750,000
Heysham	Shell	1,950,000
Ardrossan	Shell	170,000
		74,650,000

(b) NEW CAPACITY UNDER CONSTRUCTION OR PLANNED FOR COMPLETION BY 1970

Site	Owner	Capacity—long tons of crude per annum
Canvey Island	United Refineries	2,000,000
Immingham	Continental	3,000,000
Fawley	Esso	5,000,000
Milford Haven	Gulf	3,000,000
Coryton	Mobil	900,000
Killingholme (Humber)	Petrofina/Total	6,200,000
North Tees	Phillips Imperial Petroleum	4,000,000
Eastham (Mersey)	Philmac	390,000
Teesport	Shell	6,000,000
Isle of Man	I.O.M. Petroleum	500,000
		30,990,000

1.17.5. **Utilisation of Petroleum in the U.K.**

The crude petroleum refined in the U.K. in the year 1964 was effective in providing the country with the products broadly classified in Table 25, although there will have been some exportation and importation of various products in order to preserve a balance between refinery product production and product consumption.

TABLE 25

UNITED KINGDOM INLAND CONSUMPTION OF PETROLEUM PRODUCTS IN 1964

	million tons	%
Aviation fuels	2,303,000	3·8
Motor spirit	10,008,000	16·4
Kerosine	1,616,000	2·6
Automotive fuel (Derv)	3,636,000	6·0
Gas/Diesel oil	6,002,000	9·8
Fuel oil	24,753,000	40·5
Gases	1,327,000	2·2
Miscellaneous	7,440,000	12·2
Refinery consumption	4,004,000	6·5
Total consumption	61,089,000	

In the year 1964/5 the C.E.G.B. used 5·47 million tons of oil fuels for all their various purposes (see Table 28). Very little oil fuel was used by the Scottish Electricity Boards.

1.18. BRITISH STANDARD CLASSIFICATION OF OIL FUELS

1.18.1. Introduction

The relevant British Standard is B.S. 2869 : 1967, Petroleum Fuels for Oil Engines and Burners. It is recognised that many types of oil-consuming appliances which normally use a particular class of fuel, can in certain circumstances consume quite satisfactorily a fuel falling within the limits of another class or category. A typical case is the widespread use of burner fuels in the larger oil engines; a second is the occasional use of the better quality marine diesel fuel for steam-raising. Another point of interest is that the distillate fuel known sometimes as diesel fuel and sometimes as gas oil, is used both for oil engines and for burners; accordingly it is classified twice. The Standard covers only hydrocarbon oils derived from petroleum but this does not preclude the incorporation of small amounts of additives intended to improve some aspects of performance.

1.18.2. Engine Fuels

Class Al is a distillate fuel of high quality, designed primarily as an automotive diesel fuel. Class A2 is a distillate intended as a general purpose diesel fuel. Class B1 is also a distillate fuel, but is designed for the larger engines such as those in marine practice. Class B2 is also used in marine practice and the Standard allows for the inclusion of small amounts of residuum. In temperate climates these fuels can be stored at normal ambient temperatures.

1.18.3. Burner Fuels

Class C is a domestic vaporising-burner fuel of the kerosine type. Class D is to all intents and purposes identical to Class A2.

The remaining four classes E to H are essentially classified in accordance with the maximum permitted viscosity of the fuel. They are prepared by blending heavy residues from the distillation of crude oil with a distillate fraction having a flash point in excess of 150°F. Because they are blends of residuum and distillate there is a rising trend of sulphur, ash and sediment and water content as the viscosity increases, and correspondingly a falling trend in calorific value.

The names commonly in use for the British Standard classes are given below:

Class D Gas oil/Diesel fuel
E 220-sec fuel
F 950-sec fuel
G 3500-sec fuel/Bunker C
H+

+ This fuel is marketed in the U.K. only for use in power stations and consequently has no common descriptive title except perhaps "6000-sec fuel".

The Standard will provide a means of making approximate conversions to viscosities at various temperatures and, therefore, to determine the approximate preheat temperature of burner fuels required to give the makers' recommended viscosity at the burner. Viscosities in terms of commercial viscometers (Redwood, Saybolt, etc.) may also be obtained. The Standard will include a formula whereby calorific values (gross and net) can be calculated with a degree of accuracy acceptable for normal purposes, provided the specific gravity of the fuel is known, together with its sulphur, water and ash contents expressed as proportions.

S = specific gravity at 60/60°F,

x = proportion by weight of water,

y = proportion by weight of ash,

z = proportion by weight of sulphur.

Calorific value (gross), Btu/lb

$$= (22{,}320-3780S^2)(1-(x+y+z))+4050z$$

Calorific value (net), Btu/lb

$$= (19{,}945-3780S^2+1370S)(1-(x+y+z))+4050z-1053x$$

1.19. BURNER FUELS

1.19.1. Typical Characteristics

The Middle East holds the largest known reserves of crude oil in the world and formerly provided the great bulk of crudes refined in the U.K. In recent years new sources of crude oil have been discovered and developed in accordance with government policy, which requires increased diversification of supplies in order to promote security. A result of this has been changes in the properties of burner fuels which contain large proportions of residuum. In the future, changes can be expected from time to time according to the need for oil companies to vary their sources for crude.

The new sources (African) of crude oil provide more light fractions and less residuum than do the main bulk of crudes obtained from Middle East sources. They yield oil of low sulphur content but, as they contain a high proportion of waxes, the pour point of the heavy residual fuels is higher; therefore, the fuel oils need to be stored and handled at higher temperatures than those produced from Middle East sources. It is likely that in the future, fuel oils of classes F, G and H, with pour points up to 85°F will be generally marketed. The lower sulphur content of the new crudes is advantageous in connection with the drive towards clean air and in reducing acid corrosion in the cooler parts of the boiler—nevertheless the oil companies continue to seek a method of partial desulphurisation of Middle East fuel oils at a reasonable cost.

Table 26 shows typical characteristics of burner fuels supplied in the year 1964.

TABLE 26

CHARACTERISTICS OF BURNER FUELS SUPPLIED TO C.E.G.B. (1964)

British Standard Classification:	D	E	F	G	H Supplier A	H Supplier B
Viscosity, Redwood I (sec) at 100°F	34	200	900	3100	5700	4000
Equivalent viscosity, kinematic at 180°F approx. (centistokes)*		10·5	22	62	92	74
Specific gravity	0·835	0·935	0·95	0·965	0·975	0·96
Calorific value (gross) Btu/lb	19,600	18,700	18,550	18,450	18,350	18,700
Flash point P.M. closed cup °F	175	190	200	210	280	280
Pour point °F	5/10+ W/S.	10	45	50	60	70
Ash wt. %	Trace	0·02	0·04	0·05	0·05	0·03
Water vol. %	Trace	0·1	0·2	0·25	0·1	0·1
Sediment wt. %	Trace	0·02	0·03	0·04		
Sulphur wt. %	0·8	2·7	3·0	3·2	3·5	2·5

* The 1967 revision of B.S. 2869 specifies a test temperature of 82·2°C (180°F) for the determination of viscosity of classes E to H in order to ensure repeatable determinations on fuels of high pour point.
+ Oils supplied in winter have a lower pour point than those supplied in summer.

1.20. USE OF OIL FUEL BY THE C.E.G.B.

1.20.1. Lighting-up Uses

A long-standing but minor use of oil fuel in power stations has been for diesel and other oil engine-driven generators. Because oil fuels were expensive compared with coal, the first major use in the industry was for lighting-up pulverised coal-fired boilers, although some limited use had been made for lighting stoker-fired boilers. As the size of the industry has increased and as growing dependence has been placed on pulverised fuel, so has the use of oils for lighting-up purposes developed. In the earliest days lighting-up burners were no more than torches for the purpose of securing ignition, the fuels used being kerosine and gas oil. The size of lighting-up burners has progressively increased and today they are capable of being used to raise pressure and to provide as much as 15% of the total steam output of a boiler, if required. Their use has been extended to cover flame stabilisation at times when maintenance of ignition in the coal burners is difficult, either through mill loading or poor coal quality. The large boilers of today depend greatly upon the use of oil in their early commissioning periods. As the scale of oil usage for these purposes has increased it has become economic to use cheaper grades and modern stations now use B.S. Class F (950-sec) fuel.

1.20.2. Boilers converted to Burn Heavy Fuel Oil

In 1955 when oil fuel was significantly more expensive than coal delivered to the south of England, a coal shortage for the U.K. was forecast. In consequence, the Government instructed the Central Electricity Authority to make plans to convert a number of large

TABLE 27

STATIONS BURNING B.S. CLASS H (6000-sec) OIL

Station	Capacity MWso	Method of Delivery
Bankside (L.P. and H.P.)	284	Waterborne
Belvedere (L.P. and H.P.)	460	Waterborne
Barking 'C'	220	Waterborne
Littlebrook 'C'	231	Waterborne
Tilbury 'A'	348	Waterborne
South Denes	244	Waterborne
Marchwood	466	Pipeline
Poole	325	Waterborne
Plymouth 'B' (L.P. and H.P.)	206	Pipeline
Portishead 'B' (part)	122	Waterborne
Bromborough	197	Pipeline
Ince	240	Pipeline

coal-fired boilers to heavy fuel oil firing, and to arrange for long-term supply contracts with the two major oil companies operating in this country—the Esso Petroleum Company Ltd. and Shell-Mex and B.P. Ltd. The plan envisaged that by 1960 sufficient oil fuel would be used in the large oil-fired boiler plants to relieve the N.C.B. of an obligation to provide 10 million tons of coal. Before this time there was only one major oil-fired station—Bankside—and this was built despite the high cost of oil because of the value of having a generating station in the heart of London, which could operate without fear of locally increasing atmospheric pollution—washing flue gases from oil fuel being a feasible and effective proposition.

Since these stations began to burn the heaviest grade of residual fuel oil, the price has fallen. A tax was imposed on all fuel oils in 1961 to protect the coal industry against the combined effect of rising costs for coal and falling oil prices; but oil continues to be the cheaper fuel at many sites. The power stations in the 1955 oil-firing plan are shown in Table 27. They consume B.S. Class H fuels.

1.20.3. **Other Fuel Oil Conversions**

The Board has in recent years converted 31 older power stations and parts of stations from coal burning to oil fuel, in order to comply with clean air legislation in the most economic and effective manner. Such conversions have frequently also overcome maintenance and manpower problems in the power stations. The oil fuels used in the converted boilers vary according to their location, the height of chimneys and the load factor expected. The classes of oil used are British Standard D, F and G. The more expensive Class D oils have been selected for use in stations in built-up areas where a low sulphur content is important and the fact that the oil can be used cold facilitates the frequent cold starts of the true peak-load station without risk of smoke emission. Classes F and G have been selected according to the importance of sulphur content.

TABLE 28

OLDER STATIONS FIRING OIL FUELS

Using B.S. Class H (6000-sec) Oil

Station	*Capacity*
Barking 'A'	167 MWso

Using B.S. Class G (3500-sec) Oil

Station	*Capacity*
Clarence Dock H.P.	267 MWso
Kirkstall L.P. (in part)	90
Tir John	150

Using B.S. Class F (950-sec) Oil

Station	*Capacity*	*Station*	*Capacity*
Brimsdown 'A'	40 MWso	Ironbridge (part)	53 MWso
Brimsdown 'B' L.P.	107	Nechells 'A'	65
Islington (in part)	8	Southport	15
St. Pancras	17	Wallasey	21
Northampton L.P. (in part)	22	Kearsley L.P.	62

Using B.S. Classes Bl and D (Gas and Marine Diesel) Oils

Station	*Capacity*	*Station*	*Capacity*
Grove Road	57 MWso	Agecroft 'A'	55 MWso
Hastings	13	Barton L.P.	101
Luton	14	Bury	10
Taylor's Lane L.P.	10	Clarence Dock L.P.	98
Reading	10	Hartshead L.P.	32
Weymouth	4	Lancaster L.P.	8
Oxford	15	Lister Drive	49
Portsmouth L.P	16	Percival Lane 'A'	24
Salisbury	3	Ribble 'A'	41
Swindon L.P.	11	Stockport L.P.	21
Llanelly	24	St. Helens	16
Derby	47		
Halifax	37		

Table 28 lists the stations converted from coal to oil firing in recent years and classifies them according to the grade of fuel oil they were designed to use. It also includes several minor sections of stations which have used oil fuel for many years.

1.20.4. **Fuel for Gas Turbines**

A new use for oil fuel has recently been introduced by the Board in connection with aviation-type gas turbines capable of running on the normal grade of gas oil instead of the more expensive aviation fuels, which in any case are not generally available for use in land engines. There are two main purposes for the installation of gas turbine generating units. The first is to assist in meeting peak-load demands on the system. The second is to fulfil the need for standby generation for the very large generating units being or to be installed, so that they can restart whenever isolated from the Grid. In this case the gas turbine units will be smaller but can also, if required, be used to feed the Grid system. Table 29 shows the large gas turbine plant installed by March 1966 specifically to meet peak demands.

TABLE 29

Croydon 'B'	140 MW	Lister Drive	112 MW
Rye House	140 MW	Norwich	110 MW
Early	112 MW	Hastings	110 MW

Table 30 shows the use of oil fuel for all purposes in C.E.G.B. power stations during the year 1964/65.

TABLE 30

	tons
British Standard Class H Fuel Oil (6000-sec)	4,558,000
Class G Fuel Oil (3500-sec)	247,000
Class F Fuel Oil (950-sec)	45,000
Class D Fuel Oil (gas oil)	270,000
Total fuel oil used in oil-fired boilers:	5,120,000
Oil fuel used for lighting-up coal-fired boilers, flame stabilisation, etc.	321,000
Gas turbine fuel	14,000
Diesel fuel	16,000
Total oil fuel used for all purposes:	5,471,000

1.20.5. **Future Use of Oil Fuel by the C.E.G.B.**

There is no doubt that oil as a fuel for the Generating Board is attractive, being convenient and already cheap, even with the current heavy rate of taxation. The signs for the future are that the heavier grades will be able to provide a cheaper fuel than coal for many parts of the country. Consequently the Board would like to increase its use of oil fuel but the position of the coal industry in the U.K. cannot be ignored.

The following extract from the White Paper "Fuel Policies November 1967" Cmnd 3438, sets out the forward position: "In order to keep down costs and prices it is important that the electricity supply industry should be able to make use of cheap sources of primary energy. Power stations not yet started cannot at the very earliest come into comission before 1973. For future power stations, the Government have now decided that the generating Boards should base their choice of fuel on an economic assessment of the method of generation which will enable them to supply electricity at the lowest system cost consistent with security of supply and load balancing. In deciding whether to give consent to new stations, the Minister of Power and the Secretary of State for Scotland will also take into account such wider economic considerations as may be relevant.

It would be consistent with this policy for the generating Boards to seek consent in the longer term for the conversion of stations from one primary fuel to another where this would reduce system costs. The electricity industry will, however, require very large quantities of coal in the 1970's and has a strong, continuing interest in the health of the coal industry. For this reason, linked as it is with the short-term structural problems of the coal industry, the Government do not in general envisage the conversion of coal-fired stations to other fuels in the next few years, though there may be scope for some use of natural gas on a seasonal basis and there may be cases where the requirements of clean air dictate conversions from coal to other fuels.

The bulk of the oil fuel used in the Generating Board's power station boilers conforms with B.S. Class H because this is the cheapest grade available, containing as it does the highest proportion of residues from the crude distillation process. However, the Board does not restrict oil companies from supplying lighter grades at the same price if they so desire; in consequence the specifications used by the Board for heavy fuel oil are sufficiently wide to cover Class G oils as well as Class H. The specifications are wide because unnecessary limitations could increase costs to the suppliers and these would be reflected in prices.

B.S. Class H oil is currently used by the Board in 16 power stations and sections of power stations (12 sites). The total annual consumption of these particular stations is expected to decrease as new and more economical plant is commissioned. Whereas in 1964/5 they used 4·56 million tons, by the early 1970's their estimated burn is about 1·5 million tons, assuming that the current oil tax and restrictions on oil burning continues, and that the cost of coal rises only by a few per cent above present level.

In 1969, the first 500 MW oil-fired unit at the 2000 MW Fawley power station was commissioned and a second similar power station at Pembroke should commission its first unit about a year later. Each of these two oil-fired power stations should consume at least 2½ million tons per annum when in full operation. In addition there are possibilities for burning oil in similar quantity at the 2000 MW dual-fired station under construction at Kingsnorth where the first 500 MW unit should be commissioned in 1969. As indicated in the White Paper, the C.E.G.B. may well construct some further large oil-fired power plants at sites distant from sources of coal but close to or easily accessible from refineries.

Although many older stations have been converted from coal- to oil-firing, mainly in the interests of clean air, there are more which could be converted with economic advantage both from the point of view of fuel cost, improved efficiency, reduced maintenance and manpower. However, having regard to the need for this industry to ensure that its own interests in coal are safeguarded, it seems unlikely that many more stations will be converted in the next five years or so. Furthermore such stations would require their fuel mainly in the coldest part of winter when the oil companies' resources and facilities for delivery come under the greatest strain. The quantity of oil fuels consumed in older converted boilers seems likely to fall from upward of ½ million tons in 1966 to under ¼ million tons by 1970.

For the purpose of peak-load generation, the Board decided to install large gas turbine units totalling some 720 MW, the size of the individual units varying between 55 and 70 MW. These have been sited so as to give the transmission system the greatest benefit by their use.

The fuel used is B.S. Class D—more expensive than the residual fuels but it is a fuel which has the advantages that it will not foul the turbine and consequently permits a higher efficiency to be obtained; it has a low sulphur content which is important as a clean air consideration; it can be stored and used cold which facilitates automatic start-up. The capital cost of such units is low, but the running costs high, so that provided the annual load factor is of the order only of a few per cent, they prove to be a satisfactory way of supplementing the system in times of peak-load. The total annual consumption of the units installed may prove to be of the order of only 20,000 to 40,000 tons. The Board is considering the installation of more such units and the size may well be increased.

In addition, a smaller gas turbine unit (17·5–25 MW) will be installed with each of the new large boiler/turbine units of 350 MW or more. These auxiliary units will also run on Class D fuel and will provide power for starting up the steam units under emergency conditions; they may also be used as required for main generation at times of peak-load. A 1% annual load factor may be appropriate.

1.21. OIL FIRING—DELIVERY AND SAMPLING

1.21.1. Delivery

Petroleum products are produced at refining centres and transported to distribution centres by means of coastwise shipment, pipeline, rail and road. The use of pipelines and railways for transporting products is currently increasing.

Bulk supplies of oil fuel may be delivered to consumers direct from refineries and this is the case for the power stations using heavy fuel oil, which receive their supplies in ships or by pipeline. The ships used to carry heavy fuel oil to the power stations range from 500 to 5000 tons carrying capacity. The small ships are used for estuarial transport. Supplies of gas oil for use in the Board's large gas turbines are brought direct from refineries in block trainloads carrying not less than 500 tons.

Practically the whole of the other fuel oils used by the Board (as described in Section 1.20.1/4/5) are carried in road tank cars of capacities 4 to 16 tons. Vehicles to carry 22 tons are being constructed. The oil companies would prefer to provide 500-ton rail consignments to the Board's power stations but the disadvantages to the Board are that large consignments are rather slower to arrive and bigger storage is required; consequently for many purposes the Board prefers to order small quantities at a time to be delivered by road and the oil companies are, therefore, required to provide a 24-h service and to make delivery at short notice.

Oil companies need to ensure that fuel oils B.S. Classes E to H arrive for off-loading at a temperature which is sufficiently high to enable the fuel to be effectively and swiftly discharged. While most ships tanks are provided with steam heating coils which can be used if required to maintain the correct temperature, neither rail tank cars nor road tank vehicles can be heated en route. Because there can be delays on the railways, steam coils are fitted in the cars and the customer has to provide his own steam supply for emergency use. Road tank vehicles are not usually subject to serious delays and the customer is able to reject any consignment which has over-cooled.

1.21.2. Measurement

The oil companies normally sell oil fuels by the gallon (and the Government apply a tax as pence per gallon). This means that the user would check by ascertaining the volume delivered but it would also be necessary to take the temperature of the delivery and correct the ascertained volume to the volume at 60°F, in order to fit the method of billing.

Pricing by volume is satisfactory for small quantities of fuel but in the case of large quantities the Board has been successful in persuading oil companies to price their fuels on the "per ton" basis. The method of checking deliveries is the same so far as ascertaining the number of gallons supplied at 60°F, but in addition a specific gravity measurement is required so that the number of gallons per ton can be derived. Specific gravity can vary considerably from consignment to consignment, even of the same product, and the temperature of a consignment may not be even throughout. Consequently attention must be given to ensuring that the values used for specific gravity and temperature are properly representative of the whole consignment.

1.21.3. Sampling

In addition to sampling for specific gravity and temperature for the purpose of measurement checks, samples are taken in order to obtain small representative quantities from a part or whole of a quantity of oil, and are required for visual or laboratory examination. It cannot be over-emphasised that an incorrect sample invalidates all of the work ultimately done in the laboratory on the oil. The sample therefore must not contain any adventitious material or be altered in the sampling procedure by evaporation or by oxidation.

The Institute of Petroleum's handbook *IP Standards for Petroleum and its Products: Part IV: Methods of Sampling 1964* gives detailed procedures for sampling oils from cans, drums, rail and road tankers, storage tanks, ships, barges and pipelines and describes appropriate sampling equipment. Cleanliness is essential during sampling and sampling equipment and containers therefore must be chemically clean. Responsible personnel should be trained in the techniques of sampling and compositing and be skilled in the preparation of final samples for the laboratory. Because of the fire hazard, necessary safety precautions should be implemented at each stage of operation.

1.21.4. Sampling from Pipelines

Samples may be drawn continuously from pipelines by means of a drip feed system, or intermittently at timed intervals, by opening a sampling valve to give individual weighted increments. The pipeline sampling equipment consists of a $\frac{1}{2}$-in pipe which projects into the pipeline internally to one third of the pipeline diameter. The inner end is bent at 90° to face the flow of the oil stream, and a sampling cock is fitted outside. Figure 1.21.4 shows a typical pipeline sampling arrangement for continuous or intermittent sampling.

If maximum accuracy is required, a micropump may be adapted for continuous sampling.

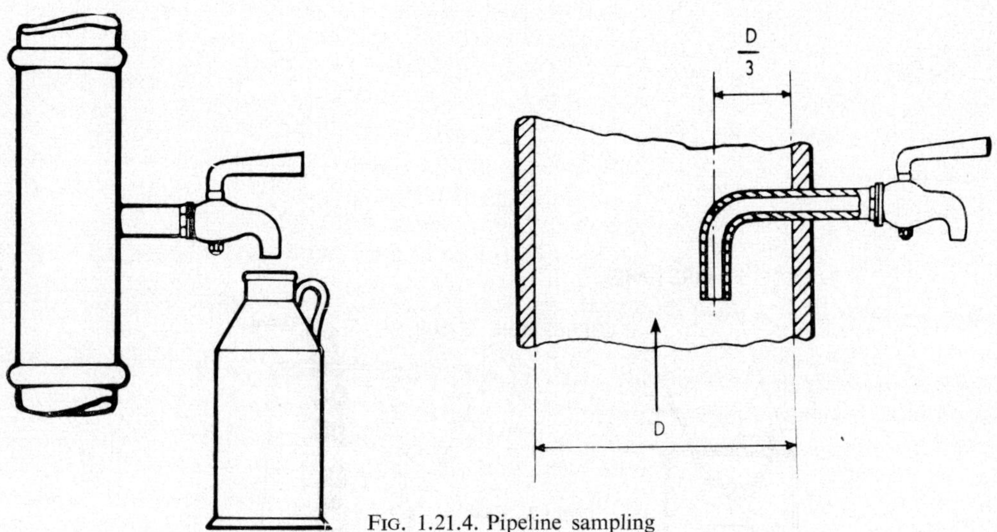

FIG. 1.21.4. Pipeline sampling

1.21.5. **Sampling from Ships**

Oil as shipped may be in several separate tanks and is sampled at the suppliers' end before despatch. On arrival at a power station wharf, each hold is sampled separately before discharge and the oil is then discharged to the station where it may again be check-sampled by a continuous sampling technique. For sampling from holds, the weighted sampling can method is conventionally used. The can, of 1–2 pints capacity and suitably weighted (Fig. 1.21.5), is fitted with a long cord attached to a stopper in its neck in such a way that the stopper can be jerked out at any desired level. Upper, middle and lower samples are taken from each hold at one-sixth, one-half and five-sixths depths, respectively, after first rinsing out the can at each depth before drawing the sample. The samples may then be composited on a weighted basis and mixed to give three final samples; one for the supplier, one for the consumer and the third retained by the consumer as a referee sample in case of dispute.

When sampling from holds, tankers and storage tanks, it is essential to sample systematically, in order, from top to bottom, so that each sample is obtained before the oil at that level is disturbed.

Where tanks are horizontal, cylindrical, or irregular in shape, the tank dip should be translated into volume from the tank calibration table, and the samples taken at levels equivalent to one-sixth, one-half and five-sixths of that volume.

1.21.6. **Sampling from Road and Rail Tank Cars**

The general principle of the one-sixth, one-half and five-sixths sampling depth procedure may be applied but it is sometimes more convenient to discharge the tankers into a site storage tank before sampling. However, the latter method has the disadvantage that the storage tank is likely to contain some oil already.

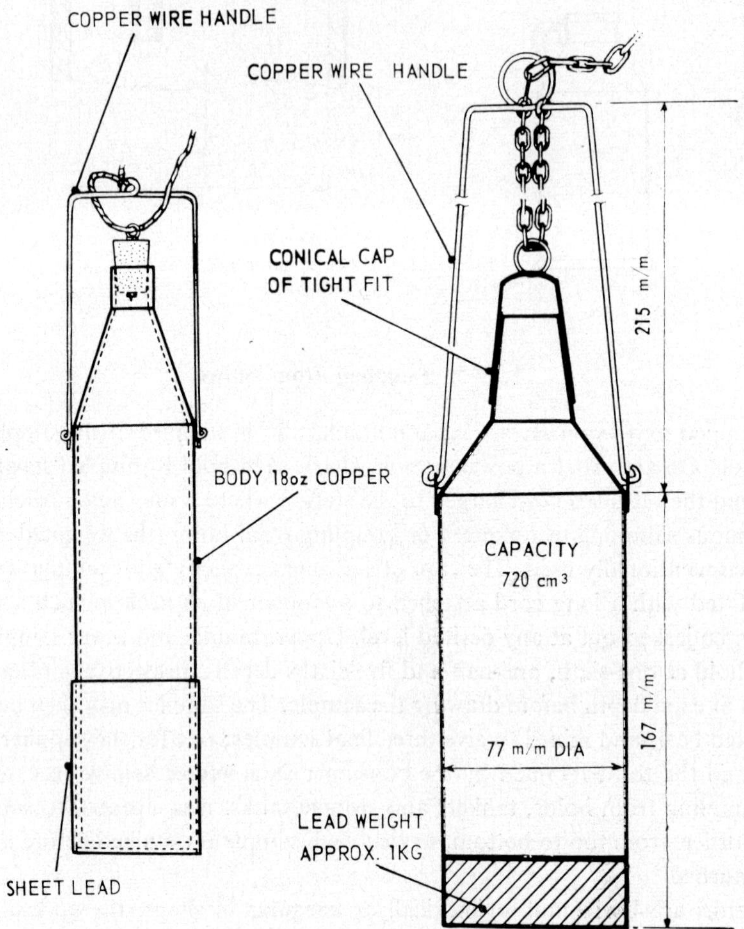

FIG. 1.21.5. Oil sampling containers

1.21.7. Sampling from Storage Tanks

A number of samples are taken from different parts of the tank and combined in suitable proportions to provide a single tank composite oil sample.

1.21.8. Sample Containers

Containers should be of glass or metal and fitted with a suitable cap or lid. If of glass, the bottles must be scrupulously clean and treatment with chromic acid followed by copious water rinsing is recommended. After washing and draining, the bottle is dried in an air oven at 105°C. If cleaning used oil bottles, precleaning with an oil solvent before the above treatment is essential. Cans are often used as sample containers and these should have external seams. Particular care is necessary in cleaning, particularly new cans, for it is essential that all traces of rust, soldering flux, etc. be removed.

1.22. TESTING

1.22.1. Safety Precautions

It is seldom necessary for regular routine tests to be made for all the main characteristics and properties of the oil fuels. However, an inherent danger of oil fuels is the fire risk. For this reason it is wise either to check every consignment before accepting delivery or, because of the delays this would involve in many cases, to take deliveries into storage tank(s) not supplying oil to the boilerhouse and which would not be used until their contents had been tested and passed as satisfactory for use. The check most desirable is the flash point test, and fuel oils and gas oil should have a closed flash point (Pensky–Martens) of not less than 150°F. This test is not simple; consequently other precautionary tests may be applied before accepting consignments. For example, smell and sight tests may give an indication of a wrong delivery; a specific gravity test by hydrometer will in some cases give an indication as to whether or not the correct grade of oil has been delivered. In addition the details of the delivery note should be carefully examined. However, none of these simple precautionary checks would provide a safeguard against receipt of the correct grade of fuel contaminated with a low flash product; therefore, the flash point is the only true indication. Oil companies are well aware of the danger and do all that is possible to ensure that correct and uncontaminated deliveries of fuel oil are made.

1.22.2. Specific Gravity

This is required when a fuel oil is sold on the "per ton" basis.

1.22.3. Viscosity

Viscosity is the main characteristic by which fuel oils are classified. This test is, therefore, necessary as a routine to see that the correct grade of oil is in use and that the temperature of preheat is satisfactory to give the correct viscosity of the fuel at the burners. The viscosity

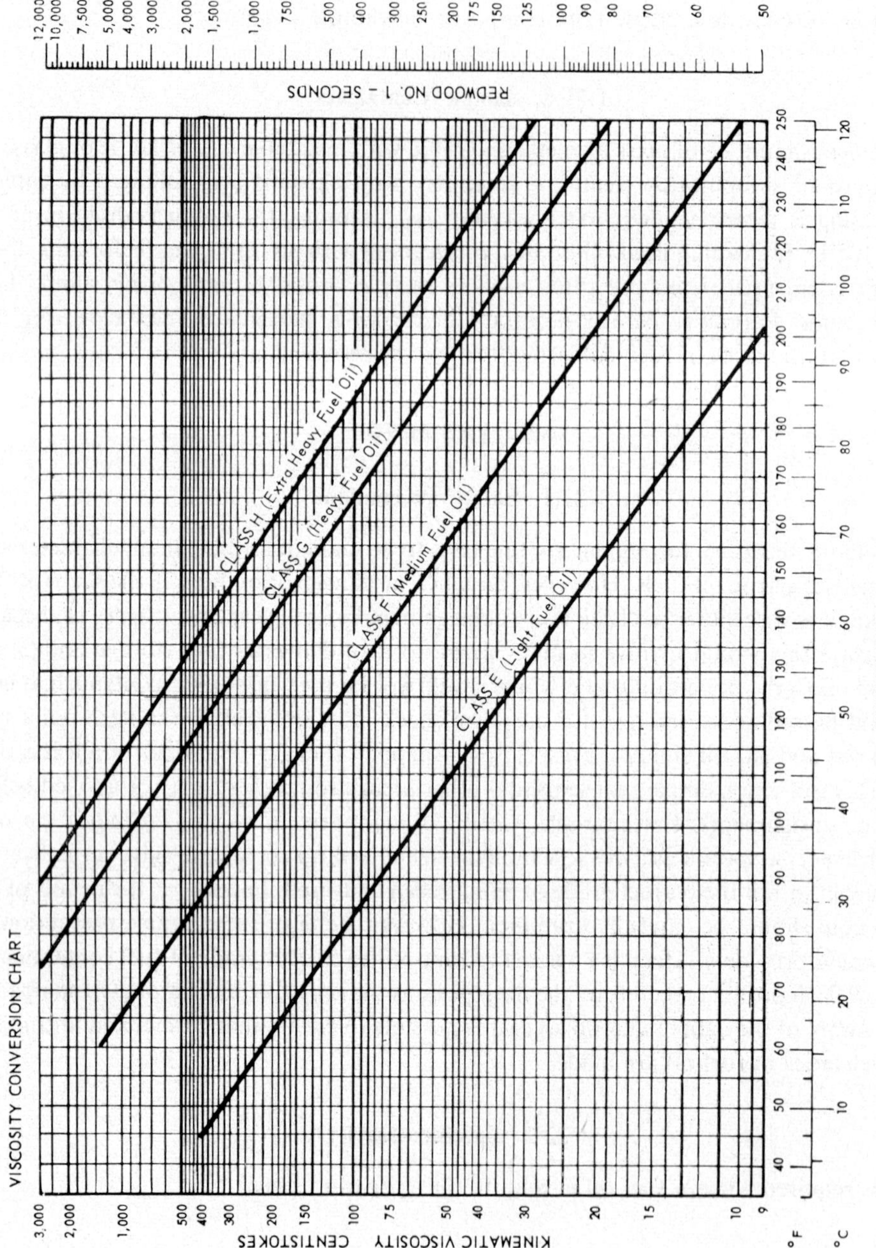

FIG. 1.22.3. Viscosity temperature diagram—fuel oils

of oil is strongly affected by temperature and diminishes with rising temperature (see Fig. 1.22.3). Burners are designed to function with oil of a specified viscosity range, and in Britain this is 12·5 to 24 centistokes (60 to 100 Redwood No. 1 seconds) at the temperature at which it is supplied to the burner.

1.22.4. **Cloud and Pour Point**

The cloud point test is used with engine oils (B.S. Classes A1 and A2 and also gas oil B.S. Class D). It is the temperature at which waxes normally held in solution, begin to solidify in small crystals causing the oil to appear cloudy or hazy. At temperatures below the cloud point these waxes will block the fine strainers used in oil engines. Such strainers may have apertures as small as 5 microns and this is the case in the Board's gas turbine units.

The pour point of an oil was intended to indicate the lowest temperature at which an oil can be pumped. It has been found that most oils can be used at temperatures considerably below the pour point, but the pumping rate of others, notably those from the North and West African crudes becomes abnormal because of thickening at temperatures above the pour point. (The Institute of Petroleum are working on the development of a pumpability test to replace the pour point test.) However, if the limitations of the pour point test are realised, it is of value in indicating the minimum temperatures for storage and handling (see Table 31).

1.22.5. **Calorific Value**

Gross calorific value is determined mainly for the purpose of calculating power station overall thermal efficiency. The determination is rarely made for checking purchases because oil companies do not normally specify the calorific value of the product. Where it is specified, it is usually only as a minimum value and values seldom fall below it. The net calorific value is required only for purposes of boiler design.

1.22.6. **Sulphur**

The sulphur content of fuels used by the Board is important if only for the large weight burned and its consequent effect upon the air throughout the country. The oil fuels used are, therefore, regularly tested to ascertain sulphur contents in order to study trends and to see that the specified upper limit is not exceeded. An additional reason for requiring to know sulphur in fuel oils used in the Board's boilers is to control corrosion in the cooler parts of the boilers, chimneys and to minimise acid soot emission.

1.22.7. **Sodium**

Sodium compounds in fuel oils, particularly in heavy fuel oils, if present in amounts greater than 100 ppm, can cause a significant degree of fouling and therefore reduce efficiency and possibly availability. It is not usual to find much sodium in the crude oils, but the quantity can be increased by contamination with sea water in the tankers which carry the crudes to the refineries. A neutralisation process carried out in the refineries, if not carefully controlled, can add significantly to the sodium content of the fuel. The quantity of sodium found in the heavy fuel oils used by the Board is generally of the order of 40 to 150 ppm (0·004 to 0·015% by weight).

1.22.8. **Vanadium**

Vanadium compounds are present in crude oils to varying degree and therefore in residual fuel oils. The highest vanadium contents are found in Venezuelan oils and the B.S. Class G/H residual fuel oils from this source can contain as much as 800 ppm (0·08% by weight). However, the normal range for vanadium in the B.S. Class H oils used by the Board is 50 to 250 ppm (0·005 to 0·025% by weight). The interest in vanadium compounds in fuel oils is centred mainly upon their corrosive properties upon boiler metals (essentially the superheater), at temperatures of 566°C and above. It is for this reason that the final steam temperature of oil-fired boilers installed or planned by the Board, does not exceed 543°C. Early work on the subject of sodium and vanadium compounds in fuel oils and their tendency to foul or corrode boilers, indicated that deposit formation is at a minimum when the ratio of vanadium/sodium (both calculated as the metal) is within the range of 4/1 to 12/1. However, it was recognised that the significance of this ratio becomes less with decreasing vanadium and sodium contents so that with low vanadium and sodium contents a ratio of 1/1 could be tolerated. The Board, in its experience of fuel oil burning, has not encountered any serious problems engendered by compounds of these elements.

1.22.9. **Ash Content**

The quantity of ash in fuel oils is very small and therefore relatively unimportant. Nevertheless, an upper limit in ash content is specified and occasional tests are, therefore, necessary to ensure that the limit is not exceeded.

The following table shows maximum percentages of ash as specified, compared with the amounts commonly found.

Ash Content of Fuel Oils (% by weight)

British Standard Class	D	E	F	G	H
British Standard—upper limit	0·01	0·10	0·15	0·20	0·20
Typically found in practice	Trace	0·02	0·04	0·05	0·05

1.22.10. **Sediment**

The specification for maximum sediment appears unnecessarily high in relation to the amount found in the bulk of oil fuels delivered. The maximum permitted sediment in the standard is 0·25% by weight for Classes F, G and H—the maximum for Class E is 0·15% and for Class D 0·01%.

The deleterious effects of sediment could be either the chokage of strainers—which are installed to protect pumps and the burner tips from blockage—or erosion of pumps, pipework and burner tips by abrasive materials.

For these reasons it is important to examine occasionally the sediment in the fuel oil, although experience has shown that abrasion does not constitute a problem and the bulk of the sediment is sufficiently fine to pass through the system into the combustion chamber. The oil companies are, of course, anxious to supply products which are trouble-free in operation.

1.23. STORAGE AND HANDLING

1.23.1. **Policy**

Oil fuels are costly to store not only because of the capital cost of the fuel lying in store (this includes tax which is paid on oil as delivered), but because the tanks necessary to hold the oil are expensive. Tanks which need a piled foundation are about 40% more expensive than those which do not. Notwithstanding the high cost of oil storage, and because petroleum is not indigenous, the Government requires many weeks' supply of finished products to be held available by oil companies in strategic positions throughout the country. Oil companies also hold on hand at the refineries, reserves of crude oil awaiting processing; in addition there is at all times a considerable quantity of crude oil on the high seas in transit to this country. It has already been mentioned that refineries are capable of varying the output of the various petroleum products; consequently the consumer is not pressed to provide storage to compensate for differences between their winter and summer use.

The Board's policy in connection with storage of oil fuel, is to carry proportionally greater stocks at those stations most important to the satisfactory operation of the system. This means that an oil-fired 2000 MW power station would be expected to have reserves of fuel available to enable it to continue operation at full output for 4 weeks in a national emergency. Stations converted to oil firing as a result of the 1955 Plan were arranged to have approximately 2 weeks' supply of fuel on hand. All gas turbine units will have a fuel storage capacity equivalent to 60 h run at full output. Stations recently converted from coal to oil firing have fuel storage capacity to an extent governed by the size and importance of the plant and also by the ease with which the supplying oil company can replenish the stocks. In no case will the storage amount to less than 18 h at full output and one week's supply at the highest expected rate of consumption would not be uncommon. The amount of fuel to be stored in reserve for the purpose of lighting-up P.F., etc., will vary according to the consumption rates expected in a week, and to the proximity to oil companies' distribution facilities. Storage ranges from much less than 1 week to 4 weeks' supply, at the highest monthly rate of consumption experienced.

1.23.2. **Practice**

As previously mentioned, fuel oils are delivered at a temperature which gives good handleability, and this presents some advantage to the customer when the turnover of his stock is rapid. For example, it is very useful in cold weather to have gas diesel oil delivered at above ambient temperature, because the user does not usually provide any heating for his equipment. B.S. Class D fuel can be stored and fired at ambient temperature but provision could with advantage be made to see that the temperature does not fall below 0°C; this can be effected by a small tank outflow heater, since it is the fuel handling system which requires protection against extreme cold.

The gas turbines have proved particularly sensitive to the presence of finely divided solid material in the gas oil supplied. This is understandable because the sole previous experience has been with aviation fuels. Aviation fuels are not viscous and have a lower specific gravity— therefore, adventitious solids settle out easily and any small amount carried over is easily separated by a filter of small area. Great precautions are taken on these lines in fuelling aircraft. It has been found that the comparatively small amount of fine solids present in gas oil (much of this comes from rusting of the consigning tanks, although the using plant tankage and pipework can contaminate the fuel) rapidly block the 5 micron aircraft-type of filters fitted in the gas turbines. Special precautions against sediment are being taken for the Board's gas turbine installations and additional larger pre-filters are being introduced to remove +10 micron material.

All petroleum products exposed (after preparation) to moist air will take up some water; the heavier the product the greater the amount. Gas oil, therefore, contains sufficient to rust tanks and pipework (as mentioned above) and sufficient also to form crystals of solid ice at temperatures below 32°F. The ice particles can rapidly block fine filters such as are used with gas turbines, for aircraft and also for land-based use.

In the case of gas turbine units which it is essential to bring on load quickly and automatically, where oil transfer pipework is not set well below ground but remains subject to the influence of the weather, it would seem essential to arrange for oil to be continuously warmed and recirculated (at times when air temperature is below freezing) through the external pipe runs and filters and back to the outlet from the storage tanks.

Residual fuel oils should be stored and handled at temperatures well above their pour-points or serious pumping difficulties may occur. Table 31 shows suitable storage and handling temperatures and in addition approximate temperatures for atomisation.

Note: The temperatures adopted should be the higher of the alternatives.

It is necessary to provide trace heating and lagging for all pipelines, filters and valves, etc., handling residual fuel oils. It is particularly important that pressure gauges and other instruments using the fuel oil as a control fluid, should be either heated or filled with a suitable

TABLE 31

STORAGE, HANDLING AND ATOMISATION TEMPERATURES REQUIRED FOR VARIOUS FUELS

B.S. Class	Storage temperature °F	Outflow temperature °F	Atomisation temperature °F
D	No preheat normally necessary	No further heating normally required	
E	45	45	140–150
F	(P* + 10) or 65	(P* + 15) or 80	190–200
G	(P + 10) or 75	(P + 15) or 100	240–250
H	(P + 10) or 85	(P + 15) or 120	260–280

* P, pour point

buffer fluid such as gas oil. The heating may be effected by means of steam or hot water or it may be electrical. Even after the boiler fuel ring main circulating oil is preheated to atomising temperature, there can be sufficient cooling in branches to burners which are shut down, to make lagging and trace heating necessary.

The oil fuel in store should never be raised in temperature to its closed flash point temperature, because this would increase the fire hazard.

1.24. OIL FIRING—COMBUSTION AND ASSOCIATED PROBLEMS

1.24.1. Combustion

As a boiler fuel, oil is easier to control than coal; it is much more homogeneous and much less likely to vary in its properties over short periods. Consequently, near-optimum boiler efficiency can be obtained in normal operation. As a fuel its hydrogen content is more than twice that of coal and, therefore, the theoretical boiler efficiency obtainable (gross C.V. basis) is lower than from coal. However, in practice, because of the points mentioned and because the boiler heat-exchange surfaces remain comparatively clean for long periods, oil-fired boilers tend to give efficiencies in operation equal to those of coal-fired boilers. It has proved possible to operate oil-fired boilers with a much smaller margin of excess air without making smoke, than has been possible so far with coal-fired boilers. With oil fuels, problems with smoke and dust emission are considerably less than with coal, and ash removal and disposal does not create difficulties in view of the small amounts present.

As compared with coal, fuel oils burn with a highly luminous intensely radiating flame, which allows a high rate of heat transfer to be attained. When burnt in boilers converted from coal-firing, this characteristic means that excessive heat is released in the combustion chamber which, together with the lower mass-flow of flue gas from oil, causes a reduction in steam temperature. Therefore, boilers should be specially designed for oil-firing.

The combustion of oil fuels takes place in the vapour phase, and it is necessary to feed the oil into the furnace in such a way that this vaporisation can be rapid and continuous. The heat used in evaporation is obtained from the flame itself and from surrounding surfaces, and to obtain the required vaporisation it is necessary to expose a large oil surface area to the flame. This is achieved by a process known as "atomisation". The burner itself must break the fuel down into minute droplets—between 30 and 150 microns—and in addition must spread and inject them into the combustion zone in an even pattern. Burners are available of intricate designs which atomise in different ways, for example by oil pressure alone; by steam or air pressure; by rotating cup. For large oil burners the oil pressure method is preferred.

On starting a new boiler from cold, atomisation may not at first be easy to achieve and carryover of oil droplets into the colder parts of the boiler can occur. There have been instances where accumulation of unburnt oil in air heaters has subsequently led to serious fires. Poor combustion can easily occur during the commissioning period of new plant, when cold starts are frequent and operating experience is limited. For this reason it is prudent to consider the use of a more expensive but more easily atomised (that is, lighter) oil during this period, until experience and confidence are gained.

1.24.2. High-temperature Fouling and Corrosion

On combustion, the metallic compounds which are constituents of residual fuels, principally sodium and vanadium, are converted to oxides and sulphates. These products have melting points in the range 1100° to 1650°F and the soft sticky particles give rise to deposits which are highly corrosive in the molten state. In the solid state they are virtually non-corrosive. The mechanism of corrosion depends on the amounts and types of compounds present and the temperature; the main effect is accelerated oxidation promoted by vanadium. Sodium oxide materially assists corrosion by forming low-melting compounds with vanadium pentoxide.

In a boiler trial at Plymouth, burning Middle East B.S. Class H (6000-sec) fuel oil, measurements of the rate of deposition of sodium and vanadium compounds on metal surfaces at temperatures between 850° and 1400°F (455–760°C), were made using an air-cooled probe. The oil contained about 100 ppm each of sodium and vanadium but the test results showed the rate of deposition of sodium to be between 4 and 50 times that of vanadium.

Sodium sulphate was the major constituent of the deposits (about 80% of the total). Below 1000°F (540°C) sodium pyrosulphate was present, whereas above 1150°F (620°C) soluble vanadium compounds were present.

To date neither fouling nor corrosion of superheaters has been a major problem in C.E.G.B. stations. In some of the trials certain additives used for the prevention of low temperature corrosion gave rise to significant superheater fouling.

1.24.3. Low-temperature Fouling and Corrosion

Sulphur on combustion, produces sulphur dioxide of which a small percentage is converted to sulphur trioxide. As the flue gases pass through the boiler and get progressively cooler, a temperature is reached at which "moisture" may form and condense on the plant. This temperature is known as the dew point. When sulphur trioxide is present, the condensation is not only water but a solution of sulphuric acid, which is corrosive. The corrosion takes the form of direct sulphuric acid attack on ferrous metal surfaces to form iron sulphates, and fouling occurs due to the accumulation of the corrosion products, ash and unburnt carbon carried over from the furnace.

The dew point of flue gases is related to the amount of sulphur trioxide present and increases with rising concentrations. Therefore, for a given final flue gas temperature the higher the dew point, the greater becomes the area of boiler surfaces exposed to acid attack.

Thus it will be seen that the acid dew point of the flue gases is the factor which controls the extent and degree of corrosion and fouling in the low temperature part of the flue gas system.

When boilers are taken off load and brought on again at intervals, as happens when a station is meeting peak loads and to a lesser degree single- or two-shifting, low-temperature acid corrosion could be a serious problem. If the problem cannot be reasonably handled in any other way, there is the possibility of using a distillate fuel of low sulphur content at times when condensation would be expected. This has the disadvantages of raising fuel costs, and separate storage and handling is needed for the additional class of fuel oil.

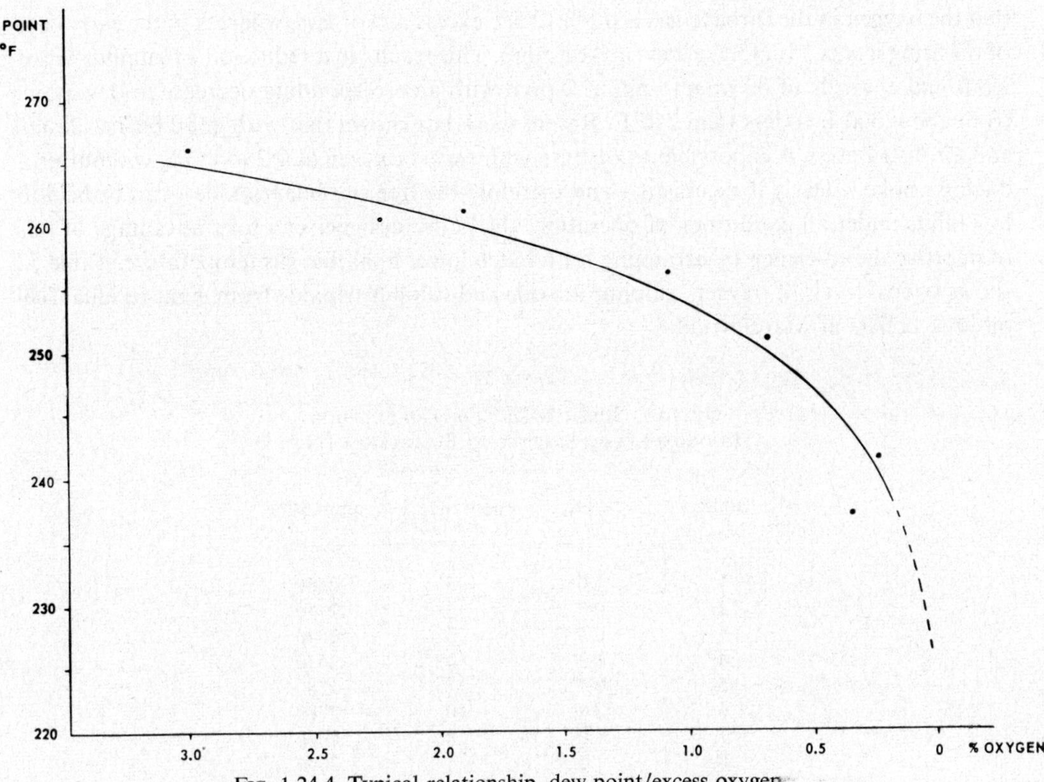

FIG. 1.24.4. Typical relationship, dew point/excess oxygen

1.24.4. Factors Affecting the Acid Dew Point

The acid dew point of a flue gas is related to the amount of sulphur trioxide present, rising as the concentration rises. Sulphur trioxide is formed by further oxidation of the dioxide, which in turn is the product of combustion of sulphur.

$$S+O_2 \longrightarrow SO_2 \quad \text{and} \quad 2\,SO_2+O_2 \longrightarrow 2\,SO_3$$

The percentage of sulphur in the fuel being burnt is thus the first factor affecting the dew point. The conversion of sulphur dioxide to trioxide may be affected by the amount of excess air and the presence of catalysts. From work done at several oil-fired stations it was shown that an increase in excess air was accompanied by an increase in sulphur trioxide detected, whilst low-temperature corrosion of boiler plant was kept to the lowest level by maintaining excess air at a minimum. This does not conclusively prove that less sulphur trioxide is formed when excess air is at a minimum. Because the quantity of carbon particles and fine dust in the gaseous combustion products increases as the amount of air is restricted, it may be that sulphur trioxide is absorbed. Furthermore, the sampling equipment for oxides of sulphur includes a filter to remove carbon and ash from the gases—this filter could therefore possibly take up sulphur trioxide.

Figure 1.24.4 shows a typical relationship between the dew point and excess oxygen in a 500 k lb/h boiler. In modern boilers it is possible to reduce the excess combustion air so that the oxygen in the furnace gas is 0·5% (2·5% excess air) or less, whereas in the early days of oil-firing it was 3% (15% excess air) or more. This results in a reduction of sulphur trioxide found upwards of 30 ppm to about 2 ppm with a corresponding decrease in dew point from about 300°F to less than 240°F. Recent work has shown that with good burner design and air distribution it is possible to operate with excess oxygen at 0·2 to 0·3% without producing smoke. Clearly if excess air—and therefore the free sulphur trioxide—can be held to low limits under all conditions of operation, the boiler designer can take advantage of this to improve the efficiency by arranging for a much lower final flue gas temperature. Table 32 shows typical levels of oxygen, sulphur dioxide and sulphur trioxide from eight residual fuel oil-fired boilers at Marchwood.

TABLE 32

OXYGEN, SULPHUR DIOXIDE AND SULPHUR TRIOXIDE LEVELS SAMPLED AT ECONOMISER INLET

Boiler	% O_2	ppm SO_3	ppm SO_2
1	1·0	3	2560
2	0·7	4	2750
3	0·5	3	2540
4	1·3	4	2630
5	0·7	7	2780
6	2·8	10	2780
7	1·0	6	2620
8	1·0	5	2690

Although excess air is a principal factor in the formation of sulphur trioxide in flue gas, it is also influenced by the presence of catalysts in deposits on the boiler surfaces and this has been observed on commissioning new oil-fired plant. With a very clean boiler, levels tend to increase after a few weeks' operation. This increase suggests that the oxidation of sulphur dioxide is catalysed possibly by iron and vanadium oxides present in the deposits on heating surfaces.

1.24.5. Control of Low-temperature Corrosion

Control of low-temperature corrosion can be exercised in the following ways:
(a) by selection of low sulphur fuels;
(b) by operational means;
(c) by use of additives;
(d) by use of corrosion-resistant materials for construction.

(a) FUEL QUALITY

The quality specifications for the bulk of fuel oil (B.S. Class G/H) burned by the Board could permit high sulphur contents (some specifications for sulphur are as high as 5%) and this could give rise to serious operational problems if the plant was single- or double-shifting. However, the trend of sulphur content in the fuel supplied has been downwards. Whereas one company supplied at about 4·7% S, their fuel later contained nearer 2·5%. The other company supplied at 4·2%, later at 3·2%.

(b) Operational

If the plant can be operated so that the flue gas temperature does not fall below the dew point until after it has passed the vulnerable parts of the boiler (i.e. air heaters and economisers), then this will considerably reduce corrosion. This can be achieved in several ways.

Attack by sulphur trioxide can be kept to a minimum by operating at the highest possible CO_2 figure (i.e. minimum excess air). The plant can be operated with a high "back end" temperature. Here the loss of efficiency involved must be compared with the saving due to the decrease in corrosion that is achieved, and any improvement made in boiler availability.

By maintaining clean heat-exchange surfaces, the rate of corrosion can be markedly reduced and this is achieved by planned off-load water washing of air heaters and economisers. It is essential, however, that uniform distribution of water is achieved over the surfaces and that large quantities are used in a short time. On one boiler where excellent washing facilities were available, it was estimated that air heater tube life was increased from two to seven years by this means.

(c) Use of Additives

Certain chemical compounds can be added to the flue gas, either to prevent the formation of sulphur trioxide or to neutralise it once formed. Solid, liquid and gaseous additives are used and may be applied at various points in the boiler gas circuit.

(i) *Solid additives.* While dispersion in the fuel oil itself would be desirable, it has not been found possible to overcome difficulties with this or with erosion and blockage of pumping equipment and burner nozzles. Solid additives are, therefore, injected into the combustion chamber in a finely powdered form using compressed air as the carrier. Figure 1.24.5 illustrates in principle, one commonly used type of additive feeder.

Materials used have included magnesium carbonate, magnesium oxide, dolomite (a naturally occurring mixture of magnesium and calcium carbonates), the metal zinc in finely powdered form, and various proprietary materials. The latter are additives containing mainly magnesium carbonate with additions of zinc, copper oxide and carbon, as a dry 200-mesh size powder.

Magnesium carbonate has been favourably reported. A cheap mineral, it allows long runs with little air heater or superheater fouling and significantly reduces the sulphur trioxide content of the flue gases.

Magnesium oxide has also been used and is fairly cheap. It is, however, difficult to handle, consistently clogging the injection system and the air heater deposits formed are difficult to remove by water washing.

Dolomite, a mixture essentially of magnesium and calcium carbonates, ground to 30-micron size, has also been reported favourably in reducing corrosion and fouling of air heaters.

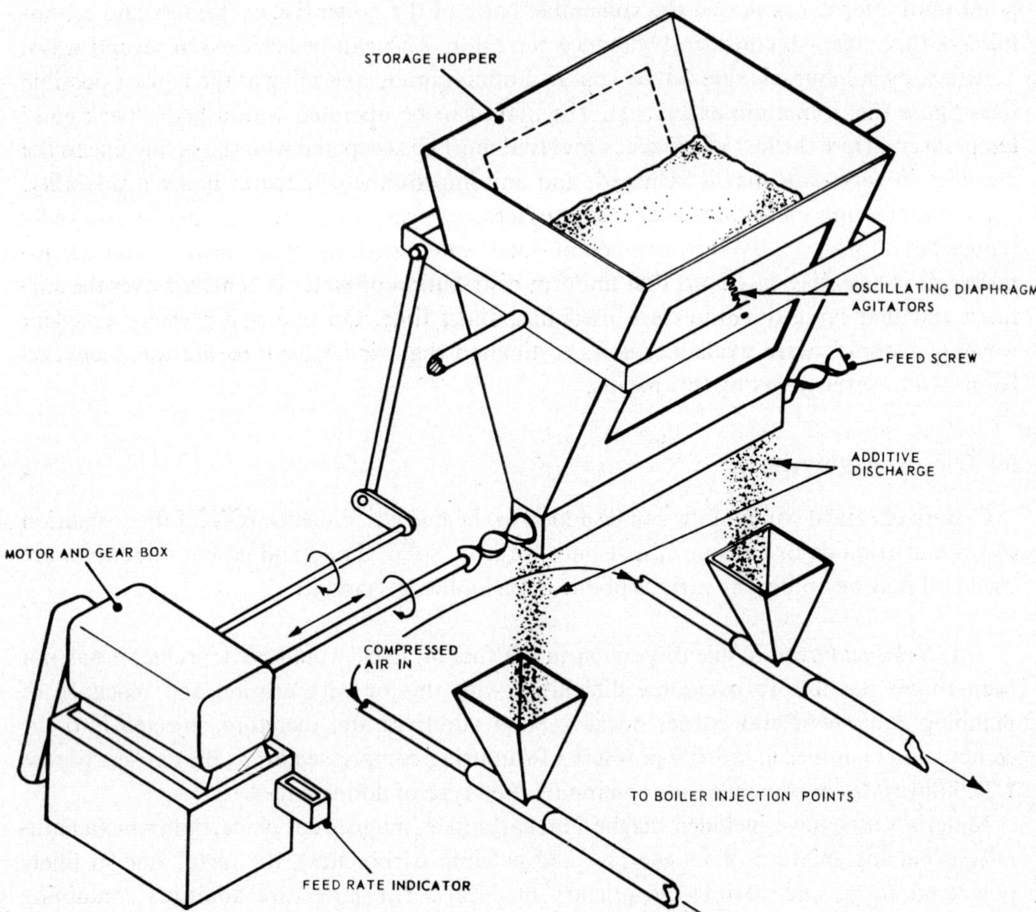

FIG. 1.24.5. Additive feeder

and for the reduction of sulphur trioxide concentration. Runs of considerable length have been achieved but superheater fouling is often significant. The cost of treatment is relatively small, and the superheater deposits formed are usually readily removed.

Zinc powder: The fine zinc particles oxidise in the combustion chamber to form a zinc oxide smoke which is capable of limiting considerably the formation of sulphur trioxide. Trials at Southport and Tilbury were quite successful in reducing the dew point to 140° and 120°F respectively. In the case of Tilbury, sulphur trioxide was eliminated completely. At Poole, it was found that daily intermittent doses of zinc powder, between 3 – 10 lb/h per boiler of 340 klb/h, for periods of 1½–3 h, were sufficient to eliminate the acid dew point for the remainder of the day's operation. Bromborough, however, reported less satisfactory results, owing to severe slagging developing in the combustion chamber after only a few days' operation, the slag so formed being found to contain a high proportion of zinc. There was also some evidence from stations that zinc additives may cause corrosion of the high temperature parts of superheaters.

Proprietary materials: Where used, these additives have been reported to have reduced corrosion of air heaters by about 50% compared with an untreated unit, and reduced the sulphur trioxide content of the gases from about 40 to 20 ppm. Long runs were found possible between cleaning but sometimes superheater fouling at the end of such runs was appreciable. Treatment using proprietary materials is more expensive than using the other chemicals.

(ii) *Liquid additives:* Tertiary amines, including the proprietary liquid inhibitor, Teramin, are by-products produced from coal tar and act as corrosion inhibitors on the metal surfaces on which sulphuric acid deposits during boiler operation. They are injected into the flue gases by air atomisers at a suitable point in the system, below 570°F (299°C). Vaporising readily in the hot flue gas stream, they ultimately condense on the metal surfaces and inhibit attack by sulphuric acid. Tertiary amines are normally used in concentrations of 0·03–0·04% by weight of the oil burned. Bankside power station has reported a reduction of 43% in air heater corrosion during inhibiting amine trials.

At Marchwood, pyridine was used in five trials but only limited reduction in dew point was achieved. Treating one half of the boiler only, with the other half untreated for comparison, it was noted that deposition was heavier on the pyridine treated half. Subsequent water washing, however, indicated the protective value of pyridine as a corrosion inhibitor, when tests showed higher metal loss from the untreated side of the plant. The protective factor of the inhibitor in these trials was estimated to be 30%.

(iii) *Gaseous additives. Carbon monoxide:* It has been suggested that carbon monoxide should in theory chemically reduce sulphur trioxide to the dioxide but trials on boilers have proved abortive. Using ten times the theoretical amount required, no significant reduction in sulphur trioxide was achieved.

Ammonia: In theory, if gaseous ammonia is injected and intimately mixed with the flue gases, the sulphuric acid in the gases will be neutralised with the formation of ammonium sulphate, if temperature, ammonia concentrations and reaction conditions are correct. Further, if the temperature is above 300°F (149°C) sulphur dioxide does not react with the ammonia.

At Bankside, corrosion rates of air heaters have been reduced by 70–80% by the introduction of gaseous ammonia between the primary and secondary air heaters. At this point, metal temperatures downstream of the injection point are below 221°C (430°F), coinciding with flue gas temperature of the order of 260°C (500°F). Work carried out earlier with gaseous ammonia at Marchwood had shown, that although acid dew points could be eliminated, rapid and serious fouling of air heaters occurred, and further, that corrosive ammonium bisulphate was the main foulant, and not ammonium sulphate. Supporting evidence for the successful use of gaseous ammonia, however, has also been reported from oil-fired power stations in France, South Australia and Malaysia. The recommended ammonia doses are 0·06–0·08% by weight of fuel burned, depending on the sulphur trioxide content of the flue gases.

(d) USE OF CORROSION-RESISTANT MATERIALS

Many alternative metallic and non-metallic materials for use in low temperature air heaters have been tested in power stations, in a search for new suitable materials to combat corrosion. For tubular air heaters, tube materials tried have included glass, aluminium, titanium and special ferrous alloys. Proprietary plastic coatings too have been applied to specimen air heater tubes, but under operating conditions these have proved unsuitable.

The most successful air heater tube materials tested to date have been borosilicate glass and a 1% copper 1% silicon grey cast iron alloy. Air heater tubes made of these materials have successfully withstood operational hazards, including thermal stress changes and water washing over long service periods. Silicon cast iron alloys, however, are difficult to cast and still present problems in manufacture, because of their inherent brittleness. They are, therefore, not readily available on the commercial market. Spun cast tubes are superior to sand cast tubes.

For the regenerative type air heaters for stations of advanced design, consideration is being given to the use of low alloy steels, of the 3% chromium type, with or without ½% molybdenum. Ferritic straight chromium (12–14% Cr), austenitic type stainless steels and mild steel with surface diffused chromium and/or silicon, have also been recommended as suitable low temperature air heater materials, but they are costly.

Commercial aluminium has proved fairly successful in trials, but resultant corrosion products, when corrosion does occur, are voluminous and cause rapid blockage. Titanium metal with its high corrosion resistance has given excellent results in station trials, but again, unfortunately its cost today is still prohibitive.

The final choice of material, therefore, must be based on an assessment of all relevant factors including cost, availability, ease of manufacture, service life and corrosion resistance.

1.24.6. **Smut Emission**

Heavy oil-fired stations sometimes exhibit the phenomenon of "fall-out", particularly during periods of inefficient combustion, starting up conditions and inadvertant changes in flue gas oxygen levels. This "fall-out" takes the form of emission from chimneys of agglomerated carbon particles or "smuts" which are usually also acidic due to sulphuric acid. The particles agglomerate up to an inch or so in size and may damage crops, buildings and various other property. The quantity and acidity of the smuts emitted can be reduced by strict control of excess air in combustion.

Much work has been done by scientific teams within the Board's oil-fired stations on the specific problem of fall-out, including the use of additives for injection after the economiser or directly into the base of the stack. Additives such as dolomite, used in the control of low-temperature corrosion, also have a beneficial effect on smut emission.

In the stations converted from coal to oil firing, electrostatic precipitators have usually suffered damage to the internal surfaces of concrete work owing to condensation of acid and water. The external application of heat-insulating material has given considerable relief to this problem, and by maintaining gas temperature, has also reduced the emission of smuts. There is also some evidence that a higher temperature may be maintained by filling the annular spaces in chimneys with a granular insulating material.

1.24.7. **Gas-side Cleaning of Oil-fired Boilers**

The fouling of oil-fired boilers derives from the ash content of the oil, from carbon and from solid additives such as dolomite which are injected into the combustion chamber. On-load cleaning resembles that for coal-fired boilers, soot-blowing supplemented by water lancing. Off-load cleaning differs only in the need to take more stringent precautions against exposing cleaners to the dust hazard. The vanadium present in oil ash can cause temporary disability after a time and the wearing of goggles, breathing masks and protective clothing is essential. To date no cases of affected persons have been reported.

1.24.8. **Ash Disposal from Oil-fired Boilers**

Although the ash content of fuel oil is relatively low, when oil is burned in quantity, the amount of ash for disposal can be important; it has been calculated that a 2000 MW station can produce about 20 tons dry, soot-free dust per week, say 80 tons if moisture and carbon are allowed for. The disposal of such material has to be carefully planned because of the possibility of vanadium leaching out of the refuse and finding its way into underground water which might subsequently be used for drinking purposes. At coastal stations it is possible to dump refuse at sea outside the 3-mile limit.

1.25. TESTING OF FUEL OILS

The fullest details for the testing of petroleum products can be found in the book *IP Standards for Petroleum and Its Products, Part I, Methods for Analysis and Testing* which is published by the Institute of Petroleum. The tests appropriate to engine and fuel oils are also to be found in the current British Standard (see paragraph 1.18.1). A brief description of the main tests normally performed at power stations is given below.

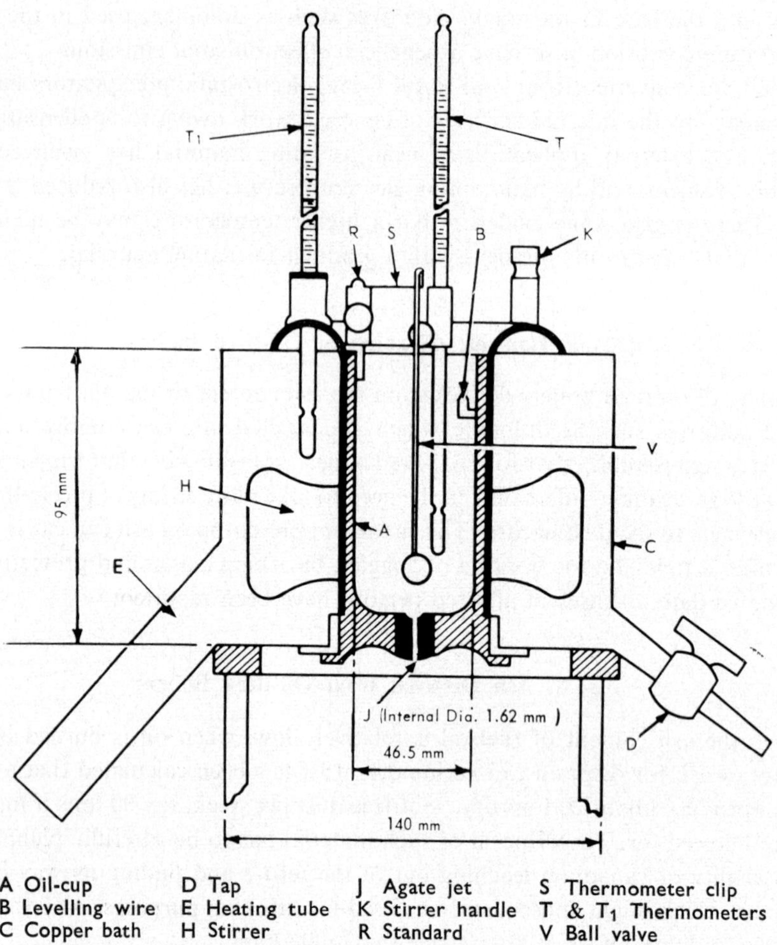

A Oil-cup
B Levelling wire
C Copper bath
D Tap
E Heating tube
H Stirrer
J Agate jet
K Stirrer handle
R Standard
S Thermometer clip
T & T_1 Thermometers
V Ball valve

NOTE: — The Redwood II Viscometer differs slightly from the No. 1 instrument in size and design.
The dimensions of the oil-cup are the same, but the agate jet is longer and has an internal diameter of 3·80 mm.
The heating bath is larger.

FIG. 1.25.1. Redwood viscometer

1.25.1. **Viscosity**

The viscosity of a fluid can be measured in absolute kinematic units (centistokes) using glass, capillary type U-tubes or by commercial viscometers such as the Redwood and Saybolt which measure in empirical units. Using a conversion table it is possible to convert kinematic viscosity to Redwood seconds. Most of the Board's power stations use the Redwood viscometers, although some, in common with the oil suppliers, have changed to the absolute method. If large numbers of samples have to be tested this latter method is preferred due to its quickness, although more expensive equipment is required.

The Redwood standard apparatus (Fig. 1.25.1) consists of a metal cup, which has a carefully standardised jet outlet that can be closed. This cup is inside a liquid bath which controls the temperature of the oil sample in the cup to close limits. The sample of oil fills the cup to a standard level and the time taken for 50 cc of oil to flow through the jet into a receiver below is recorded as Redwood seconds. There are two types of Redwood instrument. The No. I for the lighter fuel oils and the No. II for the more viscous fuels. Redwood II seconds may be converted to Redwood I seconds by means of a $\times 10$ factor. Redwood viscosities are normally quoted in terms of the No. I instrument.

1.25.2. **Specific Gravity**

The specific gravity of a liquid is defined as the ratio of the mass of a given volume of the liquid at a temperature t_1, to the mass of an equal volume of pure water at a temperature t_2. Specific gravities are most commonly reported on the basis that both t_1 and $t_2 = 60°F$.

There are three methods which can be used to determine the specific gravity of an oil fuel, the pyknometer method in which the weights of equal volumes of the oil and of water are compared, the hydrometer method which uses standard hydrometers, and the Westphal balance method. The pyknometer method gives great accuracy but is lengthy and requires great care, whereas the hydrometer method is quicker but less accurate, particularly for viscous, opaque oils such as fuel oil. The Westphal balance compares the weight of a plummet when suspended in pure water and when suspended in the sample being tested. It is quick and accurate when used for light products such as gas oil, but is unsuitable for viscous oils.

1.25.3. **Flash Point**

The flash point is the minimum temperature at which an oil gives off sufficient vapour to ignite momentarily on the introduction of a bead-sized flame, when the oil is heated at a prescribed rate in an apparatus of specified type. The introduction of the flame is made at regular temperature intervals as the oil warms up. The flash point depends on the type of instrument used and may be "open" or "closed" depending on whether the apparatus is used with or without a lid. The "closed" flash point is most commonly used because it always gives a lower flash temperature than the "open" and therefore gives a better indication of the hazards in handling the fuel.

There are two forms of apparatus, the Abel used for oils with flash points below 120°F and the Pensky-Martens for oils having flash points above 120°F. (The essential parts of a

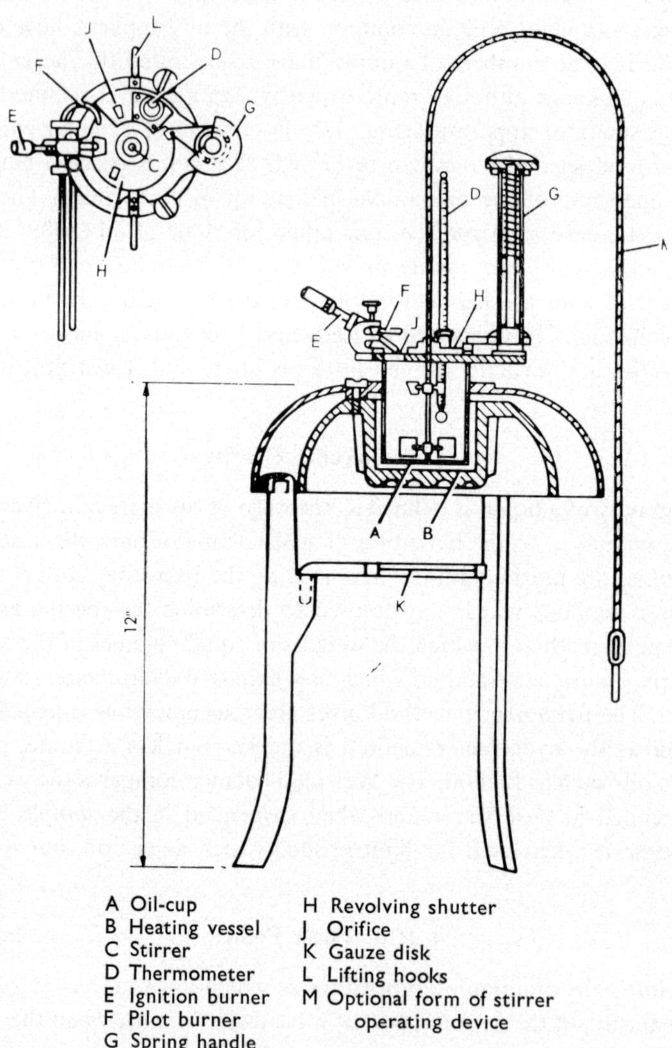

Fig. 1.25.3. Pensky–Martens flash point apparatus

flash point apparatus, which can be seen in Figure 1.25.3, are an oil cup, a cover provided with ports for the introduction of the test flame and for observation, and a means of heating the oil cup.)

For a rapid guide on whether the flash point of a consignment of oil awaiting discharge, is above or below 150°F, the "miniflash tester" is used at the Board's stations. It is an adaptation of the standard Pensky–Martens apparatus. In the miniflash tester, 2 ml of the sample are heated to 150°F in a modified oil cup and one application only of the test flame is made. If a flash occurs the consignment is suspect and further checks must be made using the standard Pensky–Martens apparatus before accepting or rejecting the oil. If no flash occurs, then the consignment can be accepted.

1.25 4. **Water Content**

Water content of 0·05% or greater is determined by the Dean and Stark method where 100 ml of the sample is shaken with 100 ml of a standard boiling-range petroleum spirit. They are boiled together under a condenser arranged so that the condensate falls into a graduated tube. The water sinks to the bottom and the amount may be read off (Fig. 1.25.4).

1.25.5. **Sediment**

Sediment in fuel oils is that material which is insoluble in benzole. Basically the oil sample is taken up in benzole and the sediment may be extracted in one of two ways—by centrifuging or by filtration.

1.25.6. **Ash—Sodium and Vanadium**

The ash content of an oil is the percentage by weight remaining after controlled ignition of the sample. The sodium and vanadium compounds present in the ash can be determined by chemical methods. At the time of writing the Institute of Petroleum have been invited to produce standard methods for determination of these elements. Methods currently in use by C.E.G.B. involve dissolving the ash in a mixture of sulphuric and nitric acids. The vanadium so taken into solution is determined photometrically as the phosphotungstate. The sodium is determined with a flame photometer.

1.25.7. **Cloud and Pour Points**

Cloud point can only be determined on oils which are transparent. The sample is cooled by immersion in a liquid cooling bath and inspected for cloudiness at intervals of 2°F. The temperature at which cloudiness first appears is reported as the cloud point.

The pour point is that temperature which is 5°F above the temperature at which the oil just fails to flow under specified test conditions. The oil is heated to a temperature above the expected pour point and is then allowed to cool. (If the oil does not solidify above ambient temperature, a cooling bath is used.) As it cools, the jar is tilted to see if the oil will move until a temperature is reached when the jar can be held horizontal for 5 sec and the oil shows no movement.

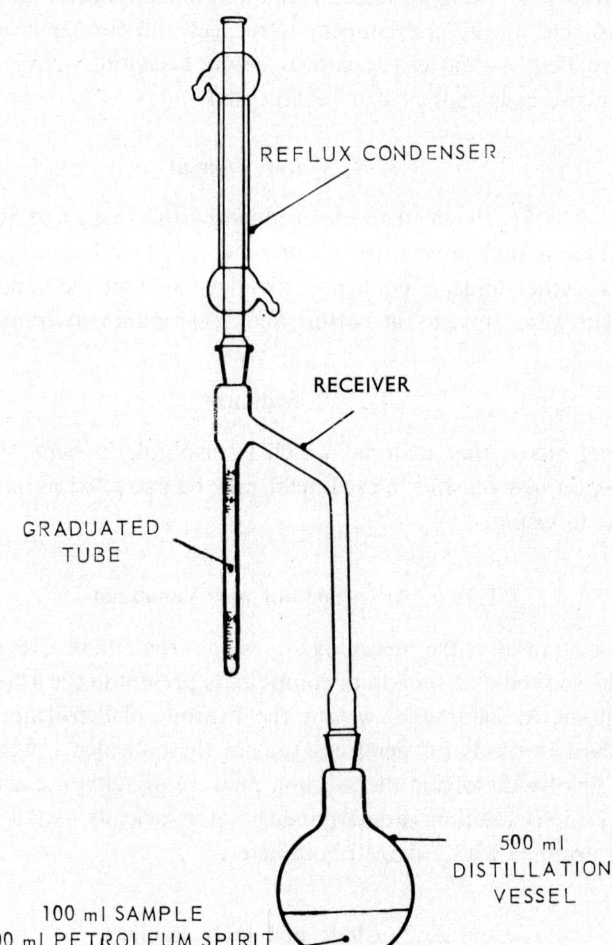

FIG. 1.25.4. Dean and Stark apparatus

1.25.8. Gross C.V. and Ultimate Analysis

Similar methods are used for these determinations, as those employed for coal.

1.25.9. Sulphur

Sulphur may be determined by the bomb combustion method or the quartz tube method. In the former, a sample of the oil is burnt in a pressure vessel (bomb) filled with oxygen under pressure. The method makes use of the same apparatus employed in calorific value determinations. A difference is that in order to absorb the sulphur oxides formed on combustion, sodium carbonate solution instead of water is put into the bomb before firing. The sulphur in this solution is precipitated as barium sulphate and weighed, and the percentage of sulphur in the original sample is calculated.

In the second method the sample is burned in a quartz tube at a temperature of 950–1000°C in a stream of purified air. The products of combustion are absorbed in a dilute solution of hydrogen peroxide and the sulphuric acid thereby formed is titrated with sodium hydroxide solution. Percentage sulphur in the sample can then be calculated.

1.26. TESTING OF FLUE GASES

1.26.1. Introduction

If near optimum conditions of boiler operation, both with regard to efficiency and availability, are to be obtained, it is necessary to make regular studies of flue gas conditions at various parts of the boiler system. In addition to the routine checking of carbon dioxide and oxygen recorders to ensure that excess air is controlled to the minimum practical amount, the following tests are necessary if corrosion and fouling of heat exchange surfaces are to be kept to a level that can be tolerated:

Dewpoint measurement and rate of acid build-up.
Corrosion probe tests.
Acid deposition rate measurement.
High-temperature deposition measurement.
Dust burden measurement.
Gas analysis.

A publication of the Boiler Availability Committee entitled *Testing Techniques for Determining the Corrosive and Fouling Tendencies of Boiler Flue Gases* describes in detail most of the measurements discussed below, and is recommended for further study.

1.26.2. Dew Point Measurement and Rate of Acid Build-up

In order to measure the dew point of the flue gases at any point, portable dew point meters have been developed. A satisfactory instrument for this work is the British Coal Utilisation Research Association (B.C.U.R.A.) dew point meter, illustrated in Figure 1.26.2A, which in

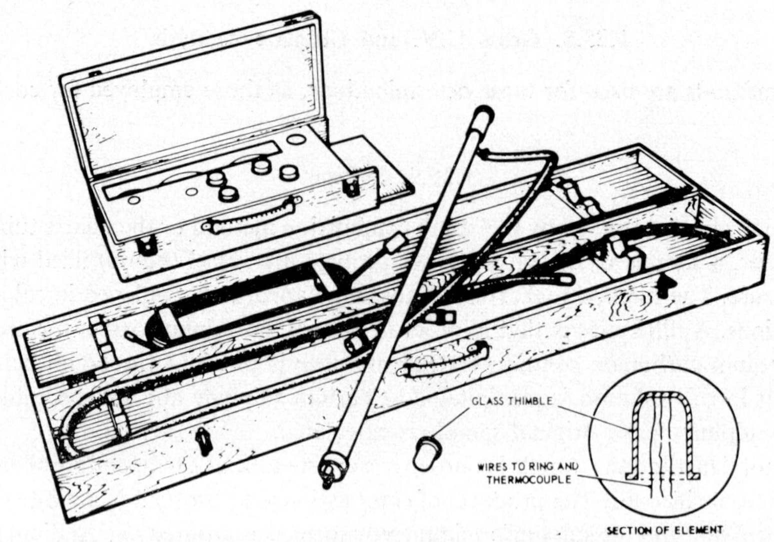

FIG. 1.26.2A. Dew point meter (flue gas testing)

principle depends on the breakdown of electrical resistance across a domed glass head, if condensation is made to occur on its surface when it is suitably placed in a flue gas path. By means of two thermocouples mounted on the head, one in the centre of the glass dome and the other attached to a ring surrounding the first, the temperature of the head can be accurately measured. The temperature of the head is controlled by directing cold compressed air on to the inside surface of the glass dome. In practice the head, mounted at the end of a metal probe, is inserted into the gas stream, where a potential of 10 V is maintained across the couples. The temperature of the glass dome is then slowly reduced by controlling the amount of air supplied until a deflection is observed on the microammeter. This temperature is the dew point of the flue gases at the test point. A dew point survey carried out under varying known boiler loadings and conditions provides very useful information for corrosion control.

Using the dew point meter and by progressively reducing the temperature below the dew point, an increasing build-up of acid is noted and this reaches a maximum at a temperature some 40–80°F below the observed dew point. The temperature of maximum build-up is then obtained by plotting rates of increase in current (in micro-amps per minute) against probe head temperature, working in 5-min periods at temperatures below the initial dew point. By this means one may obtain a guide to the temperature and corresponding rate at which sulphuric acid is likely to be deposited. Figure 1.26.2B shows a typical curve, exhibiting the relationship commonly obtained.

As a general rule results in the following ranges give an indication of possible corrosion rates:

Up to 100 μA/min—Minor
100–500 μA/min—Moderate
Over 500 μA/min—Serious

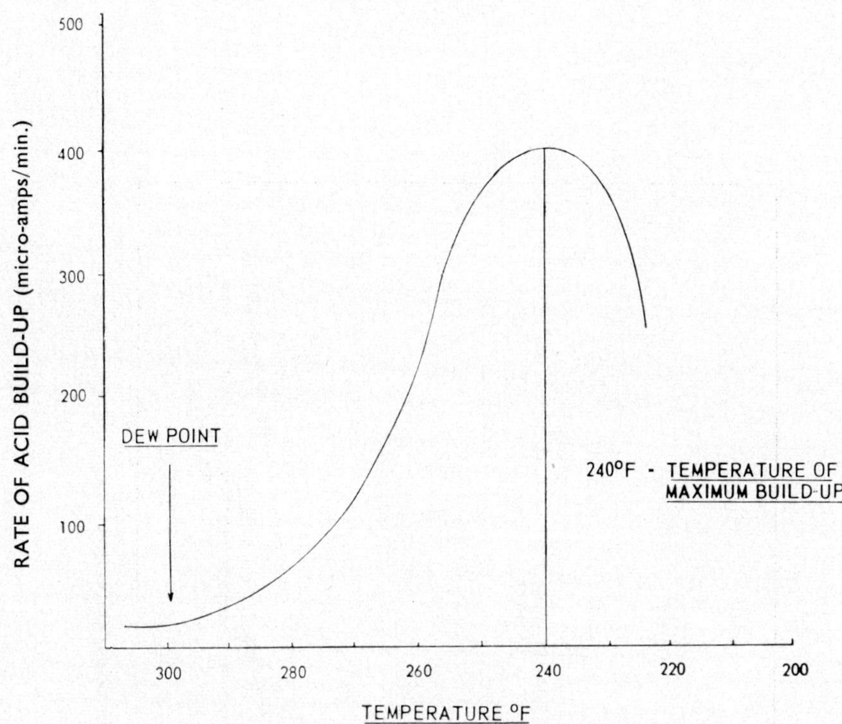

FIG. 1.26.2B. Relationship, rate of acid build-up temperature

1.26.3. **Corrosion Probe Tests**

Corrosion probe tests, using specially prepared metal heads, may also be made to assess rates of metal removal at various temperatures, and to measure direct loss of metal for any corrosive condition in the flue gas. The specially prepared metal (usually steel) head is weighed initially and then again, after thorough washing and drying, at the end of the measured exposure time. The loss in weight represents a physical measure of the corrosion which is also assessed by chemical determination of the metal (iron) in the corrosion products. Both temperature and time of exposure can be varied, to correspond with specific operational conditions. Figure 1.26.3A shows typical curves indicating the variation of corrosion with temperature for half hour exposures and Figure 1.26.3B shows the variation of corrosion with time. Figure 1.26.3C shows in section, the B.C.U.R.A. low-temperature corrosion probe.

1.26.4. **Acid Deposition Rate**

In order to give relative assessments of the acidic properties of flue gases, measurements of acid deposition rates can be determined by means of the C.E.R.L. probe. In addition to giving an indication of the potential harm to any cool metal surface exposed to the flue gas, the results are also useful in investigating the emission problem of acid smuts from chimney stacks. The construction and use of the C.E.R.L. probe is given in detail in C.E.R.L. Report number 859, dated 8 April 1959, and is shown in Figure 1.26.4.

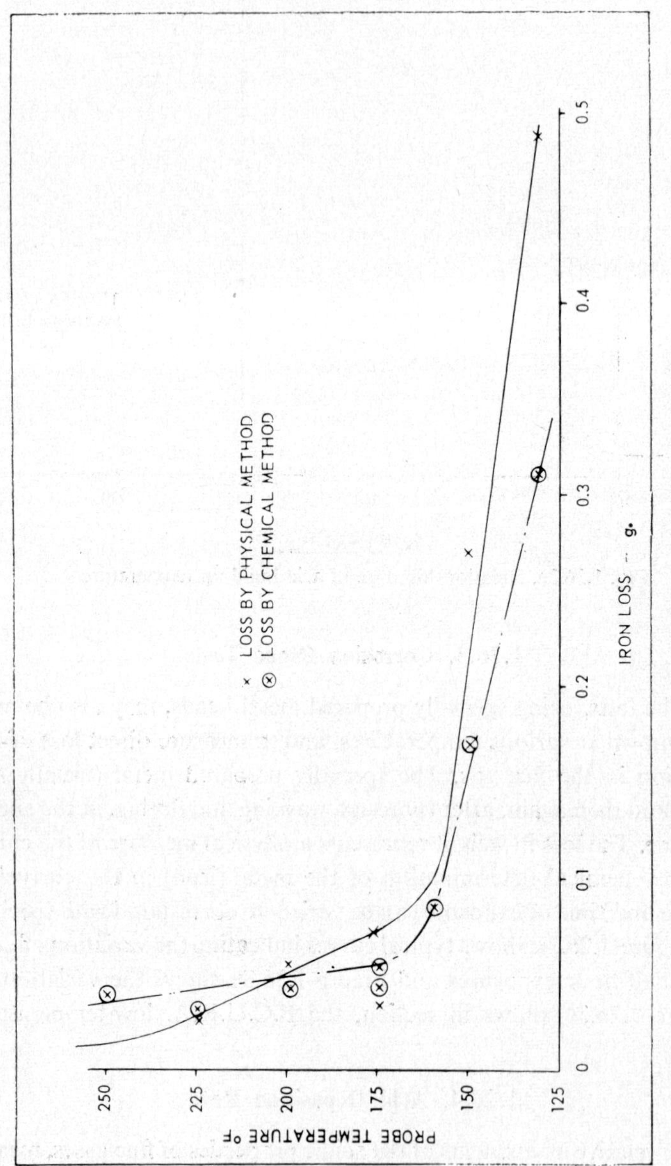

FIG. 1.26.3A. Relationships, corrosion/temperature

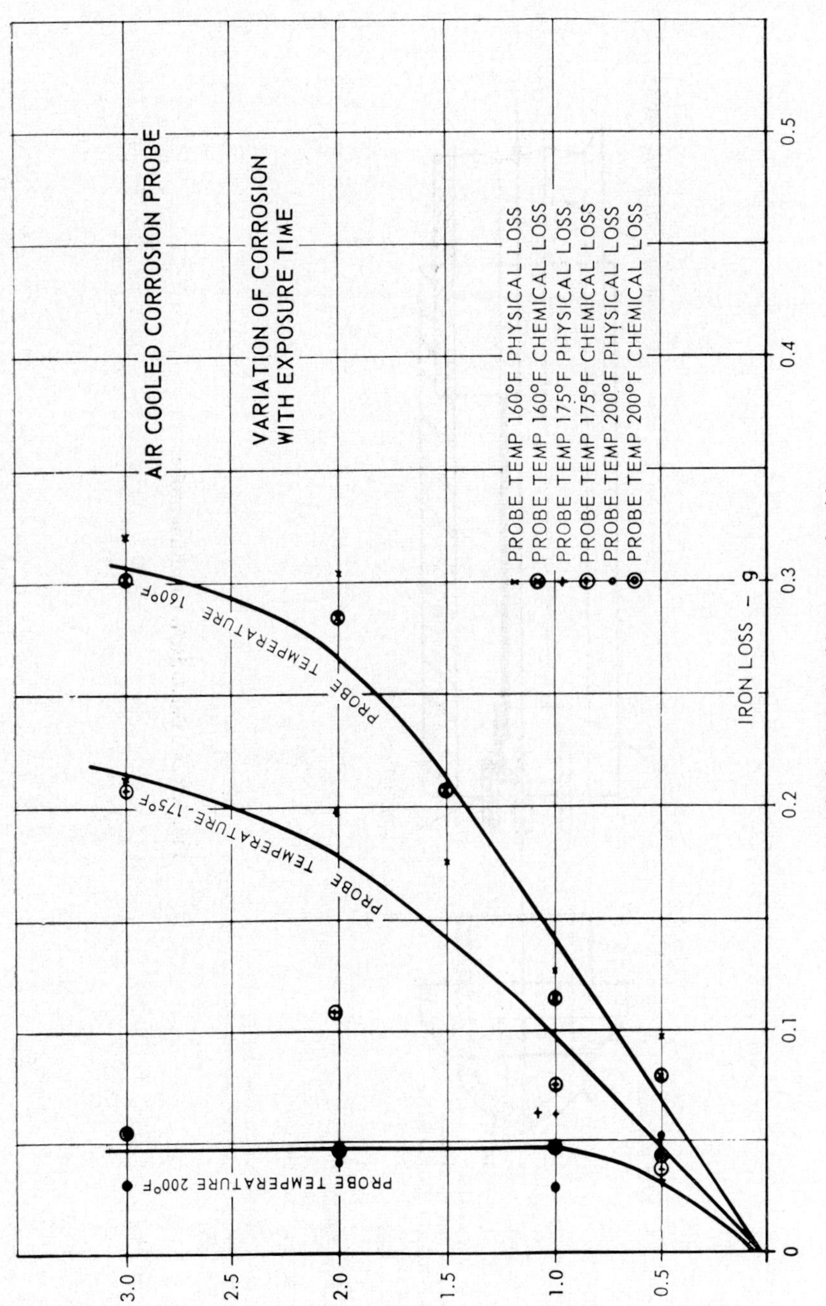

FIG. 1.26.3B. Relationships, corrosion/time

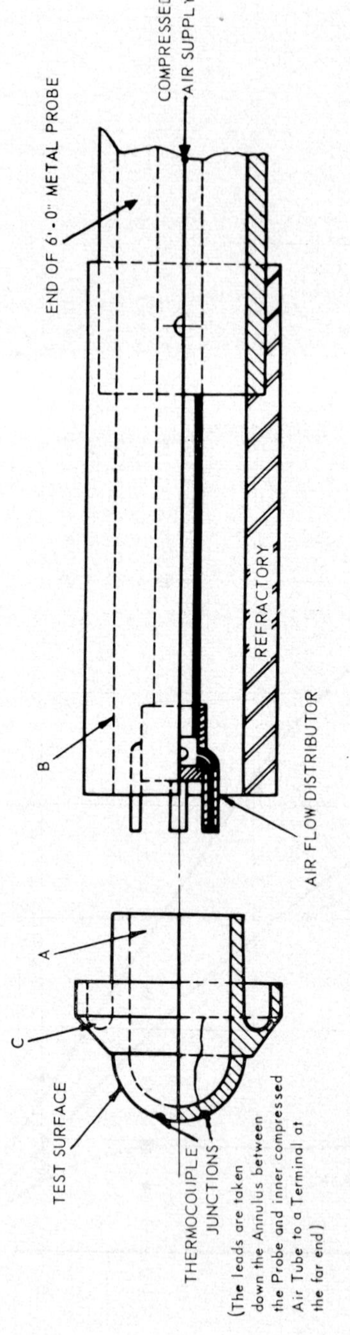

FIG. 1.26.3C. B.C.U.R.A. corrosion probe

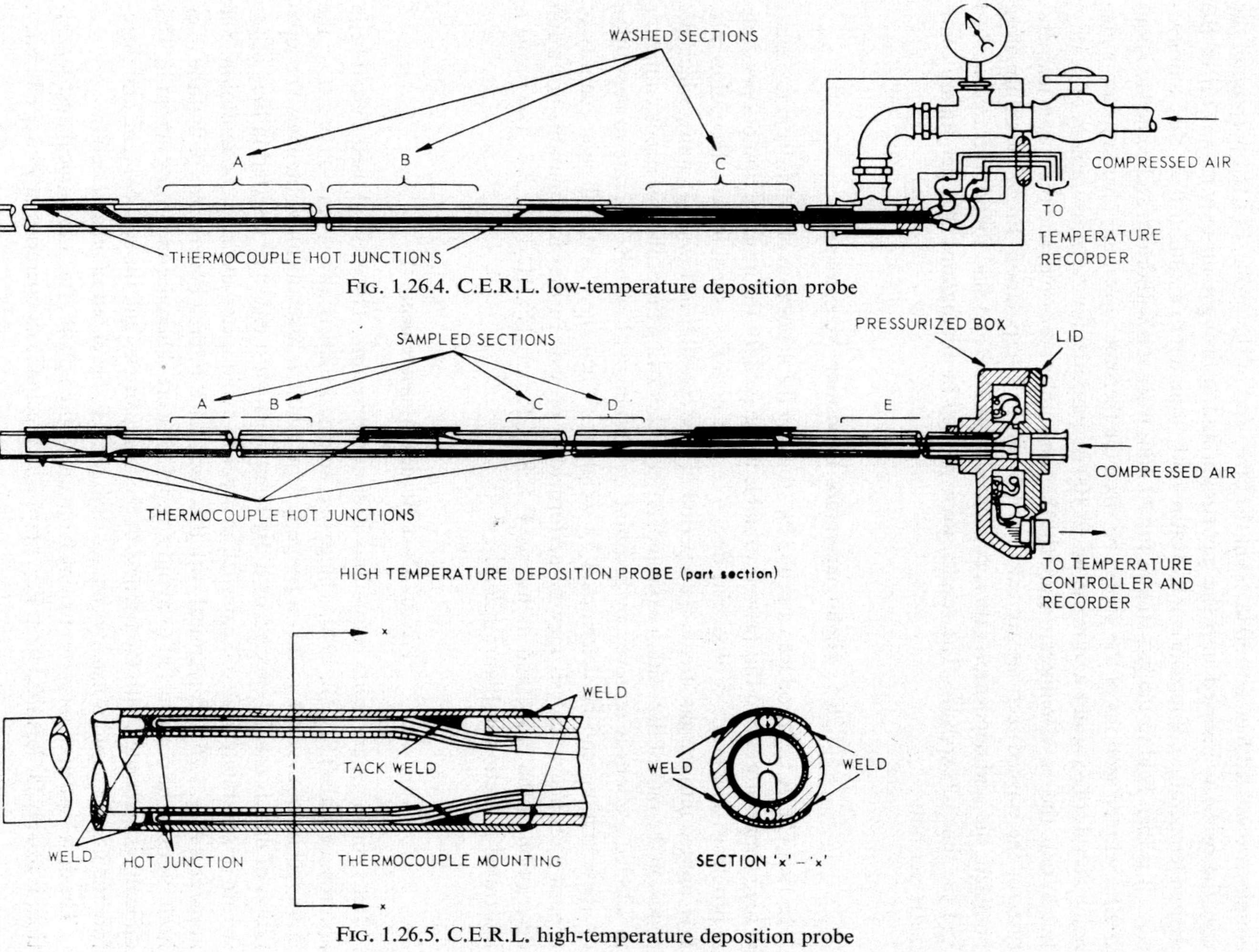

Fig. 1.26.4. C.E.R.L. low-temperature deposition probe

Fig. 1.26.5. C.E.R.L. high-temperature deposition probe

The probe, 8 ft long and of 1 in. outside diameter, is made of austenitic stainless steel and is fitted at one end with a T-piece connection to introduce compressed air. Two thermocouples are welded 6 in. and 30 in. from the other end of the probe. For the purpose of the test, the probe is inserted into the gas stream and is brought up to within 10°F of the desired temperature of measurement, before air is admitted to maintain it at this temperature. It is then held in the gas stream for 30 min, when it is withdrawn and cooled to about 200°F. Selected sections of the probe, between the thermocouples, are then washed externally with distilled water and made up to 100 cc quantities for the measurement of acidity. In addition, the mean temperatures of the thermocouples are noted, and the mean temperatures of the washed sections are calculated by linear interpolation. From the information obtained, the acid deposition rate is calculated from the total acid found, the area washed and the time of exposure. The results are expressed in micrograms SO_3 per square centimetre per hour.

1.26.5. **High-temperature Deposition Measurement**

In 1961 C.E.R.L. introduced a probe for studying the deposition of solids from flue gases at high temperatures and this proved valuable in the field for the investigation of operational problems. The probe is of similar design to that used for assessing acid deposition rate. It is inserted into the gas stream for a period of time (possibly as low as 15 min), which depends on the rate of deposition expected. After withdrawing the probe, the deposit adhering to selected sections is removed by brushing or washing with water. The rate of deposition on these sections can be determined by weighing. When there is a sufficient quantity of deposit a chemical analysis may be performed to provide further information. For a full description of the probe and its use see C.E.R.L. Report No.RD/L/R. 977. The probe is illustrated in section in Figure 1.26.5.

1.26.6. **Dust Burden Measurement**

Determinations are made before and after dust arrestors. The method used should be in accordance with B.S. 893 : 1940. The gas is sampled isokinetically at various points on the cross-section of the duct so that a properly representative assessment is obtained. The separation of dust is usually effected by a dry filter which must either be inserted in the hot gas stream or separately heated if external, in order to avoid water and acid condensation. The apparatus (Fig. 1.26.6) is identical with that used for P.F. fired boilers, except that the filter itself needs to be constructed to avoid a rapid pressure difference mounting across it. This difficulty is associated with the fineness of the dust particles and the acid content, and therefore the filter material is usually coarser than for P.F. firing but of greater depth.

The dust burden of the gases issuing from an oil-fired plant after mechanical dust separation, is usually far less than from P.F. plant with precipitators operating at 99·3% efficiency.

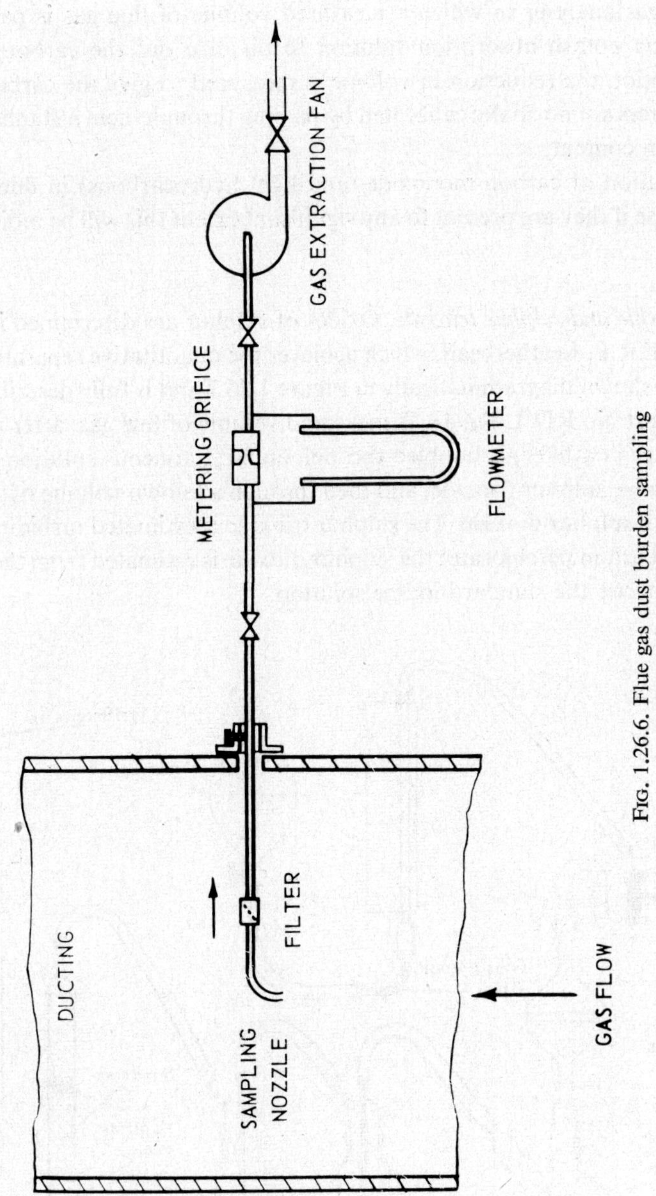

FIG. 1.26.6. Flue gas dust burden sampling

1.26.7. **Gas Analysis—Carbon Dioxide and Oxygen**

On modern boilers one or both of these gases are monitored continuously, the meters used requiring regular checking. The carbon dioxide meter can be checked by the use of the Orsat or other gas analyser in which a measured volume of flue gas is passed repeatedly through a caustic potash absorption solution to dissolve out the carbon dioxide. After complete absorption the reduction in volume is measured to give the carbon dioxide content. Oxygen meters are normally calibrated by passing through them a standard gas mixture of known oxygen content.

The determination of carbon monoxide (and light hydrocarbons) in flue gases is rarely necessary, because if they are present to any significant extent this will be indicated by smoke formation.

Sulphur dioxide and sulphur trioxide. Oxides of sulphur are determined in an apparatus developed by C.E.R.L. Leatherhead, which achieves the quantitative separation of each gas. The apparatus is shown diagrammatically in Figure 1.26.7 and is fully described in C.E.R.L. Laboratory Report No.RD/L/R/948. A measured volume of flue gas, after removal of gas-borne solid matter in a filter, is bubbled through an 80% aqueous solution of isopropanol to selectively remove sulphur trioxide, and then through a known volume of standard iodine solution to absorb sulphur dioxide. The sulphur trioxide is estimated turbidimetrically or by titration against barium perchlorate; the sulphur dioxide is estimated from the volume of gas required to decolorise the standard iodine solution.

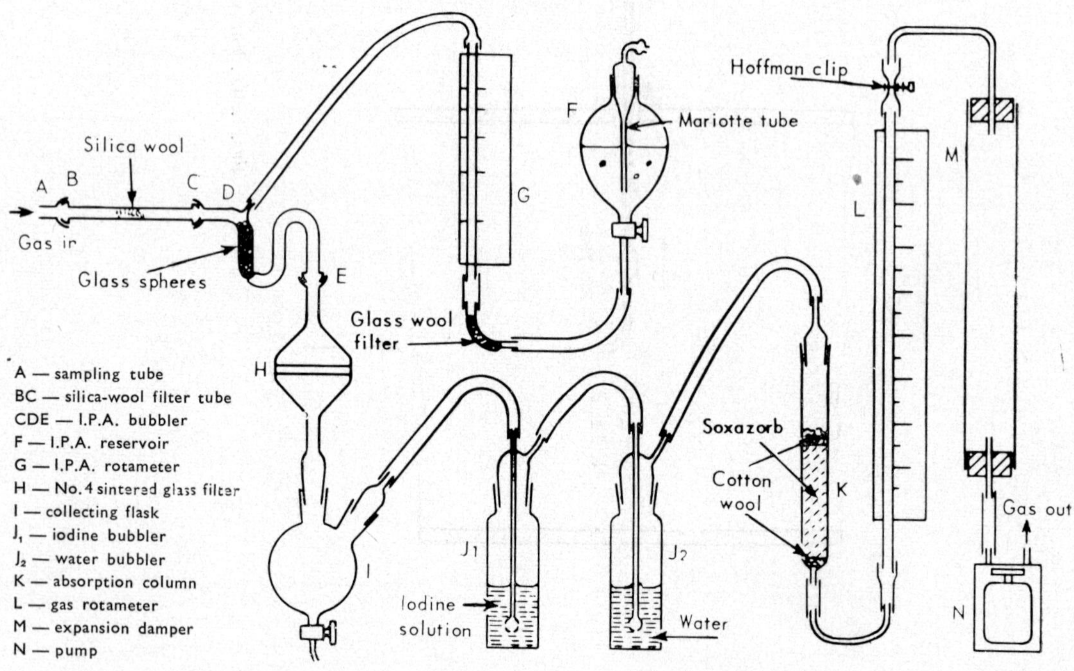

FIG. 1.26.7. Apparatus for analysis of SO_2 and SO_3

1.27. AIR POLLUTION

1.27.1. Introduction

The widespread harmful effects of atmospheric pollution have long been recognised and co-ordinated efforts to measure the extent of the principal forms of pollution have developed during the last fifty years or so. However, successful methods of pollution prevention, together with statutory enforcement, are rather more recent. A power station situated in an area where obvious pollution occurs, is often a focus of accusation or complaint and for many years the Board has paid much attention, in plant design and operation, to the elimination of air pollution.

Although pollution from industries such as those producing chemicals, bricks, cement, metals, etc., may be of major significance in particular parts of the country, combustion products from domestic and industrial fuel-burning equipment are widely distributed and form the principal sources of air pollution. Altogether, about 10 million tons of harmful pollutants are discharged into the air over Britain in a year.

1.27.2. The Nature of the Pollutants

The pollutants of common concern are solid materials in the form of discrete particles, including smoke, and gaseous products of combustion.

1.27.2.1. Solid Emissions

It is commonly observed that the larger and more dense particles of grit emitted from chimneys fall in the vicinity of the source, and thus cause a more obvious nuisance than fine dust which is distributed over a greater area. The heavier particles are composed mainly of coke and ash from coal-fired boilers, particularly stoker-fired ones, or more rarely they may be acidic "smuts" from oil-fired boilers as discussed in a previous section. The fine dust from P.F. boilers consists predominantly of spherical particles of fused ash ("cenospheres"), having diameters within a range of about 2 to 120 microns, but mostly around 20 microns.

Smoke is a suspension in air of very fine particulate matter, composed mostly of carbon with some tar and ash. The particle sizes are considerably smaller than 1 micron and smoke of the smallest particle sizes has the property of scattering visible light of short wave-length, thus giving such smoke its blue appearance. The Board and The British Coal Utilisation Research Association find it useful to adopt a terminology based on particle size range, thus "grit" has sizes greater than 76 microns, "dust" is material of 1 to 76 micron sizes, and "smoke" comprises particles of 1 micron or less. Soot may be regarded as an aggregation of smoke particles.

1.27.2.2. Gaseous Emissions

Of the numerous gaseous products of combustion discharged from chimneys, sulphur dioxide combines a sufficiently harmful effect and concentration in flue gas (500 to 3000 ppm by volume) to be singled out as the significant gaseous pollutant. In addition to the corrosive property that it has when associated with moisture, sulphur dioxide causes recognizable physiological harm to people who, for long periods of time, inhale air containing more than about 10 ppm of the gas.

1.27.3. Measurement of Pollution

1.27.3.1. Introduction

For many years the Department of Scientific and Industrial Research, D.S.I.R. (which is now a service within the Ministry of Technology), local authorities and the Board have used the British Standard deposit gauge and lead dioxide candle to measure deposited solid matter and sulphur dioxide in air, respectively. In 1961, D.S.I.R. recommended that the deposit gauge should continue in use only for local measurement of water-insoluble deposition. Formerly it was used to compare total deposition observed all over the country as well as within localised surveys. At the same time, it was recommended that a more direct method of measuring sulphur dioxide, by absorbing it in hydrogen peroxide, should displace the lead dioxide method which is not sufficiently precise.

Towards the end of 1964, the Board reconsidered the matter of routine air pollution measurements and the following policy was made.

(a) Routine surveys around new stations should continue to be planned to operate not less than two years before commissioning the first unit and until at least two years after the station has been completed.

(b) For stations of 2000 MW or more, the D.S.I.R. daily volumetric apparatus should be used in place of the lead dioxide method for the measurement of sulphur dioxide, and the C.E.R.L. directional dust pollution gauge should be used in place of the British Standard deposit gauge.

(c) For stations of less than 2000 MW, Regional management in consultation with the Planning Department should decide whether or not to change the older types of instruments in favour of the newer ones.

The British Standard deposit gauge (B.S. 1747, Part 1, 1961) and the lead dioxide sulphur dioxide gauge (B.S. 1747, Part 3, 1963), are not described in this Chapter but the student should read the standards quoted for a description of these instruments. Results from the deposit gauge represent the amount of material which falls in a month into a bowl about 12 in. diameter; the amount is commonly expressed in tons per square mile per month. The lead dioxide gauge results are in arbitrary units, also from a month's exposure, and are expressed as milligrams of sulphur trioxide produced per day, per 100 cm^2 of a standardized lead dioxide paste.

Statutory control of air pollution is administered by the Alkali Inspectorate, and although the long-established methods of measurement just referred to have served a valuable purpose in the past, they are not regarded by the Inspectorate as being adequate in relation to the very large stations recently commissioned.

Several independent studies have been published of theoretical calculations of the manner in which chimney emissions are dispersed into the air and settle to the ground; the results generally take the form of complex formulae which vary according to the theoretical basis and the meteorological variables considered. No completely satisfactory formula has been deduced, but for the ground-level effects from an effective chimney height H, discharging pollutants at a rate Q, the quantity Q/H^2 is commonly accepted as being approximately proportional to the amount of pollution produced at ground level under similar meteorological conditions. The expression "effective chimney height" is the actual chimney height increased by a factor which takes account of the gas discharge velocity, temperature and density. Table 33 shows for a wide range of station sizes that the value of Q/H^2, or relative ground-level pollution, can be kept to a fairly uniform level by designing a suitable effective chimney height. It relates to sulphur dioxide emission.

TABLE 33

Station	Capacity (MW)	Rate of SO_2 emission(Q) (lb/h)	Effective chimney height (H) (ft)	Q/H^2 $\times 10^3$
Bold 'A'	128	4600	530	16·4
Drakelow 'A'	244	6400	670	14·3
Ferrybridge 'B'	300	10,400	960	11·3
Blyth 'A'	480	15,200	850	21·0
Northfleet	720	18,400	1040	17·0
High Marnham	1000	18,200	1100	15·1
Eggborough	2000	63,500	1970	16·4

1.27.3.2. C.E.R.L. Dust Pollution Gauge

There are two principal disadvantages with the B.S. deposit gauge when it is used for monitoring power station dust emission; it is not selective of the directions from which the collected deposits originated and it is most responsive to coarse particles, having much reduced efficiency in retaining fine dust such as may be emitted from modern P.F.-fired plant.

The C.E.R.L. dust pollution gauge is designed to obviate these disadvantages and it is discussed in detail in C.E.R.L. Laboratory Report No. RD/L/R. 1069. The apparatus consists basically of a closed hollow cylinder, about 12 in. long and 3 in. diameter, having an aperture about 1½ in. wide along much of the length of the cylinder. It is mounted in an upright position with the lower end forming a container for the collected dust, which enters through the now vertical aperture.

Normally, an assembly will comprise four such cylinders mounted close together on a stand so that the four apertures face directions which are 90° apart in a horizontal plane (see Fig. 1.27.3.2). The gauge is usually exposed for one month, with one of the apertures facing the chimney or other source of dust. The material caught in each cylinder is transferred by washing it into an absorptiometer beaker. The amount of dust present is determined by placing the beaker in a photo-electric absorptiometer, which measures the amount of light obscured by the dust under standard conditions of test.

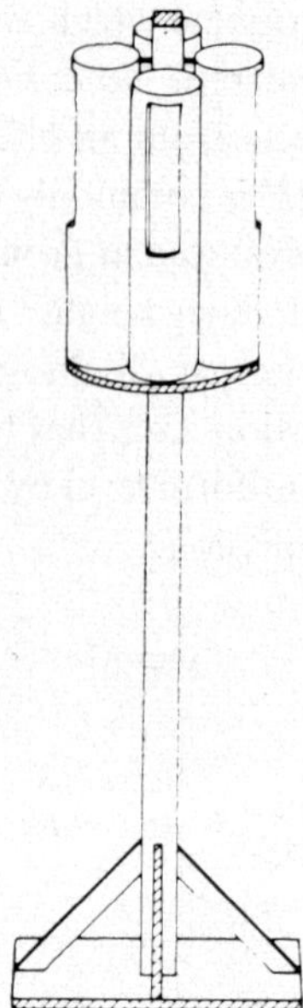

FIG. 1.27.3.2. C.E.R.L. dust pollution gauge

1.27.3.3. D.S.I.R. VOLUMETRIC SULPHUR DIOXIDE APPARATUS AND SMOKE FILTER

This apparatus is shown diagrammatically in Figure 1.27.3.3. The method depends on bubbling a measured volume of air through a dilute solution of hydrogen peroxide, which oxidises sulphur dioxide to sulphuric acid. After 24 h operation, the acidity of the solution is measured by titration with standard alkali solution, and the result is calculated in terms of micrograms of sulphur dioxide per cubic metre of air. It is possible that other acids or ammonia in the air may sometimes affect the result, but these effects are normally quite small.

It is convenient to incorporate a means to measure airborne smoke within the D.S.I.R. apparatus, and this is done by arranging for the air sample to pass first through a disc of white filter paper of standardised grade and dimensions. The stain produced on the paper by smoke is measured by means of a photo-electric reflectometer and the smoke concentration in microgrammes per cubic metre is calculated from the reading.

Where considerable numbers of D. S. I. R. sulphur dioxide instruments are in use in a survey, the instrument can be modified so that it will automatically replace the hydrogen peroxide solution and the filter every 24 h during a week, so that the processing of the accumulated daily results needs to be performed only once per week.

1.27.3.4. C.E.R.L. SULPHUR DIOXIDE RECORDER

In order to measure continuously the short-term variations in the level of atmospheric sulphur dioxide, a recording instrument was developed by C.E.R.L. in 1957. The chemical principle of operation is similar to that of the D.S.I.R. apparatus, but the sulphuric acid

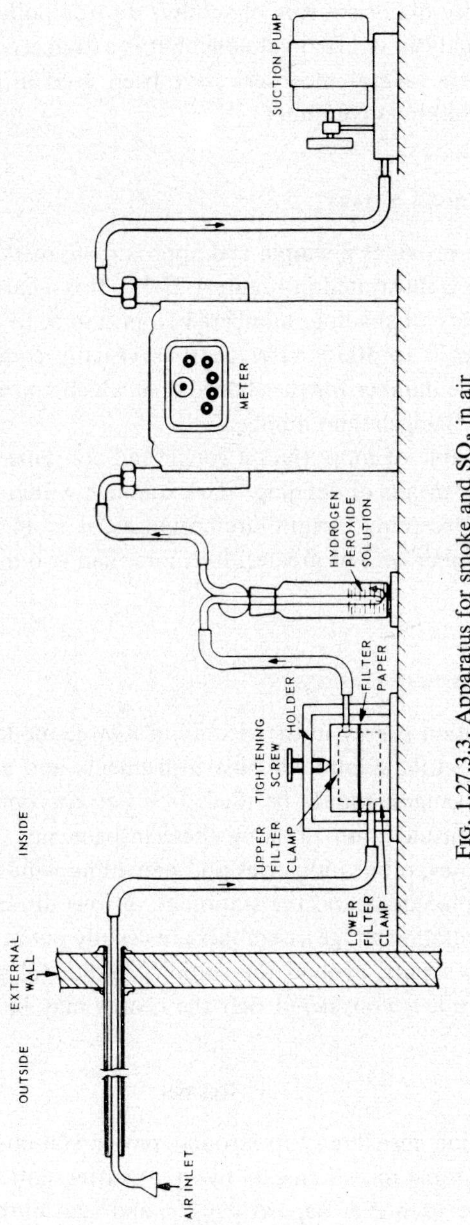

FIG. 1.27.3.3 Apparatus for smoke and SO_2 in air

produced is measured by the increase in electrical conductivity of the hydrogen peroxide solution. The variations in conductivity are displayed on a strip chart recorder, calibrated in parts per hundred million (pphm) of sulphur dioxide.

In conjunction with records of wind speed and direction, the sulphur dioxide record can be analysed to identify major sources of sulphur dioxide pollution in a given area of investigation. Such an analysis is often valuable, but is extremely laborious when undertaken manually and where several recorders have been used in surveying an area, the results need to be processed by a computer.

1.27.3.5. Ringelmann Smoke Chart

The Ringelmann chart provides a simple and approximate method of visually gauging smoke density. The chart is illustrated in Figure 1.27.3.5. It is a card which bears six equal areas of graduated intensity of shading, numbered from 0 to 5, to indicate equal intervals of light obscuration from 0 to 100%. The chart is visually compared with the smoke under observation, and the number for the shaded area which appears to match the smoke density, is quoted as the "Ringelmann number".

The method is not capable of more than a rough and convenient assessment of smoke density, but it is used as a means of defining "dark smoke", within the terms of The Clean Air Act of 1956. Thus, excepting certain circumstances, it is an offence to emit smoke having a Ringelmann number of 2 or greater, for more than two minutes in any half-hour period.

1.27.3.6. Siting of Pollution Gauges

In making an air pollution survey in the vicinity of a large modern station, it is recommended that 8 to 12 D.S.I.R. sulphur dioxide instruments and at least 4 assemblies of C.E.R.L. dust pollution gauges, should be used. It is not appropriate to discuss here all the factors taken into consideration in siting these instruments, but examples are local topography, housing estates, other industries and prevailing wind direction. The sulphur dioxide instruments are placed around the station at various distances up to 7 or 8 miles from it, whilst the dust pollution gauge assemblies are usually put at a distance of 2–3 miles, spaced approximately 90° apart, around the station. Additional gauges may be sited in particular localities where it is considered that the results may be of special significance.

1.27.4. **Results**

For comparing pollution measurements around power stations for two years before commissioning, with the same measurements two years after station construction is completed, most surveys have used B.S. deposit gauges, and lead dioxide gauges for sulphur dioxide. Stone and Clarke (C.E.G.B.) have been able to make valid comparisons for 16 surveys. They found that in 12 of these there was no significant increase of local dust pollution, while in 14 of them there was no significant increase of sulphur dioxide pollution. It must be recognised that the gauges used were of low precision in the results obtained,

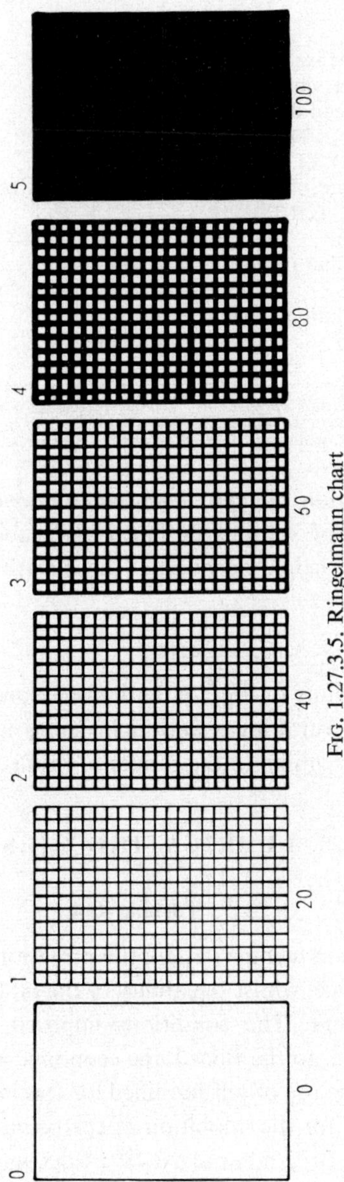

FIG. 1.27.3.5. Ringelmann chart

TABLE 34

TYPICAL RANGES OF RESULTS OF AIR POLLUTION MEASUREMENT

Item and method of measurement	Results		
	low	medium	high
B.S. deposit gauge; insoluble deposit (tons/sq. mile/month)	2	15	70
Sulphur dioxide, lead dioxide method (mg SO_3/100 cm^2/day)	0·5	2	9
Sulphur dioxide, D.S.I.R. instrument ($\mu g/m^3$, daily mean)	18	160	1600
Sulphur dioxide, C.E.R.L. recorder (pphm)	2	10–15	60
Smoke, D.S.I.R. filter ($\mu g/m^3$, daily mean)	10	200	1400
C.E.R.L. dust pollution gauge (see note below)	<0·5	1–2·5	7–15

Note: The C.E.R.L. dust pollution gauge results are given as "Ten-day percentage covering power", calculated from the measured light-obscuring value of the dust caught in the gauge. This rather arbitrary unit is such that, up to the value 0·5 the level of pollution is very unlikely to cause complaint, whilst at values above about 4 a progressive degree of nuisance is caused.

but there is now an accumulation of many such measurements, which indicates that the operation of a large majority of stations designed since about 1945 has not materially increased pollution levels in the neighbourhood of those stations.

TYPICAL RANGES OF POLLUTION MEASUREMENTS

As one would expect, the results of these pollution measurements vary widely according to locality, with low figures in rural areas and high figures in industrial areas. The above figures (Table 34) are given as a guide to the ranges of results commonly found.

1.28. LUBRICATING OILS

1.28.1. Turbine Oil

With technological advances in turbine design, the provision of clean, stable and efficient lubrication is of great importance, whilst the ability of the oil to remove heat from bearings remains a principal requirement. The conditions imposed today on turbine oils have greatly increased in severity and, to the Board, the economic penalty of oil failure is heavy. Present-day practice requires the use of oils modified for special performance by compounding with additives, particularly for the inhibition of corrosion and oxidation.

Turbine oils are classified by British Standard 489: 1955, into four main viscosity groups, namely: light, medium, heavy and extra heavy, and are specified to comply with the characteristics shown in Table 35.

B.S. 489 has adopted centistoke units of viscosity and has inserted Redwood No. 1 units for convenience. Redwood units are still quoted in many quarters and Table 36 gives the corresponding units for viscosity.

TABLE 35

LUBRICATING OIL GRADES AND PHYSICAL PROPERTIES
(B.S. 489)

Grade	Viscosity in centistokes (Redwood No. 1 seconds in brackets)			Inorganic acidity (mg KOH/g)	Minimum flash point (closed) (°F)	Maximum pour point (°F)	Maximum demulsification number (sec)	Sulphur (corrosive) (copper strip)	Rusting	Maximum total acidity (mg KOH/g)
	100°F (37·8°C)		210°F (97·9°C)							
	min.	max.	min.							
Light	29 (120)	43 (176)	4·7 (38)	Nil	330	20	300	Negative	Absent	0·20
Medium	43 (176)	60 (245)	5·5 (40)	Nil	330	20	300	Negative	Absent	0·20
Heavy	60 (245)	82 (334)	6·5 (43)	Nil	330	20	300	Negative	Absent	0·20
Extra heavy	82 (334)	110 (449)	8·2 (47)	Nil	330	20	300	Negative	Absent	0·20

For gas turbine lubrication, synthetic ester type oils have been developed to meet the higher service temperature. These, however, are not mineral oils and therefore are not covered by B.S. 489. Other modern synthetic lubricants include the polyalkylene glycols and their derivatives, and the silicones. Synthetic oils with fire-resisting characteristics are of increasing interest to the Board.

TABLE 36

EQUIVALENT VISCOSITIES

Grade	Viscosity			
	at 70°F		at 140°F	
	Centistokes	Redwood No. 1	Centistokes	Redwood No. 1
Light	125	508	12·5–18·5	60– 80
Medium	192	780	16·5–25·0	74–105
Heavy	280	1140	21·5–33·0	92–137
Extra heavy	390	1580	28·0–43·0	117–177

1.28.2. Other Lubricating Oils

Other oils in the station require occasional testing (for example, P.F. mill lubricants), but they do not need as frequent testing as turbine oils, nor as wide a range of tests. Acidity and viscosity measurements, together with the visual condition of the oil, are usually all that is required in testing these oils.

TABLE 37

CHARACTERISTICS OF C.E.G.B. SPECIFICATION OILS

	Grade	Viscosity index	Viscosity (centistokes)		Closed flash point (°F)	Pour point (°F)	Total acidity (mg KOH/g)	Saponification value (mg KOH/g)	Ash %
			100°F	210°F					
General purpose plain	3	40 (min.)	10–15	2·7 (min.)	290 (min.)	–20 (max.)	0·1 (max.)	1 (max.)	0·02 (max.)
mineral oils	5	49 (min.)	30–40	4·9 (min.)	320 (min.)	–20 (max.)	0·1 (max.)	1 (max.)	0·02 (max.)
(C.E.G.B. specification 20/6900/2)	7	59 (min.)	50–80	7 (min.)	375 (min.)	15 (max.)	0·1 (max.)	1 (max.)	0·02 (max.)
	9	44 (min.)	300–400	18 (min.)	420 (min.)	25 (max.)	0·1 (max.)	1 (max.)	0·02 (max.)
	11	82 (min.)	400–650	28 (min.)	510 (min.)	30 (max.)	0·1 (max.)	1 (max.)	0·02 (max.)
	13	91 (min.)	800–1200	45 (min.)	540 (min.)	45 (max.)	0·4 (max.)	2 (max.)	0·05 (max.)
Oxidation-resistant	E.20	75	45–60	6·5 (min.)	380 (min.)	10 (max.)	0·1 (max.)	1 (max.)	0·2 (max.)
general purpose	E.30	85	85–115	10 (min.)	390 (min.)	10 (max.)	0·1 (max.)	1 (max.)	0·2 (max.)
mineral oils	E.40	78	115–165	12 (min.)	400 (min.)	15 (max.)	0·1 (max.)	1 (max.)	0·2 (max.)
(C.E.G.B. specification 20/6900/1)	E.50	80	165–225	15 (min.)	410 (min.)	15 (max.)	0·1 (max.)	1 (max.)	0·2 (max.)
Detergent oils; crank-case grades	D.20	–	75	7·3–9·7	360 (open)	–5 (max.)	—	—	—
(M.O.D. specification 2101B)	D.30	–	150	9·7–13	390 (open)	0 (max.)	—	—	—
Extreme pressure oils	EP.80	85	32–41	—	—	—	—	—	—
(M.O.S. specification CS.3000A)	EP.90	85	224	15·6–18	—	—	—	—	—

Lubricating oils are purchased by the Board to meet their own specifications, some typical examples of which are shown in Table 37. Turbine oils are excluded from these specifications at present, as they are purchased as proprietary oils approved by the turbine makers and the Board.

1.28.3. **Additives and Inhibitors**

1.28.3.1. Introduction

To meet the more onerous conditions of service in modern machines, lubricating oils are now provided with various properties and may be regarded as "tailor-made" for their purpose. This is achieved, after refining the base oil, by compounding with it various chemicals referred to as "additives" and "inhibitors", which confer certain properties or enhance existing ones, and extend the useful life of the oil. Additives are now used to inhibit oxidation, corrosion and foaming, whilst detergents, metal passivators, emulsion breakers, viscosity index improvers, and pour point depressors may also be added for the various purposes which their names suggest.

Lubricating oils purchased by the Board are based on approved specifications, which in many cases require the inclusion of appropriate additives. For turbine oils, which are compounded to give continuous service for about twenty years, anti-oxidation, anti-foam and corrosion inhibitors are included in order to stabilise the oil and the metal surfaces of the oil circuit, and to inhibit the formation of acidity and sludge.

Additives have introduced a new industry into the lubrication field and their chemistry is complex. The major oil companies have developed a large range of excellent additives, most of which, however, are jealously-guarded trade secrets.

1.28.3.2. Gear Oil Additives

These materials increase the load-carrying characteristics of an oil to maintain lubrication under conditions of extreme pressure, and are designed to ensure non-corrosivity, good oxidation resistance, high-temperature stability, constant viscosity characteristics and good water tolerance. They are polar in nature and contain compounds of sulphur, chlorine, phosphorus and zinc.

1.28.3.3. Industrial Oil Improvers

These additives confer better adhesion to metal and are used when straight oils cause excessive waste by "drip", "fling", etc., so reducing oil losses and possible mechanical damage. The active components are high molecular weight isobutylene polymers or certain synthetic esters and are added in amounts up to 5%, according to duty.

1.28.3.4. Oxidation and Rust Inhibitors

Straight mineral oils usually lack oxidation resistance and rust inhibition in the presence of water, under modern operating conditions. An oxidation inhibitor, effective through a known temperature range, is added to stabilise the oil against oxidation and to passivate the metals which act catalytically to increase oxidation. These inhibitors maintain a low neutralisation value (acidity) of an oil for many years of service.

The rust inhibitor enables the oil to protect steel surfaces from rusting when in contact with water or moist air within the oil circulating system. By eliminating rust, which catalytically oxidises oil, a further control against deterioration is achieved. A typical oxidation inhibitor is 2–6-ditertiary-butyl-para-cresol, and it is used in concentrations of 0·2–1·0% by weight, which can be measured by chromatographic methods. Rust inhibitors include some sulpho-organic compounds and selected metal-salt-type inhibitors. These inhibitors are used in turbine, hydraulic, transformer and other highly refined industrial oils.

1.28.3.5. Detergent Additives

These are used particularly in diesel engine lubrication, to inhibit high-temperature oxidation, formation of low temperature sludge and the deposition of contaminants. Such inhibitors usually contain zinc dialkyl phosphate and sulphur, or may be a blend of organic compounds containing zinc, barium, calcium, phosphorus and sulphur, together with non-metallic organic substances.

1.28.3.6. Viscosity Index Improvers

The viscosity index of an oil indicates its viscosity/temperature characteristics. Oils thin out as temperature rises, some to a greater degree than others. Oils which "thin" considerably are termed "of low viscosity index" and those which show little change on rise in temperature are said to be of "high viscosity index". These additives are usually linear isobutylene polymers of controlled molecular weight and have the additional property of promoting engine cleanliness in intermittent service.

1.28.3.7. Pour Point Depressants

When oil is cooled it thickens until a temperature is reached when the oil becomes immobile. This temperature is the pour point of the oil and is an important property of fuel and lubricating oils. The pour point may be a function of viscosity (naphthenic oils) or due to the separation of waxy constituents (paraffinic oils). Reductions in the order of 60°F can be achieved by the addition of about 1·0% of the additive. The active component of these additives is a hydrocarbon wax–naphthalene condensate compound.

1.28.3.8. Emulsifying Agents

These additives are used in some lubricating oils to surround each particle of water by oil to prevent corrosion of metal surfaces. Typical materials used are metallic soaps, tallow, sulfinates and rape seed oil.

1.28.3.9. Anti-foam Agents

In most oil circulating systems, air tends to be entrained in relatively large quantities producing a foam, and this may be aggravated by the presence of other additives in the oil. The high polymer silicones are efficient foam suppressors and are added to the oil in quantities of about 0·003% by volume. Greater quantities are ineffective.

1.28.3.10. Demulsifiers

An emulsion of oil and water is usually caused by the presence of contaminants in the circulating oils such as rust, sludge, metallic soaps or other foreign substances and, as a rule, can be broken by heating to 170–212°F and then allowing adequate settling time. There are several types of emulsion breakers in commercial use but they are selective in action, so that each emulsion must be tested first to enable a proper choice to be made. The amount of additive required is less than 0·1% and should not be exceeded.

1.28.4. **Sampling**

The general principles outlined earlier for sampling fuel oils apply; in particular, the use of weighted sampling bottles for sampling from main oil tanks, and the use of sampling tubes for sampling from barrels. In practice, many turbines are fitted with convenient sampling points throughout the oil circulating system where samples may be drawn during operation and composited. Strict cleanliness of equipment and sample containers is essential, whatever sampling technique is used.

1.28.5. **Tests and Their Significance**

For quality and performance control the following tests may be carried out on lubricating oil. Viscosity, water content, acidity (neutralisation value) flash point, demulsification number, pour point, sulphur (corrosive), copper strip test, rust preventing characteristics, and oxidation stability. These tests are detailed in B.S. 489: 1955 and in I.P. Standards for Petroleum and its Products: Part 1, Methods for Analysis and Testing 1964; but a brief discussion of the interpretation of the tests is included below. Only the first five of the tests just mentioned are carried out as routine in most power station laboratories, although additional tests may be made if an abnormal condition occurs. The C.E.R.L. test for resistance to oxidation, or "oxidation stability" is now applied regularly (every 6 months) in many stations, and is described briefly below.

Viscosity

This is one of the most important properties of a lubricating oil. In a bearing, it is the viscosity which determines the fluid friction, the heat generated and the rate of the flow of oil under given conditions of load, speed and bearing design. Changes in viscosity indicate important changes in the oil so that monitoring is necessary for trouble-free operation. Generally, an increase in viscosity accompanies the onset of oxidation.

Water Content

The water content of lubricating oil is derived from extraneous sources, and in the case of steam turbine oil, the source is usually the turbine gland steam. The water thus admitted to the oil system is usually separated in the oil tank or the turbine oil purifier, and thus removed, but substantial in-leakage of water may aggravate any emulsion-forming tendency of an oil in service. Salt water ingress into turbine oil is particularly undesirable, since it may cause corrosion of the turbine control gear, with potentially disastrous results if the corrosion products interfere with the operation of the equipment.

Water content is usually determined by the Dean and Stark method (see Section 1.25.4), although a more sensitive chemical method is sometimes used when determining very small amounts of water. Turbine oil in good condition will contain 100–200 ppm of water, but may occasionally rise to 1000 ppm (0·1%), or more.

Total Acidity (Neutralisation Value)

The neutralisation value of an oil is a measure of its acidity. It is expressed as the weight in milligrams of caustic potash (KOH), which is required to neutralise the acid in one gram of oil. When determined by extraction of the oil with an organic solvent, the result is the total acidity; when extracted with water, only the inorganic acidity is obtained. If the oil is alkaline, the sample is neutralised with acid and the neutralisation value (negative acidity) is reported as the "strong base number".

The acidity (total) of new oils is normally within the range 0·02–0·10 mg KOH/g, or slightly higher in additive-conditioned new oils. In service, oils tend to oxidise to organic acids and this is accompanied by an increase in the total acidity. Thus for many years the acidity has been used as a guide to the rate of oxidative degradation of oils and normally, with inhibited turbine oils, there is a long period of service (the "induction period") during which the acidity may increase very slightly; after this period the rate of increase accelerates. An abnormally high rate of increase in acidity, therefore, gives cause for further testing and possibly for withdrawing the oil for treatment or replacement.

Oxidation Stability

In modern turbine oils, the oxidation inhibitor concentration tends to be maintained by the "make-up" of losses, using new oil, and the induction period is normally of many years' duration. Occasionally an oil deteriorates or breaks down with sludge formation, prematurely, and although this is usually accompanied by a rapid increase in acidity, it is desirable to find a means to anticipate such deterioration. The C.E.R.L. oxidation stability test is designed to provide advance warning of significant deterioration of oil quality.

In the C.E.R.L. test, which is discussed in the C.E.R.L. Report N. RD/L/R. 1188, a 100-g sample of oil at 150°C is subjected to oxidation under standardised conditions, and the number of hours required for the sample to absorb 300 ml of oxygen, is recorded as the test result. This result is referred to as the "T300 absorption time".

Figure 1.28.5A shows the form of change of acidity and T300 absorption time, of a new

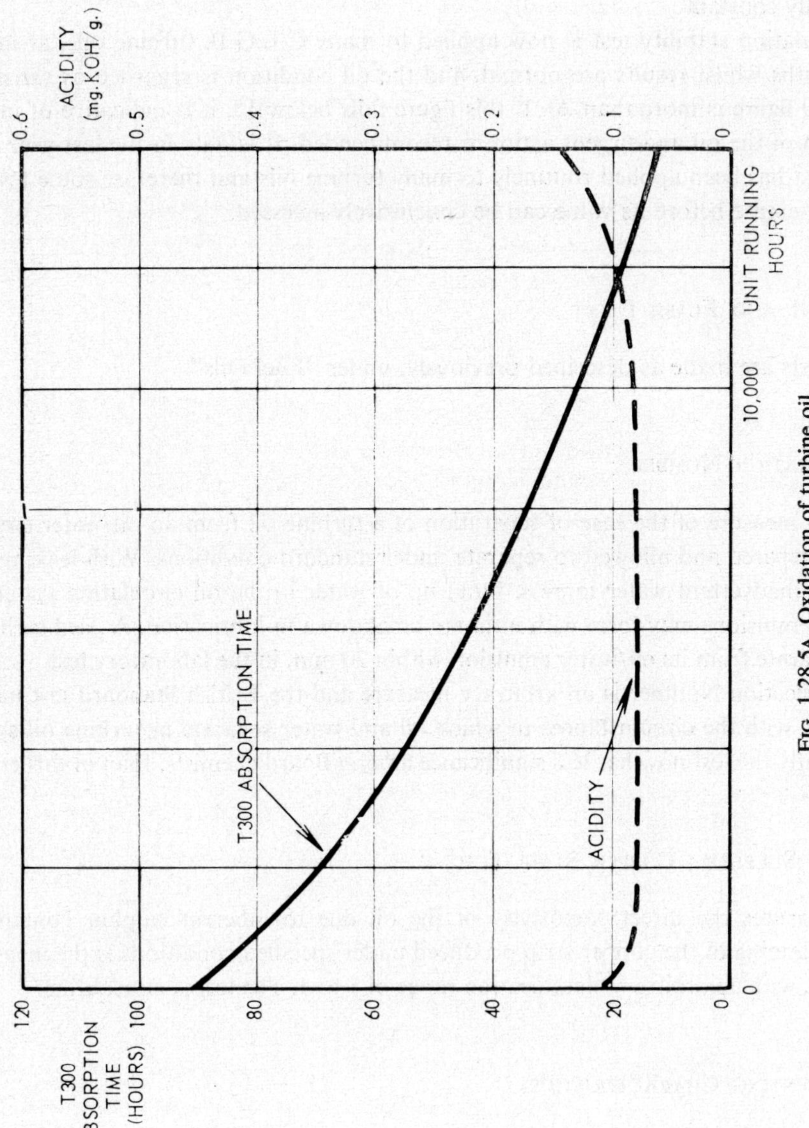

FIG. 1.28.5A. Oxidation of turbine oil

inhibited turbine oil in service. The graph represents an artificially small induction period, in order to make the significant aspects more clear, and in practice it usually requires between 50,000 and 100,000 h service before T300 has fallen to 20, or the acidity has increased significantly. Nevertheless, it can be seen that T300 is falling continuously whilst the acidity is practically constant.

This oxidation stability test is now applied to many C.E.G.B. turbine oils, at intervals of six months whilst results are normal, and the oil condition is regarded as satisfactory if the T300 figure is more than 20. If this figure falls below 15, it is indicative of incipient breakdown of the oil and urgent action is recommended. It is only in the last year or two that the test has been applied routinely to many turbine oils and therefore some few more years must elapse before its value can be conclusively assessed.

Pour Point and Flash Point

These tests are made as described previously, under "Fuel Oils".

Demulsification Number

This is a measure of the ease of separation of a turbine oil from an oil/water emulsion, specially prepared and allowed to separate under standard conditions. With leaking steam glands, or inadvertent water ingress, build up of water in the oil circulating system proceeds and emulsions may form with ultimate breakdown in lubrication. A used turbine oil should separate from its oil/water emulsion within 20 min, in the laboratory test.

Demulsification Number is an arbitrary measure and the British Standard test has little in common with the circumstances in which oil and water separate in turbine oil systems. Consequently the test now has less significance among Board chemists, than in earlier times.

Corrosive Sulphur: Copper Strip Test

This measures the direct corrosivity of the oil due to inherent sulphur content. The degree of staining of the copper strip produced under specified conditions is the measure of corrosivity, with an arbitrary classification range of 1 to 4. The test is not routine.

Rust-preventing Characteristics

This test indicates the ability of new or used oils to prevent corrosion of ferrous parts, in the presence of water. Turbine oils should give trouble-free service throughout their life and should not show any change in their ability to prevent corrosion of ferrous parts of the system by water. Water will cause corrosion if the oil does not "wet" the metal surfaces adequately. In the absence of satisfactory rust inhibitors, corrosion due to the ingress of water may produce sufficient iron oxide to choke oil passages, score bearings and block filters. This test is also not routine.

1.28.6. **Purification of Lubricating Oil**

CENTRIFUGING

In order to keep lubricating oil in good condition, regular treatment is necessary. Purification is most often carried out by centrifuge, using either single batch treatment or continuous by-pass treatment. In batch purification, the whole charge of oil is passed through the centrifuge and the treated oil is then returned to the oil system; it is, however, inconvenient in that it requires a planned outage of several days. If insoluble oil oxidation products are present the purification is done in the cold, that is, at ambient temperature, as these products are normally partially soluble at higher temperatures. In continuous by-pass treatment, purification is done during normal turbine running, the purifier continuously treating a proportion of the oil. Over-conditioning by centrifuge, however, should be avoided, in order to obviate excessive aeration of the oil.

Centrifuging submits the oil to a separating force enormously greater than that of gravity, and water and insoluble deterioration products and foreign matter are flung out of the oil. Emulsions too are broken down in the process, and water-soluble acids are conveniently removed with the discharged water. Figure 1.28.6A shows a typical De Laval centrifuge.

Centrifuges may be used as "separators" if a water seal is used. In this case, two effluents are continuously discharged; purified oil is one and water plus solid impurities and water-soluble acidic products is the other. When fitted as a "clarifier", there is no water effluent and separated solid material remains in the bowl and has to be removed at suitable intervals. Normally centrifuges in turbine installations are used as dual effluent systems, that is with separate oil and water discharges. This is particularly so with non-inhibited oils, as the scrubbing action of the water materially assists in the removal of solid matter and water-soluble deterioration products. In practice, water is normally added with the oil entering the centrifuge in order to gain these advantages. With inhibited oils, centrifuge techniques may be modified to reduce possible losses of additives in the oil.

SETTLING

In stations of low load factor, gravity settling in the main oil tank is usually sufficient to maintain the oil in a satisfactory condition. By tapping off water and deterioration products from the bottom of the main oil tank after long standing periods, the remaining oil is fit for further service. Settling can, of course, be used additionally to centrifuging.

FILTRATION

Although filtration of insulating oils has been practised in power stations for many years, this form of treatment has not been applied to turbine oil as an operational practice until recent years and to only a minor extent at present. The types of filter briefly described below can remove fine particulate matter, air and moisture more effectively than the commercial

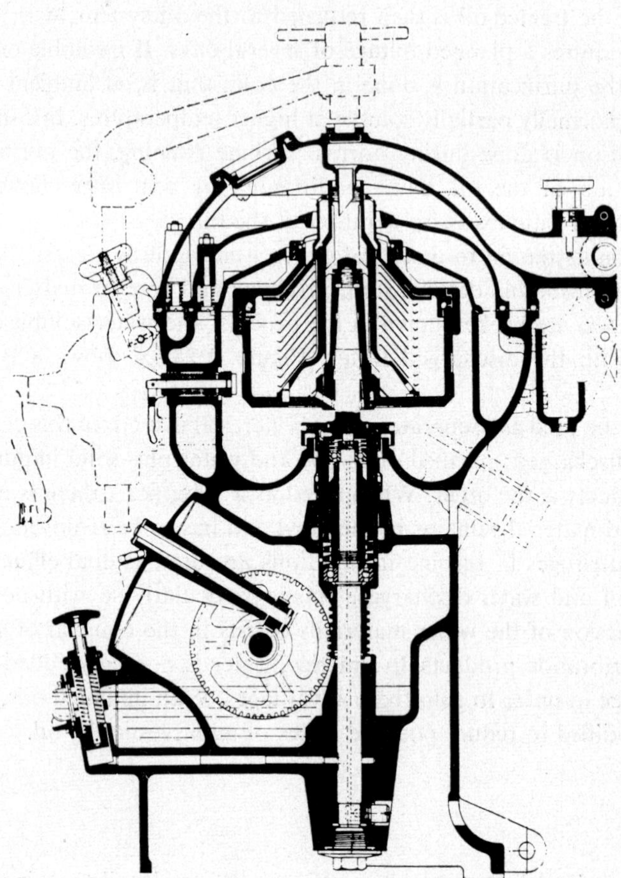

FIG. 1.28.6A. De Laval centrifuge

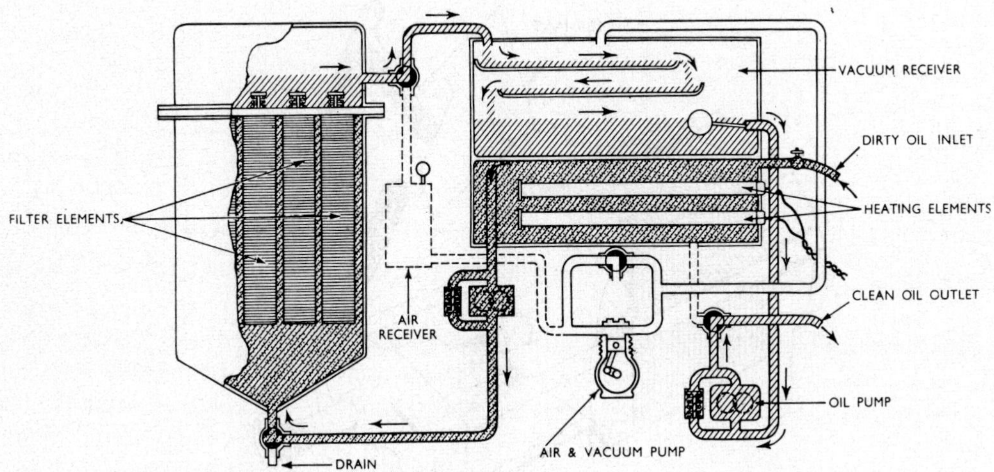

FIG. 1.28.6B. Edge type filter

centrifuge. However, it is not yet generally accepted that the consequent improvement in oil performance is sufficient to offset the rather simpler operation and maintenance requirements of the centrifuge.

Edgewise Filtration. This type of filter is exemplified by the "Streamline" filter, and the principles of operation are as follows. The oil is preheated to a maximum temperature of 180°F and pumped to a chamber containing a bank of filter packs, each pack comprising a large number of paper discs with a hole at the centre, and pressed closely together so that the assembled pack forms a hollow cylinder, about 3 in. long and $1\frac{3}{4}$ in. diameter. Filtration takes place as the oil passes from the outside of this cylinder, to the hollow core. The oil is then discharged to an evacuated chamber in which any moisture and entrained air are removed. This type of filter is illustrated in Figure 1.28.6B.

Filter Press. This consists essentially of a large number of square or circular filter papers, each supported by a metal plate having oil-ways formed in it, and the assembled plates and papers are mechanically pressed together to form a unit. The oil, which may be preheated, is pumped through the plates and papers, in parallel. In some designs, the plates have a recessed area which is filled with an absorbent filter medium, such as fuller's earth, which is able to reduce acidity and some other oil contaminants. The Menrow oil purifier is of the filter press type, and various auxiliary treatments such as deaeration, water separation, and coagulation of suspended-matter, may be included in the equipment.

It is necessary to obtain expert advice before using absorbent earths in the purification of turbine oils, because of the risks that some of the oil additives may be removed in the process. Figure 1.28.6C illustrates the filter press.

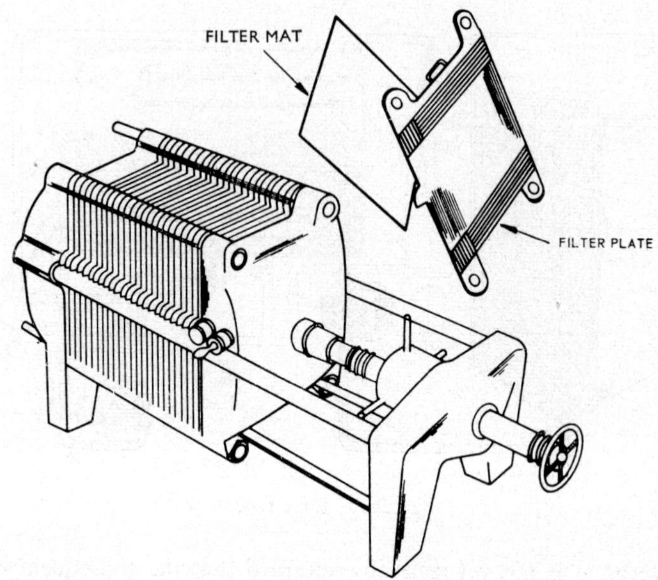

FIG. 1.28.6C. Filter press

1.28.7. **Operating Troubles in Turbine Oil Systems**

With modern additive-conditioned oils, serious operating troubles are comparatively rare and major outages from this cause have been virtually eliminated. Troubles do occasionally occur, however, and may be listed as follows:

(a) Water ingress.
(b) Oil deterioration with sludge formation.
(c) Foaming and aeration.
(d) Emulsification.
(e) High temperature of oil and bearings.
(f) Corrosion of bearing or journal metal.
(g) Excessive wear of bearings, gears, or other moving parts.

From this list, only item (a) occurs sufficiently often to warrant further discussion, although a few stations have experienced sporadic corrosion of white metal bearings. This corrosion is unexplained as yet, and is first seen as a hard black film over parts of the white metal, caused by the preferential oxidation of tin at the surface of the alloy. The trouble has occurred to a serious degree in some marine turbines.

WATER INGRESS

The common source of water which gains access to turbine oil, is turbine gland sealing steam, and in normal circumstances the water is removed from the oil by the centrifuge, by draining from the oil tank, and (as vapour) through the venting arrangements provided.

Control of gland steam quantities and maintenance of turbine glands are obvious factors in minimizing the ingress of water, and correct adjustment of the centrifuge is necessary for maximum effect in removing water.

Removal of water vapour by adequate venting is equally important, though perhaps less obvious, and modern turbines (almost all of 120 MW, or greater) have a venting system which includes provision for positive extraction of vapour from the bearing oil-return parts of the system. Regular inspection should be made to see that the venting system is functioning properly, since re-entrainment of vapour has sometimes caused water to accumulate in the oil.

1.28.8. **Fire-resistant Oils**

Introduction

Mineral lubricating oils used in hydraulic systems situated adjacent to equipment maintained at high temperatures have always constituted a considerable fire risk. Fires have in fact occurred in the past in power stations, when oils and hydraulic fluid leaking on to hot surfaces, have ignited. Even at ambient temperatures, fires have occasionally resulted from such leaks owing to sparking from commutators of adjacent electric motors. Since 1957, mineral oils used for hydraulic purposes, have generally been replaced with fire-resisting fluids. Typical power station hydraulic system applications include circulating water valve operation, boiler controls, boiler damper and sootblowing equipment, and window operating gear. Fire-resisting lubricating oils are now being introduced for use in air compressors and at one station, for use in turbine control gear where the turbine lubrication system is quite separate from the control system.

Fire-resistant lubricants may be divided into two main classes.

1. Water-containing lubricants.
2. Synthetic oils.

Water-containing Lubricants

These are either water-in-oil emulsions or water–glycol solutions of specific gravity 0·9 to 1·1, with water content varying between 35 and 60%. The water-in-oil emulsions are low viscosity oils in which water is finely dispersed by a special milling process. Emulsifying additives and oxidation inhibitors are added to form a non-oxidisable, stable emulsion.

The water–ethylene-glycol mixtures contain a polymer-type thickening agent and include additives to prevent rusting and to improve boundary friction properties.

These lubricants are not at present being used within the Board but their use in the future is under consideration.

Synthetic Oils

Two types of synthetic fluids are commercially available as fire-resisting oils; these are phosphate esters and chlorinated hydrocarbons (chlorinated diphenyls). These oils have a number of disadvantages compared with mineral oils, which disadvantages can be lessened by compounding various additives with the oils. They have low viscosity index, which may cause flow difficulty when starting under low-temperature conditions, and they have high specific gravity (about 1·15 for phosphate ester and 1·4 for chlorinated diphenyl), which may cause suction-lift difficulties and also will result in water, if present, floating on top of the oil. They also have low specific heat and powerful solvent properties on ordinary paints, varnish and rubber.

The synthetic oils are to some extent toxic, particularly at high temperatures, so that it is necessary to prevent contact with the skin and inhalation of the vapour. If spilled on very hot surfaces or flames, the oils produce copious toxic fumes and in such circumstances it would be necessary for personnel to use respirators.

Phosphate Esters

At present the Board's interest in fire-resisting lubricating oil is directed mainly at phosphate esters which, however, do have a further disadvantage in that they are prone to decompose slightly in the presence of water. This trouble is called "hydrolytic instability", and one of the decomposition products is phosphoric acid, which is likely to corrode metal parts of the oil circuits. It is possible to inhibit this effect by means of a suitable additive.

Chlorinated Hydrocarbons

These have been used successfully as electrical insulating oils, in equipment especially designed for their use. As lubricating oil and hydraulic fluid, they are usually mixed with phosphate ester in order to reduce the specific gravity, and with additives to improve other characteristics. In this form, the chlorinated hydrocarbons are widely used in industry.

Comparative Properties of Lubricants

The following relative values for cost, fire resistance, and lubricating property, have been derived by T.I. Fowle. In each category, the best material is placed highest in the list.

Relative cost:

Mineral oil	1
Water-in-oil emulsion	1·5
Water and glycol	4
Phosphate ester	5
Phosphate ester/chlorinated hydrocarbon	7

Fire resistance:

Phosphate ester	80
Phosphate ester/chlorinated hydrocarbon	80
Water and glycol	66
Water-in-oil emulsion	50
Mineral oil	3

Lubricating factor:

Mineral oil	1·0
Phosphate ester	1·2
Water and glycol	2·6

CHANGING FROM MINERAL OIL TO FIRE-RESISTING OIL

From what has been discussed above, it will be apparent that the substitution of synthetic oils in a system containing mineral oil, requires considerable attention to those parts of the system which are incompatible with the synthetic oils. For example, rubber seals and joints need to be changed, paintwork must be removed and good ventilation assured.

For turbine lubrication, a change to fire-resisting oil would scarcely be warranted on an existing machine, except in order to assess the operational problems of future designs to use such oils. These problems may be considerable, but the Board is investigating them by trials in small turbines, for possible application to future machines. Such a trial using a phosphate ester type oil, has been completed at Bedford and after some preliminary minor difficulty caused by air entrainment, the turbine operated for some six months quite satisfactorily.

1.29. GREASES

1.29.1. Introduction

A lubricating grease may be defined as a solid or semi-solid product of the dispersion of a thickening agent in a liquid lubricant, to which other additions may be made to impart special properties. They are usually soap-thickened mineral oils and are manufactured to give a variety of textures. For example, a calcium soap will give a buttery texture and a sodium soap a fibrous structure. Greases in general may be classified into five main groups.

1. Mixtures of mineral oil and solid components.
2. Heavy residual oils of the asphaltic type used straight or blended with lighter lubricating oil fractions.
3. Soap-thickened mineral oils.
4. Extreme pressure greases.
5. Roll neck greases.

The third group is the most widely used and is further classified into eight classes.

1.29.2. Lime-base Greases

Calcium soaps form the basis of the lime-base greases and range from cup greases to soft pressure-gun type greases. They are smooth, homogeneous, have a soft butter-like texture and are suitable for use only below 180°F. Lime-base greases do not dissolve in water and have, in fact, a high water resistance. Recent developments have introduced lime soap greases in which the normal water content is replaced by higher boiling solvents. These exert a stabilising effect on the grease, permitting use at higher temperatures.

1.29.3. Soda-base Greases

These are manufactured in a similar way to the lime-base greases, but as water is not used for stabilisation, they are suitable for use for temperatures up to 300°F. They are, however, water-soluble and can not be used under wet conditions. Soda-base greases have a greater stability and internal cohesiveness than lime-based greases, and thus may be used for high speed anti-friction bearings and similar applications. They are made both as fibrous and smooth greases. Mixtures of lime/soda grease with properties in between each class are made for special applications.

1.29.4. Aluminium-base Greases

Aluminium-base greases are manufactured in two grades, buttery and stringy, the former resembling lime-base greases in properties and applications. They are stable, smooth, water-resistant and can be used under wet conditions, but have poor temperature characteristics.

The stringy greases also are water-resistant and are particularly useful for lubricating slow moving gears. The aluminium greases are very adhesive to metal surfaces, providing rust protection under wet conditions but are unsuitable for high-speed applications.

1.29.5. Lithium Greases

Lithium greases are buttery, water-resistant greases and may be used over a wide temperature range. They are frequently used as multi-purpose greases to replace several greases of other types for general grease lubrication, and in many cases it is often possible to meet all the grease requirements of large industrial plants with one high quality lithium grease. Multi-purpose greases usually combine high-temperature stability with good water resistance and remain soft at low temperature.

1.29.6. Complex Greases

Simple soap greases contain only one metallic soap, whilst mixed soap greases contain soaps of two or more metals with the same or different fatty acids.

Complex greases on the other hand contain one or more conventional metal soaps with metal salts of low molecular weight fatty acids. They have some outstanding characteristics: they do not melt or lose consistency at high temperatures, they have good resistance to oxidation and water, and possess good load-carrying properties. In consequence these greases have a broader range of application than some other multi-purpose greases.

1.29.7. Fluid Greases

The stringy type of aluminium grease is sometimes made in a fluid form to permit easy dispensing from cans and similar equipment. Lime soaps are used to thicken oil.

These two types of compound are called non-fluid oils, as they do not flow under light pressures and if they ultimately work out of the bearing on to the shaft, they tend to adhere and waste is minimised.

1.29.8. Block Greases

Block greases can be handled as solids, being made from viscous mineral oils heavily thickened with metallic soap. They do not perform all the functions of an efficient lubricant as they do not carry away heat or keep out dirt. Block greases are most suitable for the lubrication of slow-moving bearings.

1.29.9. New Greases

Recent developments have introduced the Bentonite greases, in which bentonite clay is used to form a non-soap grease. Certain of the bentonite clays, when activated by suitable organic chemicals, can be dispersed in mineral oil in the form of stable gels. These gels form stable greases and combine good lubricating properties with outstanding heat resistance.

1.29.10. Graphite and Molybdenum Disulphide Greases

Under extremely heavy loading conditions at slow speeds and high temperatures, solid fillers such as graphite and molybdenum disulphide are necessary to prevent metal to metal contact and undue wear. Graphite with its laminar structure, combined with soda- or lime-base greases, has for many years been successfully used in many applications, one of its main advantages being that it forms an adsorbed film on metal surfaces which assists in retaining the lubricant. Molybdenum disulphide has similar characteristics: it is resistant to higher temperatures and is claimed to be more effective under critical lubrication conditions.

Both types of greases are not recommended for use in ball and roller bearings due to the tendency of the filler to impact, thus preventing free movement of the elements and consequent heavy wear.

1.29.11. Greases for Nuclear Applications

Greases are used for the bearings and gears of the fuel-handling machines and the control rod mechanism in nuclear power stations.

Conventional soap greases are resistant to radiation up to 10^8 rads, but as the fuel-handling machine receives a radiation dose about ten times this figure, special greases are necessary. At the radiation level of 10^8 rads, the soap structure breaks down, radically affecting the consistency. There is an equally serious objection to the use of ordinary greases in machinery associated with gas-cooled reactors, as hot carbon dioxide under pressure softens the grease by breaking down the soap structure, causing "bleeding" and failure of lubrication. The complex greases, in which the grease structure is modified, are less affected but are still not satisfactory.

Non soap-thickened greases, made with special clays which are unaffected by carbon dioxide, fulfil many of the essential requirements for nuclear purposes. These clay-thickened greases have outstanding radiation resistance, structural stability at high temperatures and satisfactory oil bleeding characteristics. A number of other non-soap thickeners have also given successful results.

1.29.12. Problems

With the wide range of greases now commercially available, lubrication problems have largely been eliminated. The main problems are in applications where operating temperatures are rising and where service periods between lubrication are increased. This is particularly so in the nuclear field where heavy demands are made on the lubricant. For example, high-temperature, gas-cooled reactors are now being designed to work with coolant outlet temperatures up to 750°C (1382°F) compared with the present maximum of 400°C (752°F), in "Magnox" type reactors.

1.29.13. Properties of Greases

The main properties of a grease of interest to the user are:

1. Consistency.
2. Drop point.
3. Resistance to shear.
4. Resistance to oxidation.
5. Resistance to water.

1.29.14. Consistency

Greases range in consistency from very soft, almost liquid greases to materials so hard that they can be handled as solids. Consistency is measured by the penetration number, that is, the distance which a standard cone will penetrate a grease under specified conditions. Thus a grease with a penetration number of 385 would be very soft and a grease of number 85 would be considered very hard. Table 38 gives the range of consistency numbers for commercial greases.

TABLE 38

NATIONAL LUBRICATING GREASE INSTITUTE CONSISTENCY CLASSIFICATION

Consistency number	A.S.T.M. worked penetration (at 77°F)	
0	355–385	Semi-fluid greases
1	310–340	
2	265–295	
3	220–250	
4	175–205	
5	130–160	
6	85–115	Block greases

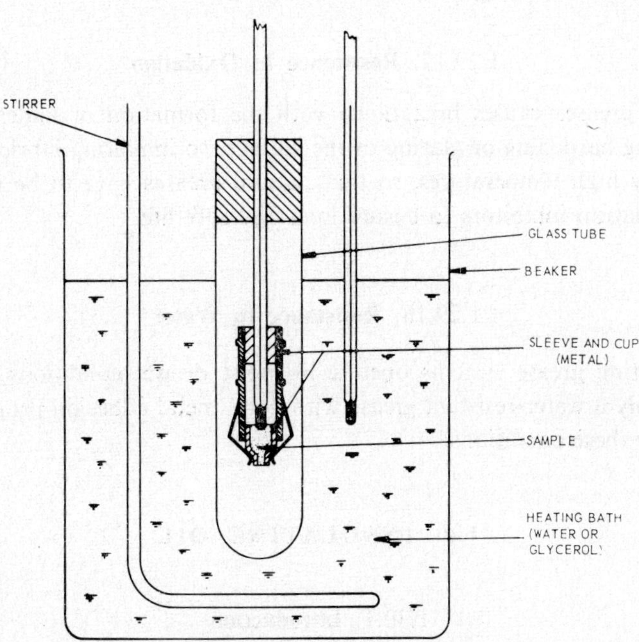

FIG. 1.29.15. Grease drop point test

1.29.15. Drop Point

The temperature at which a grease has to operate mainly determines the choice of a grease for a particular application. An indication of the maximum temperature at which a grease can be used is given by the drop point. This is the temperature, measured under specified test conditions, at which the grease begins to exude oil or to pass from the solid to the semi-solid state. In practice of course, this figure should be higher than the expected temperature of the application. Depending on the type and quality of the grease, however, some greases may be used to within 20°F of the drop point, whilst for others, the safe limit may be as low as 100°F below the drop point. The apparatus for the determination of drop point is shown in Figure 1.29.15.

1.29.16. Resistance to Shear

Greases used in anti-friction bearings are subject to considerable shearing forces which tend to accelerate breakdown. A good indication of the performance of a grease is given by the difference in consistency before and after it has been worked. A grease which will perform satisfactorily in anti-friction bearings will show little softening after working, whereas one that becomes very soft after a short period of working will not perform satisfactorily in service.

1.29.17. Resistance to Oxidation

Oxidation of greases causes breakdown with the formation of undesirable oxidation products, causing hardening or glazing of the surface, so impairing lubrication. Oxidation is accelerated by high temperatures, so that natural greases have to be improved by the addition of oxidation inhibitors to ensure long and safe life.

1.29.18. Resistance to Water

Many lubricating grease systems operate in moist or wet conditions and some, even under water. Only a water-resistant grease with good metal adhesion properties therefore, is suitable under these conditions.

1.30. INSULATING OIL

1.30.1. Introduction

In addition to electrical insulation needs, a basic requirement of an insulating oil in a transformer is to absorb heat from the core and the conductors, where it is generated, and to conduct it to the outside cooling surfaces where the heat is dissipated to the atmosphere. Insulating oil has both satisfactory specific heat and excellent electric strength.

In switchgear, the main function of the oil is to extinguish the arc formed between contacts when an electric circuit is opened. The modern tendency is to use air blast switchgear for high-voltage work, but insulating oil is still used for some large high-voltage circuit breakers, as well as for smaller units.

Special viscous oils are used as insulants in metal-clad bus-bar chambers, where exposure to high temperatures and stresses is absent.

Specially compounded insulating oils are also used as dielectrics in capacitors and in cables, for resistance to prolonged high electric stress.

1.30.2. Characteristics

British Standard B.S. 148: 1959: Insulating Oil for Transformers and Switchgear

B.S. 148: 1959 specifies that the oil to be used for transformers and switchgear shall be pure hydrocarbon mineral oil, clean and free from foreign matter likely to impair its properties, and containing no additives. By special arrangement, however, a user may request the addition of an oxidation or other inhibitor, in which case the uninhibited base oil must comply with the Standard. Oils complying are considered to be compatible with each other and can be mixed in any proportion, except in the case of inhibited oils.

The limiting characteristics of insulating oil to B.S. 148 are summarised in Table 39.

TABLE 39

INSULATING OILS

SCHEDULE OF CHARACTERISTICS (B. S. 148)

Characteristic	Limit
Sludge value (max.)	1·20%
Acidity after oxidation (max.)	2·5 mg KOH/g
Flash point (closed) (min.)	295°F (146°C)
Viscosity at 70°F (21·1°C) (max.)	37 cS*
Pour point (max.)	−25°F (−31·7°C)
Electric strength, 1 minute (min.)	40 kV (r.m.s.)
Acidity (neutralization value)	
Total (max.)	0·05 mg KOH/g
Inorganic	Nil
Saponification value (max.)	1·0 mg KOH/g
Copper discoloration	Negative
Crackle (moisture)	Shall pass test

* Approximately 151 sec, Redwood No. 1.

INHIBITORS

Inhibited insulating oils are not currently used extensively in the Board's installations, the bulk of oil supplied being straight insulating oil complying with B.S. 148: 1959. Oxidation inhibitors, however, are sometimes used in insulating oils to stabilise the oil against oxidation and to deactivate the catalytic action of metals on the oxidation process. These inhibitors are designed to hold the neutralisation value of the oil at almost its initial value for a very long service period. This period, during which formation of sludge and acidity is negligible, is called the induction period, and after this when the anti-oxidant inhibitor is depleted, the deterioration products develop at about the same rate as if no additive were present. As rust is a very active oxidation catalyst, a rust-inhibiting additive may also be incorporated in the oil.

1.30.3. **Sampling**

This is described in detail in the British Standard and the principles follow closely those previously described for fuel oil and turbine oil. Clean, accurate sampling is essential for insulating oils, owing to their extreme susceptibility to contamination by moisture and atmospheric pollution. Various sampling thiefs are available for sampling from bulk tanks or drums. Figure 1.30.3A shows a typical sampling thief which, when the valve rod strikes the bottom of the tank, automatically opens the top and bottom valves to release air and admit oil. A simple pressure thief for bulk tank sampling is shown in Figure 1.30.3B. For sampling oil from barrels, a sampling tube thief is used. These are made of clear glass or stainless steel and are designed to take the sample from within 3 mm of the bottom of the barrel. A convenient diameter is about 35 mm. A sampling tube thief is shown in Figure 1.30.3C.

For representative sampling the following minimum quantities are recommended:

Electric strength test 2000 ml

Other tests 300 ml (each)

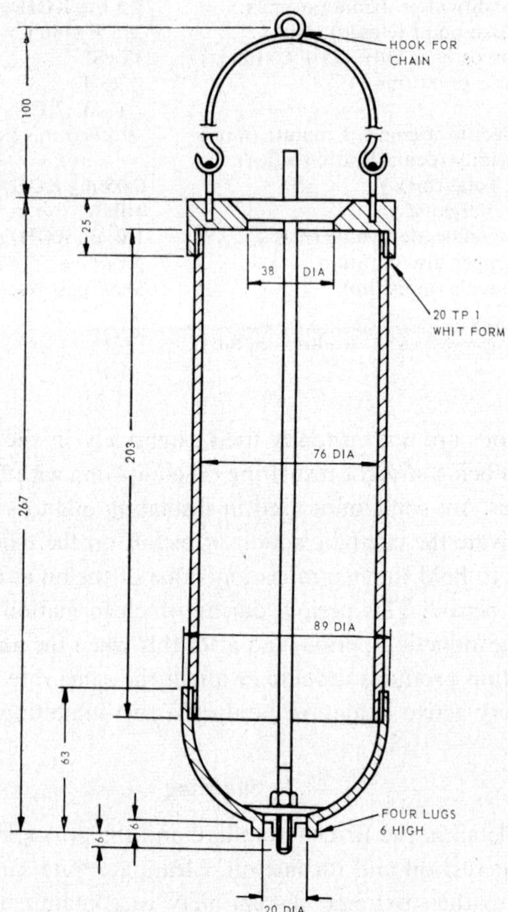

FIG. 1.30.3A. Thief dipper for sampling oil in bulk

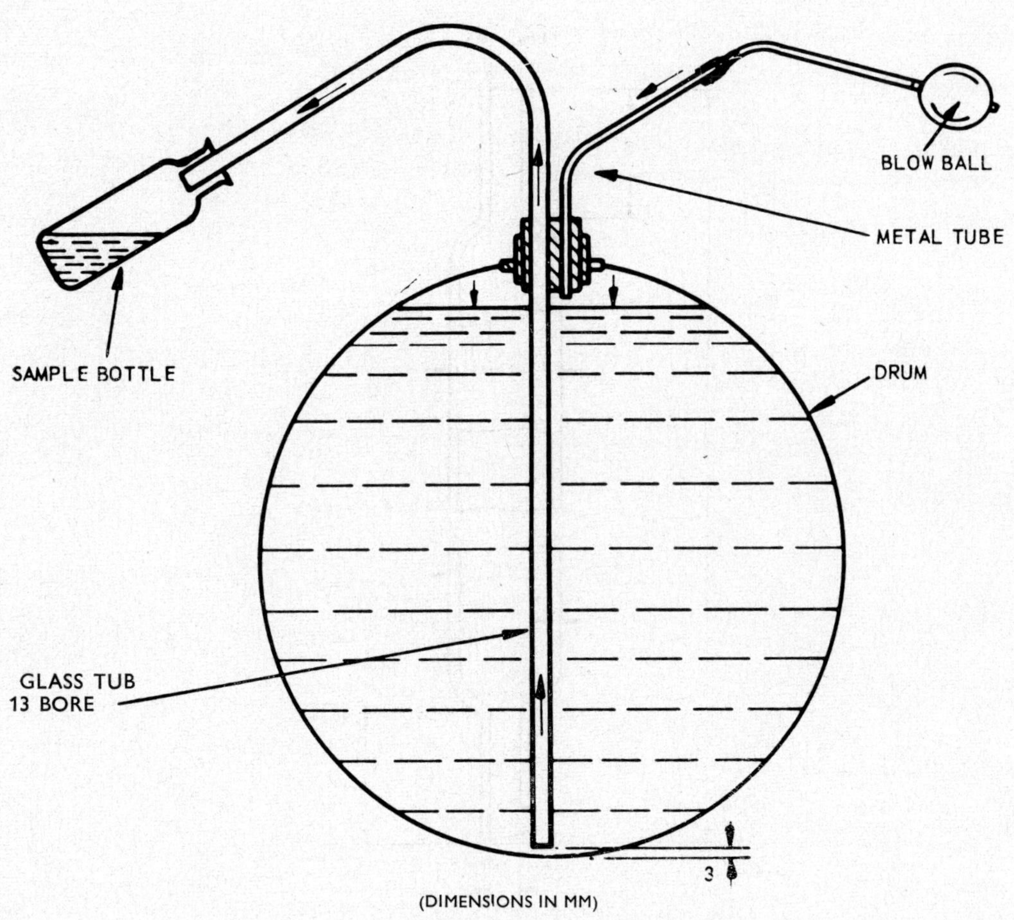

FIG. 1.30.3B. Pressure thief

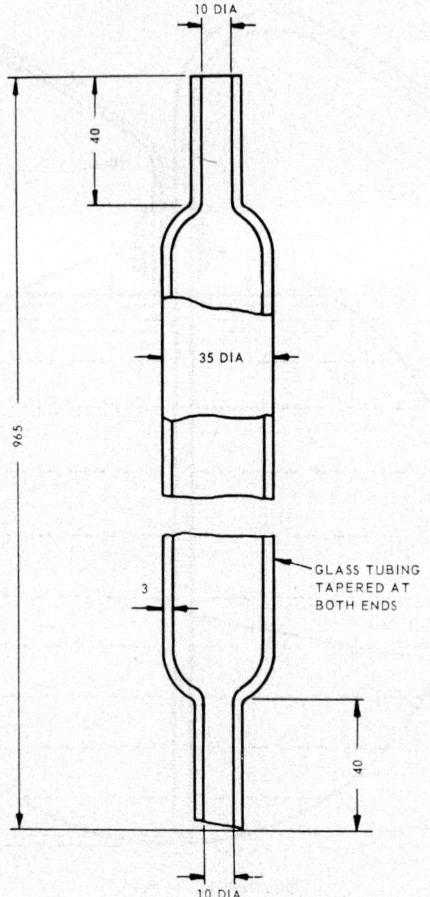

FIG. 1.30.3C. Glass sampling tube for insulating oils

1.30.4. Testing

The full range of specified tests for insulating oils include:
Oxidation test
Flash point
Viscosity
Pour point
Electric strength
Neutralisation value
Saponification value
Copper discoloration
Free water (Crackle test)
They are fully described in B.S. 148: 1959. Of these tests, flash point, electric strength, neutralisation value (or acidity), and free water, are those most commonly made as routine in the Board.

Electric Strength

The electric strength test provides a measure of the electrical insulating properties of an oil and also responds to impurities, such as water, fibres and dirt in the oil. In the test, an alternating voltage is applied to two metal spheres, submerged in a sample of the oil, using equipment of standardised dimensions and characteristics. Beginning at 15 kV, the voltage is increased to the specified test voltage (30 kV or 40 kV) in 10–15 sec and maintained at this value for 1 min. If no arcing occurs between the spheres, and it is necessary to measure the actual breakdown value of the oil, the voltage is raised by 1 kV/sec until the sample breaks down. In practice the test should be carried out in triplicate.

Acidity (Neutralisation Value)

As with lubricating oils, the development of acidity is an indication of insulating oil deterioration by oxidation. The method is similar to that used for lubricating oil except that an alcoholic solution of potassium hydroxide is used for the neutralisation instead of aqueous potassium hydroxide.

Test for Free Water (Crackle Test)

This is a sensitive test for moisture in an insulating oil and consists of heating a few millilitres of the oil in a standard dry test tube. If moisture is present an audible crackle is heard. Three separate tests are carried out and if no crackling is heard in two out of the three tests the sample is satisfactory. If unsatisfactory, the moisture may be determined quantitatively by the Dean and Stark method previously described, or by a more sensitive method.

Of the remaining tests included in B.S. 148: 1959, viscosity, flash point and pour point may be carried out as previously described for fuel oils.

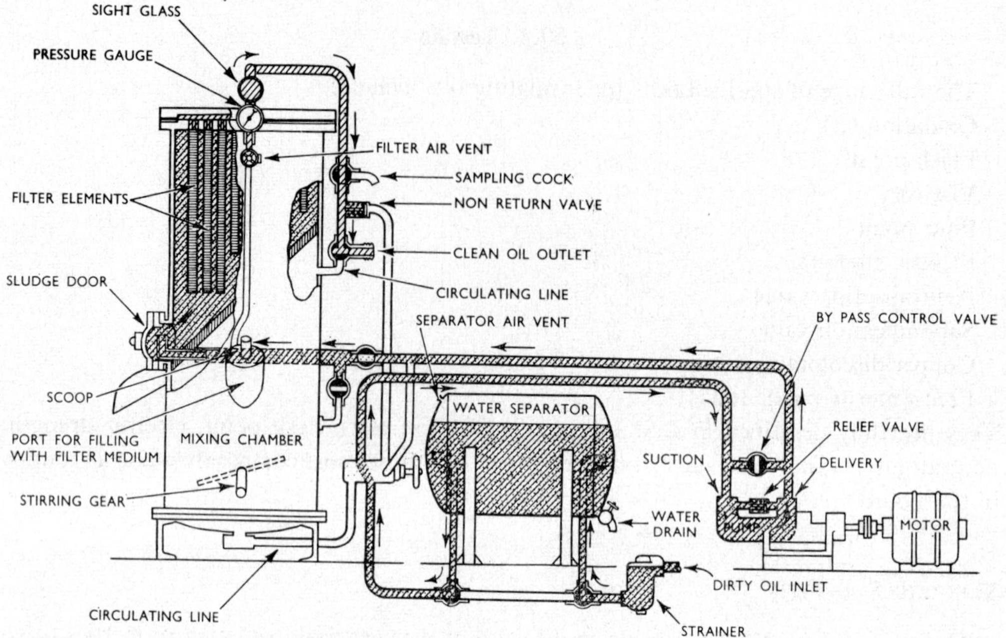

FIG. 1.30.5. Metafilter for purification of insulating oils

1.30.5. **Purification**

Extremely high standards of purity are set for insulating oils and if an oil fails the B.S. tests for crackle test or electric strength, purification is necessary before it is again fit for service. This applies equally to both new and used oils.

Filter Press and Edge Filter

These methods are commonly used for insulating oil purification and the equipment has been described earlier in this chapter, under 1.28.6.

Filter Beds

This type of filter combines mechanical filtration with absorption and a typical filter consists of metal disc filter elements mounted on a perforated tube. The incoming oil is mixed with diatomaceous earth or activated carbon, which forms the filter bed. Solids and particulate water are filtered out satisfactorily, and moisture removed by absorption. In this type of filter, the oil must circulate continuously, as a stoppage may partially breakdown the filter bed. This type of filter is illustrated in Figure 1.30.5 which shows a Metafilter.

1.30.6. Fire-resistant Synthetic Insulating Oils

Where fire hazards are particularly high, non-inflammable oils may be used. These are not mineral oils and are usually chlorinated hydrocarbons, some being mixtures of penta-chlor-diphenyl and trichlor-benzene.

The fire-resistant synthetics, while being useful in this particular field, are not superior to high-grade mineral oils in electrical properties. They suffer too from the disadvantage of forming hydrochloric acid under certain conditions, particularly when subjected to arcing, and in the presence of moisture under these conditions, serious corrosion can result.

These synthetic oils also dissolve ordinary insulating varnishes, rubber, etc., so that such materials must be excluded from equipment in contact with fire-resisting insulating oils.

1.31. BUCHHOLZ GAS ANALYSIS

1.31.1. Introduction

Almost every type of fault inside a transformer is accompanied by the generation of gas, and this is usually the first indication of a fault developing. If a fault can be detected at an early stage, serious damage to the transformer may be avoided. The Buchholz gas pressure relay shown in Figure 1.31.1 is the device used for gas detection and automatic isolation of the transformer. It is fitted to the highest point in the transformer in the connection pipe to the conservator, and is initially full of oil. If gas is generated, it accumulates and causes a float to sink which activates an alarm. If an abnormal generation of gas occurs, a flap valve is displaced which causes the relay to immediately isolate the transformer from service.

When a transformer is taken out of service following a gas generation signal, information should be sought immediately concerning the nature of the fault. Past history, rate of gas generation, whether the gas accumulation alarm or the pressure surge element has operated, and the analysis of the gas generated, will assist in indicating whether an internal fault has occurred or whether the gas accumulated is merely air. A qualitative analysis of the gas sample from the Buchholz relay will confirm a fault and generally indicate the type of fault in the transformer.

1.31.2. Analysis

The analysis is done conveniently by passing the gas from the gas relay relief cock, first through a solution of aqueous silver nitrate, then through a solution of ammoniacal silver nitrate, both made up in distilled water. The residual exit gas leaving the second solution is tested for inflammability. If the accumulated gas is air, generally indicating no fault, there is no positive colour change in the two test solutions and the exit gas is not inflammable. If, however, the gas has been formed by under-oil arcing, or by a bare red-hot metallic part under oil, a white curdy precipitate is formed in solution 1, a yellow precipitate in solution 2 and the residual gas leaving the second solution is inflammable.

If, on the other hand, the gas has been formed by breakdown of transformer insulation by overheating, a black precipitate will form in both tubes and the exit residual gas will again be inflammable. A summary of the test indications is given in Table 40.

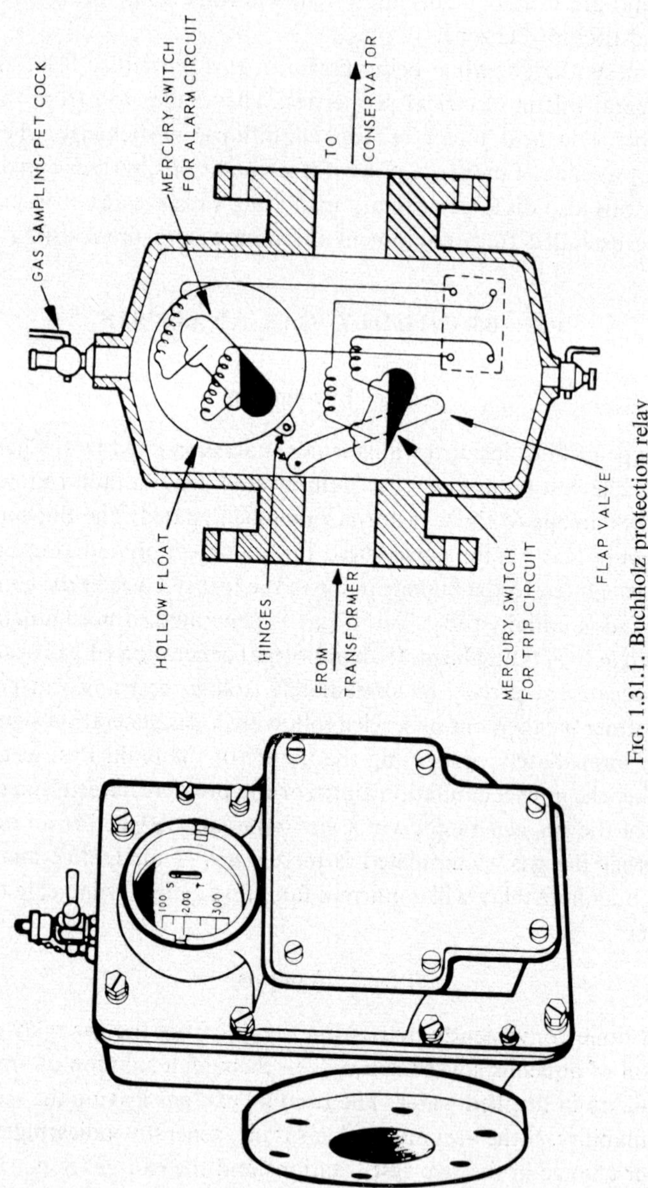

FIG. 1.31.1. Buchholz protection relay

TABLE 40

CHEMICAL TEST INDICATIONS: TRANSFORMER BREAKDOWN GASES

Solution 1 (5% silver nitrate)	Solution 2 (5% ammoniacal silver nitrate)	Exit gas inflammability	Fault indicated
No reaction	No reaction	Non-inflammable	No fault; air only present
Heavy, white curdy precipitate	Yellow precipitate	Inflammable	Under-oil arcing or under-oil hot metal
Black precipitate	Black sooty precipitate: solution opaque	Inflammable	Cellulose degradation

1.31.3. **Chemical Reactions**

REACTIONS BELOW THE OIL SURFACE

In under-oil arcing and under-oil hot metal reactions, the main gas evolved is hydrogen, together with acetylene and low-boiling point paraffins. The gas, however, producing the colour changes in both solutions is acetylene, which may be present up to 20% of the air-free volume. In solution 1, a complex of silver nitrate and silver acetylide is first formed and this changes with absorption of more acetylene to silver acetylide. In solution 2, silver acetylide is formed directly, but only after the reaction in solution 1 is completed. As both of these compounds explode violently on heating, the tubes should be thoroughly washed immediately after use.

INSULATION BREAKDOWN

During the destructive distillation of cellulose, carbon monoxide, carbon dioxide, methane and small quantities of other paraffins are given off. The identifiable gas in this case is carbon monoxide which has the property of reducing silver nitrate solutions, under certain conditions, to black metallic silver. This is the reaction which takes place in solution 2.

These tests should be carried out immediately a fault is signalled, as acetylene is very soluble in oil, and on long standing over oil may be totally absorbed. In this case no reaction would take place in either solution, but the exit gas would be inflammable. Carbon monoxide is relatively insoluble in oil and, therefore, presents no problem by delay.

1.31.4. **Test Apparatus**

The above principles have been applied to many designs of site apparatus for the rapid qualitative analysis of transformer gases. Figure 1.31.4 shows a diagram of the essential parts of a simple portable apparatus. Solutions 1 and 2 are placed in two vertical tubes

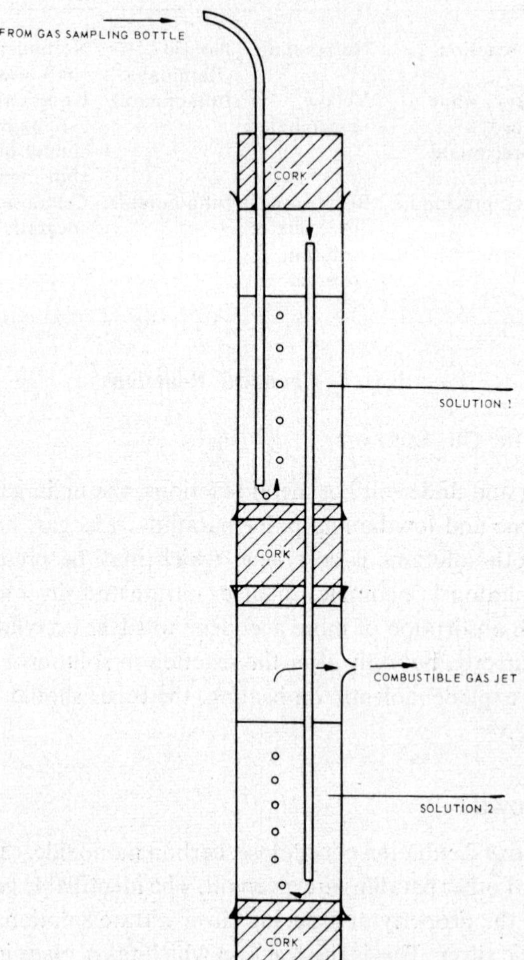

FIG. 1.31.4. Buchholz gas analysis tube

mounted one above the other and separated by a cork. Gas is forced in from the gas sampling bottle under a head of transformer oil, first through solution 1 in the top tube. It is then forced from above the level of solution 1 through solution 2. On bubbling through solution 2, it leaves by a small tapered jet situated above the liquid level of solution 2. It is at this exit jet, that the residual gas is tested for inflammability.

There are several proprietary gas test instruments now available for this work, but whatever method is used on site, it is good policy, in addition, to have the suspect gas quantitatively analysed by a competent chemist, using sensitive laboratory instruments. Such a service, however, is not always available when and where it is needed, and the simple field testing equipment provides an adequate guide, in most cases, to the cause of operation of the Buchholz relay.

EXERCISES

1. What are the principal sources of error that may occur in obtaining a representative sample of small coal, as delivered to a power station, and in reducing the sample to 8·0 oz of 72-mesh size?

Mention procedures necessary to minimise errors in sampling and sample reduction.

2. A coal sample is drawn from untreated rough slack having $1\frac{1}{2}$ in. top size with 40% passing $\frac{1}{8}$ in. square mesh B.S. sieve, and laboratory tests on the sample as prepared for analysis yielded the following results:

Moisture		8·7%
Ash		18·0%
Volatile matter		27·5%
Total sulphur		2·9%
Chlorine		0·39%
Calorific value		10,170 Btu/lb
Swelling number	$\frac{1}{2}$	
Ash fusion temperature	Deformation	1140°C
	Hemisphere	1280°C
	Flow	1310°C

Calculate the proximate analysis and calorific value on the dry basis, the as received basis, and the dry ash-free basis, given that the total moisture was 10·8% as received.

Assuming that the hydrogen content as received was 3·7%, what was the net calorific value?

Comment on the kind of difficulties, if any, likely to be encountered on burning coal represented by the above analysis, in (a) P.F. boilers and (b) chain-grate boilers.

3. What standards of coal quality do you consider the C.E.G.B. should endeavour to establish with the N.C.B., for some 4×10^7 tons of coal estimated to be required each year for the large stations expected to operate from 1967 onwards? Give reasons for your answer.

4. Discuss the causes of deposit formation and fouling of the gas-side surfaces in coal-fired boilers and in heavy fuel oil-fired boilers, with reference to the fuel constituents. How may the corrosion of superheaters be related to these problems?

5. (a) Write note on factors influencing the decision as to which grade of oil fuel should be burnt in a station which is to be converted from coal-firing to oil-firing.

(b) It is rarely necessary to carry out a full range of tests on oil fuels as delivered to converted stations. List the tests you consider important, and give reasons for your choice.

6. Give a brief account of the value and disadvantages of injecting chemicals into the furnaces or gas-passes of boilers fired with heavy fuel oil, for the purpose of controlling low-temperature corrosion. How would you test the effect of such treatment during a trial on the plant?

7. What are the most useful measuring techniques yet developed for ascertaining the contribution of modern power stations to air pollution? What further steps do you consider might be taken to reduce air pollution caused by electricity generating plant in the C.E.G.B.?

8. (a) What kinds of tests are routinely applied (within C.E.G.B.), to a modern turbine oil in service, and how are the results interpreted?

(*Note*: It is *not* necessary to describe the methods of testing.)

(b) If it were decided to substitute a phosphate-ester fluid in place of a normal mineral lubricating oil in a steam turbine, what prior attention to parts within the oil circuit would you advise, having regard to the properties of the fluid? If the oil system accumulates free water in operation, suggest methods to remove it.

CHAPTER 2

CORROSION. FEED AND BOILER WATER

2.1. INTRODUCTION

In this chapter the field of water treatment in relation to boiler operation will be reviewed. The theoretical principles of metallic corrosion will be discussed, followed by a review of boiler water and feed water treatment. Consideration will also be given to such subjects as sampling and analysis of water and steam, and chemical instrumentation.

From a study of the early history of boiler construction and operation, it is clear that two main causes of failure were present in the boiler explosions which occurred in the early years. The first cause, lack of knowledge in design, operation, and maintenance, was gradually overcome by investigation, introduction of new materials, and experience. The second, inadequate water treatment and insufficient theoretical knowledge of corrosion processes, was much slower in being attacked. It is probably true to say that fundamental knowledge of scale formation and corrosion processes, occurring in power plant systems, started to be accumulated in the 1920's and has been subjected to increasingly critical analysis in the last twenty-five years.

Although a considerable volume of knowledge and experience is now available, internal boiler corrosion remains a problem, especially in plant operating at 900 lb/in^2 and above. Whilst surveying the whole field of feed and boiler water treatment, the lesson is more concerned with plant in the pressure range 900 lb/in^2 and upwards.

In order to understand the chemical reactions which are discussed throughout the lesson, it is necessary to have an outline knowledge of the principles involved and these are discussed in sub-section 2.1.1.

2.1.1. **Chemistry of Reactions in Solution**

2.1.1.1. A Simple Model of the Atom

An atom of an element consists of a number of protons and neutrons associated together in the nucleus of the atom. Around the nucleus a number of electrons travel in orbits.

An atom of a particular element has a characteristic normal complement of electrons which is equal to the number of protons in the nucleus. When the normal complement of electrons is present, the electric charges of the protons and electrons balance each other, and the atom has no resultant electric charge.

The mass of an atom is concentrated almost entirely in the nucleus, and is approximately equal to the number of protons plus the number of neutrons. All atoms of the same element have the same number of protons in the nucleus (the atomic number) but the number of neutrons in the nucleus may differ. Hydrogen may have one of three different nuclei, which have been given the different names indicated below, but they are all in fact, hydrogen.

(i) 1 proton only = mass 1 (Hydrogen)
(ii) 1 proton + 1 neutron = mass 2 (Deuterium)
(iii) 1 proton + 2 neutrons = mass 3 (Tritium)

These forms of elements, having different numbers of neutrons in their nuclei are called isotopes. Isotopes of the same element have slightly differing atomic weights, but they have the same atomic number and the same arrangement of electrons; their chemical properties are also identical.

In an electrically neutral atom, the number of electrons must equal the number of protons. It has been shown experimentally that all electrons are identical in charge and mass, but they also have energy, and the form of this energy can vary from one electron to another within an atom. Experimental information indicates that the electrons within an atom can be divided into groups, each group consisting of electrons with similar energy levels. Various names can be given to these groups, but they may be conveniently called "shells", and there is a maximum number of electrons which can exist in a given shell.

1st electron shell—a maximum of 2 electrons
2nd electron shell—a maximum of 8 electrons
3rd electron shell—a maximum of 18 electrons
4th electron shell—a maximum of 32 electrons

The electrons can be subdivided into smaller groups, but these will not be considered here. The shells are numbered in order of their increasing distance from the nucleus, the first shell being nearest the nucleus. The higher the number of the shell the greater the energy of an electron contained in the shell.

The simplest atom is the hydrogen atom, which for a mass of 1 has one proton and one electron, and may be represented as shown in Figure 2.1.1.1(a). The oxygen atom has 8 protons in the nucleus, and has eight electrons arranged in shells, two in the first shell, and six electrons in the second shell (Fig. 2.1.1.1(b)).

An iron atom has 26 protons in the nucleus and 26 electrons arranged in four shells 2–8–14–2 (Fig. 2.1.1.1(c)).

Although atoms of each particular element have a definite normal complement of electrons, they may acquire one or more electrons, or lose one or more electrons. When an atom gains an electron or electrons, the resultant electric charge is negative; when it loses one or more electrons, the resultant electric charge is positive.

All chemical reactions are in fact interactions between the outermost electron shells of different elements, and atoms containing sufficient electrons to complete their outer shells are particularly unreactive from a chemical viewpoint, examples being helium, neon and argon, which are all inert gases.

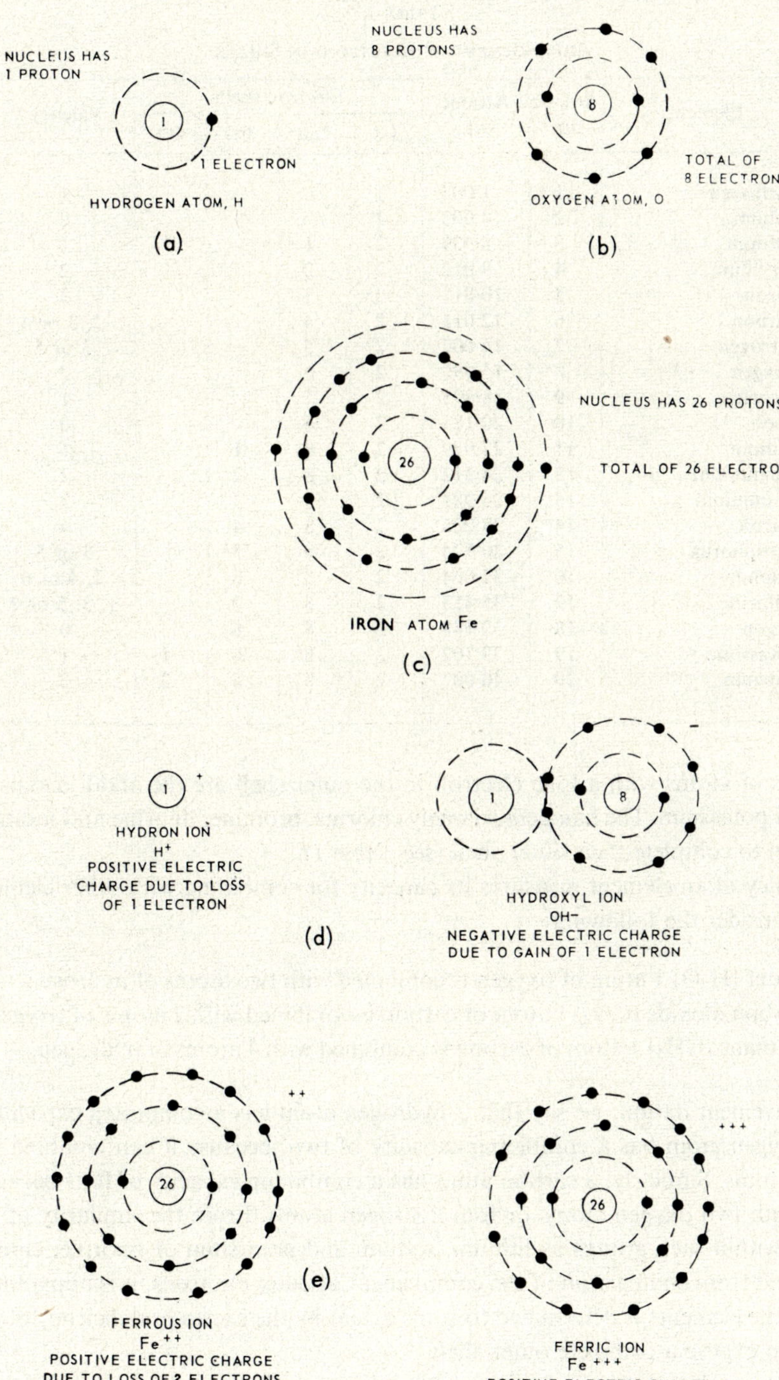

FIG. 2.1.1.1. Some atomic and ionic groupings

TABLE 1

ARRANGEMENT OF ELECTRONS IN SHELLS

Element	Atomic no.	Atomic wt.	Electron shells				Valency
			1st	2nd	3rd	4th	
Hydrogen	1	1·008	1				1
Helium	2	4·003	2				0
Lithium	3	6·939	2	1			1
Beryllium	4	9·012	2	2			2
Boron	5	10·811	2	3			3
Carbon	6	12·011	2	4			2, 3 or 4
Nitrogen	7	14·007	2	5			3 or 5
Oxygen	8	15·999	2	6			2
Fluorine	9	18·998	2	7			1
Neon	10	20·183	2	8			0
Sodium	11	22·989	2	8	1		1
Magnesium	12	24·312	2	8	2		2
Aluminium	13	26·981	2	8	3		3
Silicon	14	28·086	2	8	4		4
Phosphorus	15	30·974	2	8	5		3 or 5
Sulphur	16	32·064	2	8	6		2, 4 or 6
Chlorine	17	35·453	2	8	7		1, 3, 5 or 7
Argon	18	39·948	2	8	8		0
Potassium	19	39·102	2	8	8	1	1
Calcium	20	40·08	2	8	8	2	2

Examples of atoms with a lone electron in the outer shell are the alkali metals sodium, lithium and potassium. The halogens, namely chlorine, bromine, fluorine and iodine require one electron to complete their outer shell (see Table 1).

The valency of an element measures its capacity for combining with other elements. For example, consider the following:

(a) In water (H_2O) 1 atom of oxygen is combined with two atoms of hydrogen.
(b) In carbon dioxide (CO_2) 1 atom of carbon is combined with 2 atoms of oxygen.
(c) In methane (CH_4) 1 atom of carbon is combined with 4 atoms of hydrogen.

If, as a convenient datum, we say that a hydrogen atom has a combining capacity of one, then an oxygen atom has a combining capacity of two, because it can combine with two hydrogen atoms. Similarly, a carbon atom has a combining capacity of four because it can combine with two oxygen atoms or four hydrogen atoms. From the similarity of chemical properties within such groups as lithium, sodium and potassium or fluorine, chlorine and bromine, and from comparison of the complement of outer electrons, it is apparent that the valency of the elements is determined to some extent by the excess or deficiency of electrons necessary to expose a complete outer shell.

If the hydrogen atom loses its electron it loses a unit of negative electricity, and becomes positively charged (Fig. 2.1.1.1(d)), and in this state is known as an hydrogen ion, H^+ and is "univalent".

The iron atom can lose two or three electrons resulting in the formation of either a ferrous ion or a ferric ion (Fig. 2.1.1.1(e)), in both cases the ion is positively charged. The ferrous ion is "bivalent" and the ferric ion is "tervalent".

The concept of valency also applies to ionised groups of atoms, sometimes called acid radicles, which carry a characteristic number of negative charges; this number is the valency. Thus, the chloride (Cl^-) and nitrate (NO_3^-) ions are univalent, sulphate (SO_4^{--}) and carbonate (CO_3^{--}) ions are bivalent, and orthophosphate (PO_4^{---}) ions are tervalent.

A positively charged ion such as Na^+, Ca^{++} or Fe^{+++}, is termed a "basic radicle" and there is a notable ionised group of atoms obtained when ammonia is dissolved in water, namely the ammonium ion, which is also a basic radicle. It is formed in the following manner:

$$\underset{\text{ammonia}}{NH_3} + \underset{\text{water}}{H_2O} \longrightarrow \underset{\text{ammonium ion}}{NH_4^+} + \underset{\text{hydroxyl ion}}{OH^-}$$

It will be seen that a hydroxyl ion is simultaneously formed, rendering the solution alkaline.

2.1.1.2. Ions in Solution

Sodium chloride, or common salt (NaCl), is a simple example of an ionic type of chemical compound, and such compounds are our main concern in this lesson.

Sodium and chlorine, as elements, are both dangerously reactive substances, and when brought together will combine violently to produce sodium chloride. In this combination, the single electron in the third shell of the sodium atom is transferred to the third shell of the chlorine atom, thereby leaving completed outer electron shells to each atom. At the same time, the chlorine has gained an electron and becomes a chloride ion (Cl^-), whilst the sodium has lost an electron to become a sodium ion (Na^+), and the resultant salt or sodium chloride molecule is held close together by the attractive force of the electrical charges (see Fig. 2.1.1.2A).

When the salt is dissolved in water, which has a very high dielectric strength, this attractive force is so weakened that a large proportion of the sodium and chloride ions are able to move as separate electrically charged entities in the solution. The reactions described above may be represented thus:

$$\underset{\text{2 sodium atoms}}{2\,Na} + \underset{\text{1 chlorine molecule}}{Cl_2} \longrightarrow \underset{\text{Sodium chloride molecules}}{2\,NaCl}$$

$$\text{NaCl in water} \longrightarrow \underset{\text{Sodium ion}}{Na^+} + \underset{\text{Chloride ion}}{Cl^-}$$

It can be shown experimentally that a solution of sodium chloride in water will conduct electricity due to the ionisation just described and most other salts have a similar property. Certain substances such as hydrochloric acid, nitric acid, and sulphuric acid will also conduct electricity when they are in a solvent such as water. A solution which conducts electricity is an "electrolyte", the electric current being carried through the solution by ions. In general, the greater the concentration of ions in the solution, the greater the conductivity, thus sea water has a greater conductivity than town's water. Compounds such as alcohol and

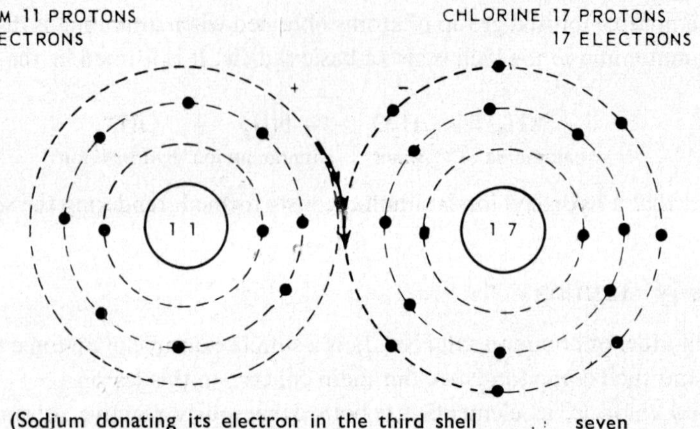

(Sodium donating its electron in the third shell, : seven electrons in the chlorine third shell, to form sodium

FIG. 2.1.1.2A. Formation of sodium chloride

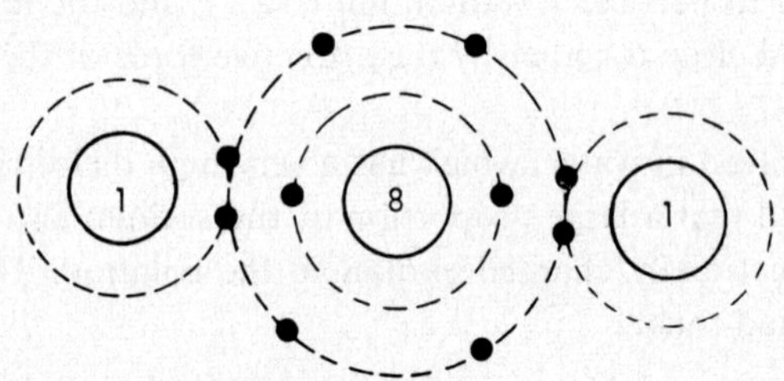

(Two hydrogen atoms share their electrons with one atom of oxygen)

FIG. 2.1.1.2B. A molecule of water

sugar, when dissolved in water, do not conduct electricity appreciably, and are called non-electrolytes.

Water is a compound of hydrogen and oxygen, and the molecule has a resultant electric charge of zero. It might, therefore, be postulated that water is not ionised. However, the purest water has a small residual conductivity of $0{\cdot}04 \times 10^{-6}$ mhos/cm, and this is accounted for by the fact that about 1 in 10 million molecules of water is ionised into hydrogen ions and hydroxyl ions (see Figs. 2.1.1.2B and 2.1.1.1(d)).

$$H_2O \longrightarrow H^+ + OH^-$$

Water molecule	Hydrogen ions	Hydroxyl ions

At 25°C the mathematical product of H^+ and OH^- ion concentrations* in water and aqueous solutions is always equal to 10^{-14}, and when each is equal to 10^{-7} as in pure water and neutral solutions, the solution is neither acidic nor alkaline. All acids contain one or more hydrogen atoms which in water provide hydrogen ions, thereby reducing the number of hydroxyl ions and, conversely, all alkalis provide hydroxyl ions in water and, therefore, reduce the number of hydrogen ions.

In order to avoid the inconvenience of negative powers of 10, Sorensen (1909) suggested a shorthand notation for the scale of hydrogen ion concentration. This entails the use of the term "pH", whereby the logarithm of the hydrogen ion concentration to the base 10, with the sign reversed, is used to express degree of acidity or alkalinity. Thus an hydrogen ion concentration of 10^{-7} is expressed as pH 7—that is, a neutral solution. An hydrogen ion concentration of 10^{-6} = pH 6—an acid solution, and an hydrogen ion concentration of 10^{-9} = pH 9—an alkaline solution. As the product of hydrogen and hydroxyl ions must be constant at 10^{-14}, the hydroxyl ion concentration can be obtained by difference, and is denoted by the term pOH. Thus where the hydrogen ion concentration is equal to pH 9, the hydroxyl ion is equal to pOH 5. Table 2 shows the relationships between hydroxyl ions and hydrogen ions in water at 25°C. It should be noted that the pH scale is logarithmic so that a decrease of pH by one unit corresponds to a tenfold increase in hydrogen ion concentration. The relationship between pH value and the acidity (or alkalinity) of a solution, is qualitatively somewhat similar to the relationship between temperature and heat.

When quoting pH values, the temperature during measurement should also be stated; where this is not done the temperature is usually assumed to be 25°C. As the water temperature increases, the pH value of the neutral point decreases, as follows:

Temperature, °C	25	50	100	200	300
pH of neutral point	7·0	6·63	6·1	5·6	5·5

Thus, pH 6·1 is slightly on the acid side of the neutral point at 25°C, but corresponds to neutrality at 100°C. The simplest way to avoid confusion arising from the contraction of the pH scale with rising temperature is to refer all pH values to their equivalent values at 25°C and this is normally done.

* The unit of ionic concentration is "gram-ions per litre", as is customary. In order to convert this unit to grams per litre, it is only necessary to multiply by the molecular weight of the ion. For example the molecular weight of the hydroxyl ion, OH, is O(16)+H(1) = 17, therefore, an hydroxyl ion concentration of 10^{-6} g-ions/l$=1{\cdot}7\times10^{-7}$ g/l.

TABLE 2

RELATIONSHIP BETWEEN CONCENTRATIONS OF HYDROXYL IONS AND HYDROGEN IONS IN WATER AT 25°C

	Concentration of hydroxyl ions in solutions, (ppm by wt)	Concentration of hydroxyl ions in g-ions per litre of solution		pOH value	Concentration of hydrogen ions in g-ions per litre of solution		pH value	Concentration of hydrogen ions in solution, (ppm by wt)	Hydrogen electrode potential (volts)
ACID RANGE	0·00000000017	10^{-14}	$\frac{N}{100,000,000,000,000}$	14	10^{0}	N "Normal"	0	1000	0·000
	0·0000000017	10^{-13}	$\frac{N}{10,000,000,000,000}$	13	10^{-1}	$\frac{N}{10}$	1	100	−0·059
	0·000000017	10^{-12}	$\frac{N}{1,000,000,000,000}$	12	10^{-2}	$\frac{N}{100}$	2	10	−0·118
	0·00000017	10^{-11}	$\frac{N}{100,000,000,000}$	11	10^{-3}	$\frac{N}{1,000}$	3	1	−0·177
	0·0000017	10^{-10}	$\frac{N}{10,000,000,000}$	10	10^{-4}	$\frac{N}{10,000}$	4	0·1	−0·236
	0·000017	10^{-9}	$\frac{N}{1,000,000,000}$	9	10^{-5}	$\frac{N}{100,000}$	5	0·01	−0·295
	0·00017	10^{-8}	$\frac{N}{100,000,000}$	8	10^{-6}	$\frac{N}{1,000,000}$	6	0·001	−0·354
Neutral	0·0017	10^{-7}	$\frac{N}{10,000,000}$	7	10^{-7}	$\frac{N}{10,000,000}$	7	0·0001	−0·413
ALKALINE RANGE	0·017	10^{-6}	$\frac{N}{1,000,000}$	6	10^{-8}	$\frac{N}{100,000,000}$	8	0·00001	−0·472
	0·17	10^{-5}	$\frac{N}{100,000}$	5	10^{-9}	$\frac{N}{1,000,000,000}$	9	0·000001	−0·531
	1·7	10^{-4}	$\frac{N}{10,000}$	4	10^{-10}	$\frac{N}{10,000,000,000}$	10	0·0000001	−0·59
	17	10^{-3}	$\frac{N}{1,000}$	3	10^{-11}	$\frac{N}{100,000,000,000}$	11	0·00000001	−0·649
	170	10^{-2}	$\frac{N}{100}$	2	10^{-12}	$\frac{N}{1,000,000,000,000}$	12	0·000000001	−0·708
	1700	10^{-1}	$\frac{N}{10}$	1	10^{-13}	$\frac{N}{10,000,000,000,000}$	13	0·0000000001	−0·767
	17,000	10^{0}	N "Normal"	0	10^{-14}	$\frac{N}{100,000,000,000,000}$	14	0·00000000001	−0·826

The "normal" or datum concentration of hydrogen ions is 1 g-ion/l of solution which corresponds to 1·008 g of hydrogen ions per 1000 g of solution.

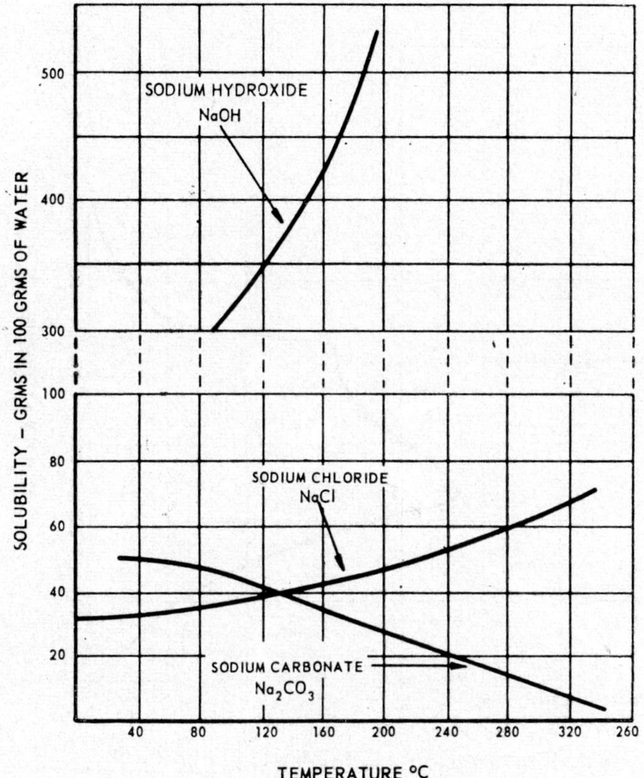

FIG. 2.1.1.3A. Solubility curves of sodium salts

In terms of corrosion processes, the importance of the pH value lies in the electrical potential associated with it. The potential of the hydrogen electrode for different concentrations of hydrogen ions or pH values is shown in Table 2. This will be referred to later.

There are two main methods of measuring pH values; these are a colorimetric method using chemical indicators (such as litmus in former times) and potentiometric methods using electrical equipment. The colorimetric method is approximate and is based on changes of colour, shown by certain organic compounds at various ranges of pH. The potentiometric method is normally used for pH determinations in power stations.

2.1.1.3. Solubility and Mechanism of Precipitation Reactions

Solutions may be defined as homogeneous systems, containing two or more distinct chemical substances. One of the substances is the "solvent", and the other substance or substances is the "solute". In aqueous solutions the water will be taken as the solvent, the substances dissolved in it being the solutes. In general, there is a limiting value to the amount of solute which may be dissolved in a given weight of water at any one temperature. When the weight of substance dissolved in the water equals this limiting value, no more will dissolve, and the solution is said to be a "saturated solution". The "solubility" of the substance in water is the weight dissolved when the substance is in equilibrium with its saturated solu-

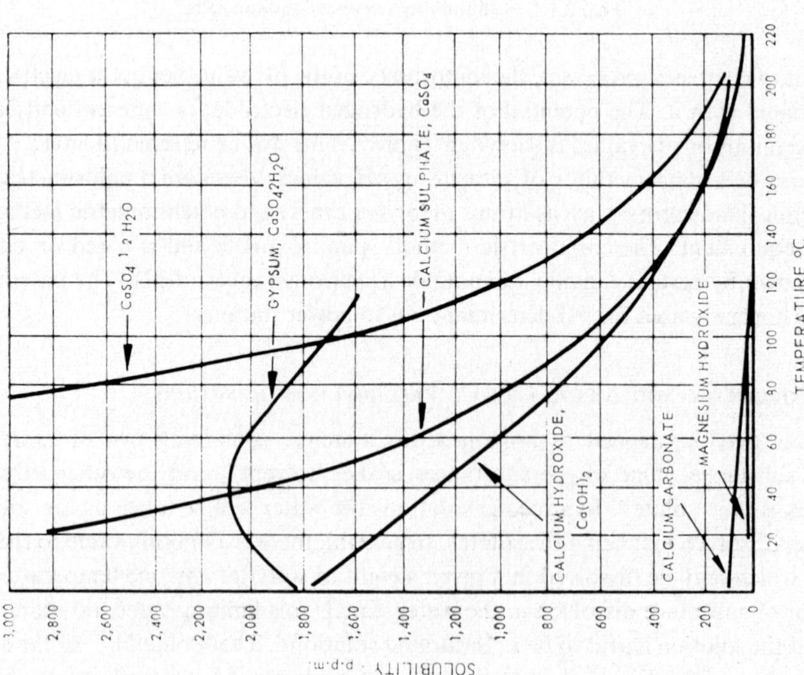

FIG. 2.1.1.3B. Solubility curves for scale-forming salts

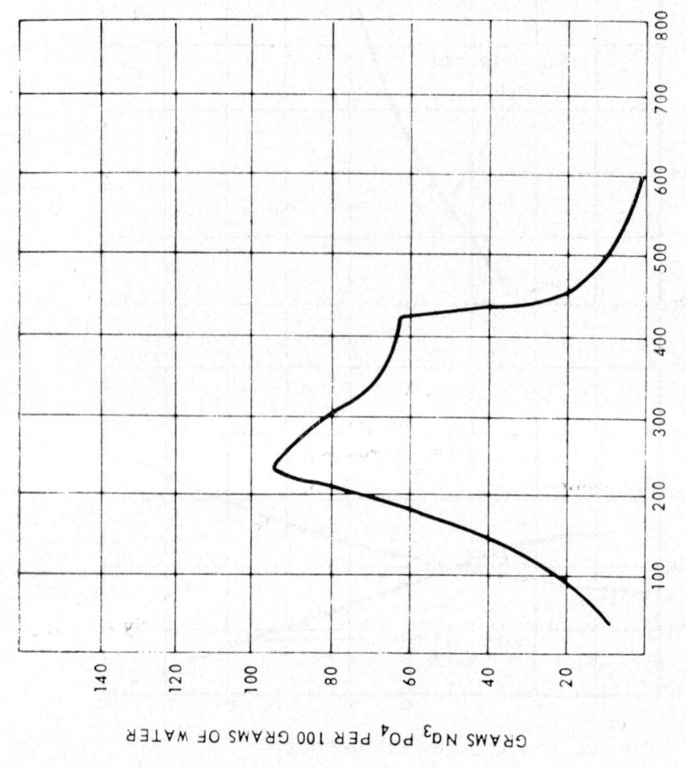

(Interpolation should not be made where abrupt changes of direction occur; these are associated with the formation of hydrates.)

FIG. 2.1.1.3C. Solubility curve for trisodium phosphate

tion. Figure 2.1.1.3A gives the solubilities of some sodium salts in water, and Figure 2.1.1.3B the solubility curves of some scale-forming substances.

The solubility of various substances in water varies widely; as extreme examples, the solubility in water at 20°C of sodium iodide is approximately 180 g/100 ml, whilst for barium sulphate it is 2×10^{-4} g/100 ml. Table 3 shows the solubilities of most of the chemical compounds concerned in water treatment technology. In many cases the solubility increases with rise in temperature, but for a significant number the reverse is true, and these are said to have a retrograde temperature/solubility relationship. Some substances, such as trisodium phosphate, show an increase followed by a decrease in solubility as boiler water operating temperatures are approached (see Fig. 2.1.1.3C). Solubility is also affected by the presence of other compounds dissolved in the solvent.

The precipitation of practically insoluble substances from aqueous solution is of major importance in all sections of power station water treatment, and a limited discussion of the principles involved is necessary. Practically all power station waters are dilute solutions of chemical compounds which are ionised to a large extent, and this fact enables approximations to be made in this discussion without significant inaccuracy.

When a sparingly soluble salt, such as calcium sulphate ($CaSO_4$) is added to pure water and stirred, the point will be reached when no more of the solid salt will dissolve, and the saturated solution is in equilibrium with the residual undissolved solid. Since even the saturated solution is dilute (0·2 g/100 ml at 20°C) it is almost completely dissociated into the separate ions Ca^{++} and SO_4^{--}, and the equilibrium may be represented:

$$\underset{\text{Solid}}{CaSO_4} \rightleftharpoons \underset{\text{Ions}}{Ca^{++}+SO_4^{--}}$$

A similar equilibrium may be represented for any sparingly soluble salt, symbolised as BA (B=basic radical, A=acid radical) as

$$\underset{\text{Solid}}{BA} \rightleftharpoons \underset{\text{Ions}}{B^{+}+A^{-}}$$

It can be demonstrated that at a given temperature the mathematical product of the concentrations of the dissolved ions is constant, and this constant is called the "solubility product". This statement may be written as

$$(B^{+})\times(A^{-})=Ks$$

where

(B^{+})=concentration of B ions in gram-ions per litre,
(A^{-})=concentration of A ions in gram-ions per litre,
Ks=solubility product of the salt "BA".

It follows that if separate solutions of quite soluble salts containing the ions B^{+} and A^{-} are mixed together, the resultant solution is likely, initially, to have the product $(B^{+})\times(A^{-})$ much exceeding Ks, in which case the excess ions will recombine to precipitate the salt BA as a finely dispersed solid.

TABLE 3

SOLUBILITY OF SOME INORGANIC COMPOUNDS IN WATER

Compound	Chemical formula	Solubility g/100 ml
Calcium bicarbonate	$Ca(HCO_3)_2$	0·166 at 20°C
Calcium carbonate	$CaCO_3$	0·00153 at 25°C
Calcium chloride	$CaCl_2{\cdot}6H_2O$	74·5 at 20°C
Calcium hydroxide	$Ca(OH)_2$	0·165 at 20°C
Calcium phosphate	$Ca_3(PO_4)_2$	0·0029 at 20°C
Calcium sulphate	$CaSO_4$	0·209 at 30°C
Magnesium carbonate	$MgCO_3$	0·0169 at 25°C
Magnesium chloride	$MgCl_2$	54·25 at 25°C
Magnesium hydroxide	$Mg(OH)_2$	0·0009 at 18°C
Magnesium sulphate	$MgSO_4$	26·0 at 0°C
Potassium phosphate	K_3PO_4	90·0 at 20°C
Sodium carbonate	Na_2CO_3	7·1 at 20°C
Sodium chloride	NaCl	36·0 at 20°C
Sodium hydroxide	NaOH	109·0 at 20°C
Sodium sulphate	$Na_2SO_4{\cdot}7H_2O$	44 at 20°C
Sodium sulphite	$Na_2SO_3{\cdot}7H_2O$	26·9 at 20°C

For example, magnesium sulphate ($MgSO_4$) and sodium hydroxide (NaOH) are readily soluble in water, and if a solution of say, 1 g/l of magnesium sulphate is mixed with another solution, say 10 g/l of sodium hydroxide, a cloudy white precipitate of magnesium hydroxide is rapidly formed in the mixture. At the moment of mixing the solutions, relatively large concentrations of the ions Mg^{++}, Na^{+}, SO_4^{--} and OH^{-+} are present together and the solubility product of magnesium hydroxide (5×10^{-11} at 20°C) is greatly exceeded, so that the excess of Mg^{++} and OH^{-} ions is removed from the solution by combining to form the magnesium hydroxide precipitate. The overall reaction here is

$$MgSO_4 + 2\,NaOH \longrightarrow Mg(OH)_2\downarrow^{\dagger} + Na_2SO_4$$

This particular reaction occurs in lime/soda water softeners (see Chapter 3) and is typical of many which arise in chemical treatment generally.

2.2. THE CORROSION OF METALS

2.2.1. Introduction

Metals are usually found in nature as compounds, principally in the form of their oxides and sulphides, and corrosion may be loosely defined as the process by which metals return to a form in which they occur in nature.

Metals mined as ores are obtained in their elemental state by a process of chemical reduction, for example, iron from iron ore in the blast furnace, or aluminium from aluminium-bearing ores in the electric furnace. In both cases considerable amounts of energy have to be expended to bring about the reduction.

† The symbol ↓ is often used to indicate a precipitated reaction product. The symbol ↑ is sometimes used to indicate that a gaseous product such as carbon dioxide, is produced.

In general, the greater the amount of energy used to produce the metal from the metallic ore, the greater is the tendency for the metal to return to a metallic oxide, by corrosion. Under normal conditions, gold, which is mined in an elemental state, does not corrode, whilst aluminium and magnesium, which require large quantities of electrical energy for their manufacture, have a strong tendency to revert to the oxides. It should be noted, however, that all metals are liable to attack by corrosion, provided the environment is sufficiently aggressive, thus gold for example, is attacked by a mixture of nitric and hydrochloric acids. Corrosion will take place when the environment is suitable to enable the metal to revert to a lower energy form, as an oxide by reaction with oxygen, or into metallic salt compounds by reaction with salts or acids.

The effect of oxygen on the surface of the metal is to produce a thin layer of oxide, which may or may not lead to destructive corrosion. In some cases the oxide film is impervious and results in a state of passivity, preventing further corrosive action.

With a basic metal such as aluminium, the metallic oxide film is easily formed, and the film is also very stable. For such a coating to give protection against destructive corrosion, it must be stable, adhere firmly to the metal, and entirely cover the metal surface with a non-porous coating. These requirements are met if the oxide coating is insoluble in the liquid in contact with the metal, and is self-repairing when damaged.

Aluminium and chromium in favourable circumstances form protective oxide coatings, whereas the oxide formed on low-alloy steel surfaces is generally porous, and does not prevent further corrosion. The reddish-brown coatings of rust which form on steel, and also mill scale formed at high temperatures, not only do not prevent progressive corrosion, but may in fact accelerate the rate of attack.

In moist environments, the corrosion of metals is essentially of an electrochemical nature, and the following sub-sections of the lesson discuss broad theoretical principles of corrosion.

2.2.2. An Electrochemical Cell

A piece of metal, such as zinc, carries no overall electric charge. If, however, it is placed in a solution containing zinc ions, say zinc sulphate, some of the zinc metal dissolves to form zinc ions, leaving electrons on the metal. Thus a potential difference will exist between the zinc metal and the zinc sulphate solution. The potential between the metal and the solution of standard strength of its ions, is called the "standard electrode potential" for the metal. In the case of zinc, the potential has a value of $-0{\cdot}76$ V.

Similarly, a copper electrode assumes a "half-cell" potential, with respect to a standard solution containing copper ions, and this is $+0{\cdot}34$ V.

If the zinc and copper electrodes are now joined externally, as in a simple type of Daniell cell (Fig. 2.2.2) a current will flow, the zinc becoming the anode, and the copper the cathode of the cell. The zinc rod dissolves or corrodes, as Zn^{++} ions, and releases electrons, which pass via the metallic path to the copper, where they are consumed in discharging cupric ions from solution:

$Zn - 2e \rightarrow Zn^{++}$ (anodic reaction—oxidation)

$Cu^{++} + 2e^{-} \rightarrow Cu$ (cathodic reaction—reduction)

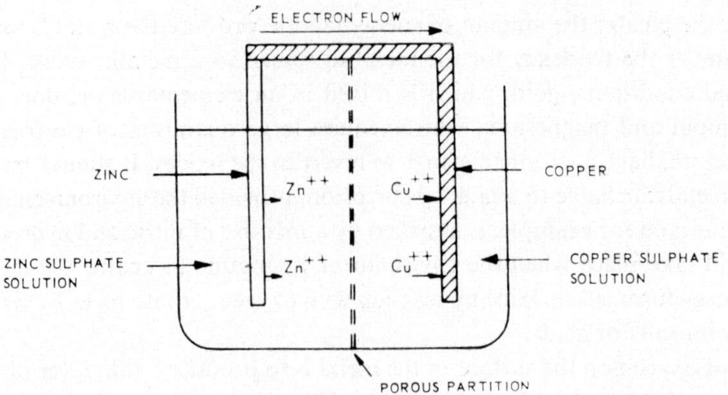

FIG. 2.2.2. Daniell cell

The rates of reaction at the anode and cathode are electrochemically equivalent, and the rate of dissolution of the zinc anode and deposition on the copper cathode are determined by Faraday's laws.

The copper electrode is the positive pole or cathode of the Daniell cell, and it has a more noble or more positive potential than the zinc anode (negative electrode) of the cell. In such a current-generating cell, the direction of flow of the "positive" current in the metallic path is from the copper to the zinc (which corresponds to the flow of electrons from the zinc to the copper); within the electrolyte, however, the "positive" current flows from the zinc to the copper, the current being carried by the appropriate ions in solution.

Some confusion arises when the operation of a current-producing cell is compared with that of a current-consuming cell in which the electrodes are connected to some outside source of potential. In the latter case the electrode joined to the positive terminal of the source of potential is the anode, and that to the negative terminal is the cathode. However, in each type of cell, "positive" current flows from the anode to the cathode through the electrolyte, and so "cations" (positively charged) move towards the cathode, and "anions" (negatively charged) towards the anode.

2.2.3. A Simple Corrosion Cell

If the solutions of zinc sulphate and copper sulphate in the Daniell cell are replaced by a solution of sodium chloride, a current will still flow, zinc ions again passing into solution, and releasing electrons which are consumed at the copper electrode in a cathodic process. (In corrosion processes, it is convenient to think of an anodic process as an electron-producing one, whereas a cathodic process consumes electrons.) The cathodic process will not, however, be the same as in the Daniell cell, as the concentration of copper ions in solution is negligible. The only positive ions or cations the solution contains are sodium ions and hydrogen ions from the water. As corrosion proceeds, a third cation—zinc ions—will appear in the solution, from corrosion of the zinc anode. Of the cations, the sodium cannot be deposited on the cathode because this would require a potential far more negative than that which even the zinc attains in this cell. The cathodic deposition of zinc ions put into

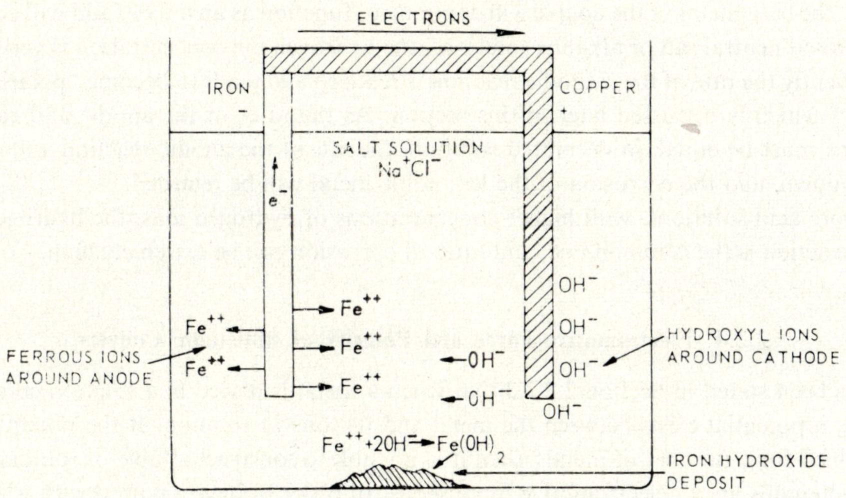

FIG. 2.2.3. A simple corrosion cell

solution at the anode would imply that the cell was in equilibrium, which is not the case, as corrosion of the zinc continues. Therefore, discharge of H^+ ions may be the only possible cathodic reaction.

$$\underset{\text{Hydrogen ion}}{H^+} + \underset{\text{Electron}}{e} \longrightarrow \underset{\text{Hydrogen atom}}{H}, \quad \text{followed by} \quad H+H \longrightarrow \underset{\text{Hydrogen molecule}}{H_2}$$

However, there is another possible cathodic process, namely, the reduction of oxygen (dissolved in the salt solution), to hydroxyl ions.

$$O_2+2\,H_2O+4e^- \longrightarrow 4\,OH^-$$

For corrosion occurring in approximately neutral, aerated solutions, the "oxygen absorption reaction" shown immediately above is the predominant cathodic reaction.

In Figure 2.2.3 the Daniell cell has been modified to show a simple corrosion cell. In the solution of sodium chloride are immersed a strip of copper and a strip of iron, joined with a copper wire. To the solution is added an indicator solution of phenolphthalein and potassium ferricyanide. With hydroxyl ions a pink colour will be given with phenolphthalein, while ferrous ions will give a blue coloration with ferricyanide.

The solution adjacent to the copper cathode becomes pink (OH^- is produced) and the solution near the anode turns blue, Fe^{++} ions in solution producing a blue colour with the ferricyanide. Both the anodic and cathodic reaction products are soluble but the ferrous and hydroxyl ions diffuse towards one another and form a precipitate of ferrous hydroxide somewhere between the anode and cathode of the cell. As the corrosion product is not formed in contact with either electrode, there is no tendency for either the anodic or cathodic process to be stifled and corrosion of the iron continues.

If the solution in the simple bimetallic couple cell is deaerated and the atmosphere above the cell contains no oxygen, then the hydrogen evolution reaction will become the cathodic process, the base metal of the couple will continue to function as an anode, and will corrode. In deaerated neutral salt or alkaline solutions, the hydrogen ion concentration is very small, consequently the rate of the cathodic reaction is reduced and tends to become "polarised"—an effect which is discussed later in this section. As the rates of the anodic and cathodic reactions must be equal, in deaerated solutions the rate of the anodic reaction will also be slowed down, and the corrosion of the less noble metal will be retarded.

In more acid solutions, with higher concentrations of hydrogen ions, the hydrogen evolution reaction is the common one, and rates of corrosion can be extremely high.

2.2.4. **Electromotive Force and Polarity of Bimetallic Couples**

It has been stated in Section 2.2.2.3 that when a metal is placed in a solution containing its ions, a potential exists between the metal and its ions in solution. If the potentials are determined for a number of metals, then it is possible to construct a table of standard electrode potentials, or an electromotive force series. In order to have a datum with which the electrode potentials of metals may be compared, the standard electrode potential of hydrogen is taken as zero. (The standard hydrogen electrode consists of a platinum electrode in acid solution containing 1 g of hydrogen ions per litre, its surface being coated with hydrogen bubbles.) From the results obtained, an electromotive force series can be drawn up as in Table 4.

The reactive or base metals which are not readily prepared in the free state owing to their strong tendency to form ions, are at the negative or "base" end of the table, for example, sodium and potassium. Conversely, the noble metals such as gold and platinum, are at the positive or "noble" end of the series.

It may be expected that if two metals in electrical contact are immersed in an electrolyte containing the metal ions, it should be possible to predict from the e.m.f. series, both the e.m.f. of the resulting cell, and also which of the two metals would function as the noble, cathodic member, and which as the base or anodic member of the cell. For the Daniell cell, it can be deduced correctly that the potential of the cell is:

$$\underset{Cu^{++}}{+0{\cdot}34\ \text{V}} - (\underset{Zn^{++}}{-0{\cdot}76\ \text{V}}) = +1{\cdot}10\ \text{V}$$

the algebraic difference of potential between the Cu^{++}/Cu electrode with respect to the hydrogen electrode, and the Zn^{++}/Zn electrode to the hydrogen electrode. It can also be predicted that the zinc will corrode and copper will be deposited at the cathode.

If the concentration of the cupric ions is reduced, the potential of the copper electrode moves in a base direction, and approaches that of the zinc electrode, the shift in potential being 29 mV for every tenfold decrease in the strength of the cupric ions. A marked reduction of the Cu^{++}/Cu potential may be obtained by addition of potassium cyanide to the copper sulphate solution to complex the cupric ions as copper cyanide, which leaves relatively few free copper ions in solution. In this case the potential of the Cu^{++}/Cu electrode is depressed such that it becomes the anode of the Daniell cell.

If the zinc/copper couple were immersed in sodium chloride solution the solution would initially be free from copper and zinc ions, and the e.m.f. series would be of little use in

TABLE 4

STANDARD ELECTRODE POTENTIALS

Element	Electrode reactions	Standard electrode potential, (E_0) Volts at 25°C, on hydrogen scale
	Noble End	
Gold	$Au = Au^{+} + e^{-}$	+1·68
Gold	$Au = Au^{+++} + 3e^{-}$	+1·42
Platinum	$Pt = Pt^{++} + 2e^{-}$	+1·20
Mercury	$Hg = Hg^{++} + 2e^{-}$	+0·854
Silver	$Ag = Ag^{+} + e^{-}$	+0·800
Copper	$Cu = Cu^{+} + e^{-}$	+0·522
Copper	$Cu = Cu^{++} + 2e^{-}$	+0·345
Hydrogen	$H_2 = 2H^{+} + 2e^{-}$	0·000
Lead	$Pb = Pb^{++} + 2e^{-}$	−0·126
Tin	$Sn = Sn^{++} + 2e^{-}$	−0·136
Nickel	$Ni = Ni^{++} + 2e^{-}$	−0·250
Cobalt	$Co = Co^{++} + 2e^{-}$	−0·277
Iron	$Fe = Fe^{++} + 2e^{-}$	−0·440
Zinc	$Zn = Zn^{++} + 2e^{-}$	−0·762
Manganese	$Mn = Mn^{++} + 2e^{-}$	−1·05
Titanium	$Ti = Ti^{+++} + 3e^{-}$	−1·63
Aluminium	$Al = Al^{+++} + 3e^{-}$	−1·67
Magnesium	$Mg = Mg^{++} + 2e^{-}$	−2·34
Sodium	$Na = Na^{+} + e^{-}$	−2·712
Calcium	$Ca = Ca^{++} + 2e^{-}$	−2·87
Potassium	$K = K^{+} + e^{-}$	−2·922
	Base End	

TABLE 5

GALVANIC SERIES OF METALS AND ALLOYS IN SEA WATER

Noble End

Titanium

Monel (67% Ni, 30% Cu, 1·2% Fe)

* Passive stainless steel:

- 13 Cr, 1 Ni
- 17 Cr
- 18 Cr, 8 Ni
- 18 Cr, 8 Ni, 3 Mo

Silver

Inconel (80% Ni, 13% Cu, 6·5% Fe)

Nickel

Copper/Nickel:

- 70/30, high iron
- 70/30, low iron
- 90/10

Aluminium bronze

Copper

Alpha brass (70% Cu, 30% Zn)

Aluminium brass

Muntz metal (60% Cu, 40% Zinc)

Tin

Lead

* Active stainless steel:

- 18 Cr, 3 Ni, 3 Mo
- 18 Cr, 8 Ni
- 17 Cr
- 13 Cr, 1 Ni

Cast iron

Mild steel

Aluminium

Zinc

Magnesium

Base End

*Passive stainless steel is covered with an oxide film and resists corrosion. Active stainless steel is of the same composition, but the oxide film has been destroyed, and the alloy behaves in a much less noble manner. Obviously, in specifying stainless steel in a sea water environment with the metals listed, careful consideration must be given to the possibility of oxide film damage or removal.

indicating the potential difference available to drive current round the circuit, although the series indicates correctly, that copper would again be the cathode of the cell. Sometimes, however, the e.m.f. series fails to predict the polarity of the electrodes if one of the metals is covered with a thin stable oxide film. For a zinc/aluminium couple, Table 4 predicts that zinc should be the cathode, but in fact the aluminium with a stable oxide film acts as the cathode, and the zinc the anode.

2.2.5. Galvanic Corrosion

In Section 2.2.4 a simple corrosion cell of iron and copper in sodium chloride was discussed. The metallic iron in this cell is converted by corrosion into oxides resembling the ores from which the metal was first extracted. The corrosion of the less noble member of a pair of metals which are joined together is called "galvanic" corrosion or "dissimilar metal" corrosion. The e.m.f. series of electrode potentials is not a reliable guide in indicating relative polarities of components of metallic couples, other than in solutions of their own ions, at a standard concentration.

For practical purposes, it is usual to take pairs of metals and alloys, and determine by observation which is anodic and which is cathodic in the solution under consideration. From the results a table can be drawn up, giving a "galvanic series" for the particular environment.

Table 5 shows a galvanic series for metals in sea water. The electrode potentials are not included, as they are dependent on the particular conditions to which the metals are exposed.

The effects of galvanic corrosion are often serious, especially in estuary and sea water power stations, and this aspect of corrosion is discussed in Chapter 3.

Comparison of Tables 4 and 5 shows that the order of the metals in the two series is not the same. Aluminium and nickel are less base in sea water than is indicated by the e.m.f. series. Titanium, which is at the base end of the e.m.f. series, behaves like a noble metal.

2.2.6. The Effect of Polarisation on Currents Flowing in Corrosion Cells

The difference in potential between the anode and cathode in a corrosion cell is the driving force of the corrosion process and the rate of corrosion will be determined by the current flowing round the circuit. Any factor that increases the resistance of the circuit or decreases the potential difference will lead to a decrease in the current flowing, and, therefore, a decrease in corrosion.

If a piece of mild steel is placed in hydrochloric acid solution it bubbles vigorously as hydrogen is evolved. Very often fractures in mill-scale provide the anodes; Figure 2.2.6A represents one such cell greatly enlarged. The ferrous ions are leaving the anode, and electrons are moving through the metal to the cathode where they neutralise hydrogen ions, enabling hydrogen atoms to be formed. Electrons are, therefore, continuously consumed at the cathode areas, and so corrosion of the anode proceeds rapidly.

If the steel was initially placed in a neutral electrolyte, such as salt solution, the hydrogen bubbles are not rapidly produced and removed as in acid. The slowly-forming bubbles

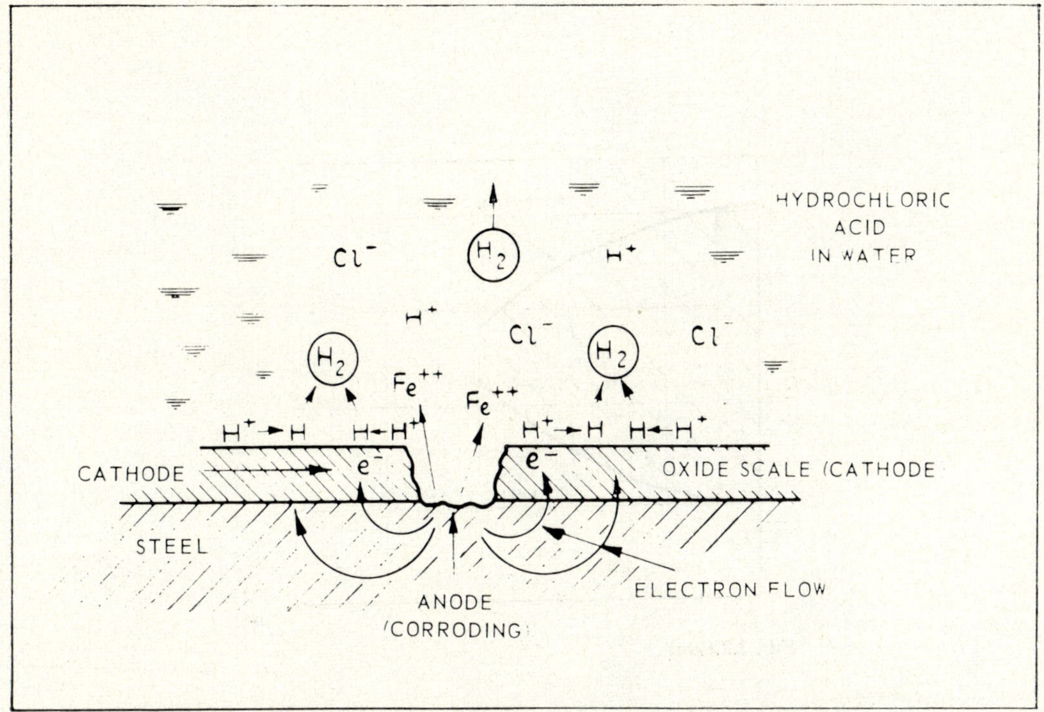

FIG. 2.2.6A. Corrosion of steel in hydrochloric acid at a break in mill-scale

coat the cathode surface, thus checking the flow of ions to and from the surface and limiting the corrosion current, hence the rate of corrosion is slower. This effect is known as "cathodic polarisation", and is a manifestation of the more general phenomena of polarisation in practical electrochemical cells.

It has been stated in Section 2.2.4 that the corrosion rate is greater in solutions containing dissolved oxygen, the oxygen being reduced to hydroxyl ions and consuming electrons:

$$O_2+2\ H_2O+4\ e^- \longrightarrow 4\ (OH)^-$$

The removal of the electrons from the cathode surface effectively prevents the accumulation of hydrogen gas at the cathode surfaces. Thus oxygen acts in this case as a "depolariser", and enables the corrosion to proceed.

If corrosion products accumulate at the anode and hydrogen at the cathode, this results in polarisation occurring with a drift in anode and cathodic potentials towards each other (Fig. 2.2.6B(a)). The anode assumes an increasingly negative potential, and the cathode an increasingly positive one. In a corrosion cell, polarisation moves the potentials closer together, and the driving force of the reaction is diminished.

The degree of polarisation of an electrode is determined by the current density rather than by the total current supplied or drawn from it, and is also dependent on the metals used for cathode and anode. Figure 2.2.6B(b) shows that whilst the cathodic polarisation for metal (A) and metal (B) increase with the increase in current density, the effect is

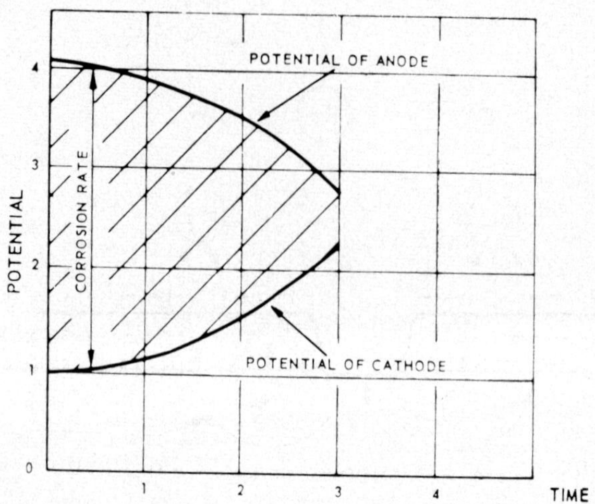

FIG. 2.2.6B(a)

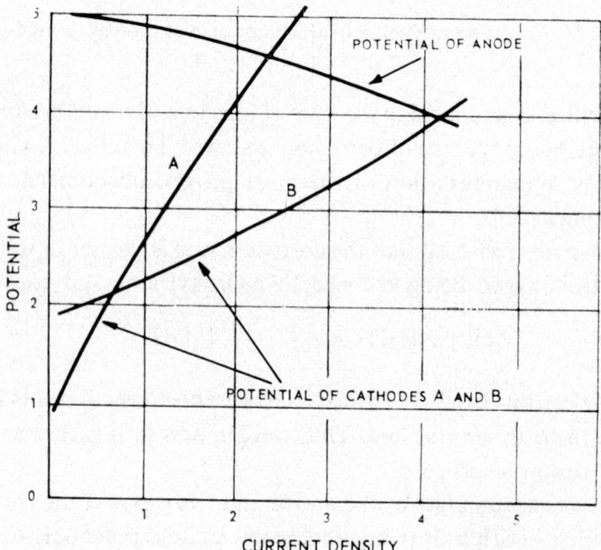

FIG. 2.2.6B(b)

FIGS. 2.2.6B. Corrosion rate/polarisation relationship

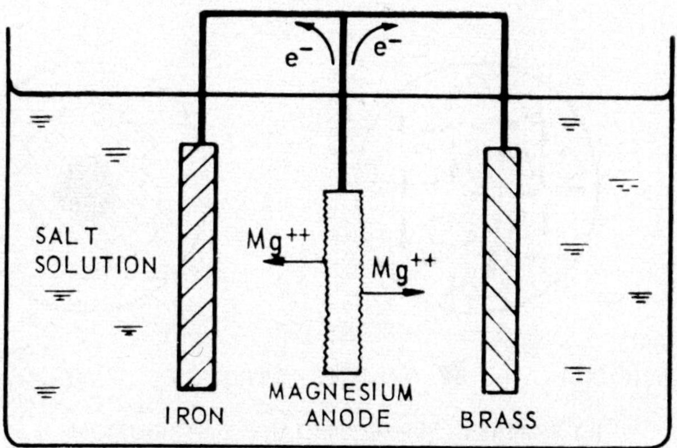

FIG. 2.2.6C. Cathodic protection of iron/brass couple

greater for metal (A). Therefore, the extent of corrosion of the anode is less with a cathode of metal (A) than for a cathode of metal (B), and is determined by the current density corresponding to the point of intersection of the cathode and anode polarisation curves.

A practical use of the matter discussed in this section is the "cathodic protection" of metals by wasting or sacrificial anodes, or by means of an impressed current.

Consider the couple iron/brass, found in power station condensers with sea water as the electrolyte. From the galvanic series Table 5, the iron will corrode. If, however, an auxiliary anode capable of readily providing a supply of electrons is joined to the couple as shown in Figure 2.2.6C, the corrosion of the iron will be much reduced. The sacrificial auxiliary electrode (zinc or magnesium) will be lower in the galvanic series than the iron and will corrode.

In practice, the most satisfactory form of cathodic protection is obtained by using an external source of direct current, in place of the sacrificial anode, with the negative pole of the d.c. source connected to the metal to be protected, and the positive pole to an electrode (usually platinized titanium), which is immersed in the electrolyte. This subject is discussed further in Chapter 3.

2.2.7. **The Corrosion of Single Metals**

In the previous sections we have discussed corrosion in terms of dissimilar metals. It is obvious, however, that it is not necessary for dissimilar metals to be in contact for corrosion to occur. The most familiar manifestation of single metal corrosion is the rusting of iron and steel objects in the atmosphere or under water.

It can be shown by experiment and measurement that this corrosion is electrochemical, and this being so anodic and cathodic areas must be present on the single metal. Earlier we showed an experiment in which ferricyanide/phenolphthalein indicator (ferroxyl) was used to follow the reactions in a bimetallic corrosion cell; this indicator solution can be used to show the presence of anodic and cathodic areas on a single metal.

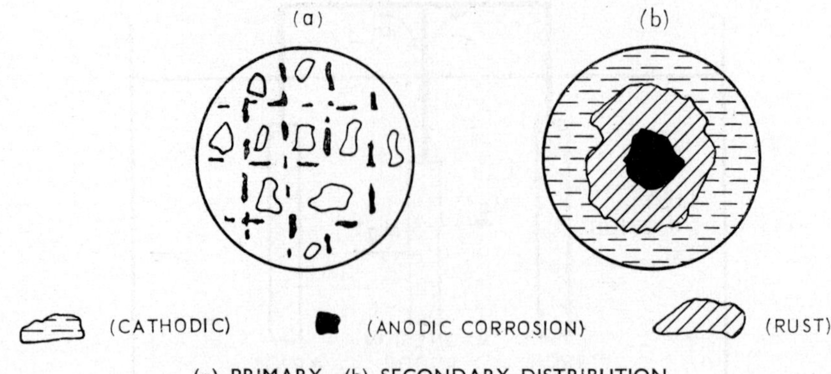

FIG. 2.2.7. Corrosion by salt solution on abraded steel

If a drop of sodium chloride solution containing a little ferroxyl indicator and saturated with air is placed on an abraded steel plate, small pink and blue spots appear over the whole drop area (Fig. 2.2.7(a)). The blue (anodic) points tend to lie along the abrasion lines, and this is the primary distribution and persists until the oxygen originally dissolved in the liquid begins to be exhausted by being consumed in the cathodic process. Replenishment of oxygen in solution from the air is most rapid at the edges where the drop is thinnest, and much slower at the centre where the drop is thick. The cathodic reaction will become predominant in the periphery of the drop, which will become pink, and if any anodic spots develop in this area the hydroxyl ion concentration precipitates the ferrous salts as fast as they pass into solution and these anodic areas will be sealed. In the centre of the drop the lower oxygen concentration will ensure that the amount of hydroxyl ion produced will be small and the anodic reaction will predominate. This is the secondary distribution shown in Figure 2.2.7(b), the centre of the drop becoming blue, the outer rim pink, and at the junction of the two zones, a membrane of rust is formed. If the salt solution is supersaturated with oxygen, the primary distribution will persist for an abnormally long time. Conversely, if the original drop of salt solution was deaerated the secondary distribution would be observed from the start.

If the experiment is allowed to continue for some time and the plate is washed off, it will be found that the metal surface under the centre section of the drop will be corroded, this being the anode, whilst the metal under the outer ring (cathodic) will be unattacked.

In the above experiment we have produced anodic and cathodic areas by abrading the oxide film, but even without this treatment, corrosion would have occurred in time. All metals after exposure to air carry oxide films, but at some points the film may be cracked, porous or thin, and at these points it will be easier for metal ions to leave the metal lattice, than at areas where the film is intact. The mechanism is similar to that depicted in Figure 2.2.6A, except that hydroxyl ions are formed at the cathode and hydrogen is not evolved. The formation of anodic and cathodic areas is also promoted by inclusions in the metal, lack of homogeneity, surface imperfections, orientation of the grain structure, localised stresses and variations in environment.

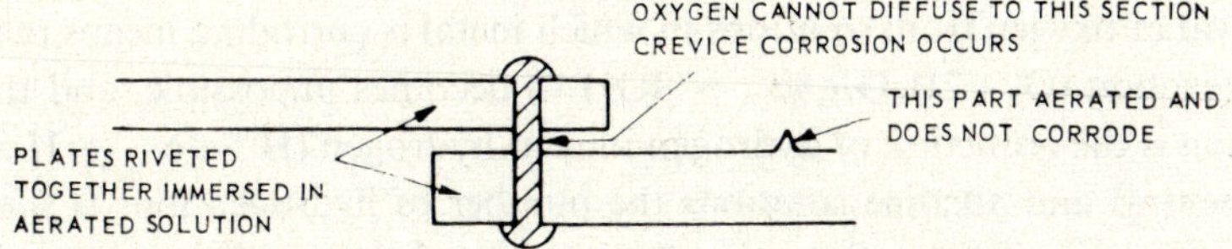

FIG. 2.2.8. Crevice corrosion taking place between riveted plates

2.2.8. **Differential Aeration**

From the previous section the student may deduce that the parts of a metal surface that are in contact with liquid of the higher oxygen concentration will become cathodic and protected from corrosion. Areas where the dissolved oxygen concentration is low will become anodic and will corrode. This is known as "differential aeration", and was first described by U.R. Evans, a pioneer in the scientific investigation of corrosion. The fact that currents flow between a well-aerated electrode and a less aerated one, can readily be demonstrated.

Corrosion due to differential aeration is a widespread and damaging process, which may occur in many situations, for example, in unprotected structures, in boiler tubes and in water tanks. It is also of practical importance in the corrosion of condenser tubes or other situations where deposits can impede access to oxygen beneath them and lead to pitting and failure. A classic example of this type of corrosion is shown in Figure 2.2.8. In this case two plates are riveted together, the joint being in contact with liquid. At the riveted joint oxygen replenishment is difficult and corrosion will take place in the crevice.

2.2.9. **Factors Influencing the Corrosion Rate**

Anodic and cathodic reactions proceed in step with each other, but the overall rate of electrochemical corrosion may be governed by either the anodic or the cathodic reaction rate.

For many base metals such as iron in aqueous solution, the rate of the cathodic reaction determines the rate of corrosion of the metal and the reaction is under cathodic control. If the supply of oxygen to the cathode is increased, the rate of cathodic reaction is raised and the anodic corrosion reaction correspondingly speeds up. A similar effect is obtained if the area of the cathode is increased, the anodic area remaining constant. If the cathodic area is constant and the area of the anode increased, the total amount of corrosion would be about the same, but the amount of corrosion per unit area of the larger anodic area would be decreased, so the intensity of corrosion would be reduced.

If the situation arises where large cathodic areas are joined to small anodic areas in an electrolyte, attack on the anodes will be intense. Thus, riveting brass or copper plates together with iron rivets will lead to rapid failure of the rivets; similarly, if steel retaining a film of "mill scale" is immersed in water or exposed to a humid atmosphere, rapid rusting or corrosion occurs. The mill scale usually has a number of defects through its thickness and a combination of large cathodes (mill scale) and small anodes (gaps in the scale) is produced, resulting in severe pitting where the mill scale is defective.

The removal of oxygen from solutions in which metal is corroding means that the oxygen absorption reaction ($O_2+2H_2O+4e^- \rightarrow 4OH^-$) becomes impossible, and the only alternative reaction is the reduction of hydrogen ions to hydrogen ($H^++e^- \rightarrow H$ and $H+H \rightarrow H_2$). In neutral and alkaline solutions the number of hydrogen ions is small, therefore the cathodic process will take place at a slow rate, and the anodic process will be equally slow. Thus deaeration of solutions will effectively reduce the corrosion rate to very low values, and this fact is the theoretical reason for deaeration of boiler feed water.

Summing up, it is evident that by having consideration for the factors we have outlined, in the design and operation of power stations, damage caused by corrosion can be largely prevented. However, unexplained forms of corrosion sometimes occur which are difficult to control, and the internal corrosion of boilers in operation is currently the most important of these. This particular problem is frequently referred to in the following sections and is discussed in some detail in Section 2.4.

2.3. FEED AND BOILER WATER

2.3.1. Introduction

The overall objective in any regime of control in power plant systems is to maintain operation at the best possible levels of availability, economy and efficiency.

To attain this objective, chemical control of the water and steam system is directed to:

(a) Prevention of corrosion in the boiler, steam and feed systems.
(b) Prevention of scale and deposit formation on heating surfaces.
(c) Maintenance of a high level of steam purity.

This section reviews the present position in relation to boiler plant, mainly at and above 900 lb/in^2, in the field of water treatment. The subject is not a static one, and this should be appreciated in reading this chapter.

2.3.2. Historical Survey

From the early days of boiler operation it was realised that many natural waters were unsuitable for direct use in boiler plant. If natural waters were used, deposits and hard scales were formed on the metal surfaces, resulting in loss of heat transfer, and eventually to metal failure through overheating. The problem was intensified with the introduction of the water-tube boiler, with its large number of relatively small bore tubes.

The first major breakthrough in removing scale-forming constituents from boiler feed or make-up water, was the introduction at the end of the 19th century of a commercial plant, based on Clark's lime-soda process of 1841. Although the lime-soda softening process removes the bulk of the scale-forming constituents, the residual hardness if fed to a boiler, may eventually result in tube blockage and failure. In many cases, however, the plant would be operated on lime-soda softened water, and taken out of service for cleaning periodically, but with loss of availability.

Experience showed, however, that the addition of sodium carbonate to the boiler water could be used to control scale deposition. The residual scale-forming salts were precipitated in the boiler as a sludge, and the solids could be removed by blowdown.

In parallel with the improvement of boiler feed water from the viewpoint of scales and deposits, it became apparent that corrosion of the metal surfaces could also occur. It was found that below a boiler pressure of about 200 lb/in^2 (200°C, 392°F) corrosion could be controlled by maintaining the alkalinity of the boiler water at 10–15% of the total dissolved solids. Even with this treatment, above 200 lb/in^2, severe local corrosion of boiler surfaces often occurred. The recognition that dissolved oxygen could cause severe local corrosion, led in the 1920's to the introduction of the closed feed system. In this system the open hot wells and cold wells disappeared, and the condenser was designed to act as a deaerator. A good design of condenser and closed feed system reduced the dissolved oxygen in the feed water to about 0·03 ppm or less.

At this stage evaporators were incorporated in the feed train, to produce a distillate for make-up purposes which contained practically no scale-forming constituents. The feed to the evaporators was given a treatment dependent on the quality of the raw water supply. With a suitable supply, internal treatment of the evaporator, rather than external softening could be used. It is interesting to note than in the late 1920's, attempts were made to use sea water in land power stations, as the feed to evaporators, but the problems of scaling, corrosion and carryover led to the abandonment of this type of feed.

Thus, as the boiler and feed system design stood in the 1920's, the problem of scale deposition was greatly reduced. Internal conditioning of the boiler water, from the scale deposition viewpoint, was based on precipitating solids resulting from condenser leakage and evaporator carryover, by added chemical reagents.

The first theoretical treatment of scale formation in boiler plant, was published by R. E. Hall in 1923–4. Hall postulated that the concept of the "solubility product" is capable of predicting whether a scale or sludge will be produced in a solution which is supersaturated with a salt. He used the solubility product to calculate how much carbonate must be added to a water containing, say, calcium sulphate, to ensure that concentration by evaporation will render it supersaturated with respect to calcium carbonate (sludge forming), before it becomes supersaturated with respect to calcium sulphate (scale forming).

About this time Hall also introduced the use of sodium phosphates into boiler water treatment. This was another step forward in water trèatment, as phosphate has considerable advantages over sodium carbonate, for boiler pressures above 200 lb/in^2.

At the end of the 1920's, boiler pressures began to increase from the general level of 200–400 lb/in^2, and it was considered that the levels of dissolved oxygen obtainable from a condenser were not stringent enough for the higher pressures. In the 1930–40's, the deaerator made its appearance in the feed train, although many later stations in the 650 lb/in^2 range operated satisfactorily on a baseload basis, without deaerators. Where additional security against corrosion due to dissolved oxygen was required, sodium sulphite was introduced as a chemical deoxygenator.

With riveted drum boilers, attention had to be paid to the control of boiler water composition, with respect to caustic cracking of seams, rivets and rivet holes. The cause of the attack is the excessive concentration of caustic soda occurring in a leak path at a seam or rivet hole, and for many years, the recommended remedy was the maintenance of a minimum ratio of sodium sulphate concentration total alkalinity. In the early 1950's Berk introduced a more scientific method of control based on maintaining a sodium nitrate/alkalinity ratio in the boiler water.

In the 1920–30's, corrosion in boiler systems was probably mainly attributable to high levels of dissolved oxygen and carbon dioxide in the feed water. With the introduction of deaerators, feed water dissolved oxygen figures fell to low levels, but corrosion of tube surfaces still occurred, as boiler pressures increased.

In 1939, Partridge and Hall in America suggested that corrosion of boiler tube surfaces could take a form other than that of pitting by dissolved oxygen. The basis of the theory was the formation of concentrated films of caustic soda (sodium hydroxide), at the boiler tube steam generating surface, and this led to the introduction of Purcell and Whirl's co-ordinated phosphate treatment (originally devised to prevent caustic cracking in riveted drum boilers), to the high pressure welded or forged drum boilers. However, the incidence of corrosion in high pressure boilers in more recent years, has given cause to avoid this type of treatment (see Section 2.4).

In the 1950's hydrazine was introduced into Great Britain as an alternative chemical deoxygenator to sodium sulphite. Sodium sulphite increases the total dissolved solids in the boiler water, with consequent heat loss in additional blowdown, and furthermore, at high pressures it can decompose to form corrosive gases in the steam. Hydrazine has neither of these disadvantages but it is more expensive than sodium sulphite.

In the field of feed water treatment, it was realised that the alkalinity of the feed water was important in preventing the corrosion of metals in the feed system, and subsequent transport of corrosion products to the boiler. The simplest method of increasing feed water alkalinity is the addition of caustic soda, on a continuous basis. Such addition will increase the caustic soda concentration in the boiler, resulting in heat loss through blow-down necessary to reduce the boiler water total dissolved solids. As caustic soda is not volatile at the pressures considered (pre-1960) it gives no alkalinity protection to the turbine condensate. In the 1950's two organic chemicals, cyclohexylamine and morpholine were introduced which, in common with ammonia, have the property of increasing feed water alkalinity. They do not increase the total dissolved solids concentration of the boiler water, and they are volatile in steam. More recent developments in boiler and feed water treatment, and the established methods which are of current interest, are more fully discussed in the following sections.

2.3.3. **Boiler Water**

2.3.3.1. Scale Formation

In a power plant system there are two principal sources of ingress for scale-forming materials.

(a) Condenser tube failure or condenser tube-end leakage, resulting in river, sea or estuarine water entering the feed and boiler system.
(b) Contamination of the make-up system. With evaporator plant, some carry-over of water as spray or foam may occur. In modern stations where evaporators are used

to produce make-up water from river or sea, the evaporator distillate will be "polished" in mixed bed ion-exchange units, to remove any impurities carried over. Where demineralisation (ion-exchange) plants are used alone to provide make-up, they normally include automatic devices to prevent dissolved impurities gaining access to the treated water.

Contamination from either of the two sources will result in the constituents of natural waters being concentrated in the boiler, with eventual deposition of scale-forming substances.

If concentration is allowed to proceed slowly then the solubility products of the least soluble constituents in the water will be approached slowly. Thus a tightly packed and adherent scale is likely to be formed on the boiler metal. An analogy of the situation is provided by the formation of stalagmites in caves in limestone rock areas. Here the deposit is formed during many years from the slow fall of drops of water, saturated with calcium carbonate, which gradually precipitates in crystalline form, due to evaporative concentration of the water.

If the solubility product of a constituent of the boiler water can be suddenly exceeded by the addition of a chemical to precipitate the least soluble salt, then the result will be a finely divided sludge, which can be removed from the boiler by blowdown.

Other factors than the speed and completeness of precipitation have some influence on the alternative formation of scale or sludge, and these include the crystalline or other form of the precipitated particles, which may or may not thus tend to grow into an interlocked crystal mass; another factor is the temperature to which any immersed metal surface may be raised above that of the water.

2.3.3.1.1. *Phosphate treatment*. In 1927, R. E. Hall introduced the use of sodium phosphate as an alternative to sodium carbonate, in the control of scale-forming compounds in boiler water.

Trisodium phosphate dissolved in water reacts with calcium and magnesium salts to form calcium and magnesium phosphates, which precipitate out as sludge.

$$2\,Na_3PO_4 + 3\,CaSO_4 \longrightarrow 3\,Na_2SO_4 + Ca_3(PO_4)_2$$

Trisodium phosphate	Calcium sulphate	Sodium sulphate	Calcium phosphate

$$2\,Na_3PO_4 + 3\,MgSO_4 \longrightarrow 3\,Na_2SO_4 + Mg_3(PO_4)_2$$

Trisodium phosphate	Magnesium sulphate	Sodium sulphate	Magnesium phosphate

Originally specific phosphate/sulphate ratios were used, but the solubilities are influenced by other boiler water salts. The approach was abandoned in the 1930's, and a sufficient quantity of soluble phosphate ion was maintained in the boiler water to precipitate a normal influx of calcium and magnesium ions into the boiler system. In the range of boiler pressure 600–900 lb/in^2, the recommended phosphate level was 20–50 ppm by weight, as trisodium phosphate.

Aqueous solutions of trisodium phosphate and disodium hydrogen phosphate are alkaline and this aspect is discussed in Section 2.3.3.2, under the heading "Alkalinity".

Where it is required to reduce the sodium hydroxide concentration of the boiler water, sodium dihydrogen phosphate or sodium metaphosphate can be used.

$$NaH_2PO_4 + 2\,NaOH \longrightarrow Na_3PO_4 + 2\,H_2O$$

Sodium dihydrogen phosphate	Sodium hydroxide	Trisodium phosphate	Water

$$NaPO_3 + 2\,NaOH \longrightarrow Na_3PO_4 + H_2O$$

Sodium meta-phosphate	Sodium hydroxide	Trisodium phosphate	Water

It will be seen that one molecule of each of the sodium dihydrogen phosphate and the sodium metaphosphate requires two molecules of sodium hydroxide to convert them to trisodium phosphate. This reduces the alkalinity of the boiler water, and leaves the phosphate available for dealing with calcium and magnesium salts.

This treatment would be useful where feed water contamination with carbonate or hydroxide was taking place, due for example, to condenser leakage or evaporator carry-over.

The metaphosphate also has the property of "complexing" or preventing precipitation of scale-forming salts such as calcium and magnesium, and was used in earlier years in the feedwater to prevent the formation of scales in the hotter sections of the feed system and economiser.

It has been found in some boilers, generally at 900 lb/in^2 and higher pressures, that loss of trisodium phosphate concentration may occur in the boiler water, when the boiler is on load. If the boiler load is reduced, the phosphate again appears in the boiler water. This phenomenon is known as "hide-out".

Schroeder (1937) showed that the solubility of trisodium phosphate reaches a maximum at 120°C, falls fairly sharply to 220°C, and rapidly above this figure (Fig. 2.1.1.1c), Potter (1964) has calculated that at boiler pressures in the region of 1800 lb/in^2 or greater, trisodium phosphate will probably reach saturation solubility, and that hide-out will inevitably occur on load. Experience at West Thurrock (2350 lb/in^2) does not confirm this, since at this station a small reserve of phosphate, 5 ppm by weight as trisodium phosphate, has been held experimentally in the boiler water without evidence of hide-out on load.

In some cases where hide-out has been detected, the trisodium phosphate has been replaced with tripotassium phosphate. The potassium salt has a higher solubility than the sodium salt, and, therefore, will not show the same phenomenon so readily. However, the advisability of substituting potassium phosphate is in some doubt, partly because it is necessary at the same time to exclude sodium ions from the water* and this is difficult, but also because hide-out is often indicative of excessive metal temperature in the boiler tubes and the potassium salt may only serve to conceal this evidence.

On-load corrosion has occurred in many of the boilers, but not all, in which hide-out of sodium phosphate has been reported. Conversely, in some cases the absence of hide-out has not guaranteed immunity from corrosion. This aspect is discussed further in Section 2.4.

* The reason for this arises from the fact that all salts dissolved in the water are ionised, and if sodium ions also gain access from slight condenser leakage or as impurity in the potassium phosphate, then hide-out of sodium phosphate can still occur by the association of sodium ions and phosphate ions.

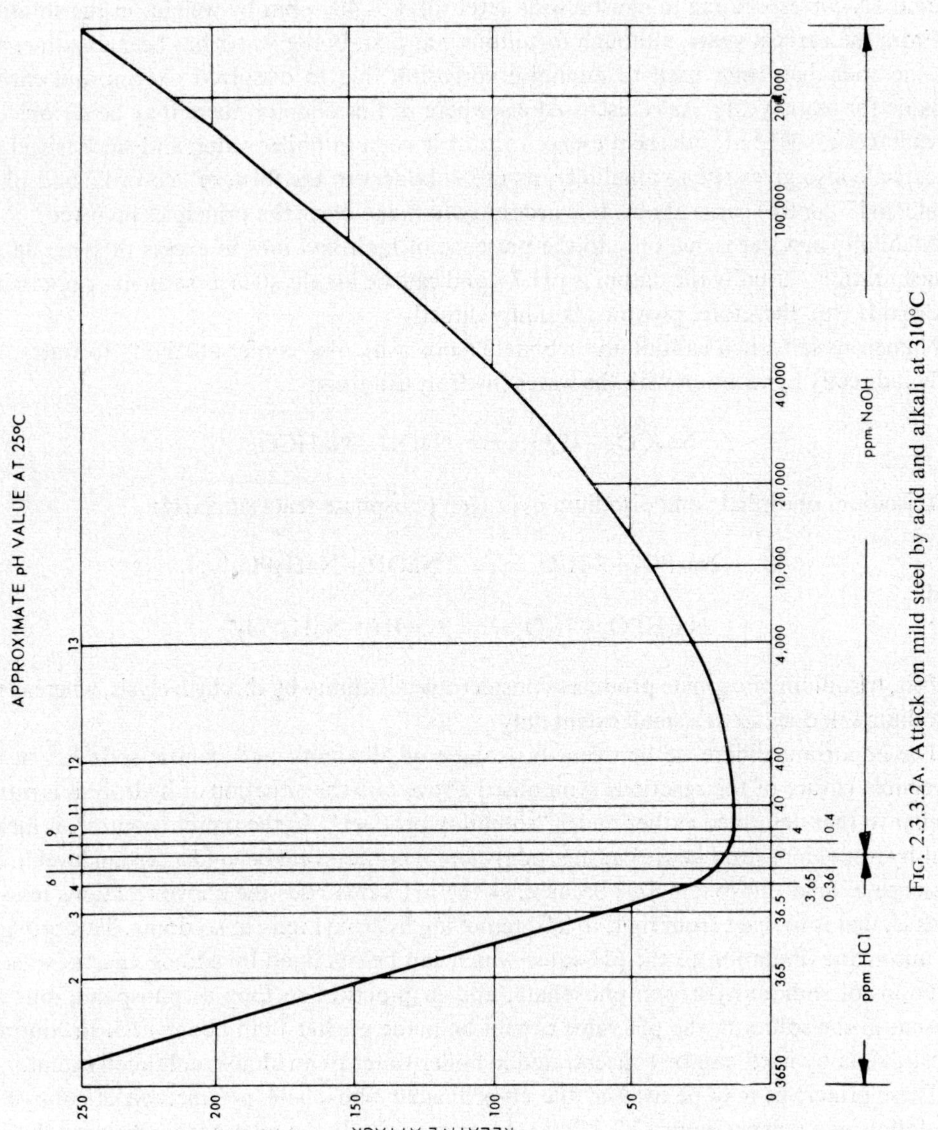

FIG. 2.3.3.2A. Attack on mild steel by acid and alkali at 310°C

2.3.3.2. ALKALINITY

The effect of acidity from hydrochloric acid and alkalinity from caustic soda, on the corrosion rate of steel at 310°C (500°F), is shown in Figure 2.3.3.2A. It is clear that the alkalinity from caustic soda, confers maximum protection to the steel at pH values (at 25°C) between 10 and 12, corresponding to caustic soda levels of 4 to 400 ppm by weight, in the solution.

From the earliest years, although fortuitously at first, boiler water has been alkaline, and caustic soda has been used to minimise corrosion due to dissolved oxygen and carbon dioxide for many years. As is discussed elsewhere in this chapter, there may be serious disadvantages associated with the presence of caustic soda in boiler water, and since trisodium phosphate also gives rise to alkalinity, its use in boilers in the form of "co-ordinated phosphate–pH" control came about. It is necessary to make clear the principles involved.

Alkalinity in water is due only to the presence of hydroxyl ions in excess of some datum concentration—usually the datum is pH 7—and caustic alkalis such as sodium or potassium hydroxide can, therefore, provide alkalinity directly.

Numerous salts such as sodium carbonate (soda ash), also confer alkalinity to water, but only indirectly by reaction with the water (hydrolysis), thus:

$$Na_2CO_3 + H_2O \rightleftharpoons NaOH + NaHCO_3{}^*$$

Trisodium phosphate and disodium hydrogen phosphate react similarly:

$$Na_3PO_4 + 2\,H_2O \rightleftharpoons 2\,NaOH + NaH_2PO_4{}^*$$

and

$$Na_2HPO_4 + H_2O \rightleftharpoons NaOH + NaH_2PO_4{}^*$$

In fact, trisodium phosphate produces considerable alkalinity by this hydrolysis, whereas the disodium salt does so to a small extent only.

The important difference between this source of alkalinity and caustic soda lies in the reversible nature of the reactions symbolised above and the situation of hydrolysis is rather similar to that described earlier under "solubility product". In the present context, it means that hydroxyl ions (and hence caustic soda) cannot concentrate beyond a certain level in the presence of acid-phosphate ions because, as the pH value rises the above reactions tend to reverse, that is to react from right to left, removing hydroxyl ions in so doing. This provides an automatic limitation to the pH-value which can be obtained by adding caustic soda to solutions of sodium hydrogen phosphate, and in practice, so long as phosphate ions are present in the solution, the pH value cannot be made greater than about 12·5. In contrast, caustic soda by itself can be concentrated in boiler water to an almost unlimited extent.

These principles may be used in the co-ordinated phosphate–pH method of control, in the following manner. Figure 2.3.3.2B(a) shows graphically the relationship between the pH value of a solution of trisodium phosphate, and the solution strength. In the area above the curve, caustic soda and trisodium phosphate are present, and, therefore, the boiler water treatment is controlled so that the measured pH value and phosphate concentration lie at

* The reactions are essentially ionic, and it should be borne in mind that the various compounds formulated (except water) are largely ionized in boiler water.

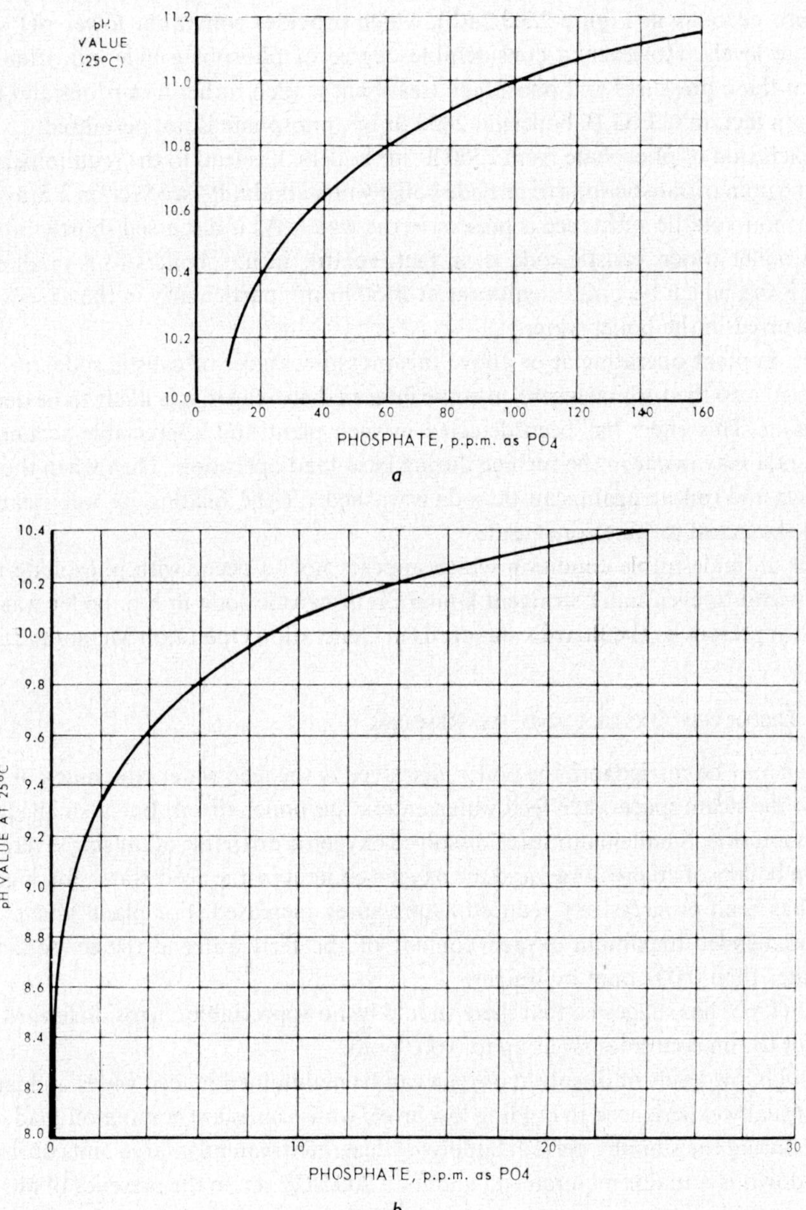

FIG. 2.3.3.2B. Co-ordinated phosphate–pH curves for medium- and high-pressure boilers

about 0·1 pH unit below the curve, in which area only trisodium phosphate and disodium hydrogen phosphate coexist. This curve is suitable for low- and medium-pressure boilers, but for pressures of about 1500 lb/in^2, or greater, the low solubility of trisodium phosphate is taken into account in Figure 2.3.3.2B(b), which provides somewhat lower pH values and phosphate levels. However, a considerable degree of phosphate hide-out often occurs in boilers at these pressures and phosphate treatment is then rather hazardous and difficult to control; in fact, in C.E.G.B. boilers at 2350 lb/in^2, phosphate is not permitted.

The exclusion of phosphate from 2350 lb/in^2 boilers, has lead to the requirement to maintain 3 to 6 ppm of caustic soda to provide boiler water alkalinity (see Section 2.5) and virtually no other non-volatile substance is present in the water. As is discussed shortly in relation to silica in boiler water, caustic soda is, in fact, volatile in h.p. boilers to a small extent (see Fig. 2.3.3.4B), which becomes significant at 2350 lb/in^2, particularly in the absence of other salts dissolved in the boiler water.

Hence, in plant operating at or above this pressure, traces of caustic soda are volatilised and carried into the turbine where, at some intermediate stages, it is likely to be deposited on the blading. This effect has been detected in such plant and appreciable accumulation of caustic soda may occur in the turbine during base-load operation. Then when the turbine is shut down and run-up again, caustic soda is washed off the blading by wet steam and appears in the initial turbine condensate.

This is an undesirable condition which appears not to occur with phosphate treatment, and may lead to even more stringent limitation of caustic soda in h.p. boiler water, than is required at present by the Board's standards in Generation Operation Memorandum No. 72.

2.3.3.3. Dissolved Oxygen and its Removal

Oxygen may be carried into the boiler, dissolved in the feed water, and much of it is liberated into the steam space when feed water enters the boiler drum, but a small quantity remains in solution. Small quantities of dissolved oxygen are capable of causing severe corrosion pitting in boilers of all pressures, and the oxygen content in the feed water for high-pressure boilers has been progressively reduced as pressures increased. For plant above 900 lb/in^2 the recommended maximum oxygen content of the feed water at the economiser inlet is not greater than 0·007 ppm by weight.

Potter (1963) has suggested that there should be no appreciable corrosion hazard in permitting plant to run continuously at up to 0·007 ppm.

Although low levels of dissolved oxygen can be maintained in feed water on load, difficulties are usually experienced in holding low levels whilst units are coming off load or starting up, and during the standby period. Supplying deaerated water for large units during start-up or shut-down is a major requirement, and for a 200 MW set, in the presence of nitrogen sealing of the vapour spaces, a total of 25,000 gallons of deaerated water has been suggested (Sherry and Gill).

The mechanism of the formation of an oxygen pit can be visualised as shown in Figure 2.3.3.3.A. Where bubble attachment occurs, a potential difference will exist on the metal surface between the oxygen-rich area under the bubble, and the surrounding areas in contact with water which will be less rich in oxygen. As a result of the current flow, between the an-

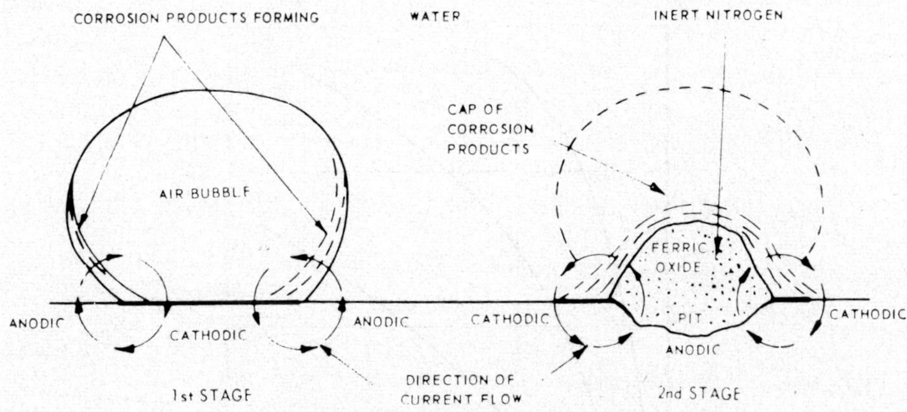

FIG. 2.3.3.3A. Formation of an oxygen pit

odic areas (low in oxygen) and the cathodic area (high in oxygen), a ring of corrosion products will build up at the anodic areas. Corrosion products will build up over the bubble by capillary attraction, isolating it from the surrounding water. As the electrolytic reaction proceeds, the cap of corrosion products around the bubble thickens, oxygen inside the bubble being used up to convert the corrosion products into a fully oxidised state leaving inert nitrogen inside the pit. When the amount of oxygen inside the bubble is less than that outside, the polarity of the cell is reversed, the area around the bubble becoming cathodic, and that beneath the bubble anodic, attack proceeding beneath the cap as long as it remains intact. The cap of corrosion products has to be sufficiently porous to allow the passage of ions, but oxygen depletion persists underneath, because the oxygen which does diffuse inwards is used up in the oxidation of corrosion products, immediately underneath the cap. Thus once a bubble of oxygen has attached itself to the metal surface, corrosion will be initiated, and will continue by the differential aeration mechanism.

Sodium sulphite. Until about 1954, in stations where physical deaeration was unable to produce a sufficiently low level of dissolved oxygen in the feed water, sodium sulphite was commonly added to the boiler water as a chemical deoxygenator. Sodium sulphite reacts with oxygen to give sodium sulphate, the reaction being rapid under operating conditions.

$$\underset{\text{Sodium sulphite}}{2\,Na_2SO_3} + \underset{\text{Oxygen}}{O_2} \longrightarrow \underset{\text{Sodium sulphate}}{2\,Na_2SO_4}$$

The sodium sulphite was normally added to the feed or boiler water on a continuous or slug dosing basis so as to produce a reserve of up to 25 ppm of sodium sulphite in the boiler water. The disadvantages of sulphite treatment have been given in Section 2.3.2, and it has now been displaced by hydrazine in most stations.

Hydrazine. In 1950, hydrazine was introduced as a chemical deoxygenator. It is a powerful reducing agent, which reacts with dissolved oxygen under boiler water conditions to pro-

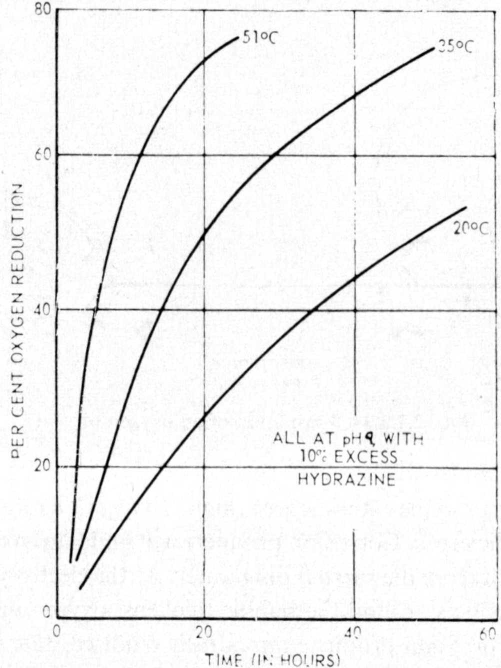

FIG. 2.3.3.3B. Effect of time and temperature on reactions between hydrazine and dissolved oxygen

duce water and nitrogen only, as follows:

$$\underset{\text{Hydrazine}}{N_2H_4}+\underset{\text{Oxygen}}{O_2} \longrightarrow \underset{\text{Water}}{2\,H_2O}+\underset{\text{Nitrogen}}{N_2}$$

Hydrazine is thus capable of removing dissolved oxygen from boiler water, without increasing the total dissolved solids concentration. Direct decomposition of hydrazine with the production of ammonia is also possible and ocurs mainly in the boiler:

$$\underset{\text{Hydrazine}}{3\,N_2H_4} \longrightarrow \underset{\text{Ammonia}}{4\,NH_3}+\underset{\text{Nitrogen}}{N_2}$$

The decomposition of hydrazine in the boiler will result in ammonia passing over with the steam to the condenser. The bulk of the ammonia in the steam will redissolve in the condensate, helping to maintain the pH value of the feed water at a satisfactory level. The reaction between hydrazine and water is dependent on time and temperature (Fig. 2.3.3.3B), and hydrazine is normally added at the extraction pump, or after the deaerator.

Experimental work suggests that the reaction between hydrazine and oxygen may be more complex than suggested by the reaction:

$$N_2H_4+O_2 \longrightarrow 2\,H_2O+N_2$$

It is known that traces of dissolved copper salts will catalyse the reaction, as is also the case in the reaction between sodium sulphite and dissolved oxygen. The speed of the hydrazine–

oxygen reaction also appears to depend on the nature of the surfaces over which the solution passes. Hydrazine in excess of dissolved oxygen in boiler water and high-pressure feed water, is capable of reducing some metal oxides to lower oxides, particularly ferric oxide to magnetite. Since ferric oxide has no protective value to boiler steel, this property of hydrazine is considered to be advantageous. Further reference to hydrazine is made in Section 2.3.4.

2.3.3.4. Silica

During the 1940's, as boiler plant design pressures were increasing, it was reported in the United States that silica deposits were being found on turbine blades. In the worst cases, the build up was such that the electrical output was reduced. Off-load techniques for cleaning the turbine blading were developed and the cause of the problem was investigated.

In a closed system consisting of water containing dissolved matter, in equilibrium with the vapour, the dissolved matter distributes itself quantitatively between the vapour and the water, in a ratio which depends on the nature of the dissolved matter and the temperature and pressure of the system. This ratio (concentration in vapour/ concentration in water) is variously called the distribution ratio or the partition coefficient, and it increases as the temperature is raised. In the case of obviously volatile substances such as ammonia or cyclohexylamine, a large proportion of the substance is present in the steam and the distribution ratio is correspondingly high; such substances are discussed later under feed water treatment.

Considering mineral substances such as silica, salt and caustic soda in boiler water, the distribution ratios at low pressure and temperature (say 400 lb/in^2, 230°C, or less) are extremely small and the amounts of these substances present in the vapour phase are too small to measure by any ordinary means. At boiler pressure of 900 lb/in^2 or above, the effect becomes significant with regard to silica (and other substances when 2500 lb/in^2 is reached) so that in high-pressure boilers silica becomes appreciably volatile in steam, or may be regarded as being soluble in it. From experimental measurements of distribution ratios by Kennedy, Table 6 below has been prepared, showing the ratio for silica in water and steam at various pressures.

Distribution ratios are affected to some extent by the presence of other dissolved matter in the boiler water, and in particular, caustic soda tends to reduce the ratio for silica. It

Table 6

Distribution Ratios of Silica between Steam and Water

Pressure, lb/in^2	Distribution ratio
1250	0·0045
1500	0·0075
1750	0·012
2000	0·02
2250	0·03
2500	0·05
2750	0·08
3000	0·16

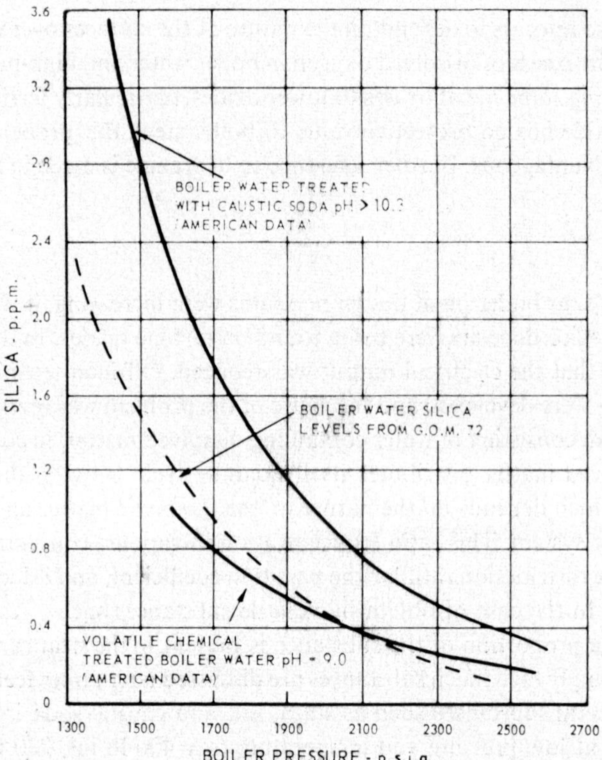

FIG. 2.3.3.4A. Permissible levels of silica in boiler water

will be apparent from Table 6 above that unless the boiler water silica content is kept to very low levels, high-pressure steam will contain more silica than is compatible with the distribution ratio corresponding to its temperature and pressure, at some point during its expansion through the turbine. At or after this point, silica will be precipitated from the steam and probably be deposited on the turbine blading, to the detriment of the performance of the machine.

It has been generally accepted that no significant deposition of silica will occur in turbines if the concentration of silica in the steam does not exceed 0·02 ppm, but this value has been chosen from experience and may need revising in due course. Figure 2.3.3.4A shows the permissible maximum concentrations of silica in boiler water at various pressures, which will limit volatile silica in steam to 0·02 ppm. It also indicates the effect of caustic soda and volatile alkalis in boiler water on the silica distribution. Figure 2.3.3.4B compares the distribution ratios of silica, caustic soda and sodium chloride at various pressures, from work reported by Styrikovich.

The solubility of silica in high-pressure steam may be increased by the presence of silica in solid form and by superheating; this would be of practical importance where there were silica-bearing deposits in superheaters or reheaters.

In h.p. plant the situation sometimes arises, usually at commissioning or after a period of shut down, when the boiler water silica concentration is higher than is compatible with the normal operating pressure. In this situation and until the silica level can be reduced, the

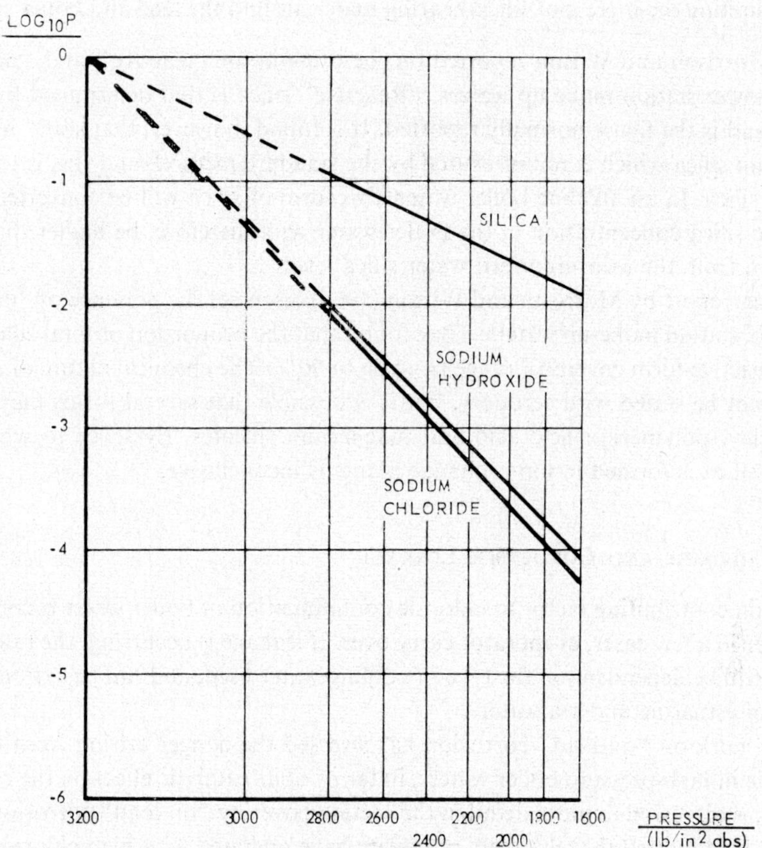

FIG. 2.3.3.4B. Distribution ratios (P) of silica, sodium hydroxide and sodium chloride, between steam and water (Styrikovich)

boiler must be operated at reduced pressure and load if turbine blade fouling is to be avoided. The permissible operating pressure for various concentrations of silica in the boiler water can be read approximately from Figure 2.3.3.4A or may be calculated from Table 6.

The concentration of silica in boiler water may be controlled in three ways:

(a) Limiting the ingress of silica to the feed water by careful control of demineralisation plant.
(b) Controlling the silica level of the boiler water by blow-down.
(c) By limiting the ingress of silica-bearing materials into the feed and boiler system.

In 1962 Morrison and Wilson reported on the presence of "reactive" and "non-reactive" silica in power station make up waters. "Reactive" silica is that determined by a standard method, and is the figure normally reported. It is found, however, that some make-up waters contain silica which is not measured by the standard method, and this is termed "non-reactive" silica. In an alkaline boiler water, this form of silica will be converted to reactive silica. The silica concentration in the boiler water will, therefore, be higher than would be anticipated from the measured feed water silica levels.

A further report by Morrison and Wilson (1965) discusses the presence of "non-reactive" silica in six station make-up waters. They found that the proportion of total silica present in the non-reactive form covered a range from 36 to 90%. The chemical nature of non-reactive silica cannot be stated with certainty, but it is possible that several forms may be present, such as clay, polymeric silicic acid and magnesium silicates. Evidence to whether "non-reactive" silica is formed in ion-exchange plants is inconclusive.

2.3.3.5. Chloride and Condenser Leakage

The main contributing factor to chloride contamination of boiler water is condenser tube leakage, or in a few cases, evaporator carry over. If leakage is occurring, the rate of chloride build-up will be dependent on the type of cooling water used, and can be extremely rapid in the case of estuarine and sea water.

Recent work on "on-load" corrosion has stressed the danger arising from the presence of chloride in high-pressure boiler waters, in terms of its catalytic effect on the rate of corrosion. This is discussed in more detail in the section covering "on-load" corrosion.

It must be admitted that numbers of boilers have operated with high chloride concentration in the past without suffering corrosion. However, in these boilers, the alkalinity levels in terms of sodium hydroxide were usually much higher than would be considered prudent for boiler operation in the range 900–2350 lb/in^2. It has been suggested by Masterson that in cases where some chloride ingress above the recommended level unavoidably occurs, it is prudent to maintain a caustic soda/sodium chloride ratio of 1·5. Thus in a high-pressure boiler with a maximum allowable concentration of 5 ppm sodium chloride, an increase to 7 ppm would entail a caustic soda level of 10·5 ppm.

A boiler water is an extremely sensitive indicator of chloride ingress because concentration in the boiler water is taking place, and it is often possible to detect a small condenser leak from the boiler chloride figures, before any rise is noted in the feed water conductivity.

For many years, a reserve of phosphate and alkalinity was carried in boiler water to precipitate calcium and magnesium arising from cooling water contamination. Depending on the size of the leak, it was often possible to run the plant until it was convenient to shut down and trace the leak. In the event of appreciable condenser leakage in high-pressure boiler plant, it is now considered that the addition of phosphate and caustic soda serves no purpose.

As an interim measure it is possible to seal small condenser leaks by adding sawdust to the cooling water, but if this expedient fails in high-pressure plant, the unit must be shut down for repair to the condenser.

With stringent limits on condenser leakage, the technique of finding leaks becomes extremely important, but at present the monitoring methods for detecting a tube failure are more sensitive than the methods of pinpointing the leaking tube. New methods of locating small leaks are being explored, but with limited success so far.

In large units double tube plates are now being installed in condensers to ensure that tube end leakage, which can be difficult to trace, does not result in cooling water contamination of the condensate. Although a majority of sources of leakage is thus made harmless, this technique does not obviate condensate contamination from tube perforations which sometimes occur along the tube, where it is not intercepted by the double tube plates.

2.3.3.6. Protection of Boilers Off Load

When a boiler is taken off load it is obviously essential to give careful consideration to the method of storage. It is in the off-load period that corrosion centres are most likely to be initiated, and these may develop into serious corrosion areas during operation.

For short periods of shut down, as in two-shift operation, it is difficult to effect a reliable chemical treatment of the boiler water for protection against oxygen ingress, but operational procedures to lessen the amount of aerated water fed to the boiler whilst taking the plant off load, can be investigated, usually with benefit. One may consider injecting additional hydrazine and a catalyst, just before an overnight shut-down, but it is difficult to ensure a uniform distribution and it would be undesirable to allow the copper or cobalt catalyst to accumulate in the boiler.

For longer periods of storage two possibilities exist, wet or dry storage. Normally, dry storage is preferred, but this may well depend on the period of storage, and also the practicability of drying out the boiler before storage. Each case has to be considered on its merits.

For short periods (a few days) of wet storage, the addition of 50 to 100 ppm hydrazine and 0·2 ppm of copper as copper sulphate may be considered adequate and the pH value should be maintained in the range 10–10·5 by caustic soda addition. If the boiler is not completely filled then a nitrogen atmosphere should be maintained above the water level. Where long term wet storage is contemplated, the advantages of sodium sulphite should be considered. Sodium sulphite catalysed with cobalt at the 0·001 ppm level reacts very rapidly with dissolved oxygen, and is a positive low-temperature oxygen remover. The reaction of oxygen and hydrazine at low temperatures is a very slow process, and complete oxygen removal cannot be guaranteed. This is of particular importance if undeaerated demineralised water has to be added to the boiler during the storage period. If sodium sulphite is used, an excess of 50 to 100 ppm should be maintained, with a pH value in the range 10·0–10·5. As with hydrazine, nitrogen sealing above the water level is required, if the boiler is not filled up.

When wet storage is practised, the addition of a pump to the circuit is desirable for slow circulation of the water throughout the system. This ensures that all sections of the boiler are receiving a supply of correctly treated water, and also enables treatment and testing of the system to be efficiently carried out.

In both methods, the storage solution will have to be drained, and the boiler refilled before returning to load.

For dry storage, the boiler is emptied whilst hot and at a pressure of 50 to 150 lb/in^2, and filled with dry nitrogen to give a slight positive pressure. Alternatively, warm dry air may be circulated through the tubes, the air being supplied through a regenerative silica-gel drier.

As a possible supplement to these methods, vapour-phase inhibitor may be used in the dry powdered state. The powder is dusted over as much of the dry surfaces as possible, using an insufflator to blow the powder down the tubes. However, it is difficult to ensure thorough distribution of the inhibitor, and the technique is not commonly used in boilers.

2.3.4. **Feed Water**

2.3.4.1. Metal Oxides

In a power plant condensing system, condensate passes through the feed train to the boiler and during its passage through the feed train, the feed water passes over a large surface area of ferrous and non-ferrous metals. (It should be noted, however, that high-pressure feed heaters now have mild steel tubes, and from 1967 onwards direct contact heaters have been used for low-pressure feed heaters.) If corrosion occurs in the feed system, ferrous and non-ferrous residues will be passed to the boiler, where they will accumulate. Iron and copper are the principal metals transferred, but nickel is also sometimes found in feed water and in boiler deposits, especially where nickel alloys are used in condensers or feed heaters. Nickel is of much significance in relation to "on-load" corrosion of boilers, and this is discussed in Section 2.4. The corrosion of metals in the feed system can be kept at a minimum by treatment of the feed water during operating periods, to maintain the pH value at not less than 8·5, and dissolved oxygen not more than 0·007 ppm by weight.

During start-up periods, it is found that the concentration of iron and copper in the feed is higher by several orders of magnitude over that obtaining in the steady operating condition. In fact, the quantities of metal oxides added during a start-up can exceed the quantity added over a large number of operating hours at full load. One method of overcoming this problem is by condensate dumping during the start-up until the concentrations of contaminating metals drop to a low level. This loss of water can cause embarrassment to make-up supplies, and also exhaust the reserve of deaerated water, resulting in dissolved oxygen being fed to the boilers. Sherry and Gill (1964) have described a method for reducing corrosion of the feed system during shut-down periods by flooding the vapour spaces with nitrogen. Condensate dumping may still be required in the subsequent start-up, but in greatly reduced quantity. Their results in a 200 MW unit are shown in Figures 2.3.4.1A and B, for copper and iron. In plant operating on a two-shift programme, nitrogen blanketing is expensive to apply.

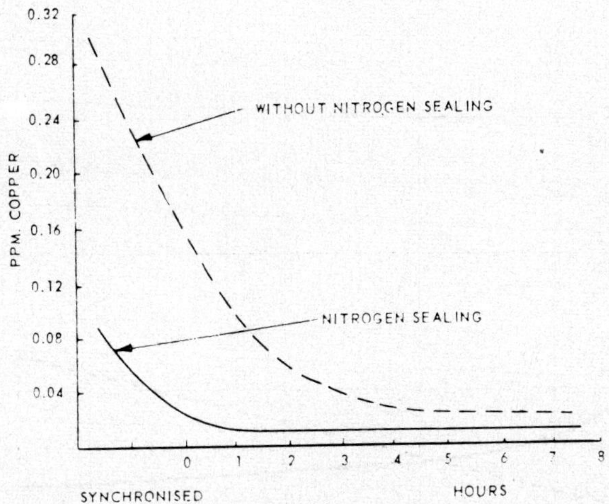

FIG. 2.3.4.1A. Copper concentration in feed water during start-up

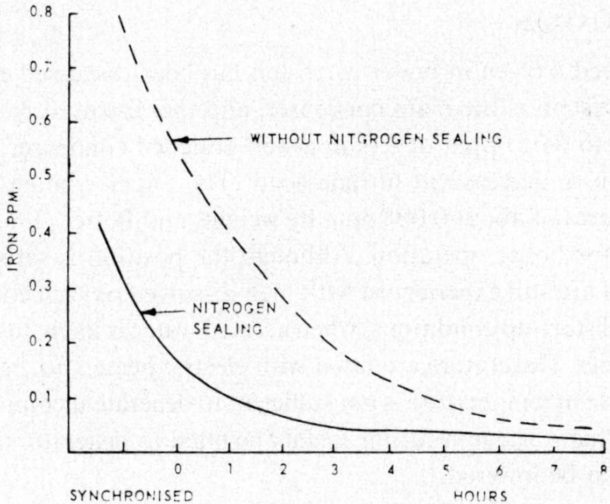

FIG. 2.3.4.1B. Iron concentration in feed water during start-up

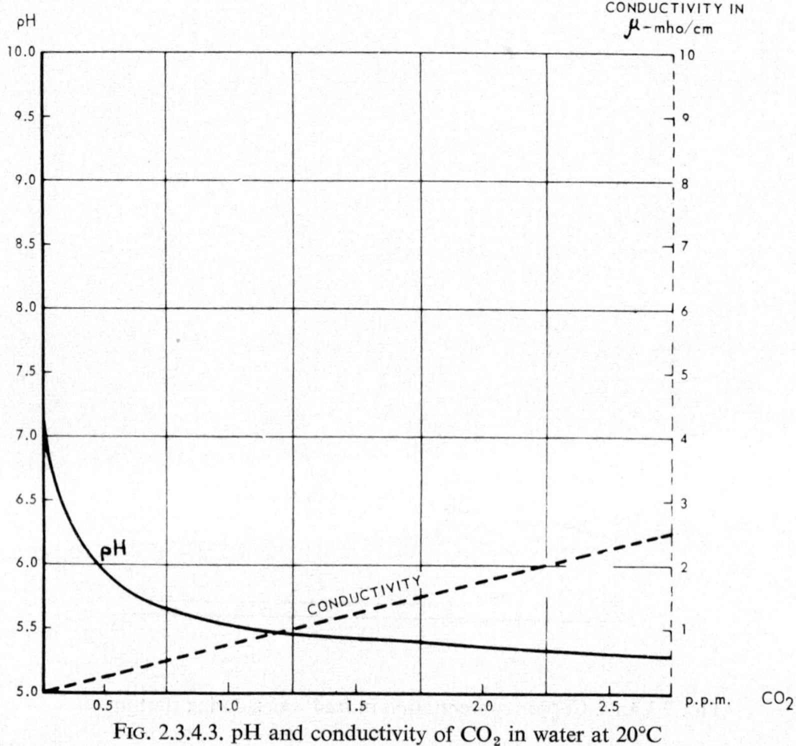

FIG. 2.3.4.3. pH and conductivity of CO_2 in water at 20°C

2.3.4.2. Dissolved Oxygen

The role of dissolved oxygen in boiler corrosion has been discussed earlier. The primary deaerator in a feed system is the main condenser, and the dissolved oxygen content of the feed can be reduced to 0·010 ppm or less in a well designed condenser, but this figure will tend to increase with a decrease in turbine load. The direct contact deaerator will give dissolved oxygen figures of about 0·005 ppm by weight, and Potter (1963) suggests that this figure is acceptable for boiler operation. Although the position is satisfactory for normal operation, difficulties are still experienced with high dissolved oxygen concentrations during boiler shut-down and start-up conditions, when aerated water is likely to be fed to the boilers in significant quantities. Deaerators are fitted with electric heaters to heat the stored water but the rate of increase in temperature is not sufficient to deaerate incoming feed water. It has been recommended that a vacuum-raising facility be fitted to deaerators, so that the boiling point of the water can be lowered.

2.3.4.3. Carbon Dioxide

Contamination of feed water with carbon dioxide can occur from condenser leakage, via carbonates and bicarbonates in the water, and from evaporator carry-over. The effluent from a mixed bed demineralisation plant should contain considerably less than 0·1 ppm by weight of CO_2. The main effect of dissolved carbon dioxide is to form carbonic acid, H_2CO_3 making the condensate acid, and very little carbon dioxide is required to cause a substantial drop in pH value, but the rate of fall in pH value decreases as the amount of carbon dioxide increases (see Fig. 2.3.4.3). In the absence of oxygen, the attack on iron by water containing carbon dioxide tends to be uniform, so that a general thinning of the metal takes place.

2.3.4.4. Treatment of Feed Water

2.3.4.4.1. *pH value*. The principal aim of feed water treatment is to reduce to the lowest possible value, the pick-up of metals by corrosion of the feed system. It is clear that this can best be achieved by adjusting the pH value of the water to give the lowest corrosion rate, and also by deaeration. Caustic soda would be a convenient reagent for pH control, but it obviously has several major disadvantages. It would greatly increase the caustic alkalinity of the boiler water, and could not be used where desuperheaters of the spray type are in operation. Spraying caustic soda into the steam would probably result in superheater and turbine blade deposition, and damage to these parts.

For boiler plant operating at 900 lb/in^2 and above, it is recommended that the iron, copper and nickel contents of the feed water should together not be greater than 0·01 ppm by weight. To achieve these figures, the pH value of the feed should not be less than 8·5 measured at 25°C, using volatile alkalis.

2.3.4.4.2. *Ammonia*. For control of the pH value, use is made of the volatile reagents ammonia and hydrazine, injected either at the extraction pump or the deaerator outlet. An increase in feed water pH value can be obtained by the addition of hydrazine. In most cases a proportion of the hydrazine is decomposed in the boiler to form ammonia which is evolved with the steam. The pH value of the feed is then largely determined by the ammonia which recirculates, and to a small extent by the hydrazine itself. In a few stations the condenser air pumps or ejectors remove much of the ammonia from the exhaust steam before it is condensed, and in such stations it is usually necessary to add ammonia as well as hydrazine.

The use of ammonia for the control of pH value has been criticised in the past, having been suspected of causing corrosion of copper-bearing alloys in the feed system, particularly where the feed water also contained more than normal amounts of dissolved oxygen. However, with very low dissolved oxygen levels present in the feed, ammonia can be successfully used, and many stations now operate with up to 1 ppm ammonia in feed water without harmful effect. Ammonia has an advantage over other volatile amines in that only 0·2 to 0·5 ppm is normally needed to raise the pH value to the required level (see Fig. 2.3.4.4.2).

2.3.4.4.3. *Hydrazine*. The property of hydrazine just discussed of providing ammonia by decomposition in the boiler, is secondary to its main purpose of absorbing residual oxygen, dissolved in the feed water. It was customary to add hydrazine to the feed water at a rate equal to twice the dissolved oxygen concentration in the water leaving the deaerator (for example, if dissolved oxygen = 0·007 ppm the hydrazine dose rate would be = 0·014 ppm), but this would not provide an adequate excess of the reagent where the oxygen content is very low and it is now recommended in Generation Operation Memorandum No. 72 (see Section 2.5) that the concentration of hydrazine at the economiser inlet should not be less than 0·02 ppm.

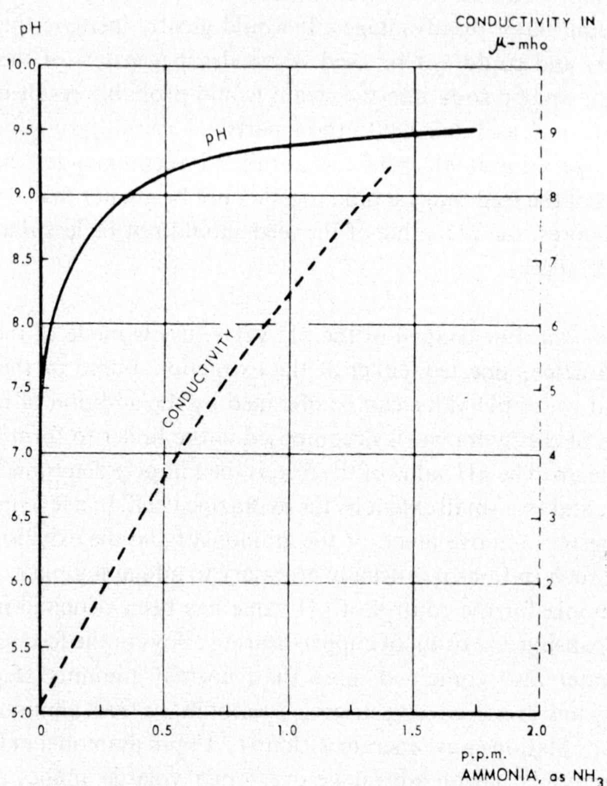

FIG. 2.3.4.4.2. pH and conductivity of ammonia in water at 20°C

Under feed water conditions, that is with low concentrations of oxygen and hydrazine, the reaction $N_2H_4+O_2 \rightarrow 2H_2O+N_2$ is fairly slow, and it is quite possible that a trace of oxygen persists in the water at entry to the economiser. This situation is not harmful, and there is evidence to indicate that an excess of hydrazine in the feed water is beneficial in that it reduces ferric oxide to the more protective magnetite, and that it may inhibit the uptake of iron and copper from the feed system. It is interesting to note that Sherry and Gill have correlated concentrations of iron and copper in the feed water at West Thurrock with hydrazine concentrations in the range 0 to 0·10 ppm hydrazine, and reported a trend suggesting that the minimum of iron plus copper occurred at about 0·05 ppm of hydrazine.

2.3.4.4.4. *Cyclohexylamine and morpholine*. For boilers operating at 900 lb/in² or less, other volatile amines may be used for pH control, and those commonly chosen are either cyclohexylamine $CH_2\cdot(CH_2)_4\cdot CH\cdot NH_2$, or morpholine $CH_2\cdot CH_2\cdot O\cdot CH_2\cdot CH_2\cdot NH$.

Ammonia is much more volatile in steam than cyclohexylamine, which in turn, is somewhat more volatile than morpholine, but all are volatilised in the boiler, and will neutralise acidity in feed water due to carbon dioxide, which may gain access to the system. By reason of the latter property, these substances are often referred to as "neutralising amines". It is argued by the proponents of cyclohexylamine and morpholine, that their reduced volatility compared with ammonia is advantageous in giving corrosion protection where condensation occurs, for example, in the turbine exhaust region, and the steam side of the feed heaters. At such points, the amines dissolve in the condensing steam more completely than does ammonia (thereby reducing losses of reagent through vents), and consequently the condensate has a higher pH value, and is, therefore, less corrosive. Cyclohexylamine and morpholine are reported to decompose at about 550°F (288°C), corresponding approximately to a boiler pressure of 1100 lb/in², but the successful use of these amines in plant operating at 1500 lb/in² has also been reported.

In choosing between cyclohexylamine and morpholine, little overall advantage attaches to either, the latter condenses out more completely, but is the weaker base of the two so that, at the point of condensation, the pH values derived from the same concentration of either amine are probably very similar. The treatment rate required for cyclohexylamine or morpholine may vary in different plant, depending mainly on the carbon dioxide content of the water, but it is generally in the range 0·5 to 3 ppm. Because of thermal decomposition referred to above, neither amine can be used economically in plant operating at 2350 lb/in², or in most plant at 1500 lb/in².

The following concentrations of the amines are required to confer a pH value of 9, in pure water free from carbon dioxide:

Ammonia	0·25 ppm
Cyclohexylamine	1 ppm
Morpholine	4 ppm

and the effect of ammonia on the pH value and conductivity of pure water, is shown in Figure 2.3.4.4.2.

2.3.4.4.5. *Filming amines.* Filming amines are wax-like organic chemicals which are almost insoluble in water but volatile in steam, and when condensed with, or dispersed in water, have the property of forming a thin water-repellant film over metal surfaces. The film thickness is of molecular dimensions and is thought to be held to the metal by physical rather than chemical bonds. For use in power stations, the material is commercially available as a concentrated aqueous dispersion, of which the main active ingredient is octadecylamine, formulated $CH_3 \cdot (CH_2)_{17} \cdot NH_2$.

In suitable conditions, feed system surfaces can be covered by a continuous film of the amine which will act as a barrier between the metal and water, and hence should greatly reduce the corrosive effect of dissolved oxygen and carbon dioxide in feed water. As may be inferred from this, the treatment rate does not depend on the concentrations of the dissolved gases in the water, and differs in this respect from the use of neutralising amines.

Thermal decomposition of filming amines in steam takes place at temperatures which probably depend on secondary factors, particularly on the presence or absence of oxygen, and from most reports they are stable at least up to 400°C in power station conditions, whilst some authors claim stability up to about 530°C.

Treatment is necessarily continuous during plant operation because the film is not very persistent under operating conditions, and the treatment rate is usually that which will give about 0·2 ppm of the amine in the condensate. For the most favourable distribution of the chemical through the condensate and vapour side of the feed system, the dispersion is injected into the turbine steam supply or the boiler drum, but it is also practicable to inject directly into the feed water.

The use of filming amines was first directed to preventing corrosion of steel in industrial plant and in low-pressure power plant feed systems, where the load factor is small. Experience in such plant in the C.E.G.B. has shown rather variable results and there have been a few undesirable side-effects which need not be discussed here, but in general the treatment in low-pressure plant is reported to be beneficial. At higher pressures (600–900 lb/in^2), octadecylamine has been tried in a few stations, including two which were experiencing corrosion of 70/30 cupro-nickel feedheater tubes, but the treatment does not appear to have shown a clear advantage.

Treatment with filming amines in high-pressure plant has also been considered, since they would appear to be of potential value in reducing the transport of iron and copper from the feed system to the boiler. The treatment is rarely applied to plant operating at or above 1500 lb/in^2, but in one C.E.G.B. station of this category, the results have been variable in regard to iron and copper concentrations in feed water.

2.3.5. **Once-through Boiler Plant**

Once-through subcritical boilers are now making their appearance in nuclear stations (Oldbury 1490 lb/in^2, Wylfa 700 lb/in^2 and Dungeness 'B' 2315 lb/in^2). For concrete reactor pressure vessels, the once-through boiler has considerable advantages over the drum boiler. It requires fewer penetrations of the reactor pressure envelope, and it does not require circulating pumps during normal operation. Figure 2.3.5 compares a once-through boiler in a nuclear station context with a nuclear drum boiler. The steam/water separator or flash vessel

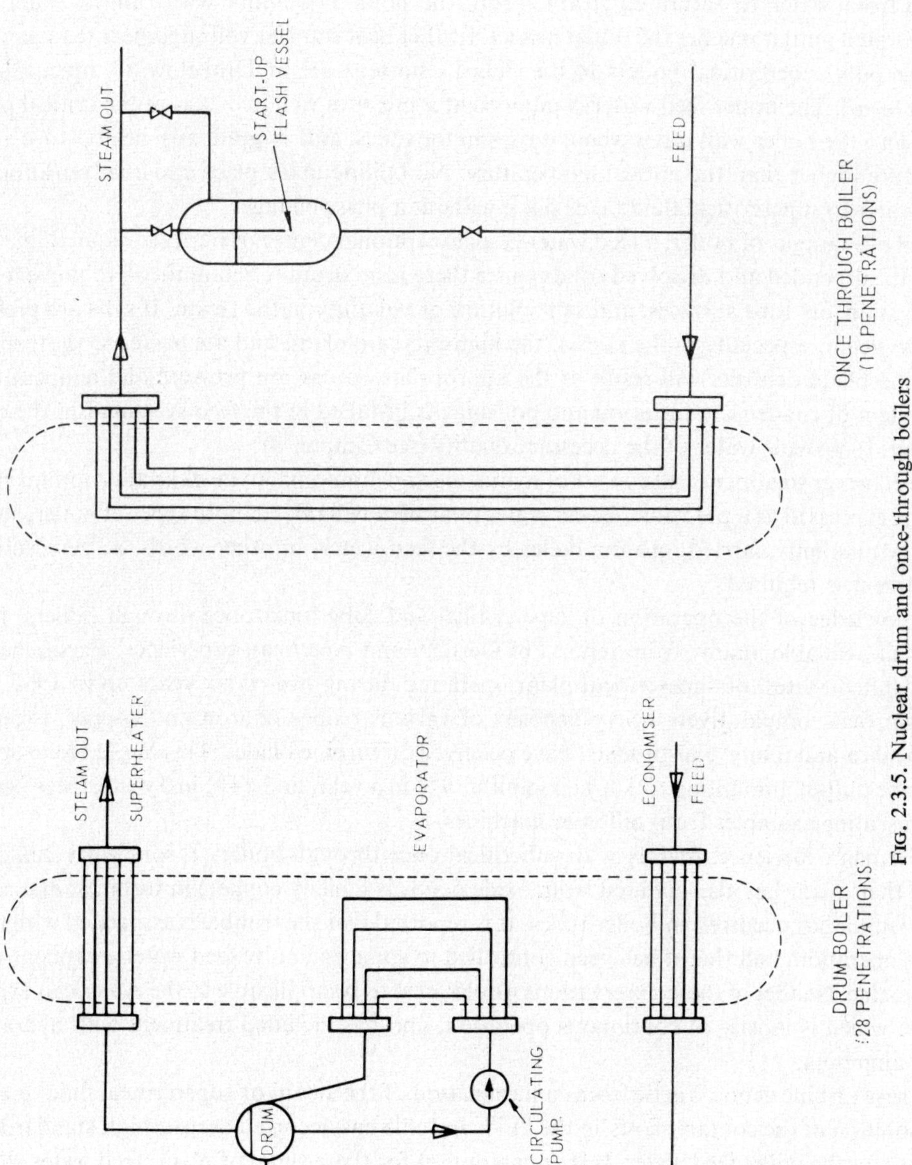

FIG. 2.3.5. Nuclear drum and once-through boilers

is used during reactor start-up to dispose of low quality steam, and is then by-passed for normal operation.

In normal operation the feed water is forced up the boiler tubes by feed pump pressure, and heated to saturation temperature. As the water continues to rise, it is heated further, converting it from water to saturated steam. From the point in the tube where all the water has evaporated until it reaches the outlet header, further heat transfer will superheat the steam.

The only supercritical boilers in the Board's stations are at Drakelow 'C' operating at 3650 lb/in^2. The boiler feed water is pumped at a pressure well in excess of the critical pressure into the boiler wall tubes where it rises in the tubes, and is gradually heated to a temperature higher than the critical temperature. No boiling takes place and the transition of the water to supercritical fluid takes place without a phase change.

In both designs of boiler, a feed water of an exceptional degree of purity is required in relation to suspended and dissolved solids (since there is no drum to retain them) to prevent salt deposition on tube surfaces, and salt solution or volatility in the steam. If salts are present in the steam, especially in the case of the high-pressure plant, and are passed to the turbine, turbine blade deposits will result at the appropriate saturation pressure and temperature. A system of condensate filtration and polishing is installed in the feed system after the condenser, to provide water of the necessary quality (see Chapter 3).

Feed water treatment will consist of hydrazine and ammonia, to give the appropriate alkalinity, expressed as a pH value. Boiler water treatment will rely, in both types of boiler, solely on the alkalinity carried into the boiler by the feed water, in other words, a "zero solids" treatment is required.

Knowledge of the operation of supercritical and subcritical once-through boilers, is at present available mainly from reports of German and American experience. Experience in the United States, of supercritical plant operating during five or six years up to 1965, has shown that comparatively heavy deposits of various oxides of iron and copper, together with silica and minor constituents, have occurred on turbine blades. This has led to losses of turbine output amounting to 1% in 6 months, 9% in a year, and 14% in 5 years, these figures representing examples from different machines.

German experience, mainly with subcritical once-through boilers, is somewhat similar in that their plant has also suffered from oxide deposits (chiefly copper) in turbines, and some deposition has occurred in boiler tubes. It is reported that the trouble is associated with two-shift operation, and that it has been controlled to some extent by feed water treatment with hydrazine. Neither of these observations would seem to be applicable to the American experience, which is mostly of continuous operation, and has included treatment with hydrazine and ammonia.

These turbine deposits arise from contamination of the steam or supercritical fluid, mainly by solution of the contaminants in the fluid, and this has occurred despite high standards of purity for the boiler feed water. It is apparent that for these types of plant, feed water standards may need to be even more stringent than Generation Operation Memorandum No. 72 currently (1965) specifies (see Section 2.5).

The vital need for feed water purity in these plants places considerable responsibility on chemical control. It is also essential that good analytical techniques are used to determine the very low concentrations of ions and suspended solids present in the feed water.

2.4. ON-LOAD CORROSION OF BOILERS

2.4.1. Occurrence

Shortly before the Second World War, when the operating pressure of large power station boilers was rising generally to about 600 lb/in^2, and in a few stations to more than twice that pressure, reports of mainly American experience indicated the occurrence of severe corrosion affecting certain regions of the boiler on the water side. This corrosion only occurred whilst the plant was steaming and since the war has become a world-wide problem of high-pressure boilers. To the C.E.G.B., on-load corrosion has cost several millions of pounds for repairs alone. The trouble very rarely occurs at operating pressures less than 600 lb/in^2 and at this pressure it is not common, whilst at higher pressures the corrosion risk appears to increase progressively.

On-load corrosion may occur at any steel/water interface where heat transfer to the boiler water takes place, thus it is most common in furnace wall tubes but it does not occur in superheaters, reheaters or non-steaming economisers. In earlier designs of boiler, where tubular attemperators were incorporated for superheat temperature control, it was not uncommon for these tubes to fail due to waterside corrosion in service.

In appearance, the corrosion may present several different forms (see Fig. 2.4.1). It appears to start as one or more localized pits which become filled with corrosion products and a considerable area of one or more tubes may be affected by the coalescence of many such pits, commonly culminating in the perforation of a tube. The area of damage varies widely; at one extreme it may be confined to the size of a penny in only one tube, and at the other extreme an area of furnace tubes, many feet in diameter, may be affected.

The corrosion products usually contain a predominating amount of magnetite and significant amounts of ferric oxide and copper oxide; they may also contain metallic copper and traces of inorganic salts such as calcium and sodium phosphates and sodium chloride. The magnetite may be present as a readily-detached shallow scale or, at high pressures more commonly, it may take the form of a thick, dense, laminated plug which adheres firmly to the corrosion cavity (see Fig. 2.4.1(b)). The morphology of the magnetite will be referred to again.

A situation sometimes arises (usually, but not exclusively, in the higher pressure groups), wherein the residual metal in the near vicinity of an area of corrosion, has become embrittled. This situation may be revealed in service by the explosive failure of a boiler tube, from which a considerable area of the wall has been suddenly ejected because of loss of ductility and strength. In this condition, the steel is described as having suffered "hydrogen damage" or "hydrogen embrittlement", since the cause may be traced to hydrogen which is produced by the corrosion process and which reacts with the pearlite in the steel (see Fig. 2.4.1(c)). This reaction liberates methane gas (CH_4), which forms voids in the metal structure and thus produces a brittle condition. Hydrogen damage does not always accompany on-load corrosion and it is uncertain what particular environment presages this kind of damage.

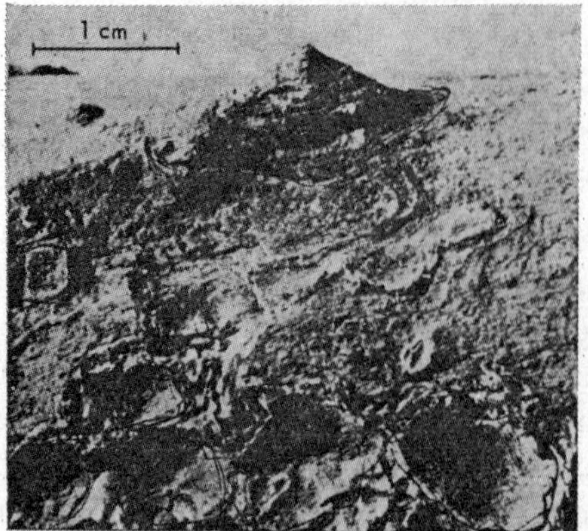

(a) On-load pitting corrosion occurring in a de-superheater operating at 900 lb/in²

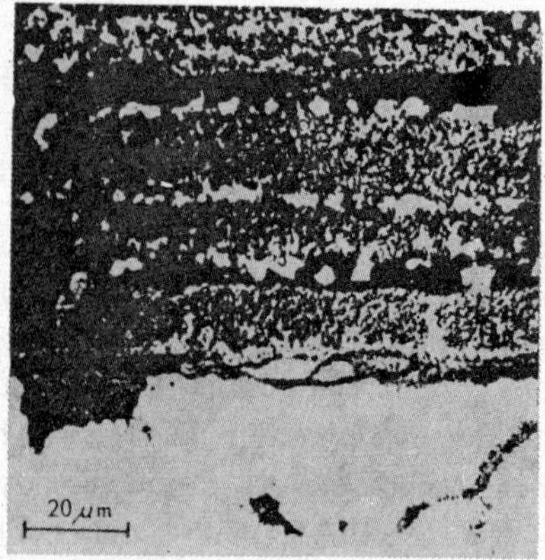

(b) Laminated magnetite at the bottom of a boiler corrosion pit.

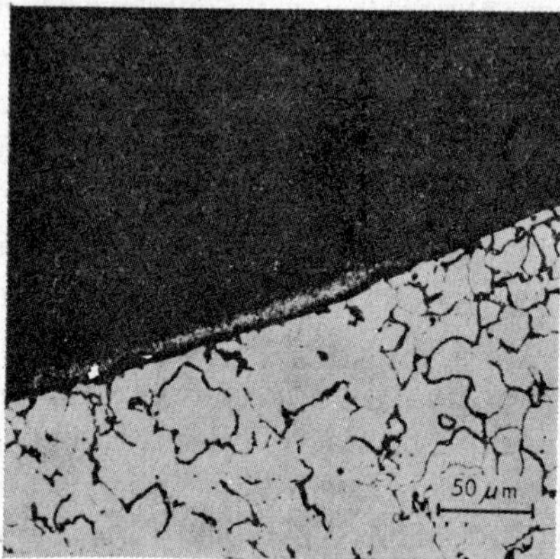

(c) Laminated magnetite and steel embrittlement at the bottom of a boiler corrosion pit.

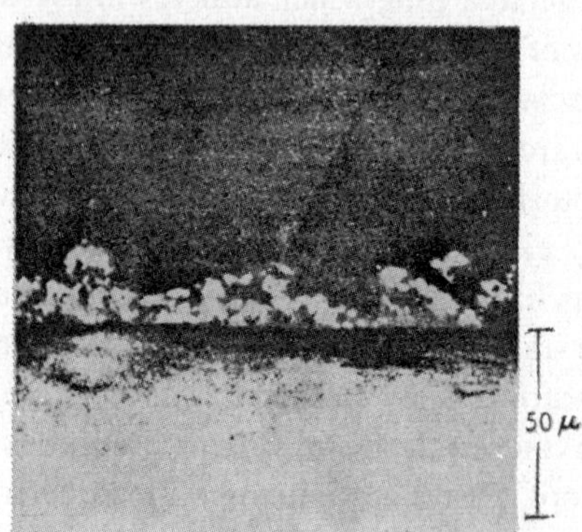

(d) Protective magnetite layer on laboratory specimen. 13% NaOH at 340°C, for 625 hours.

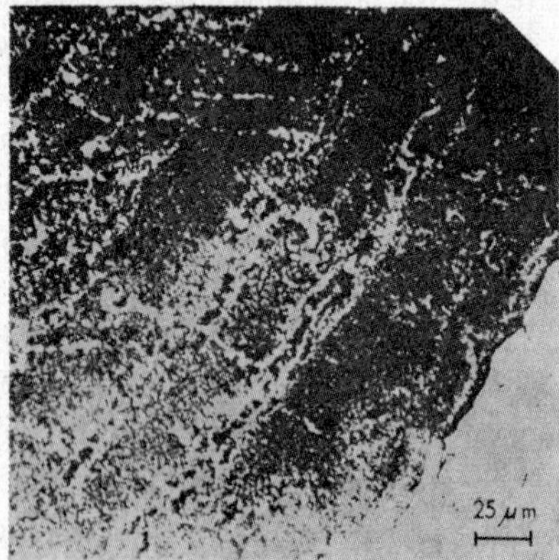

(e) Laminated magnetite and steel embrittlement resulting from corrosion of mild steel at 400°C in 0.1M $FeCl_2$ in the presence of a small amount of metallic nickel.

FIG. 2.4.1. Examples of corrosion

2.4.2. **Protective Magnetite and the Effect of Caustic Soda**

Under boiler operating conditions, steel is unstable in the presence of water and chemical reactions rapidly occur, at least initially. In the absence of dissolved oxygen, or at modern feed water oxygen concentrations of up to 0·007 ppm, the eventual stable product of these reactions is magnetite (Fe_3O_4) and the overall reaction may be represented as follows, observing that hydrogen is also produced.

$$3\,Fe + 4\,H_2O \longrightarrow Fe_3O_4 + 4\,H_2$$

Normally, an impervious and continuous film of magnetite, a few microns thick, is built up all over the steel surface and since the film prevents further access of water to the underlying metal, progress of the reaction soon ceases and the steel is protected from damage. If the magnetite film can be kept intact in boiler service, or self-repairing where local mechanical or thermal damage occurs, then freedom from corrosion is maintained and the oxide film is termed "protective magnetite". It should be remembered, however, that in contact with water and air (or oxygen), magnetite is converted into ferric oxide (Fe_2O_3) which has no protective value. This reaction is significant both to boilers in operation and those shut down, but in the former case it proceeds faster.

For many years boiler waters have been maintained alkaline with caustic soda, with pH value ranging from 10 to 12, in order to minimise corrosion due to dissolved oxygen and carbon dioxide, and for many years no harm seemed to be attributable to the practice, except perhaps in riveted-drum boilers. Thus, until recent years at least, caustic soda has almost invariably been a constituent of the boiler water with which on-load corrosion has occurred, and it is, therefore, desirable to summarise our knowledge of the effects of caustic soda solutions on steel.

Potter and Mann (1962) studied the growth of magnetite films on polished blocks of boiler steel immersed in pure solutions of sodium hydroxide (caustic soda), contained in mild steel pressure vessels. Sodium hydroxide solutions of 5% to 20% by weight (50,000 to 200,000 ppm) were first used and the work was later extended to distilled water and the temperature range 250° to 355°C. The temperature was extended to 550°C and solutions strengths of 2% to 5%, under supercritical conditions.

In the experimental ranges of 5% to 20% NaOH and 250° to 355°C, the magnetite growth rate diminished with time, according to an approximately parabolic law ($y = kt^{1/2}$), in which the coefficient, k is linearly dependent on the sodium hydroxide concentration, and exponentially dependent on the absolute temperature. In all of these experiments, magnetite was produced in two distinct layers; one was a continuous protective layer of magnetite adherent to the metal surface and the other, largely lost in the environment, was a fragmentary form of magnetite loosely attached to the protective layer, and easily detached from it (see Fig. 2.4.1 (d)). The protective magnetite did not have the laminated structure which is observed in the magnetite product of boiler corrosion.

The experimental work of Potter and Mann, using sodium hydroxide solutions of up to 20% and temperatures up to 335°C, failed to produce a damaging, or "non-protective" oxidation of boiler steel, and even up to 40% NaOH caused only a thickening of the magnetite film. However, other workers have found that in similar experiments using caustic soda solutions of greater concentrations than 40%, there was a progressive attack on mild steel and on magnetite. For example, the work of Thornhill (1961) showed that solutions in excess of 70% NaOH, attack the metal and its oxides, leaving sludge-like corrosion products, devoid of magnetite.

2.4.3. **Effect of Heat Transfer and Related Phenomena**

It is characteristic of on-load corrosion that the attack takes place where heat is transferred to the boiler water, thus the corrosion may occur on the fire-side interior of a furnace wall tube but it does not occur in boiler drums. From this observation it may be conjectured that some degree of local overheating of the metal causes the corrosion directly or otherwise, and the overheating would be enhanced by the lower thermal conductivity of the corrosion products. Experience shows, however, that even where severe corrosion has produced massive magnetite, the residual metal is rarely found to be overheated to a degree which significantly affects the metallurgical structure, that is to say it has not been heated to more than about 500°C. On this basis, a boiler tube at 2400 lb/in² gauge (350°C saturation temperature) could possibly be heated to a temperature between 350°C and say 500°C during several years of service, without subsequently showing metallurgical evidence of the fact.

In recent years, thermocouple measurements have been made in some high-merit C.E.G.B. boilers which have had corrosion failures, to investigate generator tube metal surface (fire-side) temperatures. Values in excess of the saturation temperature by 30° to 100°C have been frequently recorded, and in at least one tube, by 150°C. In this particular example (at 1650 lb/in²) the actual temperature at full load, was 477°C, and by calculation this corresponds to a localised heat flux about twice the average flux of 50,000 Btu/ft²/h.

It scarcely needs to be said that a boiler tube is cooled by water—boiling water which contains an increasing proportion of steam on moving up the tube, and which therefore, becomes correspondingly less effective as a cooling agent. It is thus reasonable to infer that in localities where the heat flux is excessive, inadequate cooling of the metal occurs accompanied by persistence of steam, which in turn, is likely to impede the flow of coolant. Actual flow measurements have shown substantial variations between similarly located furnace wall tubes, and substantial variations in the same tube, during normal operation of the boiler.

The significance of these considerations of heat flux, coolant flow and raised metal temperature, has been commonly related to the production of concentrating films of boiler water at tube metal surfaces where static or slowly-moving "slugs" of steam are generated, a condition which is sometimes referred to as "steam blanketing". Hall and others postulated that the boiler water dissolved solids, caustic soda in particular, could thus be concentrated to a sufficient degree to cause corrosion. For most of the years between 1940 and 1960, concentrated caustic soda films were widely held to be responsible for on-load corrosion. Figure 2.4.3 represents diagrammatically the relationship between tube metal temperature and the corresponding caustic soda concentration possible in the adjacent film of boiler water, initially containing 100 ppm NaOH and at 1400 lb/in².

It is common experience that corrosion occurs preferentially at sites where some constructional peculiarity exists, tending to impede water flow and raise the metal temperature, for example, near tube welds where an internal weld protrusion arises, at bends in tubes, and where lamination defects have occurred. Settlement of metal oxides and other debris on

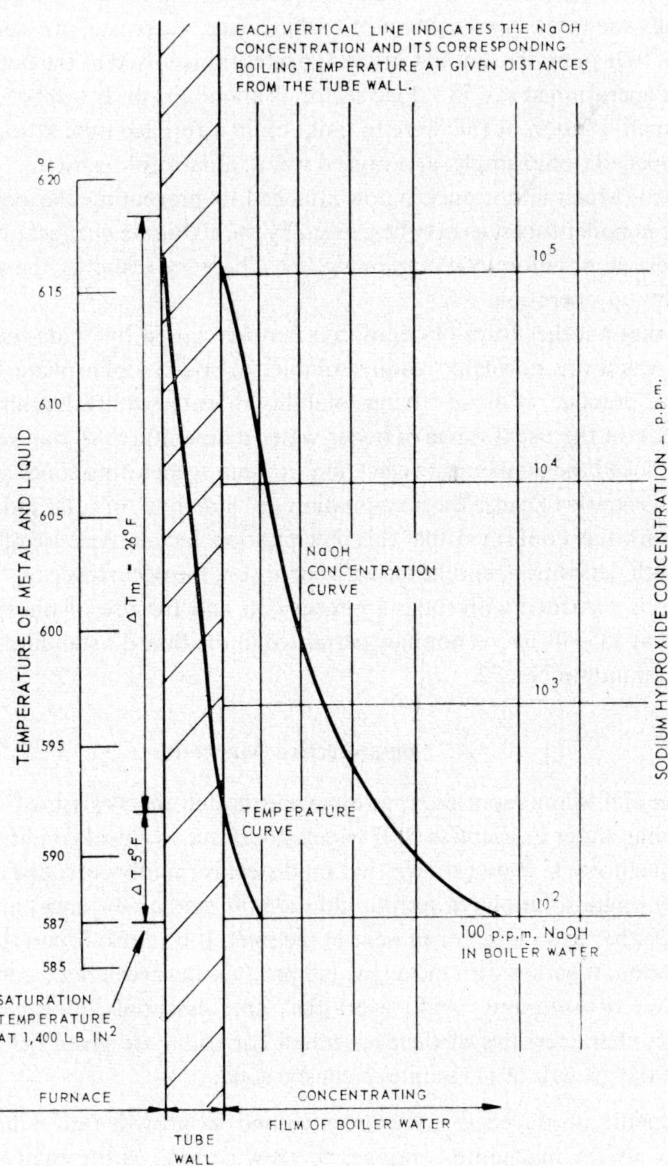

FIG. 2.4.3. Diagrammatic representation of a concentrated film of sodium hydroxide

heat-exchange surfaces, may similarly cause locally raised metal temperature and may also initiate corrosion pitting, as was discussed earlier in this chapter. The main source of metal oxides (of iron, copper and nickel in descending order of relative amounts, in most stations), is slight corrosion of the feed system and the oxides are conveyed into the boiler by the feed water. Even when the total metal content of the final feed water is controlled to an average concentration of 0·01 ppm, the amount of metal oxides transported to the boiler of a modern unit in a year of operation at say 75% load factor, is about 200 lb. It will be appreciated that if only a very small fraction of this were to build up in a furnace tube at some point where circulation is impeded (for example, at a ragged weld), a harmful restriction of heat transfer could be expected. Much importance is now attached to preventing the occurrence of insoluble deposits in boiler tubes, as may be gauged by the elaborate chemical procedures now used to clean new plant before commissioning (see Chapter 4), and by the strict control of feed water quality in operation.

It is thought that another form of deposition may be caused by "hide-out" in the boiler water, of salts which are normally readily soluble. Trisodium phosphate is the principal example in boiler practice, of a salt having solubility in water which diminishes as the temperature increases (in the usual range of boiler water temperature) so that where phosphate treatment is employed the temperature elevation in steam-generating zones may well reduce the solubility sufficiently to cause the precipitation or "hide-out" of solid sodium phosphate. Experimental work has confirmed that this precipitation occurs. Because of the association of "hide-out" with deposition, and hence with corrosion, the occurrence of "hide-out" in an operating boiler is regarded with some apprehension and the use of phosphate in boiler water treatment at 2350 lb/in^2, is not now permitted in the Board's standards of Generation Operation Memorandum No. 72.

2.4.4. **Non-protective Magnetite**

In 1963, Potter and Mann reported some quite unexpected effects arising from experiments originally involving water in stainless steel vessels under supercritical conditions. Summarising their subsequent work, it was shown that mild steel is rapidly corroded at temperatures above 200°C, by dilute solutions of ferrous chloride (an acidic substance) in the presence of traces of nickel, either in solution or present in the steel. It was also found that any chloride in an acidic solution, together with nickel, would produce the accelerated corrosion and that cobalt, vanadium or antimony could exert the same essential influence as did nickel.

The important characteristics of the accelerated corrosion are given below, and may be contrasted with the growth of magnetite in caustic soda.

(i) The magnetite produced is non-protective and its growth rate is linear with time, that is to say the magnetite continues to grow so long as the environment is maintained.
(ii) All of the magnetite produced grows at the steel surface, but it lacks cohesion and is liable to fracture where the metal contour is irregular.
(iii) The rate of growth of the non-protective oxide is many times greater than that of protective magnetite (see Fig. 2.4.4.A), and is comparable with the rates attributed to some on-load corrosion observed in boilers.

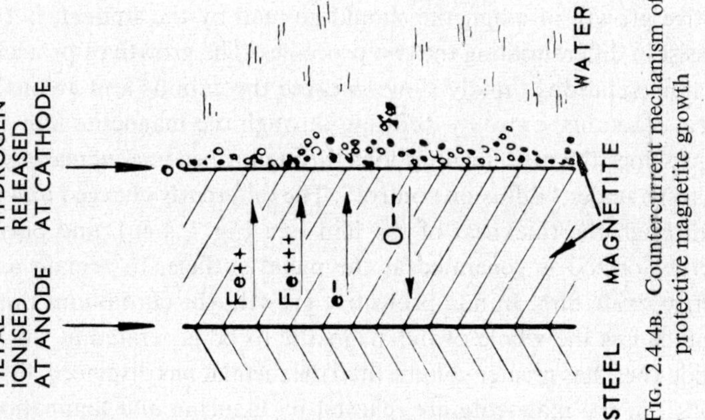

FIG. 2.4.4B. Countercurrent mechanism of protective magnetite growth

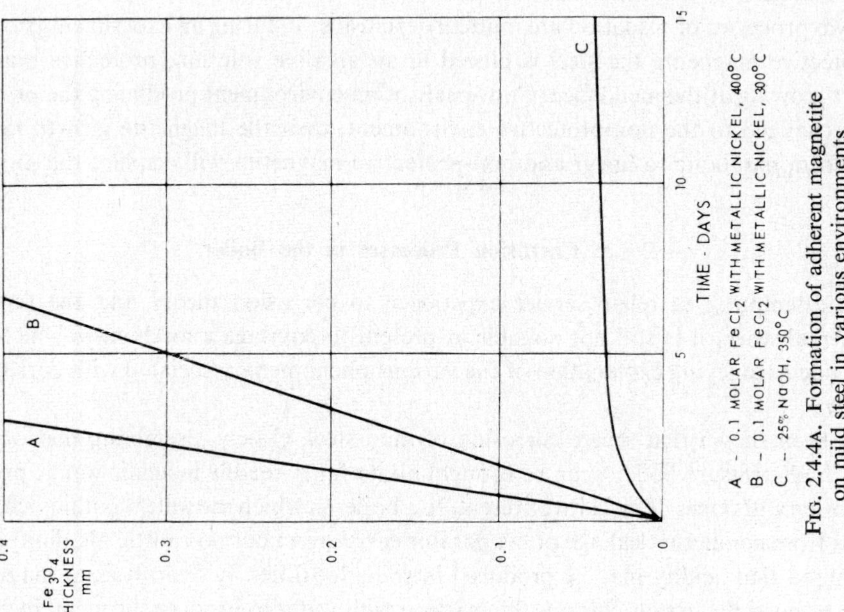

FIG. 2.4.4A. Formation of adherent magnetite on mild steel in various environments

(iv) The magnetite produced is laminated and hard, resembling that found in boiler corrosion pits, and when produced at 400°C it is accompanied by embrittlement (hydrogen damage) of the underlying steel (see Fig. 2.4.1(e)).

Dr. Potter's electrochemical interpretation of the processes of protective and non-protective growth of magnetite should be read by the student, but the following summary may assist in differentiating the two processes. The growth of protective magnetite in neutral or alkaline solutions rapidly slows because the cations and anions involved in the electrochemical reactions, move by diffusing through the magnetite film and as the film thickness grows, so does the movement of ions through it become increasingly difficult—the reaction is said to be under "diffusion control". The differently charged ions move in opposing directions through the thickness of the film (see Fig. 2.4.4B), and only about one half of the magnetite formed is generated at the metal surface, to remain as an almost impervious protective oxide film. In non-protective growth, the continuous presence of ferrous ions in solution causes the whole of the magnetite to be generated at the metal surface, but since the oxide then has greater volume than the metal it has displaced, the internal stresses thereby developed in the magnetite are relieved by fissuring and lamination of the growing oxide film. In this condition the oxidation rate is no longer controlled by diffusion since the film has been rendered porous, and it is probable that the rate is controlled by the reactions occurring at the cathode. These reactions are facilitated by the presence of nickel (or cobalt, vanadium or antimony) which are the most effective cathodes for the oxidation reaction and the oxidation of the metal is able to proceed unhindered.

The two processes of oxidation are mutually exclusive; if during an experiment producing non-protective magnetite the steel is placed in an alkaline solution, protective magnetite will start growing at the metal face. Conversely, if an environment producing the protective oxide is changed to the non-protective environment, then the magnetite growth rate will change from parabolic to linear and non-protective magnetite will displace the protective film.

2.4.5. **Corrosion Processes in the Boiler**

When attempting to relate service experience to corrosion theory and the results of experimental work, it is still not possible at present to envisage a mechanism which alone will provide a satisfying explanation of the various phenomena associated with corrosion in the boiler.

It has been shown that severe corrosion of mild steel, closely resembling that observed in many high-pressure boilers, can be brought about fairly readily in acidic water, provided that chloride and traces of nickel are present. In a boiler in which the water contains chlorides (perhaps from condenser leakage or evaporator carry-over) but no caustic alkalinity, it can be visualised that acidity may be produced in some localities by hydrolysis of magnesium chloride or possibly ferrous chloride, by reaction with water to produce the metal hydroxide and hydrochloric acid; for example:

$$MgCl_2 + 2\,H_2O \longrightarrow Mg(OH)_2\downarrow + 2\,HC$$

This reaction may well occur preferentially at the areas of high metal temperature, where concentration of the salts can occur, and in incipient corrosion pits which are prolific in producing ferrous ions. The necessary presence of nickel in the corrosion reaction (ignoring the three alternative metals because they are less likely to be present), may be accounted for as being water-borne from the feed system or as being a trace component of the boiler steel. Thus a mechanism for on-load corrosion can be postulated for a boiler water which is treated only with volatile alkaline agents such as ammonia and hydrazine, and where traces of mineral salts gain access. The same mechanism may apply even to a boiler water containing small amounts of caustic soda if there is a substantial preponderance of chloride, due for example, to cooling water ingress from condenser leakage.

There remains the problem of explaining corrosion occurring in boiler water which is comparatively strongly alkaline with caustic soda and perhaps trisodium phosphate, in which environment the corrosion mechanism just propounded is untenable, since Potter has shown that such an environment under laboratory conditions can only produce a protective oxide film on steel.

In this situation, the previously mentioned features of boiler design and construction may be crucial factors, and are invoked in the following explanation by Potter. At a steam generating surface where the local metal temperature is substantially higher than the saturation temperature, the boiler water dissolved solids will become more concentrated and at high pressures, trisodium phosphate, if present, is likely to be precipitated out of solution, aggravating the local elevation of metal temperature. On the other hand caustic soda, being extremely soluble, can concentrate to an almost unlimited extent; in fact, it would be molten and completely miscible with water in boilers of pressure more than about 1500 lb/in^2. From generator tube metal temperatures observed in recent years, it is quite feasible that, locally, the caustic soda concentration is enriched by several orders of magnitude, and at say 20% NaOH, the magnetite film could thicken to some 50 microns in a month or so. In these circumstances it is possible that local temperature fluctuations may cause fissures in the oxide film and subsequent growth of new magnetite, leading to a laminated product being formed in the corroding area. If the caustic soda were to concentrate to about 70%, it is to be expected that the magnetite would be dissolved and the exposed steel attacked, leaving little or no solid corrosion products.

As to the conditions required to cause hydrogen damage, these appear to depend largely on the rate of evolution and the relative ease with which the hydrogen evolved by the corrosion reaction can penetrate the residual metal to cause damage, or else penetrate the corrosion products to escape harmlessly with the steam and water. Where magnetite is formed under moderately alkaline conditions, any hydrogen liberated is evolved at the water/magnetite interface and is thus denied immediate access to the underlying metal; under corroding conditions at elevated temperatures, the hydrogen is evolved at or near to the metal surface and in larger amounts than in the former condition, so that there is a much greater probability that some of the gas will react with the residual steel. This interpretation suggests, in effect, that hydrogen damage may occur only where non-protective magnetite of linear growth rate, is being generated.

It is appropriate to discuss briefly the diagnostic value of the hydrogen content of the steam, in relation to boiler corrosion, since it is apparent that hydrogen is produced by the corrosion reaction, clearly so where hydrogen damage occurs, and therefore, it may be expected that the measurement of hydrogen in steam should provide a guide to the corrosion rate. Such measurement has been obtained as a continuous record in a number of stations, in boilers which experienced corrosion and in others which did not, but in general the results (usually about 0·001 to 0·01 ppm) have not shown a convincing correlation between hydrogen in steam and the incidence of corrosion, although a few investigators have reported such a correlation. It is not uncommon to find a few parts per thousand million of hydrogen in the steam from boilers which have shown no other evidence of being corroded.

Some further consideration of the circumstances of hydrogen production may explain the uncertain pattern of the results of measuring the gas. For example, it is reasonably possible that many corrosion sites do not liberate hydrogen since, in the presence of oxygen or copper oxide (which is often found in the corrosion pits), the hydrogen can be oxidised to water in a cathodic reaction. Or if one considers the background of slight but fairly uniform corrosion which normally proceeds over the whole of the boiler and feed system, this can be expected to generate at least as much hydrogen as may be produced at localised corrosion sites in the boiler, where the process is certainly more intense but involves only a very small fraction of the total area of steel exposed to water. Thus the "background" level of hydrogen in steam may be considerably greater than the contribution arising from destructive corrosion processes.

2.4.6. Prevention of On-load Corrosion

From this discussion of boiler corrosion it will be appreciated that it is the combined effect of many different factors which will decide whether any particular boiler will corrode in service or remain apparently immune for many years. The chemists cannot prescribe any regime of water quality control which will ensure freedom from corrosion in all boilers, but they do effect chemical control with the aim of avoiding, so far as is possible, those conditions which are known to encourage corrosion. This chemical control has been considered in earlier sections of this chapter and the Board's related standards for chemical control are given in the next section.

There is a substantial amount of circumstantial evidence to indicate that one of the significant factors affecting susceptibility to corrosion, is boiler design, and reference has been made to results from a number of investigations of boilers on load, which support this evidence. The design aspects involved are those which lead to generator tube metal temperatures considerably in excess of the saturation temperature, and as such are the steam/water ratio and flow rate in the generator tubes, and the rate of heat transfer in vulnerable areas. Although by design the overall heat transfer rate may be low enough to give confidence that excessive metal temperature will not occur, it has been shown that in service there may well be zones of furnace tubes where the heat transfer rate is up to three times greater than the overall average rate. In regard to coolant flow in boiler tubes, it is evident that it should not be impeded, for example by internal weld protrusions, abrupt changes in direction or the deposition of solid matter.

The manner and type of operation to which the plant is subjected is also important. Thus, boilers which have been apparently free from corrosion during some years of operation at high load factor, may develop the trouble after some months of two-shift operation—probably because of poor chemical conditions (particularly dissolved oxygen and iron, copper, etc., in the feed water) during the more frequent start-up and low-load conditions. Again, boiler operation on overload or even rapid variation of load, may sufficiently alter the mode of steam generation in a boiler tube, to produce higher metal temperature and a concentrating film of boiler water. Boilers are designed for "nucleate boiling", that is, to evolve steam at many discrete points in the tube surface; if the metal temperature is substantially raised for any of the reasons already discussed, the mode of steam generation may change to "film boiling", producing a blanket of steam over an area of tube and thus raising the metal temperature still further.

The development of persistent condenser leakage, even of small extent, is also potentially harmful, mainly because of the caustic soda and/or hydrolysable chlorides which are thereby likely to accumulate in the boiler water.

Of the chemical factors already discussed, two may be selected as being perhaps of major significance. One of these is the chloride content of boiler water, which should be kept below 4 ppm as NaCl in high-pressure boilers, since it is now known to have an essential role in at least one mechanism for the corrosion of steel, and because it is known to promote the corrosion of other metals in various environments. It is important that the chloride content of boiler water at all pressures should be accompanied by a suitable level of non-volatile alkalinity. The other factor is the controversial subject of boiler water alkalinity and we have seen that there are several important reasons for maintaining the water at the metal surface, in an alkaline condition. Nevertheless, it is quite possible that caustic soda in boiler water may concentrate in areas of high metal temperature to the extent that it will damage the protective oxide (a concentration factor of some 5000 or more), whilst trisodium phosphate is debarred from boilers at 2400 lb/in^2 because its "hide-out" property is likely to aggravate the heat-transfer problem.

These chemical factors have led some chemists to the "zero solids treatment" method of control, that is to say, to exclude all non-volatile solutes from the boiler water and to depend upon ammonia, or possibly other volatile amines, to provide the necessary alkalinity. This method of control has the serious drawback that ammonia, being so volatile, cannot reasonably be expected to persist at the steam-generating surfaces where it is particularly needed. Furthermore, the pH value of the water depleted of ammonia, will be extremely sensitive to any influence which tends to change it, such as traces of chloride and hardness which may unavoidably gain access and depress the pH value to the acid side of neutrality, with the consequent risk of incurring corrosion by non-protective magnetite formation. "Zero solids" control of boiler water is practised in a few modern stations and some have experienced corrosion—possibly because of the difficulty of ensuring that no trace of mineral salts gets into the boiler water. At present the Board accepts the view that for drum boilers, the lowest risk of corrosion lies with the presence of a small but controlled concentration of caustic soda in the overall boiler water.

Summing-up present knowledge of the feasibility of preventing on-load corrosion, it seems that in a few boilers deficiencies associated with a particular boiler design (such as have just been discussed) may be such that these boilers will corrode in spite of having been operated to provide the best known chemical control, and without any measurable extraneous contaminants in the boiler water. On the other hand, it is of great importance to realise that assiduous attention to the avoidance of all forms of feed water contamination and of other operational conditions which are known to promote a corrosive boiler environment, can maintain freedom from internal damage in a large majority of boilers.

The purification of feed water and make-up has now reached a point where it is difficult to foresee any further improvement and, from the corrosion prevention aspect, there is at least one technical disadvantage with any of the chemical treatments of boiler water which have been proposed so far. Recent experimental work has shown that alloy steels, with particular reference to chromium content, are more resistant than mild steel to conditions which promote the growth of non-protective magnetite, and it may be that developments in construction materials or in boiler design, will secure the elimination of the problem of boiler corrosion.

2.5. GENERATION OPERATION MEMORANDA NOS. 67 AND 72

2.5.1. Introduction

Two Generation Operation Memoranda have been issued for internal C.E.G.B. use, dealing with chemical standards for boiler feed water, boiler water, and saturated steam purity, for the commissioning and operation of high-pressure boiler plant in the 900–2350 lb/in^2 range. They were produced in order to give the best available advice to avoid corrosion in high merit plant, and to avoid the fouling of turbines in such plant.

2.5.2. Generation Operation Memorandum No. 67 (1st Revision December 1964)

The memorandum deals with the water treatment and control to be applied to new plant during commissioning and the period preceding the Commercial Operation Date. This date is defined as "The day immediately following a run of 72 h during which the turbo-alternator has run successfully at full load".

Sampling points and instrumentation necessary for control of start-up conditions and subsequent commercial operation are listed (Table 7), together with additional sampling points and instrumentation necessary for optimisation of chemical conditions (Table 8).

The boiler feed water, boiler water, and saturated steam standards to be achieved are the same as quoted in Generation Operation Memorandum 72 (Table 9). It is recognised that it may not be possible to achieve the standards during the commissioning period, and guidance is given in terms of flexibility in chemical control, to avoid unessential restriction of the engineering aspects of commissioning.

(a) Boiler Feed Water

Although the standards of Generation Operation Memorandum No. 72 should be achieved when the unit is fully loaded, some relaxation is allowed in relation to the limits for total iron, copper and nickel at lower loads.

The total metal concentration (Fe+Cu+Ni), ppm, multiplied by the rate of flow of feed water, shall not exceed 0·01×full flow rate, i.e. the rate of transport of metals to the boiler is never greater than the limits recommended for full load running.

TABLE 7

SAMPLING POINTS AND INSTRUMENTATION NECESSARY FOR THE CONTROL OF START-UP AND SUBSEQUENT COMMERCIAL OPERATION

Location	Type of sampler		Permanently installed instrumentation
	L.P. combined capillary/fast	H.P. combined capillary/fast	
Condenser outlet	—	—	Conductivity probes and multi-point recorder
Extraction pump discharge	√	—	* Conductivity and dissolved oxygen recorders
Deaerator outlet	√	—	Dissolved oxygen and pH recorders
Economiser inlet (at each inlet)	—	√	Dissolved oxygen and pH recorder
Boiler water (from distributor header)	—	√	* Conductivity recorder
Saturated steam at primary inlet	Conventional steam sampling equipment		* Conductivity recorder

* Measured after an ion exchange cation column in the hydrogen form.

This principle may be applied only to the concentration of iron, copper and nickel in feed water, and it is thought necessary that the total concentration (Fe+Cu+Ni) should not exceed 0·1 ppm at any load.

b) BOILER WATER

The standard for silica in boiler water should be in operation from the first time that the boiler is operated at full design pressure.

The use of phosphate is not permitted for boilers operating at 2350 lb/in^2 and above.

(c) SATURATED STEAM

If boiler water silica control is satisfactory, steam purity should meet requirements. If steam purity is not of the standard required, attention to steam purification equipment may be necessary.

(d) CONDITIONS NECESSITATING PLANT SHUT-DOWN

Whilst the memorandum admits that some relaxation of standards may be necessary during commissioning, shutting down the plant for chemical reasons may still be necessary. The chemical conditions leading to a recommendation to shut-down may include, high level of contamination of feed water with iron, copper and nickel or with dissolved oxygen, or a high chloride in boiler water.

TABLE 8

ADDITIONAL SAMPLING POINTS AND INSTRUMENTATION NECESSARY FOR OPTIMISATION OF CHEMICAL CONDITIONS AND DETAILED CHEMICAL INVESTIGATIONS BOTH DURING THE COMMISSIONING PERIOD AND THE SUBSEQUENT COMMERCIAL OPERATION

Location	Type of sampler			Permanently installed instrumentation
	L.P. combined capillary/fast	H.P. combined capillary/fast	Conventional	
Extraction pump 1st stage discharge	√	—		
Deaerator inlet (inlet to vent condenser)	√	—		
H.P. heater drains (on common line)	—			
L.P. heater drains (on common line)	√	—		
Outlet of waterside of each h.p. heater	Stubs only to be fitted suitable for capillary and fast samplers.			
Outlet of waterside of each l.p. heater				
Individual h.p. and l.p. heater drain lines				
R.F.W. tanks (on outlet from each tank)	—	—	√	
Clean drains tank return to condenser	—	—		Conductivity probes and recorder
Inter tube-plate space (on double tube plate condensers)	—	—	√	

It is stated that there are certain limits above which continued operation is considered dangerous with respect to boiler corrosion.

Firstly, running with chloride concentration in the boiler of 15 ppm NaCl for an accumulated period of one week (or proportionally shorter periods of higher levels of contamination).

Secondly, running for more than a few hours with the dissolved oxygen in the feed water at 0·03 ppm or more.

2.5.3. Generation Operation Memorandum No. 72

This Memorandum was first issued in June 1963, and the first revision in June 1964. It lays down standards to be observed during normal operation, for boiler feed water, boiler water, and saturated steam for drum-type boilers in the pressure groups 900, 1500 and 2350 lb/in^2. It will be noted that the first revision has included the 900 lb/in^2 group of boilers.

The standards are divided into two sections.

TABLE 9

CHEMICAL STANDARDS FOR BOILER FEED WATER, BOILER WATER AND SATURATED STEAM FOR DRUM-TYPE BOILERS IN THE PRESSURE GROUPS 900, 1500 AND 2350 lb/in²

Section	Units	Pressure groups lb/in²			Pressure groups lb/in²		
		900	1500	2350	900	1500	2350
		Primary standards			Advisory standards		
Boiler Feed Water							
Dissolved oxygen	ppm by wt.	≯ 0·007	≯ 0·007	≯ 0·007	—	—	—
Total Fe, Cu and Ni	ppm by wt. as Fe+Cu+Ni	≯ 0·01	≯ 0·01	≯ 0·01	—	—	—
Hydrazine (to be injected at extraction pump, determined at economiser inlet)	ppm by wt.	—	—	—	≮ 0·02	≮ 0·02	≮ 0·02
pH	at 25°C	—	—	—	≮ 8·5	≮ 8·5	≮ 8·5
Electrical conductivity (after cation exchange to H form)	μ-mho/cm at 20°	—	—	—	≯ 0·3	≯ 0·3	≯ 0·3
Boiler Water							
Caustic soda	ppm by wt. as NaOH	5–15	5–10	3–6	—	—	—
Phosphate	ppm by wt. as Na_3PO_4	—	—	—	≯ 45	≯ 10	Nil
Chloride	ppm by wt. as NaCl	≯ 6	≯ 6	≯ 4	—	—	—
Silica	ppm by wt. as SiO_2	—	—	—	≯ 5	≯ 1·5	≯ 0·2
Saturated Steam (Superheater inlet)							
Silica	ppm by wt. as SiO_2	≯ 0·02	≯ 0·02	≯ 0·02	—	—	—
Sodium	ppm by wt. as Na	≯ 0·03	≯ 0·03	≯ 0·03	—	—	—
Electrical conductivity (after cation column, H form)	μ-mho/cm at 20°C	—	—	—	≯ 0·3	≯ 0·3	≯ 0·3

(a) Primary Standards

If these standards are maintained it is believed that the boiler plant will operate at an acceptably low risk in relation to waterside corrosion.

(b) Advisory Standards

These give the best advice that can at present be given to assist in reaching or supplementing the Primary Standards. Advice is given in relation to exceptions from the standards. It is suggested that a chemical regime which does not reach the Standards laid down, but has given five years' operation free from corrosion, would be acceptable. Attention is called to the need for continual vigilance and attention to detail, in terms of water treatment and boiler operation.

The Standards discussed in Generation Operation Memorandum 72 are summarised in Table 9.

2.6. SAMPLING AND ANALYSIS OF FEED WATER, BOILER WATER AND STEAM

2.6.1. Introduction

In order to maintain the chemical standards set out in Generation Operation Memorandum No. 72, it is necessary to sample and analyse the feed and boiler water, and saturated steam at various points in the system.

Sampling and analysis of the water throughout the system is of importance to the station, and from the overall viewpoint of corrosion and water treatment throughout the Generating Board. Accurate data is required for the correlation of boiler corrosion with the analysis of the water in the system. A mass of analytical data has accumulated in the past but much of this is of doubtful value, because the accuracy of the figures is unknown or is known to be poor.

It is clear that the determination of a few parts per thousand million of contaminants such as iron, copper, and nickel requires refined analytical techniques. Precise methods of analysis have been developed and tested mainly by C.E.R.L. in co-operation with Regional R. and D. Departments, under the Analytical Working Party of the Regional Feed Water Survey.

It is relevant at this point to give a brief account of the Feed Water Survey, which is a major investigation organised on a national basis, and involves meticulous chemical analysis of feed water in twenty-four high-merit stations. The survey arose from the widespread occurrence of internal boiler corrosion in recent years, of which a general enquiry led to the conclusion that it was necessary to correlate feed water analysis with the circumstances of corrosion in the various stations. It was realised that such a correlation could not at the time be made, with any confidence in the results, because many different methods of analysis were normally used in different stations, and very little was known of the accuracy and precision of these methods, when applied to the very small quantities to be measured.

The work, therefore, commenced with the establishment by the Board's research departments of analytical methods of adequate precision for their different purposes. This was followed by a test programme for the station staffs to ascertain that the selected methods would also be satisfactorily precise when carried out in a station environment. The first main action of the survey was then achieved by carrying out measurements of all normal feed water constituents (iron, carbon dioxide, hydrazine, etc.) on thirteen occasions on two units in each of the selected stations. The remaining purpose is to correlate the measured constituents with each other, with plant design, and with plant operation. This work is being processed by computer at the time of printing this chapter, and has already provided unique information on the variability of the different parameters.

The first difficulty in starting an analysis is to obtain a representative sample, and care is required in choosing the design, location, and method of use of the sampling devices. Even when a representative sample is obtained, analytical results from the chemical analysis are subject to random errors. In order that the measurement of a contaminant or additive can be capable of detecting the smallest significant deviation from a set standard, the precision of the analytical results must not exceed given values. Therefore, analysts must obtain experimental estimates of their "standard deviations" for any particular contaminant or additive, and make regular checks to ensure their precision remains satisfactory; a statistical assessment of the results is necessary.

2.6.2. **Sampling and Analysis**

The frequency of sampling and analysis depends on the maximum time during which lack of knowledge of the concentration of contaminants or additives, is acceptable. Local conditions, such as plant design, condenser leakage, blow down and start up, will often dictate the sampling frequency. Where water is monitored by analytical instruments, the calibration of the instruments must be independently checked. Here the frequency of checking will depend on the stability of the instrument, and the magnitude of the changes required to be detected in the composition of the water.

2.6.3. **Boiler Water Sampling**

For any design of boiler the position of sampling points requires to be critically assessed during the design stage, to ensure that all circuits can be sampled. The sampling lines of suitable material should be taken to a manifold position, situated at a convenient level, usually the operating floor. Cooling coils should be of stabilised austenitic steel of the En58 type 17 : 7 Cr : Ni.

The main parameters of interest in the boiler water are the concentration of chloride, sodium hydroxide, silica and phosphate where applicable. In the absence of ingress of impurities, the concentrations in the boiler water should not change markedly from one day to the next. The primary standards, that is chloride and hydroxide, should be checked at daily intervals, or more frequently if the results are abnormal.

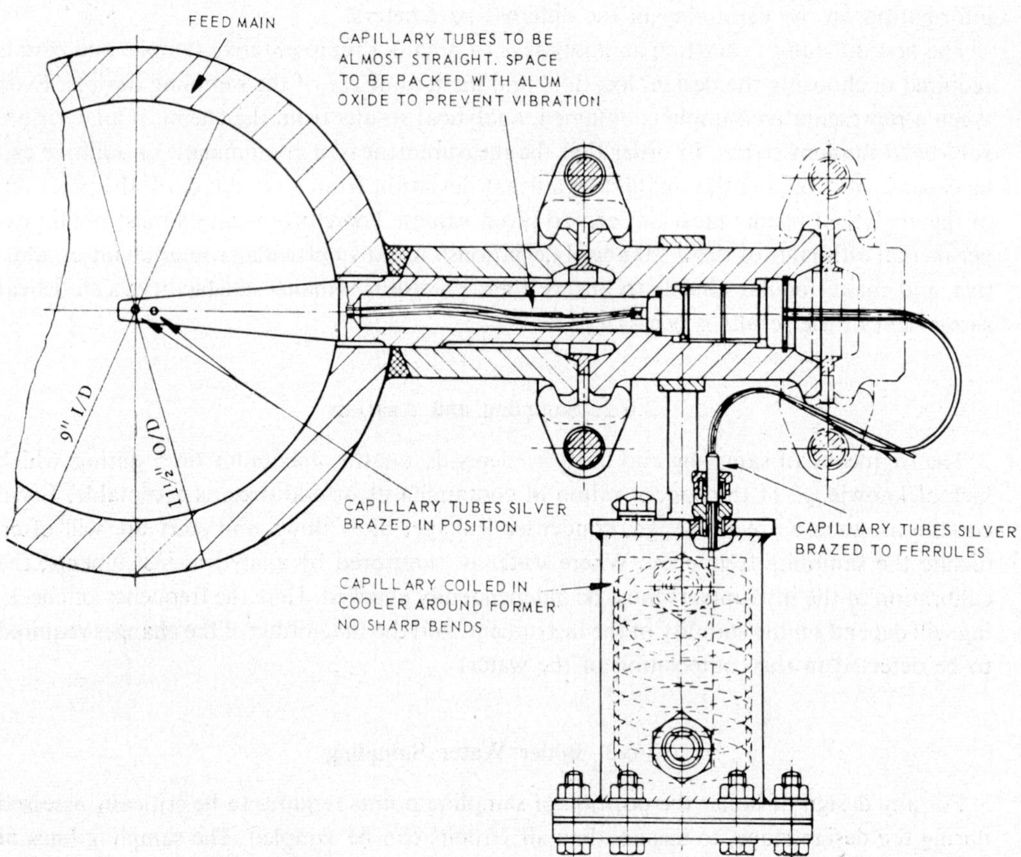

FIG. 2.6.5. General arrangement of sampling probe (high pressure)

2.6.4. Steam Sampling

The immediate purpose of monitoring steam purity is to prove the efficiency of the steam purification equipment. For this purpose continuous monitoring of the electrical conductivity of the steam, after a cation exchanger in the hydrogen form, will indicate any change in efficiency of purification. Regular checks on sodium and silica are also required for confirmation that primary standards are being maintained. It may be mentioned here that volatilised silica does not confer conductivity to the condensed steam.

Ammonia or carbon dioxide in the steam may mask the conductivity caused by boiler water carry-over, and in fairly recent years degassifying instruments have been supplied to some stations, in order to strip out dissolved gases from the condensed steam, and to present a sample of water containing only dissolved solids to a conductivity meter. Experience with these degassers has not always been satisfactory, and determination of sodium by the flame photometer is now the standard method. Considerable effort and experimentation has been expended over the years on the problem of obtaining a representative sample of steam. In the presence of solid or liquid contamination it is difficult to obtain a truly representative sample of saturated steam, and practically impossible to obtain a representative sample of superheated steam, due to the progressive tendency of liquid droplets to become attached to the steam pipe walls.

The normal method of sampling is to try to obtain an isokinetic sample, and this may be defined as "sampling conducted in such a manner that the sample enters the ports of a sampling probe at the same velocity as it flows through the pipe in the immediate vicinity of the sampling probe". The probe is provided with sampling ports of equal diameter, the total cross-sectional area of which is not greater than about half the inside cross-sectional area of the probe. The number of ports required is a function of pipe bore, and the ports must face upstream into the steam flow. The pressure drop through the probe is kept low to provide equal flow through the ports. Sampling lines and connections should be kept as short as possible, and the sample condensed under positive pressure. The material for the probe, sampling lines and condensing coil is a stabilised austenitic steel of En58 type.

2.6.5. Feed Water Sampling

Continuous monitoring of condensate for electrical conductivity, dissolved oxygen, and pH value, has been carried out for many years. However, representative samples of feed water have to be obtained for the estimation of trace quantities of metals, some of which will not be in solution. The time-honoured sampling point provided used to consist of a sample pipe which was teed into the system, and fitted with a regulating valve. The use of a valve allowed settlement of insoluble debris within the valve and this accumulated debris could be reintrained in the sample if the valve position was altered, or the flow disturbed. In addition, the sampling rate was rarely isokinetic. To overcome these objections capillary samplers are being developed for use in feed systems, and such a sampler is shown in Figure 2.6.5. In the ideal form it would be a continuous capillary of fixed bore without valves, so that the cross-sectional area of the flow path is constant throughout. The bore and length of the capillary are chosen so that isokinetic flow is achieved.

2.7. INSTRUMENTATION FOR WATER QUALITY CONTROL

2.7.1. Introduction

Recorders for electrical conductivity, dissolved oxygen and pH value, have been in use for many years, because the information provided has been of sufficient importance in maintaining plant availability and efficiency, to be continuously monitored. Since about 1955, there has been a progressive need to continuously monitor other constituents of water and steam, and this has necessitated the development of further sensitive and reliable instruments.

These additional requirements include measurement of the following items: silica, chloride, sodium, iron, copper and perhaps nickel hydrazine and caustic soda. For some of these items the development of satisfactory instruments is still proceeding, both within the C.E.G.B. and externally. One particular type (for example, the Technicon auto-analyser) may be mentioned as being of potentially multi-purpose use, since in principle it can carry out repetitively most of the analytical procedures which the analyst may perform manually on a sample of water, and which are concluded by some form of electrical measurement, the result of which may be displayed on a recorder.

Apart from the technical necessity for the instrumental recording of any particular parameter throughout the life of a station, in recent years a philosophy has been propounded which suggests that for the 2000 MW class of stations, it is of economic advantage to install a very comprehensive range of instruments for use especially during the early operating life of the station, when its principal problems should be resolved. At present the immediate difficulty concerns the lack of some of the instruments required, and the heavy maintenance requirements of some of the recently developed ones.

The remainder of this section, however, discusses the principles of the well-established instruments for recording conductivity, dissolved oxygen and pH value.

2.7.2. Measurement of Electrical Conductivity

The electrical conductance of an electrolyte is due to the ions it contains, and depends on the nature and number of ions present. In order to compare the contributions of specific types of ions to electrolytic conduction, a unit must be chosen to allow for ions carrying different numbers of electric charges (valency) and varying in conductivity with their concentration.

The resistance R, of a uniform conductor is directly proportional to its length L, and is inversely proportional to its cross-sectional area A. If the constant of proportionality is denoted by ϱ then:

$$R = \varrho L/A \tag{i}$$

If L and A are both made equal to 1, so that the conductor is a cube of 1 cm edge, then $R = \varrho$. The constant ϱ is characteristic for each conductor, and is known as the specific resistance or resistivity. It represents the resistance between opposite faces of a centimetre cube of the conductor, and has units of ohm-centimetres. In practice it is preferable to

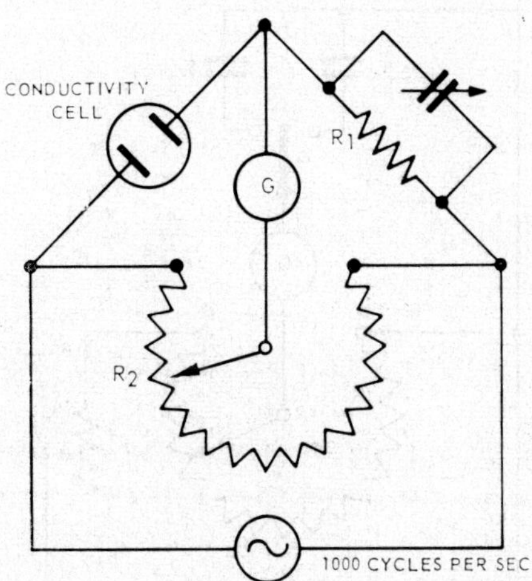

FIG. 2.7.2. Principle of measurement of electrical conductivity of solutions

employ a measure which increases in the same sense as conductance, and such a measure is the reciprocal of specific resistance, that is, specific conductance. This is denoted by K, and is expressed in reciprocal ohm-cm, or mho/cm. Most power station waters have low conductivity, so that mho/cm is an inconveniently large unit and hence it is customary to reduce this by a factor of 10^6, expressing the result as micromho/cm or μ-mho/cm.

Putting $\varrho = 1/K$ in equation (i) it follows that

$$K = L/AR$$

For electrolytes, the value of K depends primarily on the concentration of ions present, and in very dilute solutions the conductivity K has a nearly linear relationship to the concentration of dissolved salts, provided their relative proportions do not vary greatly. This is the basic reason for the use of conductivity measurement in power station waters.

Practically all measurements of electrical conductivity of ionic solutions are based on methods using an a.c. Wheatstone bridge method (Fig. 2.7.2). Considerable refinement of the basic principle is necessary before accurate measurements are possible, in terms of the bridge circuit and conductivity cell design.

The electrodes and body of the conductivity cell must be chemically resistant to the electrolyte being studied. For plant use the cells may be made of "plastic" with the electrodes embedded in the walls of the cell. The electrodes may be of stainless steel or monel metal. In some cases they are platinised with electrolytically deposited platinum black which eliminates the possibility of polarisation.

For any one conductivity cell the quantity L/A is constant. Rather than construct cells in which L and A are accurately measured, it is usual to calibrate each cell by making measurements using an electrolyte of known specific conductance. The electrolyte used is aqueous potassium chloride, the specific conductance of which is accurately known at various concentrations. The cell is standardised against a potassium chloride solution of approximately the same conductance as the electrolyte for which the cell is to be used. The conductivity of an electrolyte varies with temperature, and it is usual to report the conductivity as

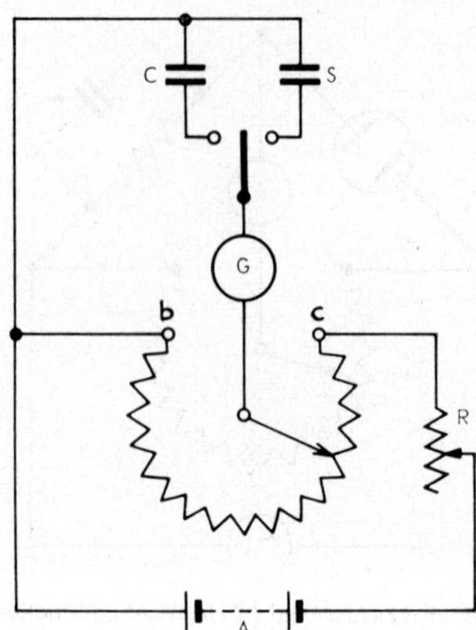

FIG. 2.7.3A. Simple potentiometer for measurement of electrode potential

μmho/cm at 20°C. Plant meters include a temperature compensating device to correct any small changes in temperature; with hot samples a sample cooler is required.

In measuring the conductivity of very pure waters as in a high-pressure boiler and feed system, it is essential to prevent the water sample being exposed to air whilst the measurement is being made. The water sample will rapidly take up carbon dioxide from the air, giving erroneous results. Thus in demineralised water and condensate, the electrodes are usually placed in the main pipework, or a closed bleed from the main pipework is taken to an enclosed test cell, by small bore sampling line.

In order to increase the sensitivity of the conductivity measurement of condensate and steam purity, it is now normal practice to pass the sample through a column of cation exchange resin in the hydrogen form, before making the measurement. The effect of this procedure is to convert all cations in the sample, into hydrogen ions. Thus salts such as sodium chloride and magnesium sulphate, which may enter the feed water through a condenser leakage, are converted into hydrochloric acid and sulphuric acid respectively, and these acids have a much higher conductivity than the salts. Carbon dioxide is unaffected by the procedure, but any ammonia, hydrazine and caustic soda are completely removed from the sample.

2.7.3. Measurement of pH Value

Table 2 in Section 2.1.1.2 shows that for each pH value, there is a corresponding hydrogen electrode potential and this potential may be measured by means of a potentiometric method. The principle is illustrated in Figure 2.7.3A. The potential difference between the two electrodes in the cell *C* (one of which is usually a reference electrode) is balanced by an opposing and accurately known potential derived from accumulator *A*. The position of balance on the uniform slide wire *bc*, is detected by galvanometer *G*, which registers a current when the opposing potentials are not exactly equal. Since the voltage from the

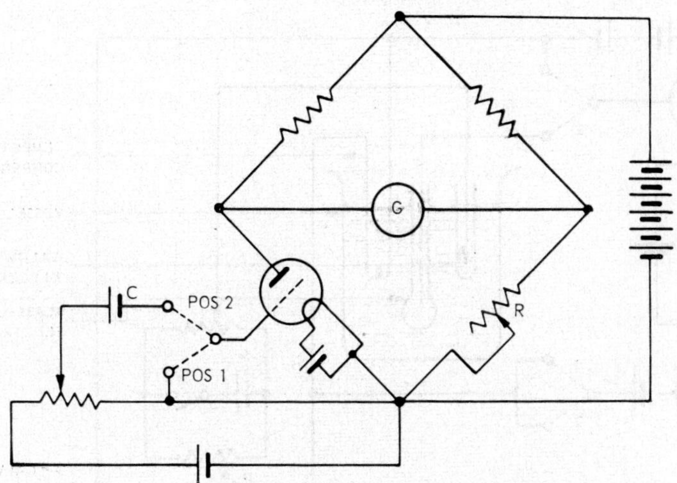

FIG. 2.7.3B. Simple thermionic valve circuit for measurement of electrode potential

accumulator is not steady over long periods of time, it is convenient to include an adjustment in the circuit to ensure that the potential fall across the slide wire *bc* is constant and is known. This standardisation is carried out immediately before each measurement by switching in the standard Weston cell (*S*), and setting the resistance *R* until the potential fall across the slide wire is at the required value.

This simple system ceases to be sensitive when the electrode system has a resistance of several megohms, in such cases as in very dilute electrolytes, for example, condensates, or where glass electrodes are used. The currents flowing are very small and difficult to measure.

These difficulties can be overcome by introducing a thermionic valve into the circuit as shown in Figure 2.7.3B. The valve through its anode and cathode is placed in one arm of a Wheatstone network, and the grid is connected to either the reference electrode or the electrode whose potential is unknown. The current which passes through the valve from cathode to anode is dependent on the potential of the interposed grid relative to the cathode. In position 1, the cell circuit is effectively disconnected from the grid/cathode circuit and the bridge can be adjusted at *R* until the galvanometer shows that no resultant current is flowing. In position 2, the potential difference in the cell system *C* is applied between grid and cathode of the valve, and the bridge is again balanced.

The circuit shown in Figure 2.7.3B is capable of further refinement, and in recent years electronic circuits have been introduced in which the zero drift of the instrument is negligible over many hours.

The most common electrode used today in pH measurement is the glass electrode, which consists of a silver–silver chloride electrode in hydrochloric acid enclosed in a glass envelope. The modern glass electrode comprises a tube blown out at one end into a bulb of approximately $\frac{1}{4}$ in. in diameter. The bulb can be reasonably thick to give mechanical strength, because modern measuring techniques can respond correctly to small signals despite the high electrical resistance of the cell. The mechanism whereby the potential of the glass electrode depends linearly on the pH of the electrolyte surrounding it, is not precisely known, but probably the glass acts as a membrane permeable only to hydrogen ions. If

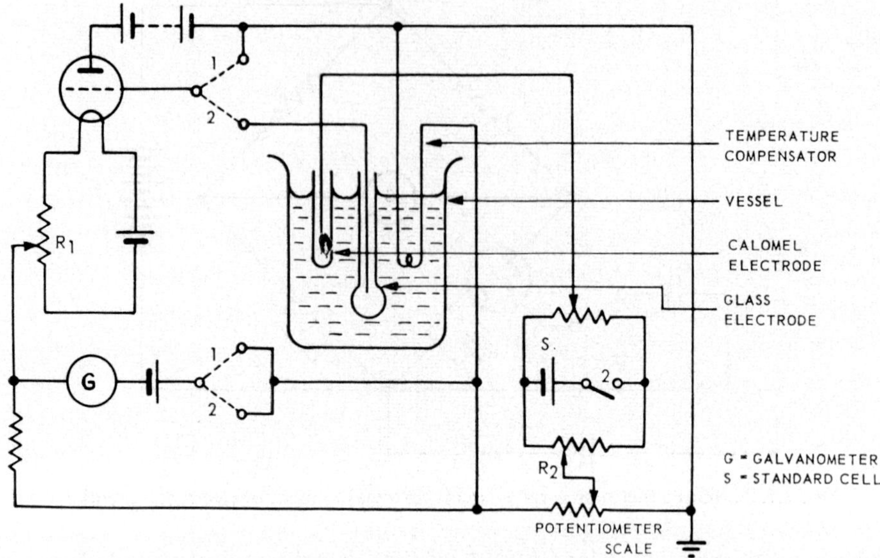

FIG. 2.7.3C. Circuit for pH measurement with glass and calomel electrodes

the hydrogen ion concentrations differ on either side of the membrane, migration of the hydrogen ions towards the more dilute solution sets up a potential difference across the membrane which can be measured.

The glass/calomel electrode system is the one in common use today. The potential of the glass indicator electrode depends linearly on the pH of the solution in which it is immersed (within the range of pH values of power station interest), the change in potential being 58 MV/pH unit at 20°C. Provided the potential difference between the electrodes is standardised with a solution of known pH value, the scale of the potentiometer may be calibrated directly in pH units.

A suitable circuit for the glass electrode/saturated calomel electrode is shown in Figure 2.7.3C. It will be noted that a temperature compensation is introduced to correct for temperature variation.

On a plant scale the electrode system is housed in a substantial compartment, and supplied with a bleed of water from the main pipe-work. As with conductivity measurements, care has to be taken to exclude air.

2.7.4. **Measurement of Dissolved Oxygen**

(a) GAS TRANSFER METHOD

One of the widely used instruments for the measurement of dissolved oxygen in boiler feed water is the Cambridge electrochemical dissolved oxygen analyser. This incorporates a gas-transfer system, whereby the oxygen in the continuously flowing water sample is scrubbed out into the gas phase by a stream of hydrogen. The hydrogen plus any oxygen is then fed into an electrochemical cell for measurement. As the electrodes are not in the sample

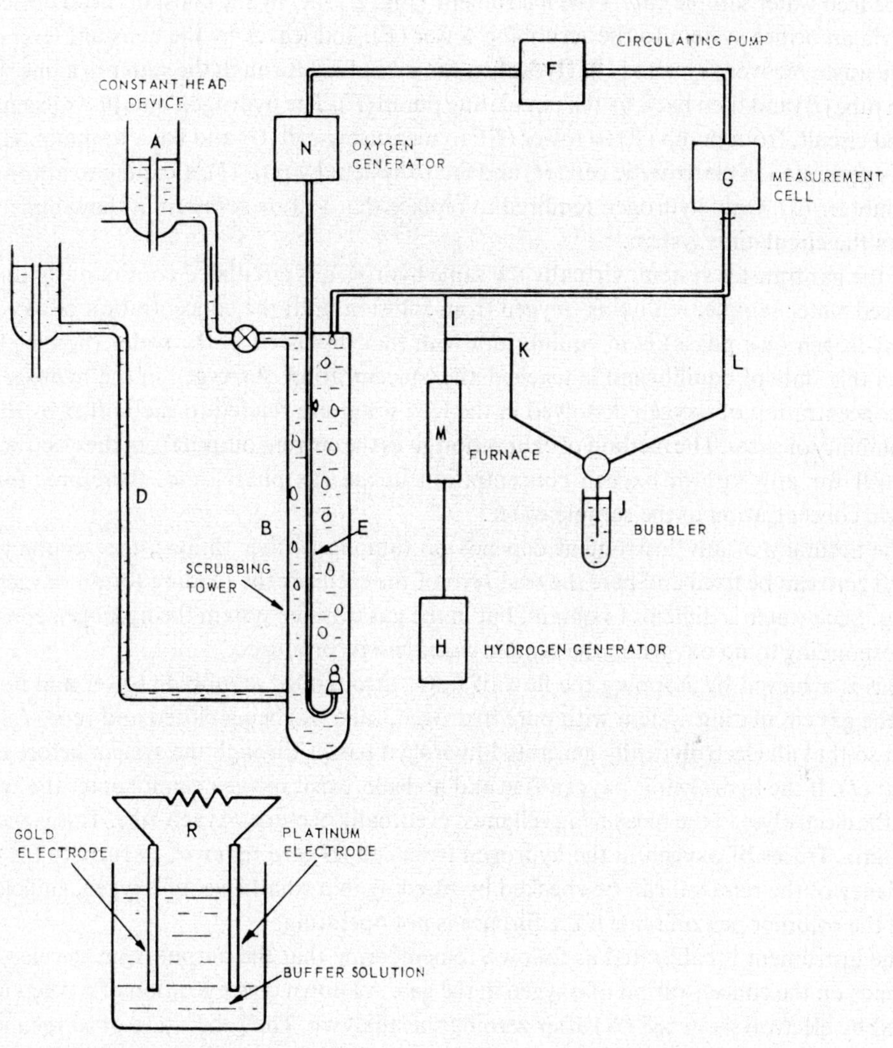

FIG. 2.7.4A. Electrochemical dissolved oxygen analyser—schematic (Cambridge)

water, they are not affected by non-volatile impurities in the sample, such as copper and iron. Volatile additives, such as hydrazine and ammonia, do not affect the oxygen readings.

The feed water sample enters the instrument (Fig. 2.7.4A) by the constant head device (*A*), and via an orifice system to the scrubbing tower (*B*), and leaves by the constant level outlet to the waste recovery system (*D*). Hydrogen gas is bubbled through the water in a fine stream from tube (*E*) and then back to the circulating pump (*F*). The hydrogen thus flows in a mainly closed circuit, from pump (*F*) to tower (*B*), to measuring cell (*G*) and back to pump. Hydrogen is generated in electrolytic cell (*H*) and fed to system by pipe (*I*), escaping to atmosphere via bubbler (*J*). Only hydrogen required to replace that lost by solution in the sample water enters the circulating system.

In the gas transfer system, virtually the same hydrogen is circulated continuously through the feed water sample, removing oxygen from solution until the concentration of oxygen in the hydrogen (gas phase) is in equilibrium with that dissolved in the water (liquid phase). When this state of equilibrium is reached, the concentration of oxygen in the hydrogen and the concentration of oxygen dissolved in the feed water are related to each other by the law of solubility of gases. The method of calibration gives the current output from the electrochemical cell for any known oxygen concentration in the gas phase, and, therefore, for any known concentration in the sample water.

The accuracy of any instrument depends on (amongst other things), the accuracy with which zero can be fixed and here the true zero of the meter is the reading for an oxygen-free water. Such water is difficult to obtain, but in the gas transfer system the hydrogen condition corresponding to no oxygen in the sample water can be produced.

This is achieved by stopping the flow of water through the scrubbing tower and flushing out the gas circulating system with pure hydrogen, tube (*K*) being closed and tube (*L*) being open so that all electrolytically generated hydrogen passes through the system before escaping at (*J*). If the hydrogen is oxygen-free and no leaks exist, oxygen cannot enter the system, and the electrolyte in the measuring cell must eventually become oxygen-free. This is the oxygen zero. Traces of oxygen in the hydrogen from cell (*H*) are removed in furnace (*M*). The efficiency of the removal can be checked by bleeding in a small level of oxygen, sufficient to send the recorder pen off scale if the furnace is not operating.

The instrument is calibrated as follows, remembering that the output from the electrodes depends on the concentration of oxygen in the gas. A known concentration of oxygen is produced by electrolysis in cell (*N*) after zeroing the analyser. The generation of oxygen in this cell is quantitative down to currents below 50 μA (equal to 0·001 ppm oxygen in water). The rate of introduction of hydrogen into the system is proportional to the current through (*H*) I_J, and the rate of introduction of oxygen is proportional to the current through (*N*) I_B. The concentration of oxygen in the hydrogen, which is determined by the ratio of the two currents, is $I_B/2I_J$, a ratio which can be adjusted to give any desired concentration in the gas phase.

Leakage of oxygen into the gas circulating system from the surrounding air will cause all readings to be high. Since it is difficult to design and maintain absolutely leakproof equipment a leak-detecting system is incorporated.

The electrode system consists of a gold electrode and platinum electrode mounted in a buffer solution, the electrodes being connected through a resistance. The platinum electrode is coated with platinum black.

On the platinum black surface, hydrogen gas in solution is converted to hydrogen atoms and thence goes into solution as hydrogen ions leaving behind two electrons per molecule.

$$\underset{\text{Molecule}}{H_2} \longrightarrow \underset{\text{Atoms}}{2H} \longrightarrow \underset{\text{Ions}}{2H^+} + \underset{\text{Electrons}}{2e^-}$$

The platinum electrode with hydrogen gas is regarded as a source of electrons at constant potential. The surface of the gold electrode has an affinity for hydrogen atoms. Hydrogen ions which strike the surface of the gold electrode accept electrons which have come from the platinum electrode, become hydrogen atoms, and adhere to the gold surface. As soon as the gold surface is covered with hydrogen atoms, the electrode is polarised. When oxygen molecules in solution strike the polarised gold surface, they react with the highly active hydrogen atoms to form water leaving a space on the gold surface into which more hydrogen ions are discharged, allowing current to flow from the platinum black through the resistance (R).

Since the rate at which oxygen atoms strike the surface is directly proportional to the oxygen concentration in solution, then the current through (R) is also proportional to the oxygen in (G).

When installing a dissolved oxygen analyser of any type in a feed system, careful consideration should be given to the position of the instrument in relation to length of sampling line, and the type of material used for the line. The levels of dissolved oxygen to be measured are small and incorrect readings will result if precautions are not taken.

The analyser should be placed as close to the sampling point as possible, and if a cooler is used this should be positioned immediately after the sample point.

The material of the sampling line is of importance; all plastic and rubber tubing allow atmospheric oxygen to diffuse into the water sample, with the possible exception of nylon. All common metals will adsorb oxygen, and increasing temperature increases the adsorption rate. Thus if mild steel tubing is used without a cooler at a high-temperature sampling point, oxygen will be lost to the steel, giving a low test result. This loss of oxygen will be enhanced if the sample contains hydrazine. For sampling lines, stainless steel is preferred.

(b) Direct Method

One other well-known analyser for the measurement of dissolved oxygen is the Wallace and Tiernan instrument. Whilst the method of measuring the dissolved oxygen is fundamentally similar to that used in the Cambridge analyser, the means of achieving the final measurement differ.

The Cambridge analyser uses a gas transfer system to scrub the oxygen out of the sample, the measurement of oxygen being carried out in a separate cell. In the Wallace and Tiernan analyser, the oxygen remains in the sample water, and is measured *in situ*, in the electrode cell. Calibration of the analysers is also different, in that the oxygen required for calibration is produced in the case of the Cambridge instrument by electrolysis of an independent solution, and in the Wallace and Tiernan instrument by electrolysis of the sample water.

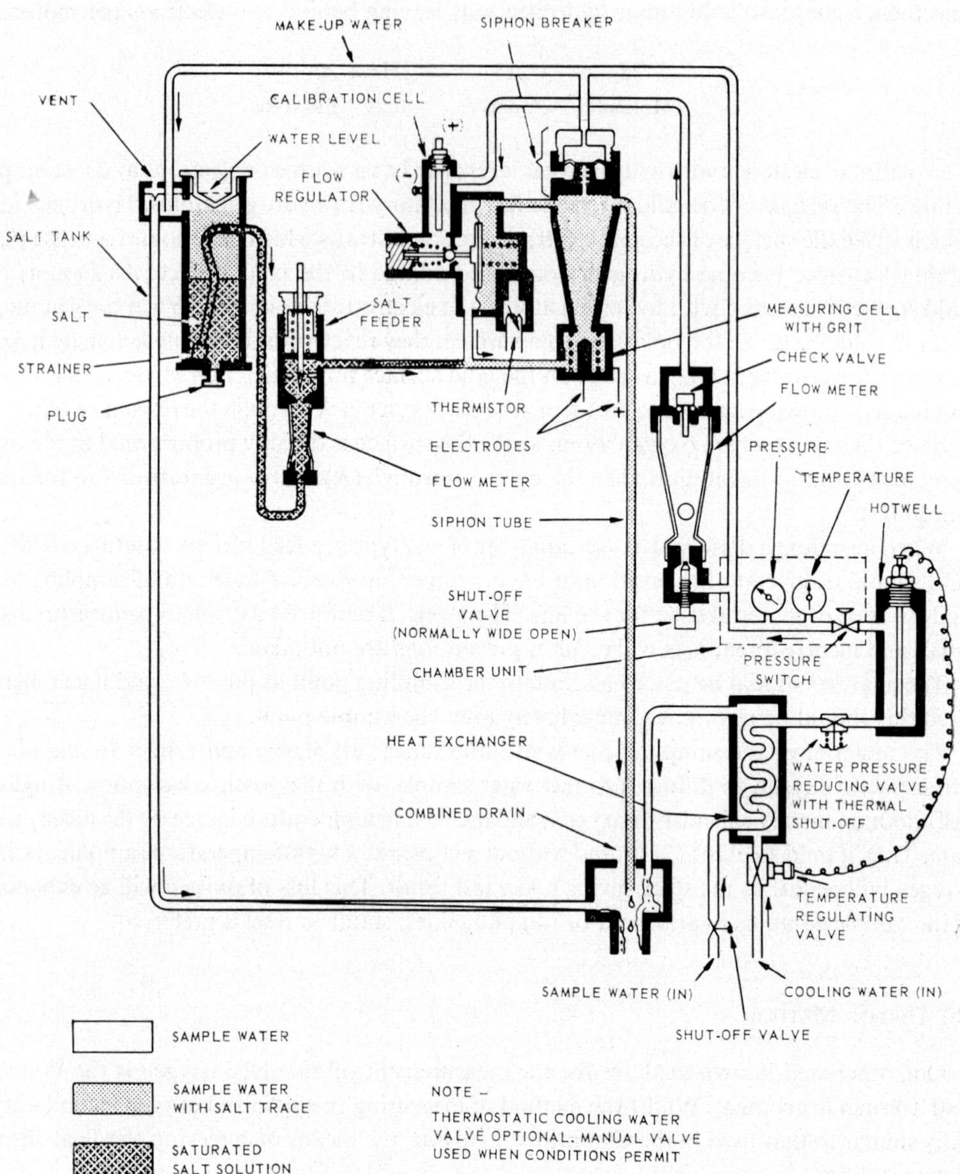

FIG. 2.7.4B. Electrochemical dissolved oxygen analyser (Wallace and Tiernan)

The Wallace and Tiernan analyser (see Fig. 2.7.4B), consists essentially of an electrode system through which the sample water is passed, together with a means of injecting a small quantity of saturated sodium chloride into the sample water. The electrodes consist of platinum for the anode, and cadmium for the cathode. A boiler feed water has a high resistance due to its low conductivity, and this would result in non-linearity of the oxygen/current curve at higher oxygen levels. In order to overcome this, the conductivity of the sample at the measuring cell is increased to 800–900 units by the addition of sodium chloride. The variables which can affect the current output of the cell in addition to dissolved oxygen and conductivity are velocity and temperature, and these are controlled by flow meters, pressure switches and thermistors.

Sample water passes through the shut-off valve, through the cooling coil and pressure-reducing valve, which is fitted with a thermal shut-off. The water flows through a thermostatic bulb container, which is connected to the cooling water inlet valve of the cooling coil, thus regulating the flow of cooling water. After passing through a flow meter, the water flow is carried through the calibration cell into the measurement cell. Before the water enters the measurement cell, a small volume of saturated sodium chloride is pumped into the sample. In the measurement cell the electrodes are maintained in a clean condition by recirculation of grit over the surfaces. Associated with the cell is a thermistor which can apply a further temperature compensation if required. The current output from the measuring cell can be displayed on an indicator or recorder.

SUGGESTIONS FOR FURTHER READING

Corrosion

Among the classic works on corrosion are two books by U. R. Evans.

1. *An Introduction to Metallic Corrosion*, London, Edward Arnold & Co. 1948.
2. *The Oxidation and Corrosion of Metals*, London, Edward Arnold & Co. 1960.

The first book covers the whole field of corrosion and may be read with profit by all. The second book gives the more recent developments in corrosion theory and practice up to 1960. It is a monument to a lifetime's work in the corrosion field by Dr. Evans.

A fairly recent book on the principles of electrochemistry is that by E. C. Potter.

Electrochemistry: Principles and Applications, London, Cleaver–Hume Press Ltd., 1956.

A simple short account of the principles of aqueous corrosion, with many photographs may be obtained from The Mond Nickel Co. Ltd., Thames House, Millbank, London, S.W.1., this is entitled "Corrosion in Action". A section on corrosion is included in the Pelican Series:

Metals in the Service of Man, W. Alexander and A. Street, Penguin Books Ltd. 1954.

Boiler and Feed Water

There is no publication which deals with modern power station practice in this field. An I.C.I.-sponsored publication covers the whole field of water treatment for varying purposes in general terms.

Industrial Water Treatment,
Hamer, Jackson and Thurston,
Butterworth, London, 1961.

The following papers cover the field of water side corrosion in relation to the C.E.G.B.

1. Water-side Corrosion in Boilers,
Report on Symposium held at Central Electricity Research Laboratories, Leatherhead.
12–13 February 1964.
C.E.G.B., London.
2. Supplement No. 1 to the Report on Symposium held at Central Electricity Research Laboratories, Leatherhead.
12–13 February 1964.

Both these reports give an extensive list of papers on water-side corrosion.
Of the many papers published by C.E.R.L., the following are of particular interest.

1. The Circumstances and Prevention of Internal Boiler Tube Corrosion, by E. C. Potter, No. RD/L/R/ 1215.
2. Large-scale Steam Generation at Steel Surfaces—a Physicochemical Enquiry, by E. C. Potter,
No. RD/L/R/1270.
3. The Fast Linear Growth of Magnetite on Mild Steel in High Temperature Aqueous Conditions, by E. C. Potter and G. M. W. Mann.

G.M.W. Mann has also published in *Chemistry and Industry*, 1964, pp. 1584–1590.

EXERCISES

1. The conductors used for the overhead transmission system usually consist of a galvanised mild steel core (to provide tensile strength), closely surrounded by strands of aluminium, and in coastal areas the conductors sometimes corrode and bulge. Assuming that sea spray is an essential part of the trouble, explain the course of the corrosion process, using the galvanic series given in Table 5. Why would the substitution of steel-cored copper conductor be particularly unwise in this environment?

In heavily industrialised inland areas, the same kind of aluminium conductor is affected by a rather different form of corrosion which is largely confined to the outer aluminium strands. Give an explanation of this form of corrosion.

2. What forms and locations of corrosion could be expected to occur in station circulating water systems, where the water is:

(a) sea water;
(b) an inland river with entrained mud and silt; or
(c) an estuary much depleted in dissolved oxygen.

Give reasons for your answers in each case.

3. If a sample of turbine condensate containing a negligible amount of any impurity were exposed to the atmosphere, explain what changes in electrical conductivity and pH value you would expect to occur in the sample. What difference in the pattern of these changes would you expect, if the sample contained about 0·2 ppm of ammonia as the only impurity initially.

4. Describe briefly the significance of silica in boiler/turbine units operating at pressures greater than about 900 lb/in^2.

Using Table 6 (page 203), calculate for pressures from 1250 to 2500 lb/in, the silica concentrations (ppm) in the water at each pressure, which are in equilibrium with 0·02 ppm of silica in the vapour. Plot these concentrations against pressure, on linear graph paper, and superimpose on your graph the uppermost curve shown in Figure 2.3.3.4A. Give reasons to account for any lack of coincidence of the two curves.

5. Where a feed water is contaminated with sea water, the following chemical reaction will occur in a boiler water which has an alkalinity maintained with caustic soda.

$$MgCl_2 + 2\,NaOH \longrightarrow 2\,NaCl + Mg(OH)_2\downarrow \qquad (1)$$

The boiler contains 100 tons of water, and has a caustic soda concentration of 5 ppm. How many pounds of magnesium chloride ($MgCl_2$) will react with the caustic soda. Assuming the sea water contains 3·4% by weight of total dissolved solids, and the magnesium chloride represents 10% of the total dissolved solids, how many gallons of sea water would contain the magnesium chloride calculated from equation (1)?

For the calculation use the following atomic weights. Magnesium (Mg) 24, chlorine (Cl) 35, sodium (Na) 23, oxygen (O) 16 and Hydrogen (H) 1.

Write down an equation which shows the course of events after the loss of caustic soda in equation (1).

6. Discuss the reasons for feed water treatment.

In addition to feed water treatment, what other steps can you suggest which would reduce the concentration of contaminants in the feed to the boiler?

7. Why is it that the avoidance of insoluble deposits in boilers is considered to be important in the prevention of boiler corrosion? What parameters of feed water quality are related to the formation of insoluble boiler deposits and in what way do they influence the formation?

8. Discuss the form and nature of the products of corrosion found in corroded boilers and in laboratory experiments on steel in caustic soda solutions, in relation to the hypothesis that caustic soda is responsible for boiler corrosion.

CHAPTER 3

WATER TREATMENT PLANT: COOLING WATER SYSTEMS

3.1. INTRODUCTION

The availability of a suitable supply of water, both for cooling purposes and for boiler feed make-up, is one basic requirement in siting a power station. In this sense water may be regarded as a raw material for the power generation industry.

In the early days of the industry, water treatment was unknown and the available water was fed directly to the boiler without treatment. This often led to scaling, corrosion and sludge deposition troubles and to overcome these, largely by a process of trial and error, various internal boiler water treatments were developed. These boiler water treatments were based on readily available materials such as soda ash, starch and tannins and sometimes helped to ameliorate the problems. With the development of larger units and progressively increasing steam temperatures and pressures, a more positive method of chemical control was required and this could best be achieved by removing the scale-forming constituents from the raw water before feeding it to the boiler. Precipitation softeners, employing lime and soda, became an item of power station plant and several different types were evolved.

Later, base-exchange softeners became available employing, at first natural or synthetic zeolites, later sulphonated coal and since about 1950 synthetic ion exchange resins developed for complete water demineralisation have been used.

In parallel with these developments in water softening came the demand for make-up water of comparable quality to power station condensate, that is virtually complete removal of salts and dissolved gases, and evaporators became an essential part of the water treatment scheme for make-up. Within the past twenty to thirty years, the synthetic ion exchange resins, referred to above, have made possible the complete removal of salts without evaporation. Demineralisation, as this process is known, is now the preferred alternative to softening followed only by evaporation.

This chapter is concerned with all these aspects of water treatment for boiler feed make-up water and in addition, with the chemical treatment and chemical control of cooling water systems. The emphasis is on modern power station practice, but in order to get into perspective the very rapid advances of the past decade, some of the early developments are discussed in some detail.

TABLE 1

COMPOSITION OF VARIOUS SOURCES OF WATER

		(A) Industrial rain water	(B) Rural rain water	(C) Plymouth surface water	(D) Newton Abbot surface water	(E) Hinkley Point surface water	(F) Manchester corporation water	(G) Padiham water	(H) Widnes water	(I) Portishead surface water	(J) Portishead well water	(K) London supply water	(L) Ipswich supply water	(M) Tilbury sewage effluent	(N) Sea water
pH		4·1	7·0	7·0	8·3	8·0	7·3	6·3	7·1	7·6	6·8	7·8	7·1	7·8	–
Total hardness	ppm as $CaCO_3$	24	40	17	25	45	17·6	38	154	244	333	270	401	452	6450
Alkalinity	ppm as $CaCO_3$	Nil	31	10	15	20	10	10	142	199	285	190	305	380	120
Acidity	ppm as $CaCO_3$	4	Nil	Nil	Nil	Nil	Nil	Nil	Nil	Nil	Nil	Nil	Nil	Nil	Nil
Equivalent mineral acidity (E.M.A.)	ppm as $CaCO_3$	43	26	48·5	–	40	18	37	67	146	189	156	208	943	30,100
Conductivity	micromho/cm at 20°C	–	–	55	88	127	51	95	–	390	550	620	880	2000	–
Total dissolved solids (T.D.S.)	ppm	40	60	44	–	–	36·6	–	222	280	390	400	600	1585	35,000
Calcium	ppm as $CaCO_3$	18	34	10	–	35	14·2	22	95	214	299	255	364	364	1020
Magnesium	ppm as $CaCO_3$	6	6	7	–	10	3·4	16	59	30	34	15	37	88	5410
Sodium and potassium	ppm as $CaCO_3$	10	15	–	–	32	–	–	65	15	22	76	107	700	23,900
Soluble iron	ppm as Fe	–	–	0·4	0·08	Nil	–	–	0·04	–	–	–	–	0·3	–
Suspended iron	ppm as Fe	–	–	–	–	–	–	–	–	–	–	–	–		–
Ammonia	ppm as $CaCO_3$	5	1	–	0·18	0·41	0·07	2·18	Nil	–	–	1·5	0·03	107	–
Chloride	ppm as $CaCO_3$	15	15	20·5	18	14·5	9·9	13	35	30	24	56	104	564	27,280
Sulphate	ppm as $CaCO_3$	28	11	28	–	12·7	5	15·5	16	31	29	77	84	318	2820
Phosphate	ppm as $CaCO_3$	–	–	–	–	–	–	–	–	–	–	3	–	30	–
Nitrate	ppm as $CaCO_3$	–	–	–	2·8	1·4	1·6	8·5	15·7	–	–	20	20	28·6	–
Nitrite	ppm as $CaCO_3$	–	–	–	Nil	Nil	1·2	–	Nil	–	–	–	–	1·8	–
Fluoride	ppm as $CaCO_3$	–	–	–	–	0·24	0·39	–	–	–	–	–	–	–	–
Free carbon dioxide	ppm as CO_2	–	–	5	Nil	0·7	0·4	–	–	6	37	–	38	–	–
Silica	ppm as SiO_2	–	–	4	–	4	2·2	5·5	16	8	6	3	23	97	–
Anionic detergent	ppm	–	–	–	–	–	–	–	–	–	–	0·14	–	10	–
*Fulvic acids	ppm	–	–	1·5	1·7	0·9	–	–	–	2·7	0·4	2·7	less than 0·05	–	–
*Oxygen demand from $KMnO_4$ (4 h at 27°C)	ppm as O_2	–	–	2·2	1·0	0·5	0·06	0·36	0·4	–	–	–	–	23·8	–
*Oxygen demand from $KMnO_3$ (30 m at 100°C)	ppm as O_2	–	–	–	–	–	–	–	–	–	–	3·8	–	–	–
Colour (Hazen units)		–	–	25	7	less than 3	–	–	–	–	–	–	–	–	–
Other salts	ppm	–	–	–	–	–	–	–	–	–	–	–	–	–	120
Biochemical oxygen demand (B.O.D.)	ppm O_2	–	–	–	–	–	–	–	–	–	–	–	–	46	–

3.2. SOURCES AND QUALITY OF RAW WATER

The composition of water from different sources varies widely both in the amount of dissolved salts and in the dissolved gases which it contains. In addition, surface waters, that is water from rivers and lakes, etc., usually contain suspended matter and often contain organic matter in solution and in suspension, derived from either decayed plant material or sewage.

In recent years the increasing use of synthetic detergents, some of which are not readily destroyed in sewage treatment processes, has resulted in measurable amounts of these chemicals being present even in public supply water.

In Table 1 the main chemical characteristics of a few sources of water are given.

It will be noted that even rain water, which is often thought to be relatively pure, can contain significant quantities of dissolved salts. This is because the moisture, as it precipitates, dissolves material from atmospheric pollution emitted by industries, particularly coal- or oil-burning furnaces and domestic coal fires. Also due to the acidic nature of industrial atmospheric discharges, rain water in industrial areas may have a low pH and be potentially corrosive.

Most waters can be treated to make them suitable for cooling services, the purpose of any treatment employed being to minimise the risk of fouling or corrosion of the heat-exchange surfaces, or corrosion of the associated plant through which the water passes. The essential requirement for cooling services is an abundant supply of water and, for example, a modern 2000 MW station with a once-through cooling system, requires about 60 million gallons per hour. There are no rivers in this country with a minimum fresh water flow of this magnitude; the River Thames at Teddington Weir has, at times, a fresh water flow of less than 200 million gallons per day. For this reason, stations of this size employing once-through cooling water systems can only be sited on large estuaries or at coastal sites.

With recirculating cooling water systems, scaling problems become more severe due to the concentration of the salts present in the water. These aspects will be considered in greater detail in Section 3.8.1.

Sewage effluent is already in use at a few power stations to provide the make-up for cooling tower systems but creates special problems due to the presence of significant amounts of phosphates and detergents. In some parts of the country where supplies of water for cooling purposes are inadequate, the use of sewage effluent may have to be extended. Sewage effluent has one particular virtue, namely that the quantity available is assured. To cater for this increasing use of sewage effluent, various treatments are being examined and developed. Work in this country is being conducted by the Water Pollution Research Laboratory of the Ministry of Technology and particular mention should be made of the process under examination for the removal of detergents from water, by air foaming.

Very pure water containing no more than a scarcely-measurable trace of dissolved salts is required for boiler feed make-up purposes. The cost of preparing this pure water, which may at times amount to some 500,000 gal/day for each 1000 MW of plant installed, will in general increase in proportion to the total amount of dissolved solids (T.D.S.) which the raw water contains. Thus whenever possible, water having a low T.D.S. is used. Where the local water undertaking can meet the additional demand this water is the first choice, since it will certainly have a relatively low T.D.S., usually less than 500 ppm, and often has received some form of treatment.

In this country, there are no nationally recognised standards for drinking water but most, if not all, public supplies are bacteriologically safe, over 90% of them being chlorinated. The Public Health Act and Waterworks Clauses Act requires that all waterworks shall provide a pure and wholesome water and this is usually taken to mean not injurious to health, but opinions vary widely as to what is acceptable in terms of chemical composition.

Since 1958 the World Health Organisation has published International Standards for Drinking Water and these are considered to be minimal standards which are within the reach of all countries throughout the world. The chemical aspects of the 1964 International Standards are summarised in Table 2. In addition the World Health Organisation have recognised that certain countries and regions of the world have the economic and technical capability of attaining a higher standard and in this connection publish European Standards for Drinking Water.

TABLE 2

INTERNATIONAL STANDARDS FOR DRINKING WATER 1964 (WHO, Geneva)

Substance	Maximum acceptable concentration	Maximum allowable concentration
Total solids	500 mg/l	1500 mg/l
Colour	5 units	50 units
Turbidity	5 units	25 units
Taste	unobjectionable	–
Odour	unobjectionable	–
Iron (Fe)	0·3 mg/l	1·0 mg/l
Manganese (Mn)	0·1 mg/l	0·5 mg/l
Copper (Cu)	1·0 mg/l	1·5 mg/l
Zinc (Zn)	5·0 mg/l	15 mg/l
Calcium (Ca)	75 mg/l	200 mg/l
Magnesium (Mg)	50 mg/l	150 mg/l
Sulphate (SO_4)	200 mg/l	400 mg/l
Chloride (Cl)	200 mg/l	600 mg/l
pH range	7·0–8·5	6·5 – 9·2

COMPOUNDS AFFECTING THE POTABILITY OF WATER

Substance	Maximum allowable limit
Total dissolved solids	1500 mg/l
Iron	50 mg/l
Manganese (assuming that the ammonia content is less than 0·5 mg/l)	5 mg/l
Copper	1·5 mg/l
Zinc	1·5 mg/l
Magnesium plus sodium sulphate	1000 mg/l
Alkyl benzyl sulphonates (ABS: surfactants)	0·5 mg/l

COMPOUNDS HAZARDOUS TO HEALTH

Substance	Maximum allowable limit
Nitrate as NO_3	45 mg/1
Fluoride	1·5 mg/1

TOXIC SUBSTANCES

Substance	Maximum allowable limit
Phenolic substances	0·002 mg/1
Arsenic	0·05 mg/1
Cadmium	0·01 mg/1
Chromium	0·05 mg/1
Cyanide	0·2 mg/1
Lead	0·05 mg/1
Selenium	0·01 mg/1

Note: mg/l is equivalent to ppm.

Returning to the consideration of the sources of water used for boiler feed make-up, it is sometimes possible to provide suitable water from a private well on the site, but frequently the raw water is obtained from a fresh water river or canal and requires some pretreatment. At a few estuary and coastal stations no fresh water of a suitable quality is available and then as a last resort, sea water or estuary water provides the starting point.

The use of sewage effluent as a source of water for boiler feed make-up water may have to be considered. It can certainly be rendered suitable for evaporation but there are at present difficulties in the utilisation of ion-exchange treatment for sewage effluent.

In order to understand the mechanism of the treatments employed to render the various sources of raw water suitable for the power generation cycle, it is useful to discuss its chemical composition in more detail.

3.2.1. **Chemical Composition**

The major constituents of all natural waters consist of the salts of sodium, potassium, calcium and magnesium, together with bicarbonate, carbonate, sulphate, chloride and nitrate ions. Other constituents present, usually in low concentrations but which may create special problems, are silica, hydrogen sulphide, ammonia, organic matter (particularly substances known as fulvic acids), detergents, phosphates, and dissolved gases. Silica occurs in some water in several different forms some of which escape detection by the normal chemical test employed ("non-reactive" silicon), and phosphate can occur as complex phosphate in which form it not only escapes detection by the normal test used for orthophosphates but there is some evidence to suggest that it can interfere with water-softening reactions. Sea water contains traces of most of the soluble salts and small concentrations of nitrates, ferrous iron and manganese are found in certain waters.

Calcium bicarbonate, usually included in the temporary hardness or alkaline hardness, is unstable and on heating breaks down, depositing calcium carbonate. Some calcium and magnesium salts are sparingly soluble so that if the water is evaporated or concentrated they crystallise out as hard scale.

The equivalent mineral acidity (E.M.A.) is a measure of the total concentration of salts of strong acids (sulphuric, hydrochloric and nitric acids, usually) present in the water.

Carbon dioxide in the form of carbonates and bicarbonates is referred to in water treatment as *combined carbon dioxide*. Natural waters also contain a small concentration of carbon dioxide dissolved in the water and this is known as *free carbon dioxide*.

The other salts present in the water create no particular problem but since they are electrolytes they will, if present in sufficient concentration (for example sea water), allow galvanic action to take place between dissimilar metals, resulting in dissolution of the more anodic metals.

3.3. BOILER FEED WATER MAKE-UP

Reference has already been made to the fact that ultra-pure water is required for boiler feed water make-up in a modern generating station. A typical specification for a new high-pressure station may call for the following guaranteed quality:

Conductivity (before and after passage through a cation-exchange column in the hydrogen form)	Not greater than 0·10 micromho/cm (corrected to 25°C)
Sodium	Not greater than 0·015 ppm as Na
Silica (including non-reactive silicon)	Not greater than 0·02 ppm as SiO_2 (with 95% confidence limits)

The quality of the water required is almost that of absolutely pure water which has a conductivity of 0·056 micromho/cm at 25°C and 0·038 at 18°C and there is, therefore, little room for improvement on the quality specified above. The real problem is to ensure that this quality is consistently maintained, particularly with regard to silica.

Two alternative processes are used to prepare this water and these processes are illustrated diagrammatically in Figure 3.3. The essential difference is that the main salt removal stage is evaporation in scheme (A) and demineralisation in scheme (B).

The choice of process employed depends partly on the quality of the available water supplies. As a general rule it is uneconomic to use demineralisation for waters containing more than about 1000 ppm T.D.S. In making this generalisation it is assumed that, as with most waters, a high percentage of the 1000 ppm T.D.S. consists of calcium and magnesium salts, particularly bicarbonates. A conventional demineralisation plant as illustrated in

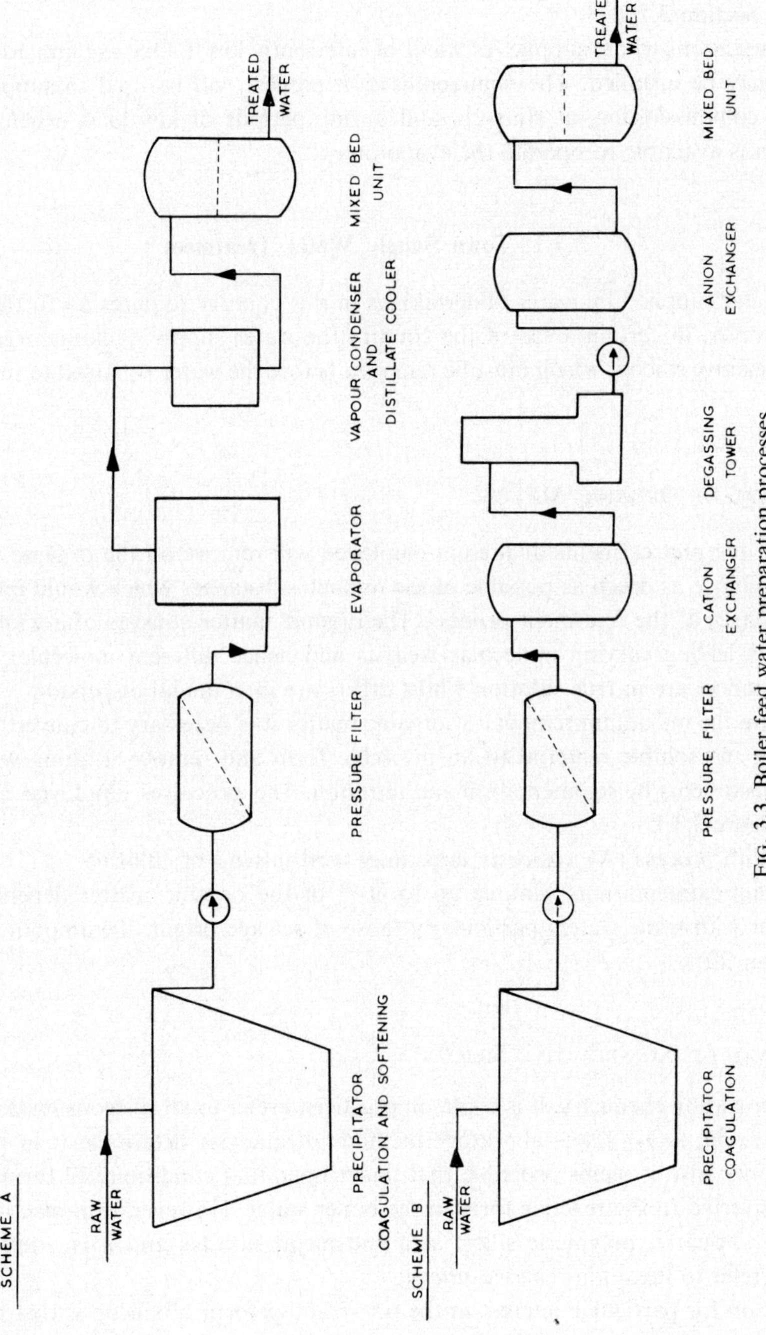

FIG. 3.3. Boiler feed water preparation processes

scheme (B) is unlikely to be economic with a high T.D.S. brackish water, derived from sea water, when sodium chloride will be the major constituent. These aspects will be discussed further in Section 3.7.

At a few stations where supplies of water of sufficiently low T.D.S. are limited, both types of plant may be installed. The demineralisation process will be used to supply make-up water for commissioning, at start-up, and during periods of low load, when insufficient bled steam is available to operate the evaporators.

3.3.1. Town Supply Water Treatment

Most water supplied by water undertakings in this country requires no further pretreatment; however, in certain areas of the country the water supply contains organic matter and non-reactive silicon, which must be removed before the water is passed to the treatment plant.

(a) Removal of Organic Matter

None of the pretreatments at present employed will remove all the organic matter; the aim is to remove as much as possible of the organic substances which would interfere with the later stages of the treatment process. The organic matter consists of a wide variety of compounds having varying molecular weights and hence different molecular sizes; thus some substances are in true solution whilst others are in colloidal dispersion.

To ensure the maximum removal of organic matter it is necessary to convert as much as possible of the soluble material to an insoluble form and remove it along with the less soluble constituents by sedimentation and filtration. The processes employed are summarised in Figure 3.3.1.

Note: With process (A), ozone is sometimes used instead of chlorine.

These pretreatments may remove up to 80% of the organic matter depending on its nature, but with some waters particularly those of sewage origin, the amount may be no higher than 20%.

(b) Removal of Non-reactive Silicon

In power station chemistry, it is common practice to refer to all siliceous material as silica (silicon dioxide, SiO_2) since laboratory methods of analysis determine it in this ionised reactive form; also it seems probable that under operating conditions all forms of silicon will be converted to the reactive form in the boiler water. However, non-reactive silica includes clays, quartz, polymeric silicic acid and metal silicates and it is, therefore, more precise to refer to it as non-reactive *silicon.*

The reason for particular interest in the non-reactive form of silicon is that it may pass undetected through a demineralisation plant and so enter the feed system. There is also some evidence to suggest that some of the reactive silicon present in the raw water can, under certain circumstances, be converted to the non-reactive form during passage of the

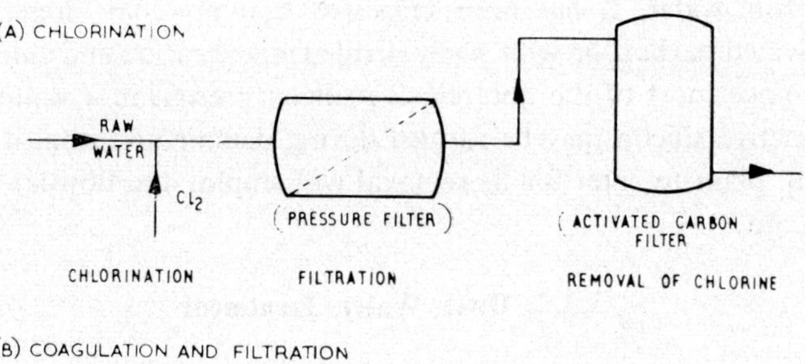

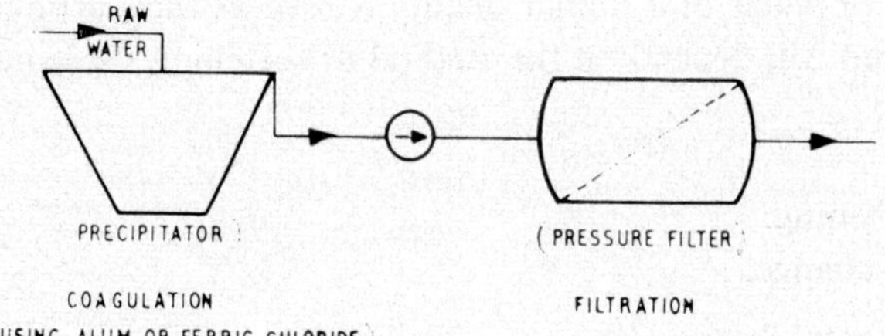

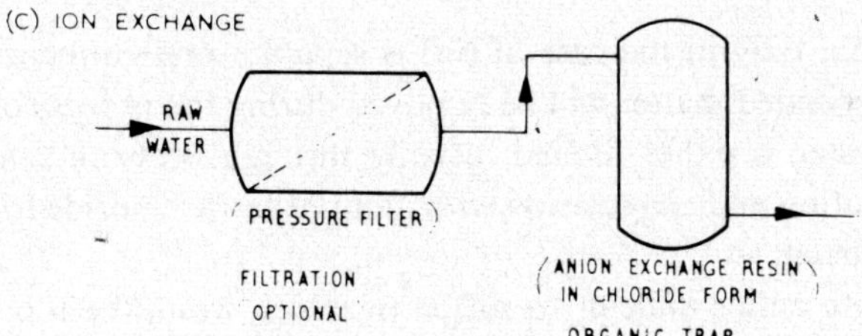

FIG. 3.3.1. Treatment for removal of organic matter

water through a demineralisation plant. This can lead to unexpected increases in the boiler water silica content.

A method of analysis is now available for determining the total silicon content of a water and by determining the reactive silicon content, the non-reactive silicon content can be obtained by difference. Techniques are also being developed for the determination of the different forms of silicon present in water.

There is as yet no experience in this country with methods suitable for removing non-reactive silicon from water. It has been suggested that pre-coat filters, especially filters coated with activated carbon or with finely-divided mixed cation and anion resins (Powdex process), can remove most of the non-reactive silicon present in a water. Because of the fear that non-reactive silicon may be formed during demineralisation, it is probable that the first C.E.G.B. plant to cater for its removal will employ the Powdex process after the final mixed bed unit.

3.3.2. **River Water Treatment**

Where river water or water of a similar quality is used as the starting point, the type of pretreatment employed will depend on the method of softening subsequently used, that is, one of the following:

(i) Lime/soda softening.
(ii) Lime—base exchange.
(iii) Blend process.

Apart from chlorination, only in the case of (iii) is separate pretreatment necessary since with (i) and (ii) any suspended matter will be removed during the process of softening. For a blend process the water is either filtered, usually through pressure sand or anthracite filters, or else sedimentation and coagulation with alum or ferric chloride followed by pressure filtration, is employed.

With waters which are either difficult to soften or which would be too costly to soften because of a high magnesium content, such as estuary waters, the treatment employed before evaporation consists of chlorination, coagulation, sedimentation and filtration, neutralisation of the alkalinity or temporary hardness with acid, degassing to remove carbon dioxide, and pH correction with caustic soda to produce a slightly alkaline water. The chlorination is used to prevent precipitated sludge from polluted water becoming septic, which could result in it being resuspended in the water by gas produced by proliferating bacteria.

The type of precipitator employed is usually similar to the sludge-blanket softener illustrated in Figure 3.4.2B. Sludge is drawn off continuously from the bottom of the precipitator as a suspension containing approximately 3% solids. If it cannot be returned to the river, it is usual to concentrate the sludge before disposal, usually in some form of settlement tank or pit, the separated water being returned to the precipitator. In this way a sludge containing about 17% solids can be produced and from a 1000 MW estuary station this may amount to about 40 tons of sludge per day.

3.4. WATER SOFTENING

The object of water softening is to remove the calcium and magnesium salts which would cause scale deposition, particularly in evaporators.

3.4.1. Lime/Soda Softening

In lime-soda softening, advantage is taken of the fact that calcium carbonate and magnesium hydroxide are sparingly soluble in water and can be precipitated from solution by the following reactions:

$$\underset{\text{Calcium bicarbonate}}{Ca(HCO_3)_2} + \underset{\text{Lime}}{Ca(OH)_2} \longrightarrow \underset{\text{Calcium carbonate}}{2\,CaCO_3\downarrow} + 2\,H_2O \quad (1)$$

$$\underset{\text{Calcium sulphate}}{CaSO_4} + \underset{\text{Soda}}{Na_2CO_3} \longrightarrow \underset{\text{Calcium carbonate}}{CaCO_3\downarrow} + \underset{\text{Sodium sulphate}}{Na_2SO_4} \quad (2)$$

$$\underset{\text{Magnesium bicarbonate}}{Mg(HCO_3)_2} + \underset{\text{Lime}}{2Ca(OH)_2} \longrightarrow \underset{\text{Magnesium hydroxide}}{Mg(OH)_2\downarrow} + \underset{\text{Calcium carbonate}}{2\,CaCO_3\downarrow} + 2\,H_2O \quad (3)$$

$$\underset{\text{Magnesium chloride}}{MgCl_2} + \underset{\text{Lime}}{Ca(OH)_2} + \underset{\text{Soda}}{Na_2CO_3} \longrightarrow \underset{\text{Magnesium hydroxide}}{Mg(OH)_2\downarrow} + \underset{\text{Calcium carbonate}}{CaCO_3\downarrow} + \underset{\text{Sodium chloride}}{2NaCl} \quad (4)$$

It will be seen in the above reactions that the precipitating agents are calcium hydroxide (lime or more correctly hydrated lime, $Ca(OH)_2$), and sodium carbonate (soda-ash or soda, Na_2CO_3). These are the reagents normally used in water softening but in fact any soluble carbonate or hydroxide could be used. However, in removing temporary hardness, lime is a particularly suitable reagent since the only soluble product of the reactions (1) and (3), is water and this therefore results in a reduction of the T.D.S. of the water.

At 0°C the solubility of calcium carbonate is 15 ppm and magnesium hydroxide 17 ppm. In the presence of a small excess of carbonate and hydroxide ions it is usually possible to reduce the combined residual hardness to below 20 ppm as $CaCO_3$ by cold lime-soda softening, the normal excess being:

20 ppm hydroxide alkalinity as ppm $CaCO_3$
40 to 60 ppm carbonate alkalinity as ppm $CaCO_3$

In considering the chemical reactions taking place in water softening, it is assumed that if the calcium and magnesium content is in excess of the bicarbonate concentration (bicarbonate alkalinity), then all the bicarbonate alkalinity is associated with calcium and magnesium and in this form is known as the *temporary hardness* or *alkaline hardness*. The excess of calcium and magnesium must then be associated with the chloride, sulphate and other anions present and this is known as the *permanent hardness* or *non-alkaline hardness*.

Temporary hardness or bicarbonate alkalinity	Permanent hardness	
	Calcium chloride	$CaCl_2$
Calcium bicarbonate $Ca(HCO_3)_2$	Calcium sulphate	$CaSO_4$
	Calcium nitrate	$Ca(NO_3)_2$
	Magnesium chloride	$MgCl_2$
Magnesium bicarbonate	Magnesium sulphate	$MgSO_4$
$Mg(HCO_3)_2$	Magnesium nitrate	$Mg(NO_3)_2$

Note: In water analysis, it is normal to report all the main constituents in terms of the equivalent concentration of calcium carbonate (ppm $CaCO_3$).

At higher temperatures the solubilities of both calcium carbonate and magnesium hydroxide are reduced and it is therefore possible to soften water to a lower residual hardness, using smaller excesses of hydroxide and carbonate alkalinities, if some form of heating is employed. As a corollary to this, poor water softening often occurs in periods of low ambient temperature so that with softeners sited outdoors, improved performance can often be achieved if heating is available during the winter months.

The reactions represented by equations (1) to (4) take place fairly rapidly since they are ionic reactions. However, the particles of calcium carbonate and magnesium hydroxide first formed, will be of near colloidal dimensions and would pass through the various types of filter employed. Over a period of time the precipitated material grows into larger crystals by a process of redissolving and recrystallisation and at normal ambient temperature, in about 4–6 h, crystals of sufficient size and density will have formed and will sink to the bottom of the reaction vessel.

(a) Use of Coagulants

If the small crystals formed in the early stages of the softening reaction can be induced to agglomerate into larger particles, the rate of sedimentation can be increased thus reducing the burden on the filter. This is achieved in water softening by adding small quantities (usually about 20 ppm) of the salts of polyvalent metals such as iron and aluminium. In the alkaline soft water these precipitate a bulky gelatinous floc of metal hydroxides which carry a small positive charge. These flocs agglomerate the calcium carbonate and magnesium hydroxide crystals, partly by pure physical entanglement and partly because the crystals carry a small negative charge.

Activated silica and starch are also used, often in conjunction with alum. Silica in the form of a gel has been found particularly useful on certain waters, notably those low in magnesium.

More recently organic chemicals having high molecular weights and with molecules having many ionisable groups, have been used successfully as coagulants. These chemicals are known as polyelectrolytes and many different types are available. They have the advantage of being effective at a much lower concentration (usually 1 or 2 ppm) but at the present stage of development they are very expensive and are not always as effective as the conventional coagulants.

To obtain the maximum effect, aluminium sulphate is added to the raw water after the softening chemicals. Alternatively sodium aluminate may be used and frequently this is premixed with the lime and soda as a combined softening charge.

(b) Effect of Nuclei

If suitable nuclei are present in the water the crystals of calcium carbonate and magnesium hydroxide will grow preferentially on them, resulting in more rapid growth of crystals large enough to settle. Thus the time required for the softening reactions to be complete may be considerably reduced. The nuclei may be either added material of suitable size, such as sand, or alternatively the previously precipitated softening sludge may be recycled. This reduces the time required for softening to an hour or less, thus enabling the size of the reaction vessel for a given output to be reduced proportionately.

3.4.2. **Types of Lime and Lime/Soda Softeners**

To illustrate the various possibilities in greater detail, the following three main types are more fully described:

(a) "Conventional" cold lime-soda softener.
(b) Sludge blanket lime-soda softener.
(c) Catalyst lime softener.

(a) Conventional Cold Softener

A typical softener of this type is illustrated in Figure 3.4.2A. In this case it consists of a tall cylindrical tank constructed of steel or concrete and a central cylindrical downcomer which terminates some distance above the bottom of the outer vessel; this serves as a mixing and reaction compartment. Raw water and softening chemicals enter at the top of this compartment and a coagulant, if added separately, enters a short way down. To ensure adequate mixing a stirrer is often included. The water flows downwards through the central chamber and then upwards through the outer annular space. As it does so the precipitated hardness falls into the dead space at the bottom, concentrates into sludge and is drawn off either intermittently or continuously.

In some of the earlier softeners (as illustrated) the upper part of the outer compartment contained a wood-wool filter through which the outgoing water passed. In other versions, filtration is done externally in pressure (sand or anthracite) filters. For either of these the aim is to produce a virtually clear water before it enters the filter, in order to ensure a long filter life. This is achieved partly by the correct choice of coagulants and by designing the softener to ensure that the flow rate is sufficiently low and the retention time sufficiently long, so that even under conditions of low ambient temperature softening is complete.

Addition of softening chemicals. The solubilities of the two softening chemicals are very different, sodium carbonate (soda ash) having solubility of about 17% at 20°C whereas calcium hydroxide (lime) will only dissolve to the extent of less than 0·2% at 0°C and also

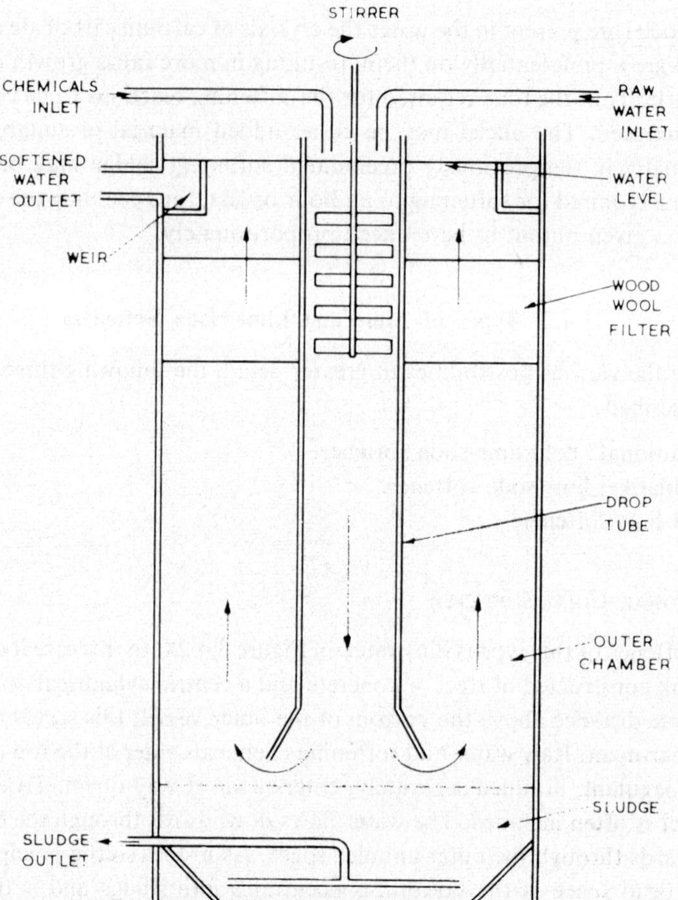

FIG. 3.4.2A. Conventional lime/soda softener

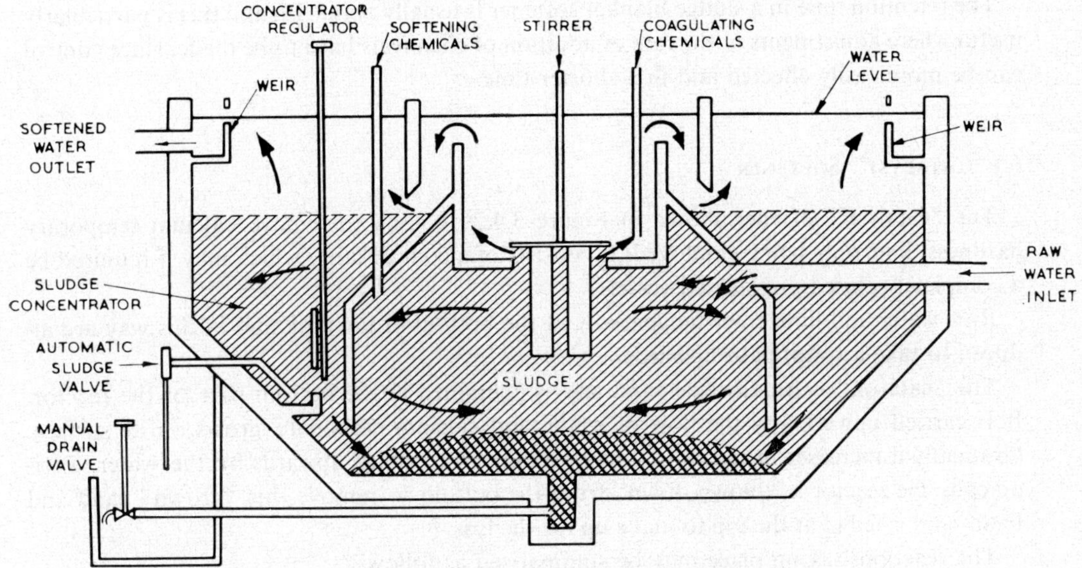

FIG. 3.4.2B. Sludge blanket lime/soda softener

exhibits retrograde solubility. Therefore, if the chemicals are added separately, soda ash is added as a solution and lime usually as a suspension. Only in some small softeners is lime solution used and in this case a lime saturator is employed in which part of the incoming raw water passes through a vessel containing powdered lime.

Where the composition of the raw water is fairly constant it is convenient to feed the two chemicals mixed together in suspension.

Control of the rate of addition of softening chemicals is effected by either volumetric apportioning gear regulated by the raw water flow or by using variable stroke metering pumps.

(b) SLUDGE BLANKET SOFTENER

A softener of this type is illustrated diagrammatically in Figure 3.4.2B. It employs the principle referred to earlier of resuspending a proportion of the precipitated sludge in the raw water, providing nuclei on which the freshly precipitated hardness can crystallise.

Raw water is first mixed with the softening chemicals and some precipitated sludge, the mixing and re-entrainment of the sludge being assisted by the central stirrer. The mixture then flows upwards through the sludge zone into the central reaction chamber at a rate many times that employed in conventional softeners, usually about 20 ft/h, and at this stage is mixed with the added coagulant.

If the flow rate, the rate of addition of chemicals and the rate of sludge removal are correctly adjusted, the softening reactions should be completed within the sludge zone, the precipitated material remaining in this zone so that the sludge level is held constant and the water rising in the outer chamber is clear.

The retention time in a sludge blanket softener is usually about 1 h and this is particularly useful where adjustments in the rate of addition of chemicals have to be made, since control can be more easily effected and in a shorter time.

(c) "Catalyst" Softener

The "catalyst" softener shown in Figure 3.4.2B is used to remove calcium temporary hardness, lime only being employed. It consists of a conical tank, which may if required be a completely closed pressurised vessel.

Raw water and lime enter tangentially at the bottom of the tank and in this way are induced to take a spiral path upwards.

The "catalyst" is usually fine sand and is contained in the bottom part of the reactor. It is carried upwards by the flow of water and calcium carbonate grows on its surface. Gradually it increases in size until it can no longer be carried upwards by the water. Periodically the reactor is "blown down" from the bottom to remove this "grown" sand and fresh sand is fed in at the top to make up for the loss.

The reactions taking place may be summarised as follows:

$$Ca(HCO_3)_2 + Ca(OH)_2 \longrightarrow 2\,CaCO_3\downarrow + 2\,H_2O \tag{5}$$

$$Mg(HCO_3)_2 + Ca(OH)_2 \longrightarrow CaCO_3\downarrow + MgCO_3 + 2\,H_2O \tag{6}$$

$$MgCO_3 + CaCl_2 \longrightarrow CaCO_3\downarrow + MgCl_2 \tag{7}$$

Note: If sufficient lime is added, magnesium hydroxide will be precipitated:

$$MgCl_2 + Ca(OH)_2 \longrightarrow Mg(OH)_2\downarrow + CaCl_2 \tag{8}$$

Since magnesium hydroxide will not grow on sand, it will be carried forward and overload the filter, if one is fitted.

In order to calculate the amount of lime required it is usually assumed that all the bicarbonates are associated with calcium, the magnesium with the sulphate and chloride, etc. and in order to ensure that magnesium hydroxide is not precipitated, slightly less than the calculated amount of lime is used. Obviously in competing chemical reactions of this sort, success depends largely on the concentration of the various constituents present. With a water having a high magnesium content (above about 40 ppm as $CaCO_3$) some precipitation of magnesium hydroxide is likely to occur and for this reason the catalyst softener is seldom used for these waters.

The advantages of the catalyst softener are:

1. Compact size.
2. It can be pressurised so that softened water can be fed direct to a filter without additional pumps.
3. The precipitated hardness is in a granular form having a low moisture content (that is, little water is wasted) and which can be easily disposed of.

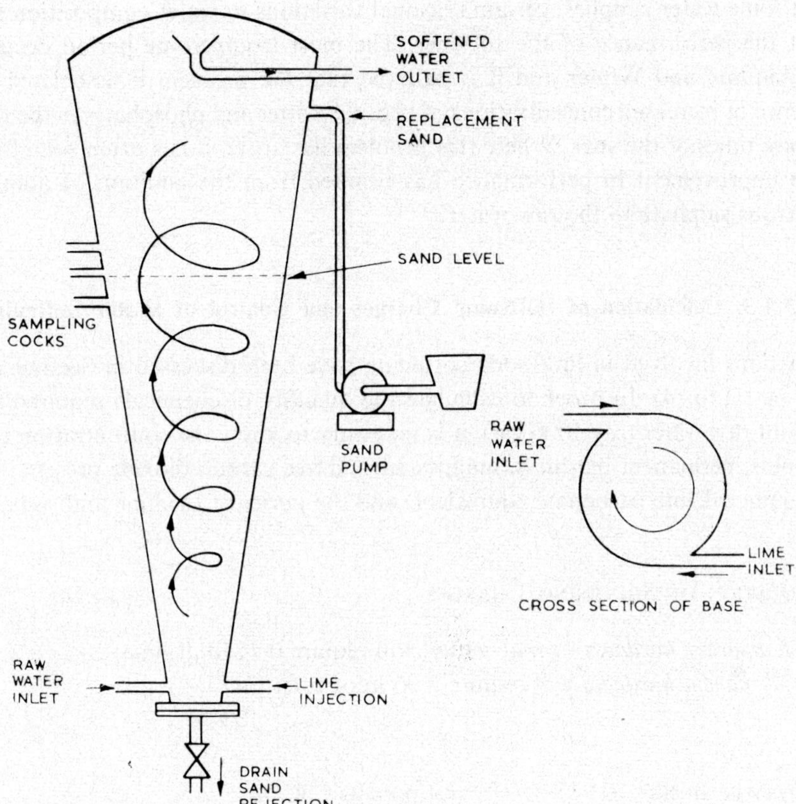

FIG. 3.4.2C. Catalyst lime softener

The disadvantages are:

1. Normally only calcium temporary hardness is removed, although it is also possible to add soda and remove calcium permanent hardness.
2. Under low-temperature conditions "sand growth" is very slow with the result that filters become overloaded by the quantity of calcium carbonate carried forward to them.
3. With some water supplies, certain seasonal variations in water composition adversely affect the performance of the softener. The most troublesome period occurs in the late Autumn and Winter and it is believed that the problem is associated with the presence of increased concentrations of organic matter and phosphates in the raw water at these times of the year. Where this problem has arisen it has often been found that some improvement in performance has resulted from the addition of about 5 ppm of ferrous sulphate to the raw water.

3.4.3. Calculation of Softening Charges and Control of Water Softening

The reactions involved in lime soda softening have been discussed in Section 3.4.1 and in equations (1) to (4). In order to calculate the quantity of chemicals required to soften a quantity of raw water (say 1000 gal), it is necessary to know the concentration of temporary hardness, permanent hardness, magnesium and free carbon dioxide present in the raw water (as ppm calcium carbonate equivalent) and the purity of the lime and soda.

(a) Calculation of Softening Charges

1 lb of *temporary hardness* or *magnesium* will require 0·74 lb of lime
1 lb of *permanent hardness* will require 1·06 lb of soda

Let

Temporary hardness	$=A$ ppm as $CaCO_3$
Permanent hardness	$=B$ ppm as $CaCO_3$
Magnesium	$=C$ ppm as $CaCO_3$
Free carbon dioxide	$=D$ ppm as $CaCO_3$
Purity of lime	$=E\%$
Purity of soda	$=F\%$
Final hydroxide alkalinity required	$=G$ ppm as $CaCO_3$
Final carbonate alkalinity required	$=H$ ppm as $CaCO_3$

Then to soften 1000 gal of raw water of the above quality, will require

$$\text{LIME} = \left[\frac{A+C+D+G}{100}\right]\times 0{\cdot}74\times\frac{100}{E}\ \text{lb} \tag{9}$$

$$\text{SODA} = \left[\frac{B+G+H}{100}\right]\times 1{\cdot}06\times\frac{100}{F}\ \text{lb} \tag{10}$$

Using equations (9) and (10) the quantity of lime and soda required to soften each 1000 gal of raw water can be calculated. The tanks used to store the charge of lime/soda suspension will hold sufficient suspension to treat several thousand gallons of water so that in practice the appropriate multiple of the amounts of lime and soda, calculated as above, are weighed out, made up into a suspension (between 5% and 10% solids) in a mixing tank and then transferred to the storage tank.

(b) Control of Lime/Soda Softening

During operation, samples of the softened water leaving the softener are taken and analysed to check that softening is being carried out correctly. The soft water will normally be considered satisfactory if it conforms to the limits given in Table 3 which follows.

TABLE 3

CONTROL OF LIME-SODA SOFTENER

Total hardness (calcium + magnesium) = 10 to 20 ppm as $CaCO_3$
Hydroxide alkalinity = 20 to 30 ppm as $CaCO_3$
Carbonate alkalinity = 40 to 60 ppm as $CaCO_3$
Bicarbonate alkalinity = NIL

It will be noted that the alkalinity of the softened water is referred to as *hydroxide* or *carbonate*, whereas previously in discussing the composition of raw water, reference was made to *bicarbonate* alkalinity or temporary hardness. To determine the form or forms of alkalinity present in a water and also to estimate the concentration of each type, a filtered sample of water is titrated stepwise against standard acid solution. The first step involves titration to an end-point determined by phenolphthalein indicator (pH 8·3) and the second step to an endpoint determined by methyl orange indicator (pH 3·8 to 4·5). The values of the total titrations to these end points are then converted to their equivalent calcium carbonate concentration in ppm and are known as the:

P value—Phenolphthalein alkalinity
M value—Methyl orange alkalinity
of the water.

In water-softening calculations, alkaline salts such as phosphates and silicates are usually ignored and it is considered that the alkalinity of the water is due to the presence of the following.

TABLE 4

INTERPRETATION OF ALKALINITY

	P and M values
(a) Bicarbonates only	$P = 0$, M only
(b) Bicarbonates + carbonates	M greater than $2P$
(c) Carbonates only	$M = 2P$
(d) Carbonates + hydroxides	M less than $2P$
(e) Hydroxides only	$M = P$

TABLE 5

CONTROL OF LIME-SODA SOFTENER

Composition of softened water				Action necessary
Hardness	Bi-carbonate	Carbonate	Hydroxide	
High	Nil	Normal	Normal	Slightly increase rate of chemical addition to softener and recheck after suitable interval.
High	Nil	Normal	High	Reduce amount of lime and soda in softening charge and recheck after suitable interval.
Normal	Nil	High	Normal	Reduce amount of soda in softening charge and recheck after suitable interval.
High	Nil	High	High	This is probably due to raw water temperature being too low or due to the presence of interfering substances in the raw water (See Section 3.2.1 and 3.2.2.4.)
High	Present	High or normal	Nil	Increase amount of lime in softening charge.

Where only one type of alkalinity is present, as in (a), (c) and (e) in the above table, the concentration of this type of alkalinity is equal to the M value.

If *carbonate and bicarbonate are both present* (M greater than $2P$) then:

The *carbonate* concentration will be equal to $2P$.

The *bicarbonate* concentration will be equal to $(M-2P)$.

If *carbonate and hydroxide are both present* (M less than $2P$) then:

The *carbonate* concentration will be equal to $2(M-P)$.

The *hydroxide* concentration will be equal to $(2P-M)$.

Since there should be no bicarbonate alkalinity in the softened water, $2P$ should always be slightly greater than M.

If the hardness, hydroxide or carbonate alkalinity falls outside the ranges given in Table 3, then it is important to first check that the softener is functioning correctly and that the correct amounts of chemicals are being fed to the softener. If everything is in order and the chemical composition of the raw water is unchanged, then the action indicated in Table 5 should be taken.

In practice unless softening is very unsatisfactory, it is best to make only small adjustments and then, after noting their effect, to make further small adjustments. Since with some softeners the retention time is long, some time will elapse before the effect of any adjustments is observed. For this reason it is essential to analyse the raw water sufficiently frequently to enable prompt adjustments to the chemical charge to be made as soon as any changes occur in the water composition.

3.4.4. **Base-exchange or Zeolite Softening**

The previous section considered various precipitation processes for the removal of unwanted cations (calcium and magnesium) from the water. In this section a process known as *base-exchange softening* which does not involve precipitation, will be described.

(a) The Development of Base-Exchange Softeners

Chemical reactions of the type

$$AX + BY \rightleftharpoons BX + AY \tag{11}$$

where AX, BY, BX and AX are salts, are very familiar and take place whenever two salts are dissolved in water. Where none of the reactants or products undergo physical change, it is not possible to regard any particular cation as being associated with any particular anion. Thus if small but equivalent amounts of sodium chloride and potassium nitrate are dissolved in water, the solution will be identical with a solution obtained by dissolving similar equivalent amounts of potassium chloride and sodium nitrate in water:

$$NaCl + KNO_3 \rightleftharpoons KCl + NaNO_3 \tag{12}$$

If one of the products of the reaction is sparingly soluble, only slightly dissociated (ionised), or is gaseous and hence is lost from solution, the equilibrium of the reaction will be displaced to favour the formation of this product (see also Chapter 2), as in the following reactions:

$$NaOH + HCl \longrightarrow NaCl + \underset{\text{Only slightly dissociated}}{H_2O} \tag{13}$$

$$CaCl_2 + Na_2CO_3 \longrightarrow 2\,NaCl + \underset{\text{Only slightly soluble}}{CaCO_3} \tag{14}$$

$$Na_2CO_3 + 2\,HCl \longrightarrow 2\,NaCl + \underset{\text{Gaseous}}{CO_2} + H_2O \tag{15}$$

These are all reactions occurring between salts which ionise in solution. If instead of a soluble salt, one of the reactants was a solid insoluble compound containing labile cations and if the solid compound had different affinities for different cations, then the exchange of cations between salts in solution and the solid compound is possible as in the following reaction:

$$\underset{}{CaCl_2}+\underset{\text{Solid}}{Na_2Z} \rightleftharpoons \underset{\text{Solid}}{CaZ}+2\,NaCl \qquad (16)$$

Na_2Z and CaZ represent the solid compound in the sodium and calcium forms respectively. This is an example of a *cation-exchange* reaction and the extent to which it will occur will depend on the concentrations of calcium and sodium ions in solution at equilibrium and the relative affinity of the solid compound for sodium and calcium ions.

The first examples of cation exchange were discovered in the early part of the nineteenth century during investigations concerning the way in which soluble manures are retained for long periods in the soil, intead of being washed out by the rain. In 1845 M. S. Thompson showed that when a solution of ammonium sulphate, $(NH_4)_2SO_4$, is passed through a bed of soil, the solution emerging from the bottom of the bed contains calcium sulphate, thus:

$$(NH_4)_2SO_4+\underset{\substack{\text{Soil in}\\ \text{calcium form}}}{CaZ} \rightleftharpoons CaSO_4+\underset{\substack{\text{Soil in}\\ \text{ammonium form}}}{(NH_4)_2Z} \qquad (17)$$

During the latter half of the nineteenth century it was shown that these soil reactions were brought about by certain minerals present in the soil, the minerals being complex compounds of alkali and alkaline earth metals, silicon and aluminium, known as alumino-silicates. The minerals exhibiting the highest capacity for ion exchange are known as *zeolites* or *greensands*. Also over this period synthetic zeolites were prepared and were found to have greater ion exchange capacity than the natural minerals.

It was shown that zeolites can be used to effect many different cation-exchange reactions and when, for example in equation (16), all the zeolite has been converted to the calcium form and can no longer exchange calcium ions from solution, it can be restored to the sodium form by treating with a strong solution of a sodium salt and can then be used to carry out further cation exchange. Thus:

$$\underset{\substack{\text{Zeolite in}\\ \text{calcium form}}}{CaZ}+2\,NaCl \rightleftharpoons \underset{\substack{\text{Zeolite in}\\ \text{sodium form}}}{Na_2Z}+CaCl_2 \qquad (18)$$

The reaction given in equation (16) is in fact water softening, since calcium (and magnesium) are removed from solution and replaced by sodium. It is known as *base-exchange* or *zeolite* softening. In 1905, R. Gans made the first practical use of the process to soften water on an industrial scale.

These zeolite materials are unstable in acid or alkaline conditions so that their use is limited to approximately neutral waters (pH 7–9). In 1934, O. Liebknecht showed that when certain types of coal are treated with sulphuric acid, an ion-exchange material is produced, known as *sulphonated coal*, which has a greater exchange capacity than natural or synthetic zeolites and which is stable over a much wider pH range. Since these materials could be used under conditions of low pH, they could be converted to the hydrogen form and used to exchange cations for hydrogen ions, thus

$$\underset{\substack{\text{Sulphonated}\\ \text{coal in}\\ \text{hydrogen form}}}{CaCl_2 + H_2Z} \rightleftharpoons \underset{\substack{\text{Sulphonated}\\ \text{coal in}\\ \text{calcium form}}}{CaZ + 2\,HCl} \quad (19)$$

In 1935, B. A. Adams and E. C. Holmes prepared the first synthetic ion-exchange resins, which were condensation products of formaldehyde and polyhydric phenols. They not only produced cation-exchange resins having superior properties to sulphonated coals but also anion-exchange resins or rather, resins capable of absorbing strong acids from solution, that is to say, they behaved as weak bases, thus:

$$HCl + \underset{\substack{\text{Resin in}\\ \text{free-base}\\ \text{form}}}{R} \longrightarrow \underset{\substack{\text{Resin acid}\\ \text{complex}}}{R \cdot HCl} \quad (20)$$

These resins, when exhausted, can be regenerated or converted back to the free-base form by treating with a solution of any alkali, thus:

$$R \cdot HCl + NaOH \longrightarrow R + NaCl + H_2O \quad (21)$$

By passing water through the two kinds of resin in series, complete demineralisation of water by ion exchange became possible and the first commercial demineralisation plant was made by The Permutit Company Limited in this country, in 1937.

In 1944, D'Alelio produced a new type of ion-exchange resin in bead form, based on *polystyrene*, which had much better characteristics than the earlier resins and are now used almost exclusively in demineralisation plants in the C.E.G.B. The latter aspects are outside the scope of base-exchange softeners and will be discussed in greater detail in Section 3.6.1.

Returning to the consideration of base-exchange softeners, these have been employed extensively since the early work of Gans and many are in service in the C.E.G.B. Inorganic zeolites are no longer used and frequently the sulphonated coal materials are replaced by the more efficient synthetic resins, when renewal becomes necessary due to degradation and loss of exchange capacity.

Salt solution (NaCl) is used to regenerate the exchange material and since this is a relatively inexpensive chemical, the cost of water softening by base exchange compares favourably with lime-soda softening. The base-exchange process has the particular advantage that, during regeneration, the calcium and magnesium are removed from the exhausted exchange material in solution and there are, therefore, no sludge disposal and filtration problems.

(b) Description of Plant

A typical base-exchange softener is illustrated in Figure 3.4.4. It consists of a cylindrical steel pressure vessel with dished ends. In the bottom of the shell is an under-drain system which serves to collect the softened water during the softening run, distribute the back-

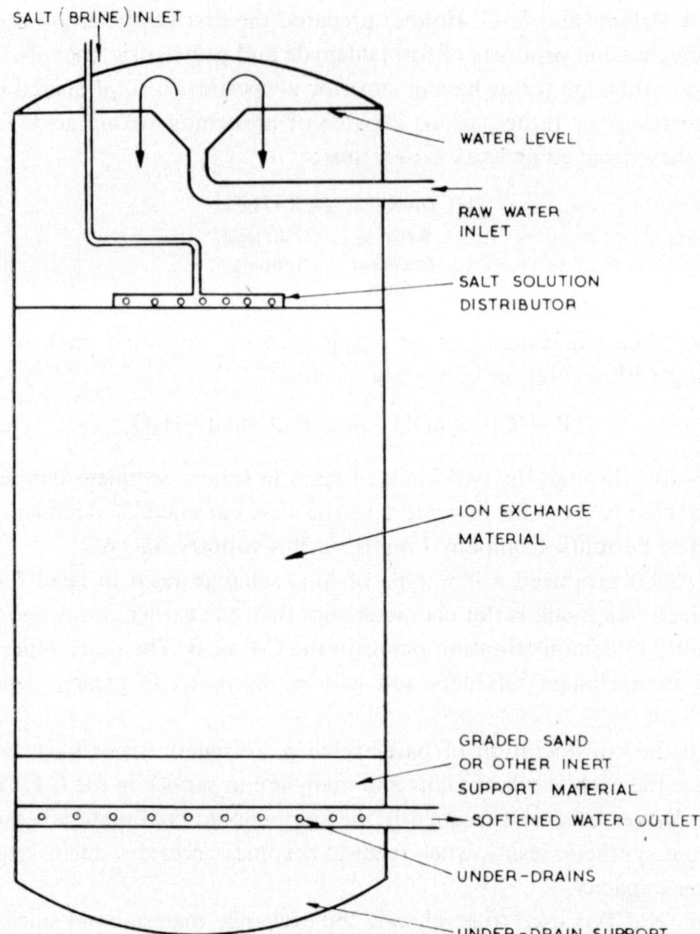

FIG. 3.4.4. Base-exchange softener

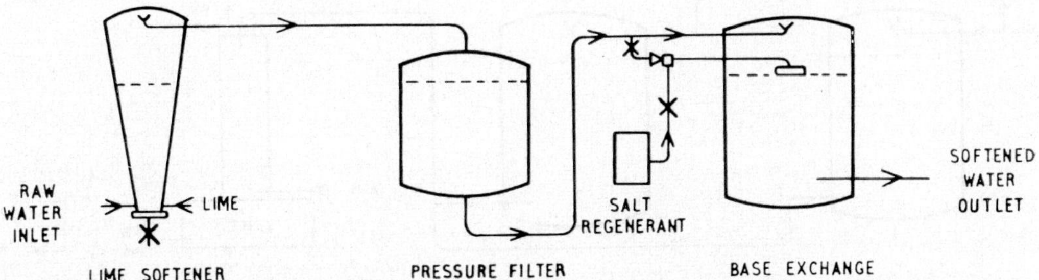

FIG. 3.4.5. Lime/base-exchange softener

wash water during backwashing operation and remove the salt and rinse water in the regenerating and rinsing operations.

Above the underdrain system is a graded gravel bed and above that the base exchange material. The gravel acts as a support for the base-exchange material and helps to ensure distribution of the water and salt solution during operation, backwashing and rinsing.

In the upper part of the shell is the water collection system, which serves during softening and rinsing to deflect and distribute the incoming water and to collect the backwash water and discharge it to waste.

A short distance above the bed of exchange material is the brine distributing system which must deliver the brine evenly over the whole area of the bed, in order to regenerate as completely as possible. Obviously the more efficient the distributor is, the less salt is required for regeneration.

The salt solution used for regeneration is prepared in a simple salt saturator.

Most types of water can be satisfactorily softened by base exchange, providing the sodium content is not too high. Soft water with a consistently low residual hardness (about 2 ppm or less) can be produced even from waters of variable quality. With waters containing suspended matter, prior filtration with or without coagulation is necessary in order to ensure that the exchange material does not become fouled.

3.4.5. **Lime/base-exchange Softener**

With a base-exchange softener, all of the free and combined carbon dioxide (see Section 3.2.1) in the raw water is carried forward into the soft water. If the raw water is first partially softened in a catalyst lime softener, to remove the temporary hardness and the remaining hardness (permanent) removed by a base-exchange softener, the final softened water will have a low carbon dioxide content and a low residual hardness (see Fig. 3.4.5).

This combination provides a very compact water softening plant wich can be made fully automatic in operation and will satisfactorily treat raw water prior to evaporation. The main difficulties arise when the lime softener is operated at low temperatures and where trace constituents of the raw water delay "growth" of the sand (see Section 3.4.2).

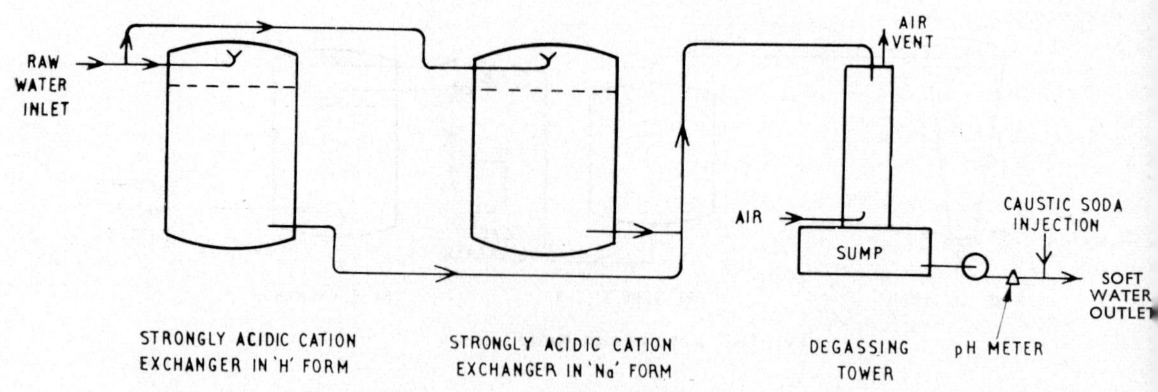

FIG. 3.4.6. Blend softener

3.4.6. Blend Softener

In Section 3.4.4 reference was made to the possibility of using both the sulphonated coal materials and the synthetic cation-exchange resins in the hydrogen form (regenerated with acid instead of salt), to effect exchange of cations in water for hydrogen ions. The following are some of the reactions which can occur:

$$H_2R + 2\,NaCl \rightleftharpoons Na_2R + 2\,HCl \quad (22)$$

$$H_2R + CaSO_4 \rightleftharpoons CaR + H_2SO_4 \quad (23)$$

$$H_2R + Ca(HCO_3)_2 \longrightarrow CaR + 2\,H_2O + 2\,CO_2 \quad (24)$$

where H_2R represents the cation exchange resin in the hydrogen form and CaR and Na_2R represent the cation exchange resin in the calcium and sodium forms.

The sulphates, chlorides and nitrates are thus converted into the corresponding, mineral acids. Bicarbonates are converted to carbonic acid (H_2CO_3) which, since it is unstable, breaks down to carbon dioxide and water.

If part of the raw water is passed through a hydrogen exchanger and the remainder through a base-exchange exchanger or sodium exchanger and the effluent is remixed, free mineral acid from the one can be used to neutralise the bicarbonate alkalinity produced by the other. The remixing reactions are:

$$NaHCO_3 + HCl \longrightarrow NaCl + H_2O + CO_2 \quad (25)$$

$$2\,NaHCO_3 + H_2SO_4 \longrightarrow Na_2SO_4 + 2\,H_2O + CO_2 \quad (26)$$

$$NaHCO_3 + HNO_3 \longrightarrow NaNO_3 + H_2O + CO_2 \quad (27)$$

By "scrubbing out" in a degassing tower, the carbon dioxide produced in the combined process, a water having a low total carbon dioxide content and a low residual hardness can be produced.

In practice, it is usual to adjust the flow to the two units so that the combined treated

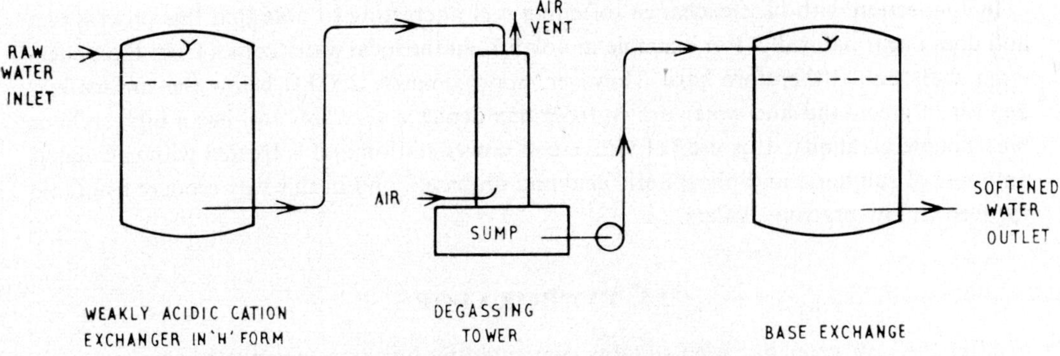

FIG. 3.4.7. Starvation/base-exchange softener

water is slightly alkaline. Alternatively the final treated water may be dosed with sodium hydroxide to raise the pH value.

A typical blend softener is illustrated in Figure 3.4.6.

3.4.7. Weakly Acidic Cation Exchange/base-exchange Softener

Certain synthetic cation-exchange resins have carboxyl (—CO_2H) active groups which will, when regenerated in the hydrogen form, only exchange cations associated with weak acid radicals, such as bicarbonates. These resins are known as weakly acidic cation-exchange resins, and the Permutit Company coined the name "Starvation" for this kind of exchange process. Since during cation exchange the bivalent ions, calcium and magnesium, are exchanged in preference to univalent cations, sodium and potassium (see Section 3.6.3), when a water having a total hardness greater than its bicarbonate alkalinity, is passed through a weakly acidic cation-exchange resin, the resin behaves as if the bicarbonate alkalinity was all associated with calcium and magnesium. The result is that all the temporary hardness is converted to carbonic acid, thus:

$$Ca(HCO_3)_2 + H_2R \longrightarrow CaR + 2\,H_2O + 2\,CO_2 \qquad (28)$$

If after treatment in a weakly acidic cation-exchange unit, the water is degassed and then softened in a base-exchange softener, the result is again a water having a low residual hardness and a low carbon dioxide content.

A plant of this type is illustrated in Figure 3.4.7.

Weakly acidic cation-exchange resins have the advantage that they require less excess acid for regeneration (that is they have a high regeneration efficiency) and they are regenerated with more dilute acid so that they require less rinse water than the strongly acidic resins.

3.4.8. Other Softening Processes

Other combinations of the above processes have been suggested, for example:

(i) Hydrogen exchanger → degasser → neutralisation with caustic soda.

(ii) Sodium exchanger → anion exchanger in the chloride form (both units being regenerated in series, using brine, and if required both resins can be combined in one unit).

However, these have had little application so far in power station water treatment.

In connection with base exchange softening it is interesting to note that this process can and does occur naturally. For example in north Kent the local water comes from the underlying chalk and is therefore hard. However, approximately 2000 ft below the surface is a deposit of greensand and water drawn from this depth is very soft and has a high sodium bicarbonate alkalinity. It is used at Gravesend power station and is treated with calculated amounts of sulphuric and phosphoric acid and degassed, and in this way rendered suitable for feed to low-pressure boilers.

3.5. EVAPORATORS

After the raw water has been suitably pretreated by filtration, coagulation, softening or a combination of these processes, the next stage in preparing boiler feed make-up water for medium to high-pressure plant, is evaporation.

Evaporation will often produce a distillate sufficiently pure for medium-pressure plant, but for high-pressure plant additional treatment is required which will be described later.

The various types of evaporators used in power stations have been described in another section of the course, but the following paragraphs are concerned with the chemical aspects of evaporator operation and maintenance.

The choice of bled steam or live steam evaporators is largely a question of economics; however, the characteristics of the water to be evaporated should influence the design of the evaporator employed. When the water is virtually completely softened, as for example by a base-exchange process, there is little risk of scaling and any type of evaporator should give trouble-free operation. With lime-soda softening it is seldom possible to reduce the hardness to below 10–15 ppm and scaling of tubular preheaters and evaporators occurs, the rate of scaling being dependent on the detailed design of the evaporator and the success attained in water softening. Some raw waters are difficult to soften and low residual hardness cannot be achieved. With these waters, evaporators in which the scale forms on the outside of the tubes are easier to clean mechanically. Moreover much of the scale falls off during normal operation or can be cracked off by the deliberate application of thermal shocks.

Sea water or estuary water contains a high magnesium salt content in addition to the temporary hardness and it is not therefore economic to soften this water by the processes described. Acid treatment with hydrochloric acid is often used to remove bicarbonate alkalinity and the main scale-forming constituent then remaining is calcium sulphate. When sea water is evaporated to approximately two-thirds of the original volume, the solubility of calcium sulphate (anhydrite) is reached, but fortunately it only crystallises slowly from solution and little scaling takes place. If concentrated to one-third volume it becomes saturated with respect to calcium sulphate subhydrate ($CaSO_4xH_2O$, where $x=0$ to $\frac{1}{3}$) and in this form is rapidly precipitated if concentration is continued. For this reason sea water evaporators are operated with a concentration factor of about 2 and under these conditions little deposition of calcium sulphate in either form will occur.

A typical modern bled steam sea water evaporator is illustrated in Figure 3.5B. The evaporator forms an integrated part of the feed system (Fig. 3.5A) and incorporates comprehensive heat recovery systems in order to achieve a high thermal efficiency. In particular a

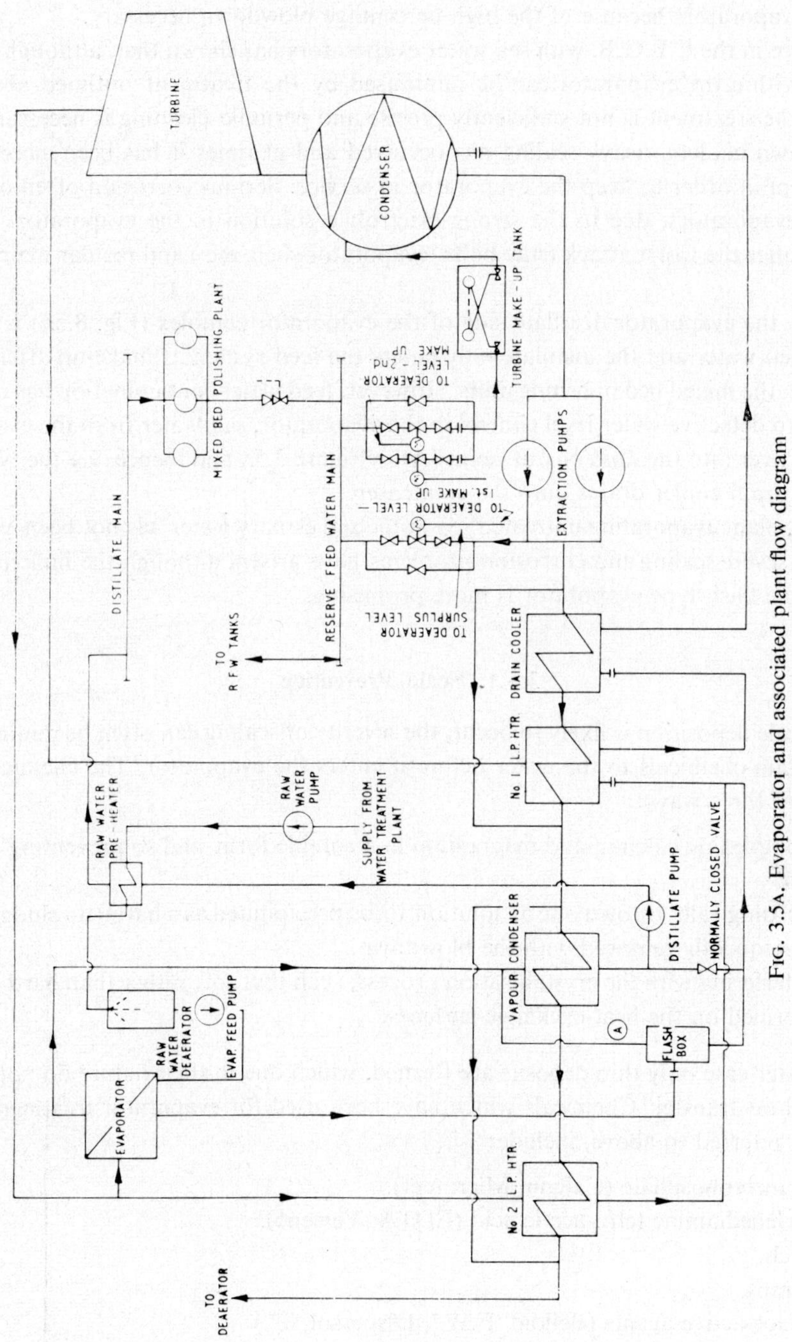

FIG. 3.5A. Evaporator and associated plant flow diagram

cooler is used to recover heat from the blowdown and this is economically essential with sea water evaporators because of the high percentage blowdown necessary.

Experience in the C.E.G.B. with sea water evaporators has shown that, although the rate of scaling within the evaporator can be minimised by the treatment outlined above, the control of the treatment is not sufficiently precise and periodic cleaning is necessary. With the blowdown coolers, heavy scaling has occurred and at times it has been necessary to by-pass them in order to keep the evaporator in service. Serious corrosion often occurs in sea water evaporators, due to the strong electrolyte solution in the evaporator. Ferrous materials suffer the worst attack (stay bolts, evaporator shell, etc.) and regular maintenance is necessary.

In theory the evaporator distillate side of the evaporator complex (Fig. 3.5A) is isolated from the feed water and the distillate only enters the feed system as make-up after further treatment in the mixed bed polishing units. However, feed water contamination has occurred when, due to defective water level control in the evaporator, salt water from the evaporator has carried over into the *flash box* at point (A) in Figure 3.5A and thence, via the No. 1 l.p. heater and drain cooler drains, into the condenser.

C.E.G.B. plant evaporating untreated sea water and estuary water has not been very successful and severe scaling and corrosion problems have arisen, although the limited experience with the flash-type evaporator is more promising.

3.5.1. Scale Prevention

Where scale deposition is likely to occur, the severity of scaling can often be minimised by adding certain chemicals to the water before it enters the evaporator. The chemicals used act in one of three ways:

(a) By complexing calcium and magnesium in a soluble form and so preventing precipitation.
(b) By causing salts thrown out of solution to be precipitated as a harmless sludge, which is subsequently removed with the blowdown.
(c) By interfering with the crystallisation process, such that soft rather than hard deposits are formed on the heat-exchange surfaces.

In the latter case only thin deposits are formed, which due to their nature do not materially affect heat transfer. Chemicals which have been used for evaporator treatment of the three types referred to above, include:

(i) Polymetaphosphate (Calgon, Micromet).
Ethylenediamine tetra-acetic acid (EDTA, Versene).
(ii) Starch.
Tannins.
(iii) Surface-active agents (Belloid 'F.W.', Dispersol 'T'.)

The choice of treatment chemical, if any, can only be determined by conducting plant trials. Different waters respond differently to the additives, as do evaporators of different design.

Solid materials, such as sand and sawdust, added to the evaporator may help to reduce the amount of scaling, by providing additional nuclei on which crystallisation can take place, although this is seldom practised nowadays.

3.5.2. Evaporator Blowdown

It is necessary to limit the concentration within the evaporator body to ensure that the less soluble constituents are not precipitated from solution and to prevent priming due to the concentrated solution of salts, which would lead to a deterioration in the quality of the distillate.

For most town's main supplies a 5% to 10% blowdown is adequate whilst with sea water, a blowdown of approximately 50% is necessary. This is usually carried out continuously, the blowdown rate being controlled by a fixed orifice. The chemist regularly checks the water in the evaporator body and arranges for additional blowdown if necessary.

Sometimes, even with the correct blowdown, foaming and priming occur particularly where the raw water contains detergents, and it is necessary to use chemical additives known as antifoams. These are usually high molecular weight polymeric organic chemicals, variously described as polyoxides and polyamides. They are relatively expensive chemicals and to be effective it is necessary to maintain a concentration of about 20 ppm within the evaporator body. It may be possible to recover part of the cost, since with antifoams the amount of blowdown can often be reduced.

3.5.3. Evaporator Cleaning

Where, in spite of the best possible chemical treatments scaling does occur, a stage will eventually be reached where heat transfer is materially affected and the evaporator will no longer produce the required output.

With coil-type evaporators, cleaning can often be done manually, but with tubular evaporators this may be very time-consuming and it is necessary to seek a chemical method of cleaning. If the scale contains 20% or more calcium carbonate, or other alkaline constituents, it is usually possible to remove it with dilute inhibited acid. If the scale is essentially calcium sulphate, acid cleaning is not very effective and although other cleaning agents such as EDTA, have been used, these are very expensive and slow in action so that it is necessary to revert to manual cleaning. For further information on these processes, see Chapter 4.

3.5.4. Evaporator Distillate Quality

The quality of the evaporator distillate will depend upon:

(a) The concentration of dissolved salts in the water in the evaporator body.
(b) The foaming characteristics of the raw water.
(c) The design of the evaporator.
(d) The efficiency of the gas venting arrangements.

Where the raw water is softened town supply water, or water of similar quality, the electrical conductivity of the distillate may vary from 0·8 micromho/cm to 3 micromho/cm, the higher value being often due to inefficient venting of CO_2.

With sea water the following analysis is typical of the distillate quality.

Conductivity	11·6 micromho/cm
Silica	0·011 ppm as SiO_2
Ammonia	0·22 ppm as NH_3
Carbon dioxide	6·4 ppm as CO_2
pH	5·3

The persibileble maximum dissolved salt concentration in the evaporator body is determined by the quality of the raw water and its scaling tendencies. For sea water, the conductivity of the water in the evaporator body will be in the range of 65,000 to 90,000 micromho/cm and for town water 6000 to 8000 micromho/cm. It can be seen, therefore, that the amount of priming and carry over is very small indeed.

3.6. DEMINERALISATION

Using synthetic ion-exchange resins similar to those described earlier, it is possible to remove all of the ionisable salts in a fresh-water supply and produce a water at least as pure as that obtained by softening and distillation.

The strongly acidic cation-exchange resins, similar to the synthetic base-exchange resins but in a hydrogen form, will exchange hydrogen ions for other cations. The result is that neutral salts are converted to their corresponding acids, a process known as salt splitting. Thus:

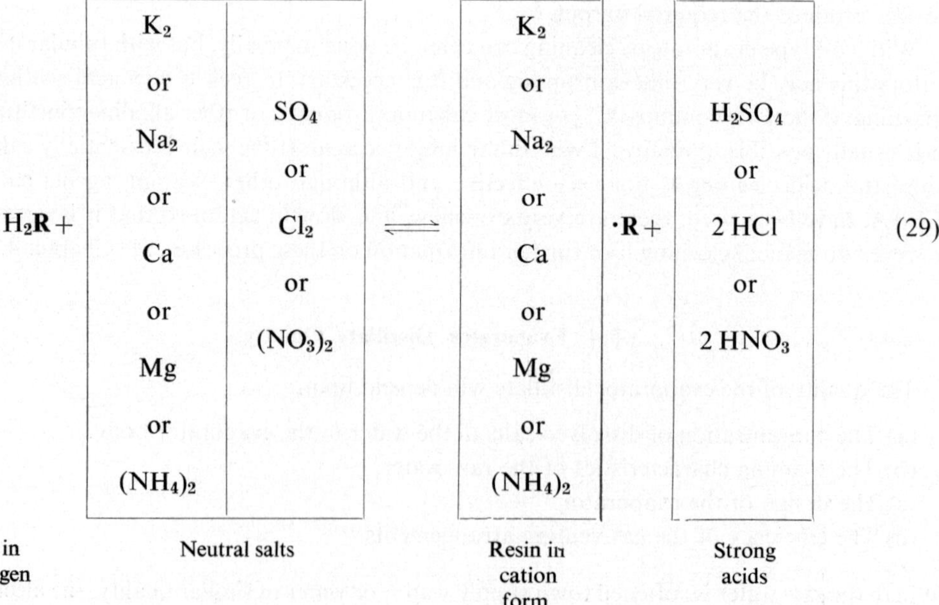

The strongly basic anion-exchange resins, when they are regenerated with strong bases, such as sodium hydroxide (NaOH), will exchange other anions for hydroxyl ions $(OH)^-$. If, therefore, the water after treatment in a cation exchanger is further treated in an anion exchanger, the salts originally present will be converted to water. Thus:

$$R(OH)_2 + \begin{bmatrix} H_2SO_4 \\ \text{or} \\ 2\,HCl \\ \text{or} \\ 2\,HNO_3 \end{bmatrix} \rightleftharpoons \begin{bmatrix} SO_4 \\ \text{or} \\ Cl_2 \\ \text{or} \\ (NO_3)_2 \end{bmatrix} \cdot R + H_2O \quad (30)$$

Resin in hydroxyl form — Strong acids — Resin in anion form — Water

3.6.1. Synthetic Ion-exchange Resins

These are organic substances of high molecular weight made by either condensing together or polymerising certain simple molecules. Chemically, the basic material is similar to many of the so-called "plastics" in common use today, but it is treated during manufacture to confer ion-exchange properties.

(a) Cation-exchange Resins

The first synthetic cation-exchange resins were formed by condensing tannins or phenols with formaldehyde and sulphonating (reacting with sulphuric acid) the resulting resin. It was produced in large masses which were crushed and screened to yield a granular material of the required particle size, and materials of this type are still in use.

Later methods of manufacturing cation-exchange resins which have improved stability, particularly towards oxidising agents, involved the sulphonation of a polymer of styrene known as *polystyrene**, thus:

$$n\left[\begin{matrix} CH{:}CH_2 \\ | \\ C_6H_5 \end{matrix}\right] \xrightarrow{\text{Polymerisation}} -CH(C_6H_5)-CH_2-CH(C_6H_5)-CH_2-CH(C_6H_5)- \xrightarrow{\text{Sulphonation}} -CH(C_6H_4SO_3H)-CH_2-CH(C_6H_4SO_3H)-CH_2-CH(C_6H_4SO_3H)- \quad (31)$$

Styrene — Polystyrene sulphonate

* Polystyrene is an example of a "linear polymer", since the basic structure of its molecule is a long chain of carbon atoms linked together.

If a small amount of divinyl benzene is included in the reaction mixture, a cross-linked co-polymer is formed which can again be sulphonated to yield a cation-exchange resin. Thus:

$$n\left[\begin{array}{c}CH:CH_2\\ |\\ \bigcirc\\ |\\ CH:CH_2\end{array}\right] + m\left[\begin{array}{c}CH:CH_2\\ |\\ \bigcirc\end{array}\right] \longrightarrow \begin{array}{c}—CH—CH_2—CH—CH_2—CH—\\ \quad | \qquad\qquad | \qquad\qquad |\\ \bigcirc \qquad\quad \bigcirc \qquad\quad \bigcirc\\ \qquad\qquad\quad |\\ —CH—CH_2—CH—CH_2—CH—\\ | \qquad\qquad\qquad\qquad |\\ \bigcirc \qquad\qquad\qquad \bigcirc\end{array} \qquad (32)$$

Divinyl benzene — Styrene* — Divinyl benzene/polystyrene copolymer

By varying the amount of divinyl benzene, the degree of cross-linking can be controlled and this affects the pore size and the stability of the resins, the more highly cross-linked resins having smaller pore size and being more resistant to temperature above ambient.

This copolymerisation can be conveniently carried out by suspending the mixed liquid monomers in water and as they solidify, hard transparent spherical beads of predetermined sizes are formed which require no further crushing and screening.

Although the degree of cross-linking can be controlled, the actual sites of cross-linking will occur in random fashion throughout the polymer chain so that we can only refer to the average pore size of the resin. By causing the linear polystyrene type of polymer to undergo self cross-linking, it is claimed that it is possible to produce a cross-linked polymer having sensibly constant pore size.

The sulphonated resins formed in this way are strongly acidic cation exchangers, that is they are capable of salt splitting. Weakly acidic cation-exchange resins can be produced by introducing a carboxylic group (—CO_2H) in place of the sulphonic group (—SO_3H), but they are produced more easily by copolymerising methacrylic acid with divinyl benzene directly:

$$x\left[\begin{array}{c}CH_3\\ |\\ C:CH_2\\ |\\ CO_2H\end{array}\right] + y\left[\begin{array}{c}CH:CH_2\\ |\\ \bigcirc\\ |\\ CH:CH_2\end{array}\right] \longrightarrow \begin{array}{c}CH_3 \qquad\qquad\qquad\qquad\qquad CH_3\\ | \qquad\qquad\qquad\qquad\qquad |\\ —C—CH_2—CH—CH_2—C—\\ | \qquad\qquad | \qquad\qquad |\\ CO_2H \qquad \bigcirc \qquad CO_2H\\ CH_3 \qquad\quad | \qquad\quad CH_3\\ | \qquad\qquad\qquad\qquad\qquad |\\ —C—CH_2—CH—CH_2—C—\\ | \qquad\qquad\qquad\qquad\qquad |\\ CO_2H \qquad\qquad\qquad\qquad CO_2H\end{array} \qquad (33)$$

Methacrylic acid — Divinyl benzene — Copolymer

The product is an active ion-exchange material, without further treatment.

* A molecule of styrene consists of a benzene molecule, $\begin{array}{c}CH\\ HC \quad CH\\ HC \quad CH\\ CH\end{array}$ in which one of the hydrogen atoms has been replaced by a "vinyl" group, —$CH:CH_2$. Thus styrene is formulated $C_6H_5.CH:CH_2$ or, more graphically $\bigcirc$—$CH:CH_2$ and here and in the formulae which follow, a hexagonal outline represents the residuum of a benzene molecule.

(b) Anion-exchange Resins

If basic groups are introduced into the polystyrene resins instead of the acidic groups, anion-exchange properties are conferred on the resins. The basic group may be derived from ammonia or an amine, and in order to facilitate the introduction of the basic group into the polymer, the latter may first be produced to contain, for example, chloro-methyl groups, $-\overset{|}{\underset{|}{C}}-CH_2Cl$. Then on reaction with ammonia or an amine, basic groups are built into the resin structure, and in the following summary of such reactions, the symbols R, R_1 and R_2 may be chosen from several subsidiary simple organic groups.

(a) With ammonia, NH_3; primary amine basic groups, $-\overset{|}{\underset{|}{C}}-CH_2-NH_2$, are produced.

(b) With a primary amine, $R-NH_2$; secondary amine basic groups, $-\overset{|}{\underset{|}{C}}-CH_2-N\begin{matrix}R\\H\end{matrix}$, are produced.

(c) With a secondary amine, $\begin{matrix}R\\R_1\end{matrix}>NH$; tertiary amine basic groups, $-\overset{|}{\underset{|}{C}}-CH_2-N\begin{matrix}R\\R_1\end{matrix}$, are produced. (34)

(d) With a tertiary amine, $\begin{matrix}R\\R_1\end{matrix}>NR_2$; quaternary ammonium salts,

$$\left[-\overset{|}{\underset{|}{C}}-CH_2-\overset{R}{\underset{R_2}{\overset{|}{\underset{|}{N}}}}-R_1\right]^{+}[Cl]^{-},$$

are produced.

The resins containing primary, secondary and tertiary amine groups behave as weak alkalis and will only exchange, or (more correctly) form, acid salts, with strong acids. The quaternary ammonium groups are strongly basic, similar in strength to strong alkalis, and will exchange even weak acids such as carbonic acid and silicic acid or, in effect, carbon dioxide and silica in water.

Anion exchange materials are regenerated with alkalis. The weakly basic resins are usually regenerated with sodium carbonate (soda ash or soda) but almost any alkali will serve. The regeneration reaction is represented by the following equation:

$$\underset{\text{Resin–acid complex (exhausted resin)}}{2\,R.HCl} + Na_2CO_3 \longrightarrow \underset{\text{Free base resin}}{2\,R} + 2\,NaCl + H_2O + CO_2 \quad (35)$$

The strongly basic resins require a strong alkali, such as sodium hydroxide (caustic soda) for regeneration, since the process involves the exchange of anions for hydroxyl ions, thus:

$$\underset{\text{Exhausted resin in chloride form}}{R.Cl} + NaOH \longrightarrow \underset{\text{Regenerated resin in hydroxyl form}}{R.OH} + NaCl \quad (36)$$

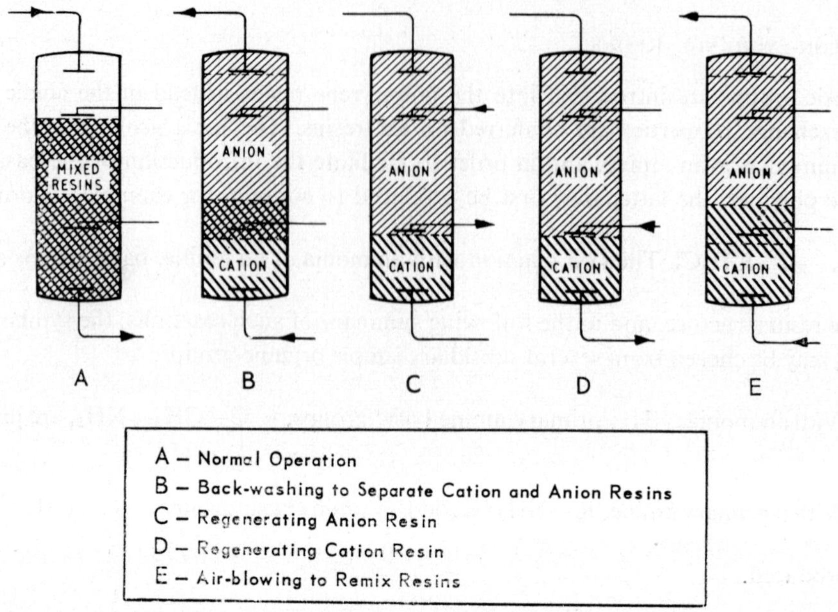

FIG. 3.6.1. Mixed bed ion-exchange unit—operation and regeneration

In demineralisation plants, the cation- and anion-exchange resins are often contained in separate columns or units and the design of these ion-exchange units is similar to that of the base-exchange unit described in Section 3.4.4. The complete demineralisation plant contains one or more of each of the two types of units, the water passing through them in series.

(c) Mixed Bed Exchange Units

The mixed bed unit is a single column or unit containing both cation- and anion-exchange resins intimately mixed together. When water is passed through such a unit, it comes into contact alternately with grains of cation and anion resin, so that the water is subject to an almost infinite number of demineralisation stages. In operation it behaves like a large number of two-stage demineralisers in series, with the result that it will produce a final water which is neutral and has a very low residual dissolved solids content.

The conditions existing inside the mixed bed unit at different stages of operation and regeneration are illustrated in Figure 3.6.1. During normal service, its mode of operation is similar to other ion-exchange units, in that water enters the top of the unit and leaves at the bottom. In order to regenerate the two different resins, it is necessary to have a more complex pipework and internals arrangement than for the single resin units, and these aspects will be described in Section 3.6.2.

(d) Strong and Weak Resins

A brief description of the chemical structure of ion-exchange resins has been given and some of the basic aspects of their use in the demineralisation of water have been mentioned. It is clear from the discussion so far that there is a wide variety of resins having different chemical structures, but in practice only a relatively small number find use in water treatment. The four major groups of resins, strong cation, weak cation, strong anion and weak anion-exchange resins, are fairly clearly defined by their different ion-exchange properties, but within each group, subtle changes in chemical and physical characteristics can yield a resin "tailor-made" for a particular ion-exchange application.

Weak resins have only a limited capability for ion exchange and since strong resins will perform all of the functions of weak resins, there would appear to be little point in using them. However, in commercial operation it is possible to reduce the cost of demineralising water by employing a combination of the two types of resin. In general it may be said that weak resins have a higher exchange capacity than strong resins and hence for a fixed output of treated water between regenerations, a smaller volume of weak resin will be required; also, weak resins can be regenerated with a smaller excess of chemicals. The higher exchange capacity helps to reduce the capital cost of the demineralisation plant and the better regeneration efficiency helps to minimise running costs, so that by employing combinations of weak and strong resin-exchange units (a multi-stage plant), the overall cost is kept to a minimum using the strong resin units to perform only those functions which cannot be performed by weak resin units.

The data given in Table 6 illustrates these differences in exchange capacity and regeneration efficiency.

TABLE 6

RESIN-EXCHANGE CAPACITY AND REGENERATION EFFICIENCY

Type of resin	Approximate regeneration efficiency %	Approximate exchange capacity kgr $CaCO_3/ft^3$
Strong cation	60	42
Weak cation	90	76
Strong anion	30	28
Weak anion	60	40

Weak resins are regenerated with more dilute regenerants than are strong resins, thus in a multi-stage demineralisation plant the "spent" regenerant from the strong exchange resin unit can often be used to regenerate the weak resin unit; this is an additional way of reducing operating costs and is known as *series regeneration*. For example, if the strong resin unit is regenerated with 2% sulphuric acid, the "spent" regenerant leaving the unit will still contain about 1% sulphuric acid. Weak cation resins are usually regenerated with 0·8% to 1% acid so that the "spent" regenerant from the strong resin can be diluted slightly and used to regenerate a weak cation resin, if one is used.

Within the broad classification of strong and weak resins, the strong anion-exchange resins are further subdivided into two types, known as Type I and Type II. Both have quaternary ammonium active groups, but in Type I the groups attached to the nitrogen atom are usually alkyl groups (C_nH_{2n+1}) such as methyl ($—CH_3$) and ethyl ($—C_2H_5$), whereas in the Type II resins one of the groups is an alkanol group such as ethanol ($—C_2H_4OH$).

Type I	Type II
$\overset{+}{—N}\begin{smallmatrix}CH_3 \\ —CH_3 \\ CH_3\end{smallmatrix}$	$\overset{+}{—N}\begin{smallmatrix}CH_3 \\ —CH_3 \\ C_2H_4OH\end{smallmatrix}$

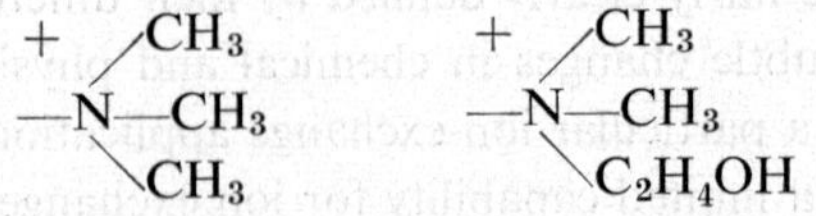

Both are strongly basic resins, that is they are capable of "salt splitting" neutral salts, but Type II is slightly less basic and has a higher regeneration efficiency than Type I. Type II resins are not as efficient at removing silica from water and suffer more degradation at higher temperatures than Type I, but where silica removal and temperature are not a problem, it is more efficient to use Type II.

3.6.2. Regeneration

The process of regeneration for all ion-exchange units is similar and follows the description given for base-exchange softeners in Section 3.4.4. The resin is first backflushed, which removes any suspended matter filtered out of the raw water during operation, and also causes the resin to increase in apparent volume.

Next the regenerant is admitted above the resin and flows downward through the bed. Finally the bed is rinsed again by downward flow to remove excess regenerant, the rinse water being run to waste until the quality is satisfactory for the unit to be returned to service.

(a) Cation Units

Sulphuric acid is usually used to regenerate cation-exchange resins as it is the cheapest acid commercially available in Great Britain. Since calcium sulphate, which may be formed as the resin is regenerated, has a low solubility (about 0·2%) and may otherwise be precipitated on the resin, it is usually necessary to keep the strength of the sulphuric acid below 2%. For raw waters with very low calcium content, sulphuric acid strengths of up to 5% may be used. Table 7 gives the maximum concentration of sulphuric acid recommended for the regeneration of the strong cation-exchange resin, Zeo-Karb 225.

In demineralisation plants, the strong cation-exchange resins are normally regenerated with about 2% sulphuric acid, but with weak resins a high level of regeneration can be achieved with more dilute acid and strengths of 0·8% to 1% are frequently used for these resins.

Hydrochloric acid may be used at almost any strength because there is no risk of precipitation. Strengths of up to 15% have been employed, but 2% to 4% is most common.

(b) Anion Units

Weakly basic anion-exchange resins can be regenerated with weak alkalis such as sodium carbonate. The regeneration process is in essence the neutralisation of an acid and not true ion exchange, since it involves the conversion of the resin–acid complex to the free-base resin (see equation 35). Other alkalis which are used include sodium hydroxide and ammonia solution. Sodium carbonate and sodium hydroxide are usually used as 4% solutions, but

TABLE 7

MAXIMUM STRENGTH OF SULPHURIC ACID FOR REGENERATION

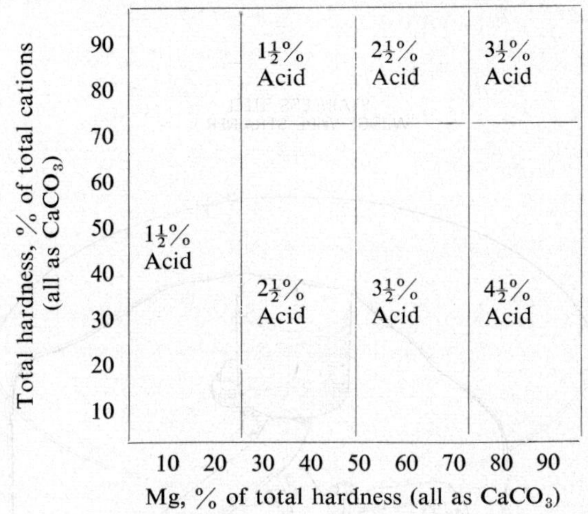

satisfactory regeneration can be achieved with more dilute solutions. If sodium hydroxide is used as the regenerant, the same solution can be employed to regenerate a strongly basic resin and a weakly basic resin in series.

Strongly basic anion-exchange resins require a strong alkali to regenerate them to the hydroxyl (OH^-) form and 4% sodium hydroxide solution is normally used.

(c) MIXED BED UNITS

In regenerating the mixed bed, the first stage is to separate the two resins into two discrete zones. Fortunately the resins employed have different densities, so that the initial backwashing causes the resins to separate into two layers, the lighter anion resin being at the top. To facilitate regeneration of the resins a central combined collector and distributor is provided at the interface of the two layers (see Figs. 3.6.1 and 3.6.2A). Caustic soda is introduced at the top and the spent regenerant runs to waste from the central collector. The rinse water follows the same path. Acid is then introduced at the central distributor, passes down through the cation resin and is run to waste from the bottom of the unit. This is followed by rinsing as for the anion resin. Finally the two resins are again intimately mixed by passing low-pressure air upwards through the unit and after a final rinse it is again ready for service.

The efficiency of regeneration in a mixed bed is never as high as in separate units, since separation of the two resins is never quite complete and there will always be some interference at the interface. For this reason a mixed bed unit requires more resin than two separate units and the running cost is higher, but as mentioned earlier it can produce almost complete de-ionisation of the input water.

(d) RESIN TRANSPORT REGENERATION

In water treatment demineralisation plants, the water flow rate through the resin bed, the depth of resin and the size of the resin granules employed, are mainly determined by the kinetics of the ion-exchange reactions (see Section 3.6.3 for further details). The design para-

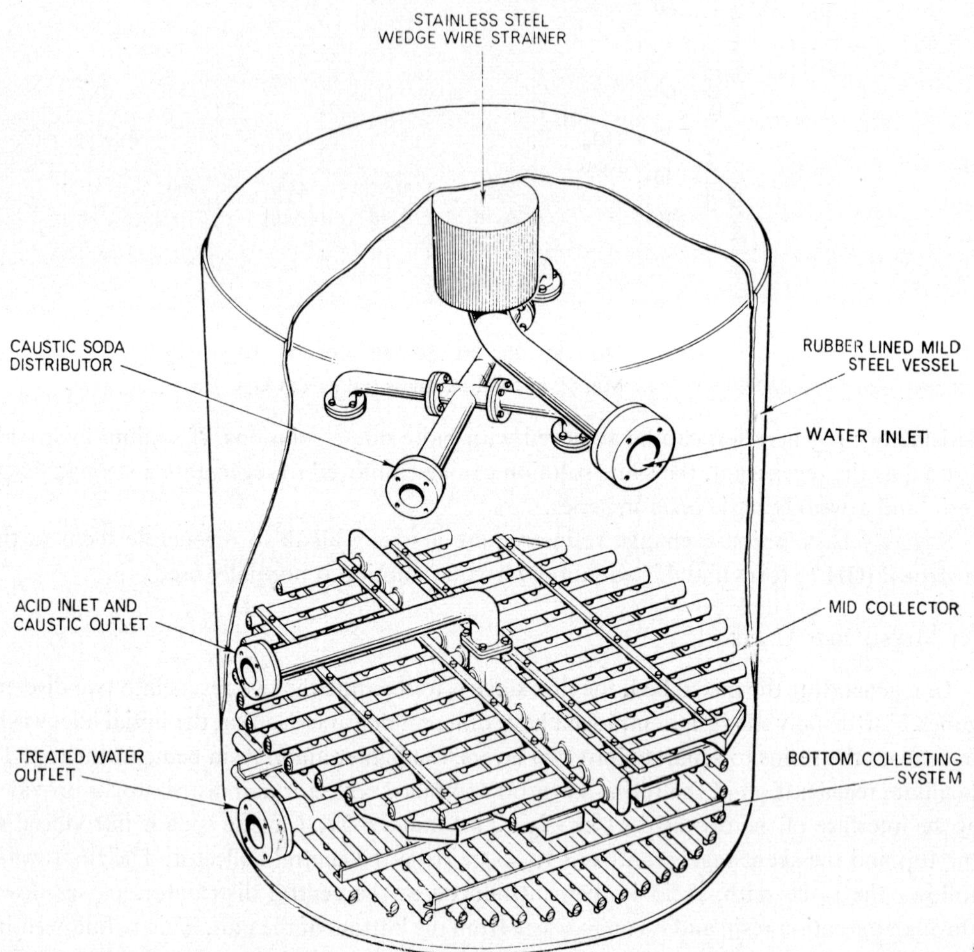

FIG. 3.6.2A. Permutit mixed bed unit (Reproduced by kind permission of the Permutit Co. Ltd.)

meters chosen ensure that, with a raw water having a T.D.S. of perhaps several hundred ppm, virtually complete demineralisation results. A water treatment plant for a 2000 MW station will have a through-put of about 45,000 gal/h and units to handle this through-put, with perhaps two complete demineralisation streams operating in parallel, can be designed to acceptable dimensions.

Mixed bed units, or sometimes cation-exchange units, are also used to remove traces of contaminants (metal corrosion products, silica and cooling water leaks) from feed water, particularly in feed systems associated with once-through boilers. This is one method of "condensate polishing", which will be discussed in Section 3.6.3. The condensate polishing ion-exchange units form an integral part of the feed system and may have to handle from 25% to 100% of the full feed water flow. For a 500 MW unit this will mean treating between 80,000 and 320,000 gal/h and with the flow rates normally used in ion exchange, would require a number of ion-exchange units operating in parallel.

These condensate polishing units have only to remove traces (less than 1 ppm) of contaminants from the feed water and reduce the concentration of these contaminants to an acceptable level (see Section 3.6.3). From this aspect, therefore, much higher water flow rates through the resin bed are acceptable and hence fewer units are required. With normally designed units, particularly with mixed beds, the laterals and distributors necessary for regeneration of the resin impose a serious resistance to water flow, but they must be sufficiently robust to withstand the pressures acting on them. These engineering considerations impose an upper limit on flow rate which is below that imposed by ion-exchange kinetics.

To achieve even higher flow rates, it has been necessary to simplify the internal design of the units so as to minimise resistance to water flow and with these simplified units regeneration *in situ* is no longer possible. Regeneration of the exhausted resin is therefore carried out in a separate unit of normal design, to which the resin is transferred hydraulically. This process is known as *resin transport* regeneration.

The water flow rates acceptable for a mixed bed unit with different methods of regeneration are given in Table 8.

TABLE 8

ACCEPTABLE WATER FLOW RATE FOR A MIXED BED UNIT WITH DIFFERING METHODS OF REGENERATION

Mixed bed units	Flow rate gal/ft² resin/min
1. Make-up water treatment plant	4 to 7
2. Condensate polishing with *in situ* regeneration	up to 25
3. Condensate polishing with external regeneration	up to 60

For C.E.G.B. plants the flow rate is normally limited to 25 gal/ft² of resin per minute, with the proviso that where two exchange units are operated in parallel, the rate may be increased up to 40 gal/ft² of resin whilst one unit is out for regeneration.

The use of resin transport regeneration in condensate polishing units has an additional advantage in that there is no longer a risk of strong regenerants accidentally contaminating the feed system.

(e) Regeneration Equipment

The principal reagents used for regenerating ion-exchange resins are sodium hydroxide, sulphuric acid and sometimes hydrochloric acid.

Caustic soda used in other kinds of water treatment is normally purchased in solid form (anhydrous flakes), but for the larger quantities used in demineralisation plants it is usually more convenient to purchase it as a 46% solution (liquid caustic soda).

Sulphuric acid is purchased as a 96% solution and at this strength it can be safely stored in mild steel vessels.

Hydrochloric acid is supplied as a 28% to 32% solution, the lower strength being normally supplied during summer months in order to reduce the nuisance due to hydrochloric acid vapour. It is very corrosive towards mild steel and all steel surfaces must be protected with a suitable lining such as rubber.

The storage tanks are surrounded by bund walls, which form a collecting tank sized to contain the whole contents of the storage vessel in the event of a major leak developing. Where necessary the storage tanks are lagged and provided with some means of heating to ensure that the chemicals do not freeze during periods of low ambient temperature. Of the three principal regenerants mentioned, only liquid caustic soda and sulphuric acid are at all likely to give trouble by freezing at winter temperatures, and Figures 3.6.2B and 3.6.2C show freezing points at various concentrations of these chemicals.

Using these liquid regenerants, the regeneration equipment employed at most C.E.G.B. stations is similar and a typical arrangement is illustrated in Figure 3.6.2D. It consists of a closed measuring tank into which the concentrated regenerant is drawn from the bulk storage tank by suction. The measured quantity of regenerant is then run into a dilution tank to obtain a suitable strength. Finally the regenerant is further continuously diluted to the required strength and introduced into the unit to be regenerated by means of water ejectors. Whenever possible, the use of pumps to handle regenerant chemicals is avoided in C.E.G.B. plants, because it is considered safer to avoid the pressurization of these potentially dangerous liquids.

3.6.3. Demineralisation Processes

The type of demineralisation process chosen for a power station will depend on four main factors:

(a) The quality of the raw water.
(b) The degree of de-ionisation required, that is, the quality of the final treated water.
(c) The capital cost.
(d) The running cost.

Although a large number of different processes is possible, the nine schemes illustrated in Figure 3.6.3A represent some that are commonly in use in C.E.G.B. power stations. Schemes

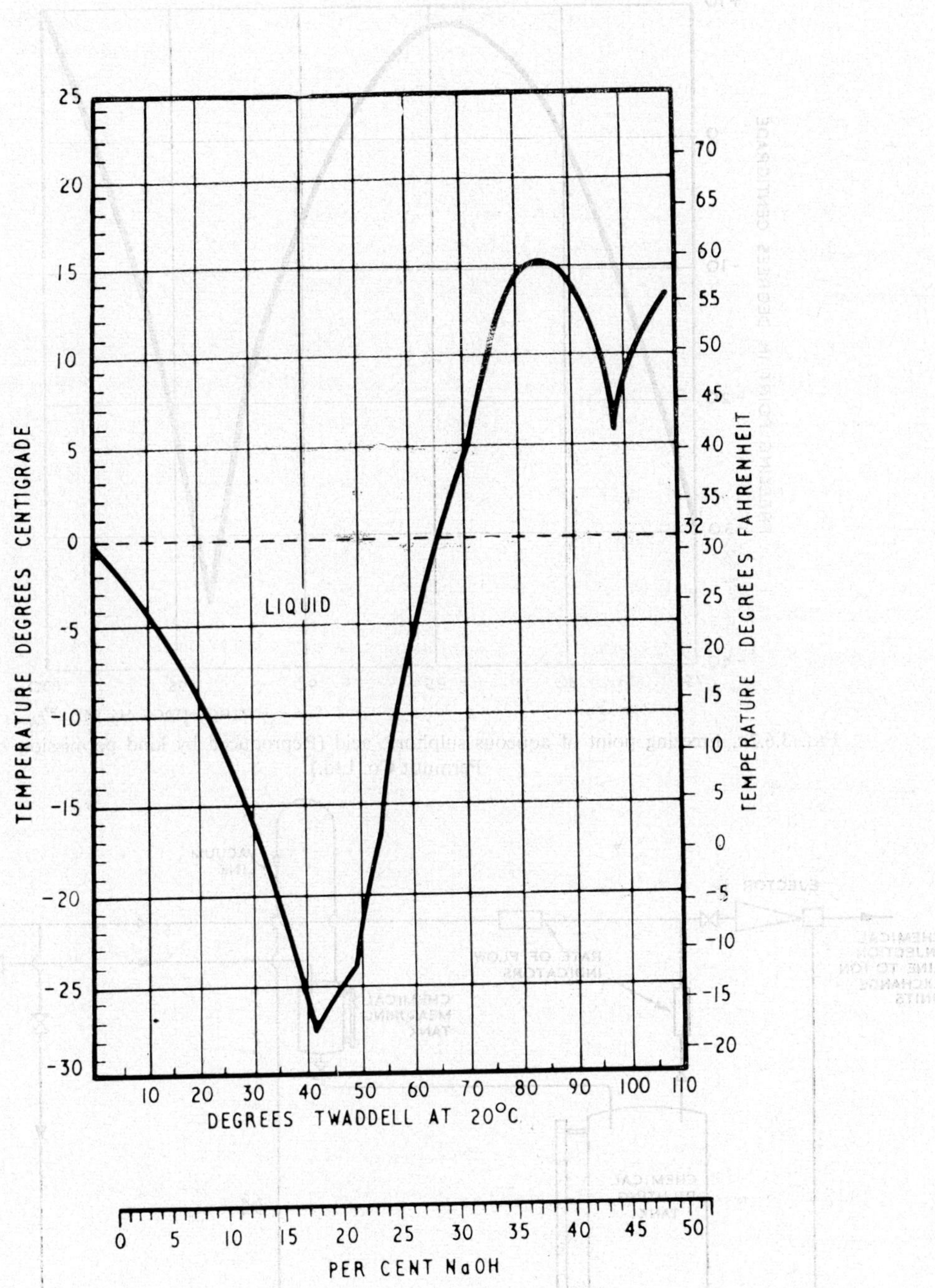

FIG. 3.6.2B. Freezing point of aqueous caustic soda solutions. (Reproduced by kind permission of the Permutit Co. Ltd.)

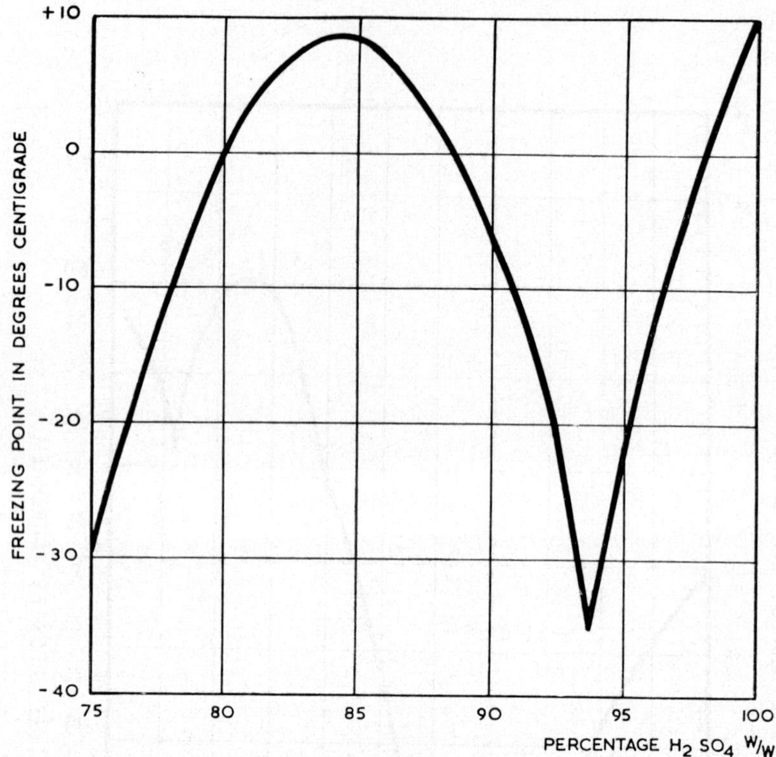

FIG. 3.6.2C. Freezing point of aqueous sulphuric acid (Reproduced by kind permission of the Permutit Co. Ltd.)

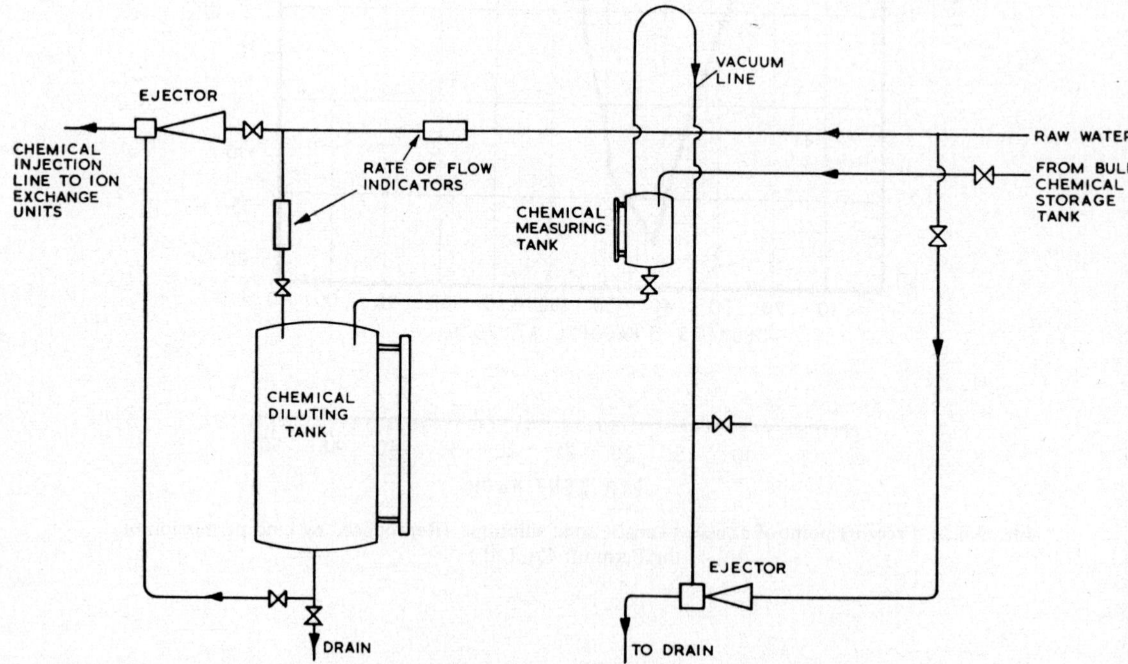

FIG. 3.6.2D. Regenerant dilution and injection equipment

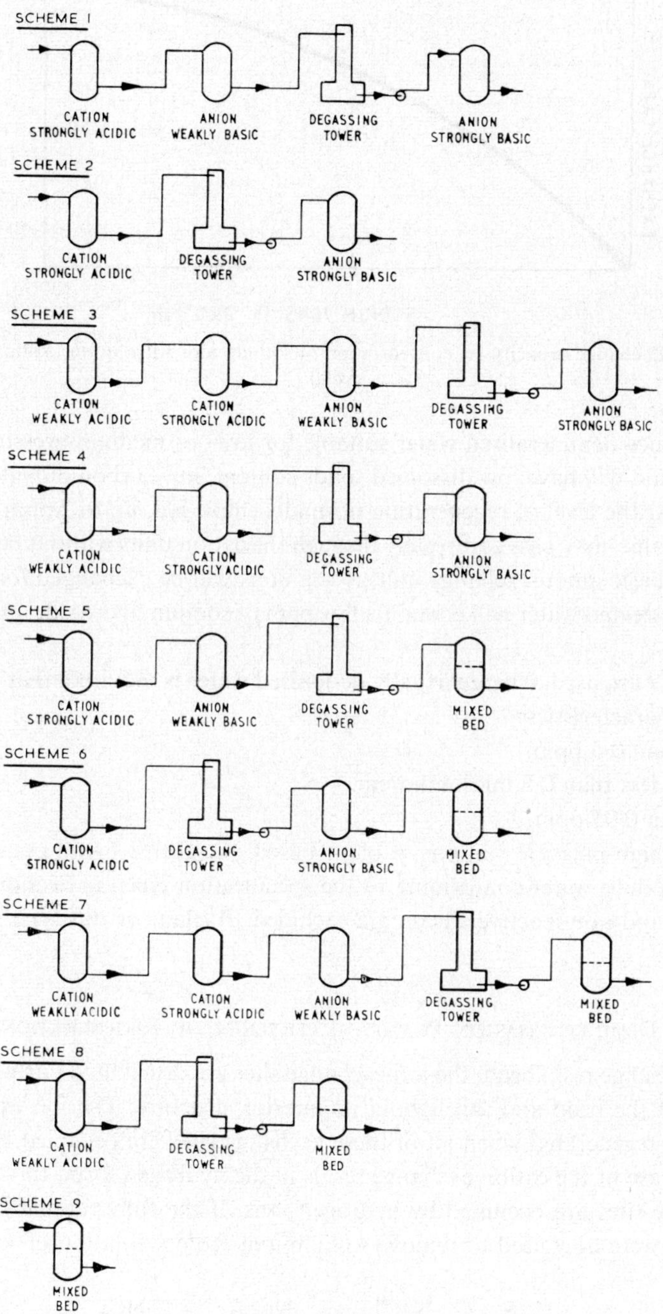

FIG. 3.6.3A. Demineralisation plants

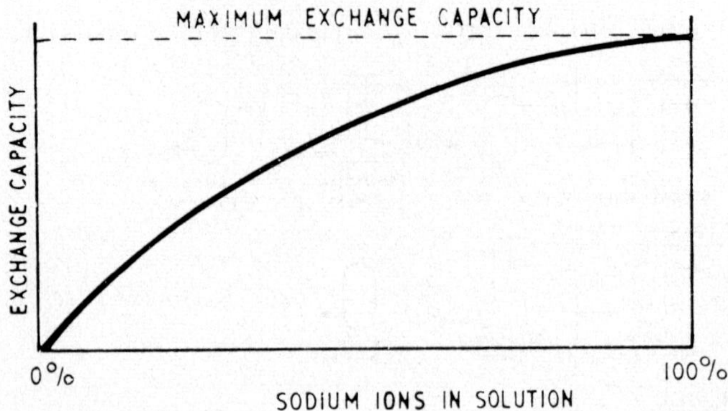

FIG. 3.6.3C. Exchange capacity vs. concentration of sodium ions in solution (sodium/hydrogen cycle)

1 to 4 will produce demineralised water suitable for low- or medium-pressure plant (up to say 600 lb/in^2) and will have low dissolved solids content, low carbon dioxide and silica less than 0·1 ppm. At the level of regeneration normally employed, up to 2 ppm of salts, particularly sodium salts, may pass completely through the cation units without being exchanged. In the strongly basic anion-exchange units, the anions will be exchanged for hydroxyl ions so that the final treated water will contain a few ppm of sodium hydroxide and will therefore be alkaline.

Schemes 5 to 9 are used, where virtually de-ionised water is required, that is water having the following characteristics:

T.D.S. less than 0·5 ppm.

Conductivity less than 0·5 micromho/cm.

Silica less than 0·05 ppm.

For modern high-pressure stations, a plant based on scheme 6 will commonly be used and this will produce water conforming to the specification given in Section 3.3 providing organic matter and non-reactive silicon are excluded. A plant of this type is illustrated in Figure 3.6.3B.

(a) DESIGN OF DEMINERALISATION PLANTS—EFFICIENCY OF REGENERATION

In an ion-exchange resin bead, the ion-exchange sites are distributed fairly evenly over the whole surface of the bead and throughout its interior structure. The ion-exchange resin is said to be fully regenerated when all of these exchange sites are occupied by the preferred ion and in the case of the cation exchange resins in the hydrogen form, this would be when all the exchange sites are occupied by hydrogen ions. If the fully regenerated resin in the hydrogen form were now used to remove sodium ions from a solution of sodium chloride, thus:

$$\underset{\text{Resin in hydrogen form}}{H^+R} + \underset{\text{Sodium chloride}}{NaCl} \rightleftharpoons \underset{\text{Resin in sodium form}}{Na^+R} + \underset{\text{Hydrochloric acid}}{HCl} \quad (37)$$

the resin could be regarded as being completely exhausted, when all of the ion-exchange sites are occupied by sodium ions. The amount of sodium ions removed from solution under these conditions would be the maximum exchange capacity of the resin for sodium in the sodium/hydrogen cycle.

These ion-exchange reactions are reversible, as is indicated by the sign $\rightleftharpoons$ in the equation above, and tend towards an equilibrium which is governed by the relative concentrations of the ions in solution which are taking part in the reaction. In equation (37), for the forward reaction (from left to right) the driving force derives from the concentration of sodium ions in solution. Similarly the driving force for the reverse reaction will be dependent on the concentration of hydrochloric acid produced or, more correctly, the concentration of hydrogen ions released to the solution by the resin. For any given initial concentration of sodium ions, the reaction will tend towards an equilibrium and only when a considerable excess of sodium ions are present will it proceed to completion. This is illustrated in Figure 3.6.3.C.

Selectivity of cation-exchange resins. The relative concentrations of sodium and hydrogen ions present in solution at equilibrium is a measure of the selectivity of the ion-exchange resin for sodium and hydrogen ions. The selectivity of an ion-exchange resin varies for different ions and the ratio of the selectivity of a resin for an ion, divided by its selectivity for a reference ion, is known as the resin's "selectivity coefficient". In Table 9 (below), which is

TABLE 9

SELECTIVITY OF CATIONS IN THE HYDROGEN CYCLE OF AMBERLITE IR 120 CATION EXCHANGE RESIN (ROHM AND HAAS CO.)

Cation	Selectivity coefficient vs. Hydrogen ion
Li^+	0·8
Na^+	2·0
K^+	3·0
NH_4^+	3·0
Mg^{++}	26
Ca^{++}	42

taken from the Rohm and Haas publication, Amber Hi-Lites No. 86, some figures are given for the selectivity of the cation exchange resin IR-120 in the hydrogen cycle for cations normally encountered in water treatment.

Exhaustion of ion-exchange resin beds. In an operating ion-exchange plant, the ion-exchange units consists of a vertical column containing the resin bed, through which the water is passed, usually in a downward direction. In order to understand the principles of operation of the ion-exchange column, it is useful to consider a cation-exchange column with the resin initially completely regenerated in the hydrogen ion form and to examine the effect of passing through it a reasonably dilute solution containing only sodium chloride. At first ion exchange will take place in the uppermost layer of resin, so that the solution passing to the layer below will contain a mixture of sodium chloride and hydrochloric acid.

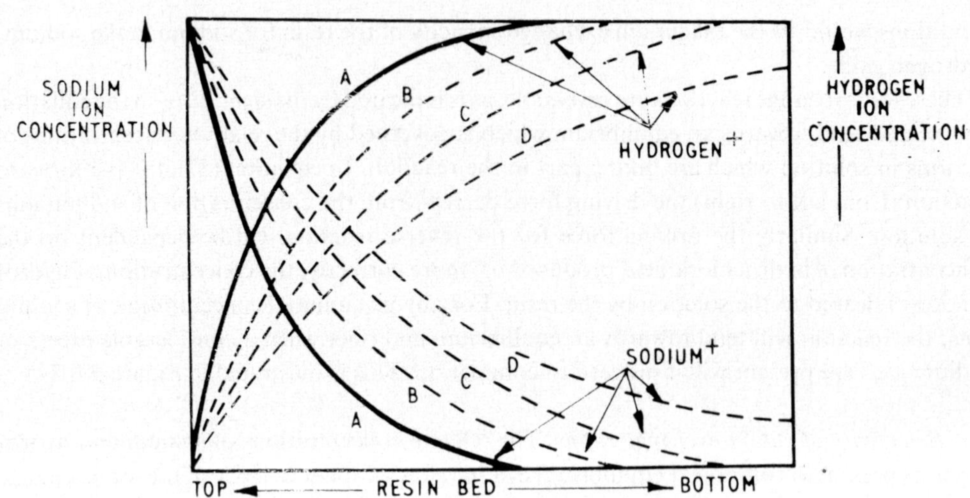

FIG. 3.6.3D. Changes in ion concentration within the resin bed

The exchange process will continue progressively down through the resin bed until at some stage virtually all the sodium ions have been exchanged for hydrogen ions. The solution passing through the remaining resin and out of the column is then almost entirely hydrochloric acid. The conditions existing within the resin bed during this initial phase of operation are given by the solid lines (A) in Figure 3.6.3D.

As operation continues the upper layers become exhausted, that is all the exchange sites become occupied with sodium ions and the exchange reactions take place progressively further down the resin bed. After some period of operation, the conditions within the resin column may be represented by the broken curves B. Curves C and D represent later stages of operation and curve D indicates that at this stage in time the solution passing from the resin column contains a measurable concentration of sodium ions in solution. For operational purposes, the resin is said to be exhausted with respect to the sodium/hydrogen cycle, when the concentration of sodium ions in the solution leaving the column exceeds some predetermined value, as at point W at time t_1 in Figure 3.6.3E. Obviously the resin is not completely exhausted until point X at time t_2, when the concentration of sodium ions in the solution leaving the column is equal to the initial concentration. The area above the curve from t_0 to t_2, is a measure of the maximum exchange capacity of the resin and the area above the curve from t_0 to t_1 is a measure of its operational exchange capacity. Similarly the area below the curve is a measure of the amount of sodium slip occurring, that is the quantity of sodium passing completely through the resin bed.

Regeneration. If at point W the resin bed is regenerated with hydrochloric acid, initially the upper part of the bed will be almost completely in the sodium form, whilst the bottom layers will consist of a mixture of resin in the sodium and hydrogen form. If the flow of acid is continued for a sufficiently long period of time, the resin can be completely regenerated, that is converted to the hydrogen form. The composition of the spent regenerant leaving the column will follow a curve similar to that given in Figure 3.6.3F where complete

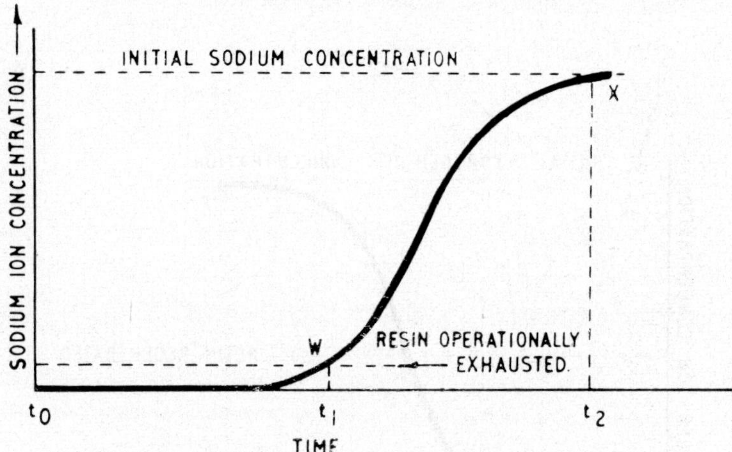

FIG. 3.6.3E. Composition of solution from a fully regenerated cation unit in the sodium/hydrogen cycle when treating sodium chloride solution

regeneration is represented by point Z. Again in operational plant, for a fixed amount of acid flowing for a fixed time, the level of regeneration achieved will be less than the maximum as, for example, at point Y.

The regeneration efficiency, that is the ratio of the quantity of acid usefully employed divided by the total quantity of acid passed through the resin bed, is given by the area above the curve from t_0 to t_1 divided by the total area from t_0 to t_2 expressed as a percentage. Some typical data for the regeneration efficiency for different types of ion-exchange resins are given in Table 10.

During the early stages of regeneration, the sodium ions removed from the upper layers of the bed and replaced by hydrogen ions will be absorbed by the resin in the bottom layers of the bed, which are still in the hydrogen form. Thus unless complete regeneration is carried out, at the end of the regeneration cycle the upper layers of resin will be in the hydrogen form but the bottom layers will contain some resin in the sodium form.

In the following operational cycle, during the early stages, these lower layers of resin will be exposed to hydrochloric acid produced in the upper part of the resin bed. Because the selectivity coefficient for sodium, referred to earlier (Table 9) is low, a reverse exchange reaction can take place between the sodium ions on the resin and the hydrogen ions in the water, resulting in leakage of sodium into the water leaving the resin column. As the cycle continues, a stage will be reached at which most of the sodium has been removed from these lower layers and the water leaving the column is virtually free of sodium ions. The sodium on concentration in the water leaving the column will then remain at a low level until the resin is exhausted when the sodium concentration will again rise. These changes which take place in the composition of the water leaving a cation resin column operating on the sodium/hydrogen cycle are given by the solid line in Figure 3.6.3G.

The exchange reactions for the sodium/hydrogen cycle described above assume that a sufficient depth of cation exchange resin is present in the column, relative to the flow rate and that a sufficiently high regeneration level has been employed. In operating plant treating waters with a high concentration of sodium ions, it is not normally economic to have a suffi

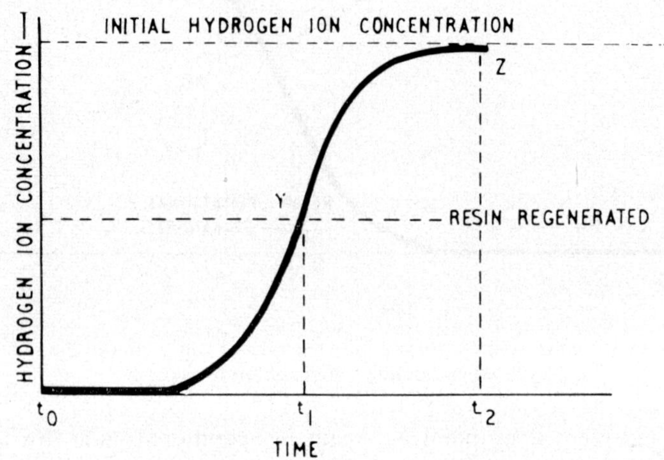

FIG. 3.6.3F. Composition of spent regenerant during regeneration cycle

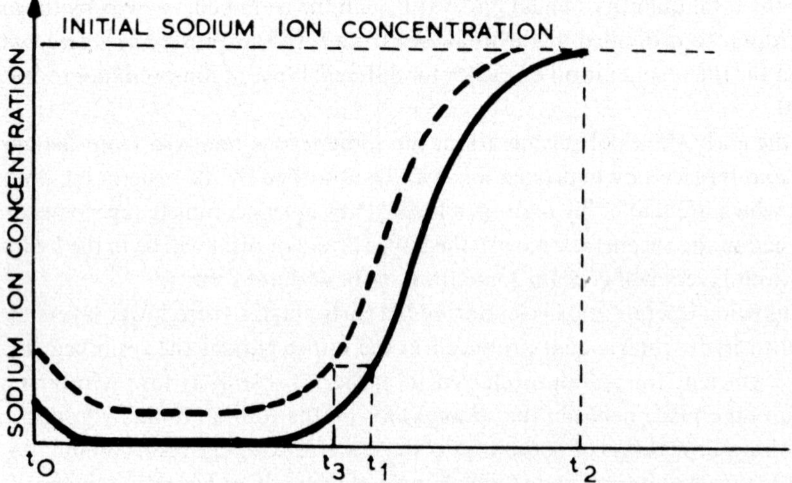

FIG. 3.6.3G. Composition of water leaving an operational cation unit

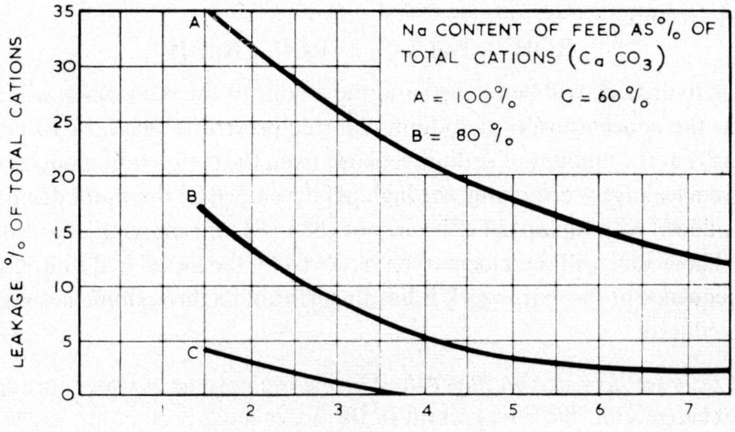

FIG. 3.6.3H. Factors affecting cation leakage (influent alkalinity 50% of total cations) (Reproduced by kind permission of the Permutit Co. Ltd.)

cient resin depth to carry out all the cation exchange in one column, and the depth of resin and the regeneration level employed assume a certain percentage of sodium leakage as in the broken line in Figure 3.6.3G.

The problem of cation slip or leakage is mainly confined to the univalent ions, sodium and potassium, since as was seen from Table 9, the bivalent cations calcium and magnesium, have much higher selectivity coefficients. Some idea of the relationship between sodium concentration in the raw water and % cation leakage can be gained from Figure 3.6.3H, which is for Permutit Zeo-Karb 225 cation exchange resin employed in the hydrogen cycle and regenerated with hydrochloric acid.

Selectivity of anion-exchange resins. So far the discussion in this section has been concerned with cation-exchange resins but a similar situation arises with strongly basic anion-exchange resins. Different anions have different affinities towards anion-exchange resins and the affinities of the anions which are of interest in water treatment can be arranged in the following order:

OH^-	$<$	Cl^-	$<$	NO_3^-	$<$	HSO_4^-
Hydroxyl		Chloride		Nitrate		Bisulphate

Effect of sodium leakage on plant performance. Returning to the problem of sodium leakage, any sodium not exchanged in the leading cation exchange unit of a demineralisation plant will have to be removed by a cation resin in a later stage of the plant, such as a mixed bed unit. If the unit following the cation stage contains a weakly basic anion-exchange resin, this will absorb the strong mineral acids and the water leaving this unit will contain sodium chloride together with the anions of weak acids which are, in effect, carbon dioxide and silica. These remaining ions will then be removed by the final mixed bed, if one is fitted. If the anion unit contains a strongly basic anion-exchange resin, the chloride ions will be exchanged for hydroxyl ions, thus:

$$ROH^{-}+NaCl \rightleftharpoons RCl^{-}+NaOH \quad (38)$$

Hence, sodium hydroxide will be formed and the anions of the weak acids will be removed, providing that the concentration of sodium chloride present is small, or to put it another way, providing that the amount of sodium leakage from the cation unit is slight. If, however, excessive sodium leakage is occurring, the high pH developed by the water due to the sodium hydroxide produced will impair the efficiency of the resin for removing weak anions, particularly silica. These ions will then have to be removed by the mixed bed unit, if one is fitted, and as a consequence of the extra work it has to perform, its throughput between regenerations will be reduced.

Effect of resin particle size on flow rate. During the passage of water through an operational ion-exchange unit, the ions present in the water must contact ion-exchange sites on the resin beads in the regenerated form, for exchange to take place. The exchange reactions take place first at exchange sites on the surface of the bead and the exchanged ion then moves inwards by a process similar to diffusion, releasing the surface sites for further exchange. The rate of diffusion is fairly rapid and it is only of concern in the exchange processes, when a high water flow rate is employed. One way of minimising the effect of diffusion would be to reduce the particle size of the resin, thus making available a greater number of exchange sites on the surface of the resin. Unfortunately the smaller the particle size, the greater is the resistance to water flow, so that for practical purposes it is normal to employ resins having a particle size of 0·3 mm to 1·2 mm across, and to design the plant such that the water flow rate is not greater than 2 gal/ft^3 per minute of resin and between 4 and 7 gal/ft^2 per minute of resin.

In the final stages of a design it is necessary to decide on the volume of resin required for a given throughput between regenerations, and the regeneration level to be employed. Some data for the four main classes of resins are given in Table 10.

TABLE 10

CHARACTERISTICS OF ION-EXCHANGE RESINS

Type	Regenerant	Regeneration level (lb/ft^3 of resin)	Regenerant efficiency (%)	Exchange capacity after regeneration (kgr $CaCO_3$/ft^3 of resin)	Maximum exchange capacity (kgr $CaCO_3$/ft^3)
Strongly basic anion resin	NaOH	1	68	6·0	26·2
		2	42	7·4	
		4	31	11·0	
Weakly basic anion resin	NaOH	2	97	17·0	43·6
		4	80	28·0	
		6	55	28·5	
	Na_2CO_3	4	87	23·0	
Strongly acidic cation resin	H_2SO_4	6	57	24·0	48·0
		10	45	31·2	
Weakly acidic cation resin	H_2SO_4	2·1	95	14·0	20·0

FIG. 3.6.3J. Demineralisation plant at West Thurrock

This data illustrates the point made earlier that an increase in regeneration level increases the operational exchange capacity of the resin but the relationship is not linear. To provide the exchange capacity required by employing a high level of regeneration results in high operating costs. Conversely, using a larger volume of resin with a low level of regeneration increases the capital cost of the plant. The correct choice can only be made when the planned operational life and the expected utilisation of the plant are known.

(b) INSTRUMENTATION AND CONTROLS

Some idea of the complexity of the instrumentation and controls required for a modern demineralisation plant can be gained from Figures 3.6.3J and 3.6.3K, which show some of the control panel at West Thurrock generating station.

The instrumentation and controls used have two main functions. Firstly, to monitor and control the plant performance during normal operation and thus determine when regeneration is required, and secondly to enable regeneration to be carried out efficiently. The effluent produced during regeneration usually requires some chemical treatment before it can be disposed of to a river or sewer, and this requires additional instrumentation.

During normal operation the following instrumentation is used.

(i) *Electrical*. To monitor the performance of pumps and blowers and to indicate which equipment is in service.

(ii) *Flow rate*. These are often of the "rotameter" type and are fitted at the inlet and outlet to each unit to indicate the flow rate of water through the unit.

FIG. 3.6.3K. Control and instrument panel—West Thurrock demineralisation plant

(iii) *Water meters.* These are integrating meters and record the total quantity of water passed through the unit. In a demineralisation plant where all the units in a single stream are regenerated at the same time, it is sufficient to employ two water meters, one at the inlet and the other at the outlet of the plant. The difference in the quantity of water recorded by the two meters is then a measure of the water run to waste during regeneration.

(iv) *Pressure gauges.* These are fitted at the inlet and outlet of each unit and are mainly used to indicate when the resin has become fouled, or when some other defect has occurred which is restricting water flow, such as an excess quantity of fine material arising from disintegration of the resin beads.

(v) *Conductivity meters.* The production of acid as water passes through the cation-exchange units and the removal of this acid by passage through the anion exchange units both cause a change in the conductivity of the water. This is shown diagrammatically in Figure 3.6.3L.

When a resin bed is approaching exhaustion, the conductivity of the water leaving the unit will approach the conductivity of the water entering the unit. For the three main types of unit the changes in conductivity that is observed as each resin becomes exhausted are:

CATION UNIT—fall in conductivity
ANION UNIT—rise in conductivity
MIXED BED—rise in conductivity

Thus by employing conductivity meters at the outlet from each unit, it is possible to establish when regeneration is required.

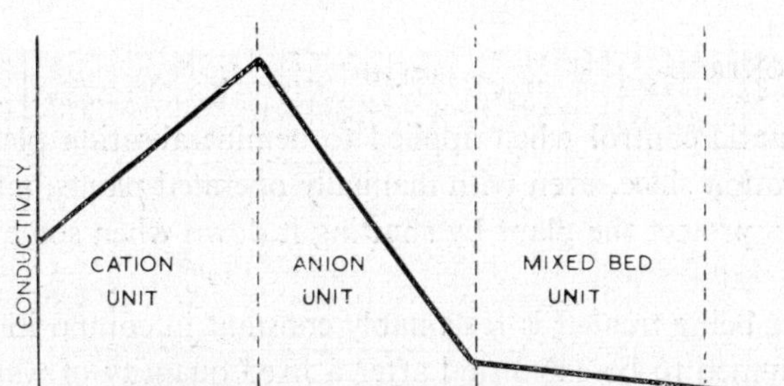

FIG. 3.6.3L. Change in conductivity during passage through plant

Note: Since the selectivity coefficient of anion exchange resins for silica is much less than for other anions, silica will appear in the final treated water before the other anions, as the resin becomes exhausted. As the silica only influences the conductivity of the water to a very small extent, the increase in silica will precede the increase in conductivity. This is known as *silica breakthrough.* If a low final silica concentration is required, the regeneration of the final mixed bed must be governed by silica concentration and not by conductivity.

(vi) *Silica meters.* These are often used to monitor the silica concentration of the final treated water, but for very low silica concentrations, a high sensitivity is required and this is at present difficult to achieve reliably with commercial meters.

(vii) *Level gauges.* Normally all enclosed tanks are fitted with level gauges or contents indicators.

(viii) *Alarms and controls.* Where the plant is under manual control, additional control is often incorporated to stop and possibly to restart the plant if some abnormal condition occurs. This overriding control may be actuated by high and low water levels in the raw water break tank, in the degassed water sump and in the treated water storage tank and by a change in conductivity in one or other of the conductivity meters, which indicates that the unit is exhausted. The plant may be arranged to shut down automatically when a fixed quantity of water has been passed through the units. This is achieved by employing preset contacts on the integrating water meters.

All of these abnormal conditions may be used to give a signal which may operate a visible or audible alarm at some remote location.

When regeneration is being carried out, in addition to the above instruments flow meters are usually employed to monitor the flow of regenerants to the units and to monitor the flow of water during rinsing and backwashing the resin bed.

The spent regenerant and rinse water are usually collected together in an effluent tank. Sufficient acid or alkali is then added to produce a neutral effluent and after thorough mixing the effluent is discharged to waste. A pH meter is usually employed to ensure that the effluent is correctly neutralised.

The ion-exchange units are rubber-lined steel vessels and to observe the resin, particularly during backwashing and regeneration, sight glasses are fitted at the resin/water interface. With mixed bed units an additional sight glass is fitted at the anion and cation resin interface.

(c) Automatic Control

The term automatic control when applied to demineralisation plants usually refers to automatic regeneration since, even with manually operated plants, automatic controls are usually included to protect the plant by shutting it down when some abnormal condition arises.

Where the water being treated is reasonably constant in composition, the ion-exchange resins can be assumed to be exhausted after a fixed quantity of water has been treated. Automatic regeneration may therefore be programmed to take place after a fixed time interval in service, or after a fixed quantity of water has passed through the plant, as measured by an integrating water meter. The backwashing, regeneration and rinsing stages are then carried out in the correct sequence and for the required period of time, and are controlled by process timers.

If, however, the raw water composition is not constant, then the regeneration cycle can be initiated by a change in the quality of the final treated water. It is usual to employ a conductivity meter, having preset contacts, to detect this change in quality.

With base-exchange softeners a residual hardness detector is used, of which there are a number of different types available.

(d) Condensate Polishing Units

The final mixed bed unit in the make-up water treatment plant of a modern high-pressure station is in effect a "polishing" unit, since its duty is to remove the last traces of dissolved salts and carbon dioxide from either a demineralisation stream or an evaporator distillate. A typical arrangement is illustrated in Figure 3.6.3M which shows the make-up system at Ironbridge "B". It will be noted that all make-up enters the feed system via the mixed bed units and the condenser, and that all water taken into the feed system from the reserve feed water storage tanks, is normally passed through these same mixed bed units and thence to the condenser.

This arrangement ensures that no contaminated make-up water enters the feed system, but it cannot be used to remove contaminants from the feed water, such as corrosion products or ingress of cooling water. If such contamination occurs, the level of contamination can only be reduced by rejecting some of the feed water and replacing it with fresh make-up water. Obviously, it is wasteful to discard this rejected water since it contains only a trace of salts, and to obviate this a method of reclaiming it is employed at some stations, the contaminated condensate being discharged to a *drains recovery system*. The "dumped" condensate, together with boiler blowdown water and superheater drains water, is collected in a drains recovery tank and after cooling is passed through the water treatment plant mixed bed units and in this way is returned to the system. Cooling is necessary because ion-exchange resins, in particular anion-exchange resins, are damaged and are liable to contaminate the water with resin breakdown products, if operated at temperatures above 50° to 60°C for long periods. Since the water from the reserve feed water tanks and from the drains recovery tank may contain solid corrosion products which would foul the resin bed, the water is passed through a filter before entering the mixed bed unit.

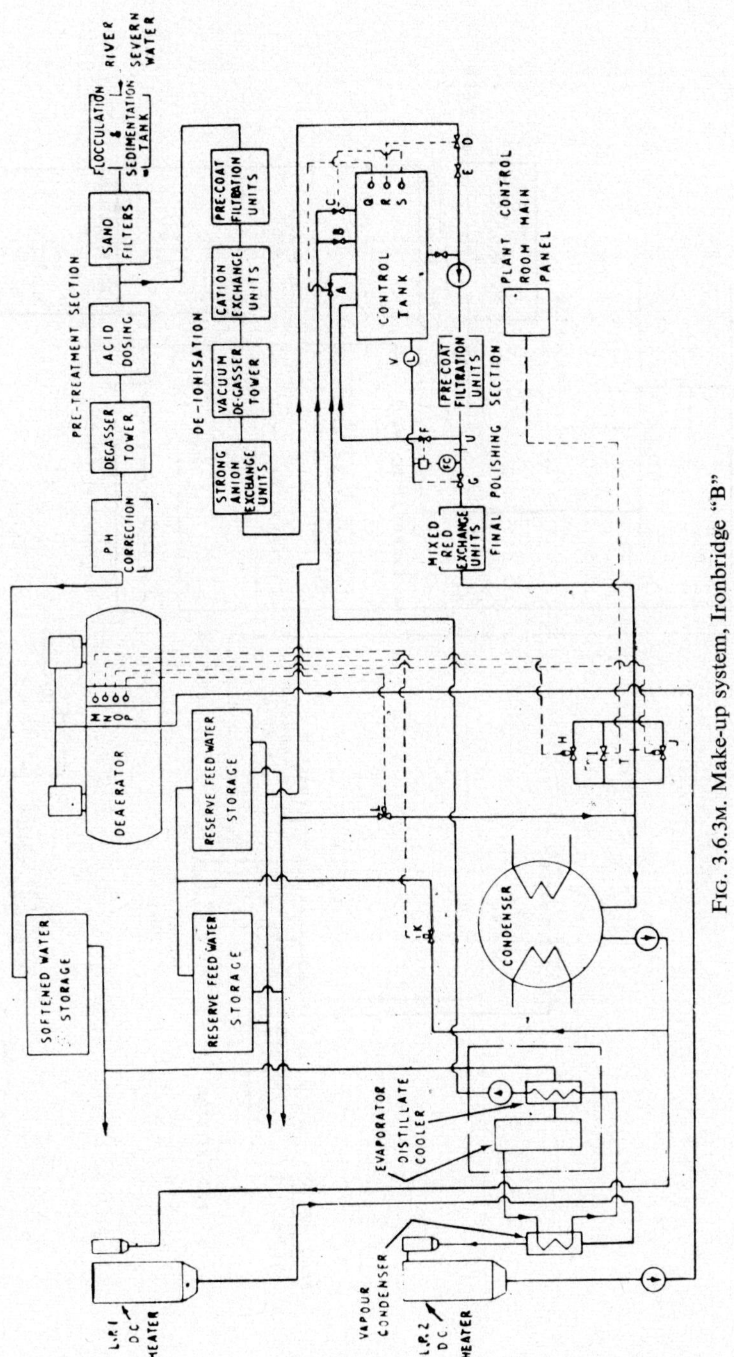

FIG. 3.6.3M. Make-up system, Ironbridge "B"

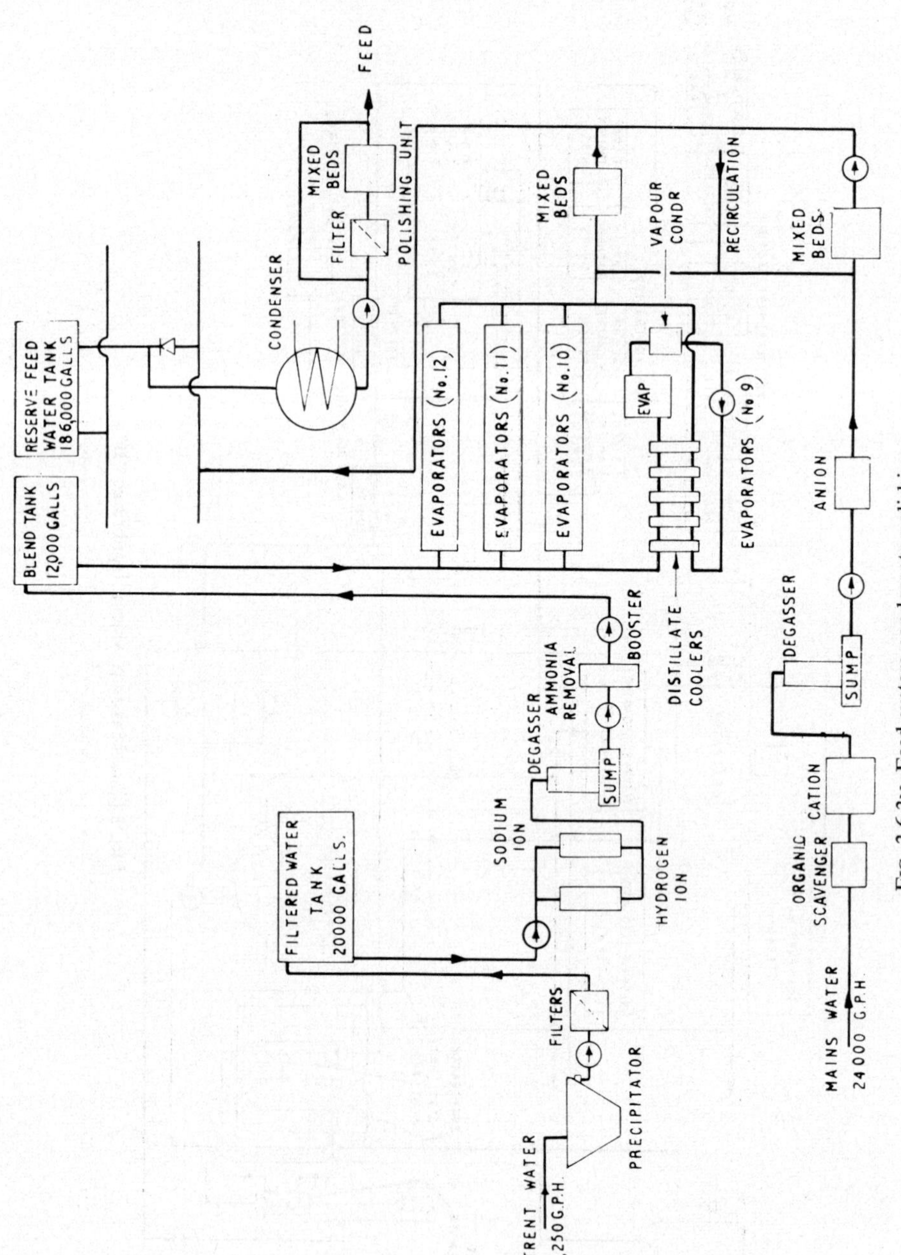

FIG. 3.6.3N. Feed system condensate polishing

The removal of contaminants by integrating condensate polishing equipment into the feed system was discussed in Section 3.6.2. A typical installation is illustrated in Figure 3.6.3N and includes an in-line filter and a mixed bed unit employing resin transport regeneration. One type of filter consists of a sand support material precoated with a cellulose filter medium, but precoated candle and cartridge types can also be used. Because of the temperature limitation, this type of plant can only be used in the low pressure section of the feed system and any corrosion products produced at a later stage in the feed system will still be carried forward to the boiler.

In C.E.G.B. stations it is intended to employ in-line condensate polishing only in units having once-through boilers. It is anticipated that the level of contamination, due to corrosion products, will be highest during "start-up" and as it is envisaged that each unit will be brought on load in turn, only sufficient condensate polishing capacity will be provided to handle one quarter of the full feed flow of one unit at any one time. Thus for a station having four identical units, once all of the units are on load, the capacity will be sufficient to treat approximately 6% of the feed flow for all of the units continuously, or up to 24% of the feed flow of one unit in the event of serious contamination occurring.

The C.E.G.B. specification for condensate polishing units requires that the water after treatment shall conform to the following quality:

TABLE 11

Iron	—not greater than 0·005 ppm (measured with a standard deviation of 0·001 ppm).
Copper	—not greater than 0·0015 ppm (measured with a standard deviation of 0·0003 ppm).
Reactive silicon	—not greater than 0·01 ppm (measured with a standard deviation of 0·005 ppm).
Sodium	—not greater than 0·005 ppm (measured with a standard deviation of 0·001 ppm).
Conductivity	—not greater than 0·08 micromho/cm at 25°C before and after passage through a cation exchanger in the hydrogen form.

Powdex process. A process has been developed in the United States of America, known as the "Powdex process", by which the ion exchange and filtering duties are carried out in a single unit. The unit usually comprises a number of filter medium support cartridges or candles, mounted in a single filter unit, and this arrangement provides a large filtering area in a compact unit. The filtering medium employed is a mixture of anion and cation exchange resins having very fine particle size (90% less than 325 U.S. mesh sieve size). The mixed resin is used as a pre-coat of about $\frac{1}{8}$ in. in thickness on the filter supports and this permits flow rates of up to 5 gal/min per square foot of filter surface.

A typical filter is shown in Figure 3.6.3P.

It is claimed that the rate of ion exchange when using the Powdex resin is about 100 times that obtained with normal ion-exchange resins and also that the operating exchange capacity is approximately double that of normal resins. For these reasons even with this thin layer of resin, good condensate polishing is said to be achieved and filter runs often have to be

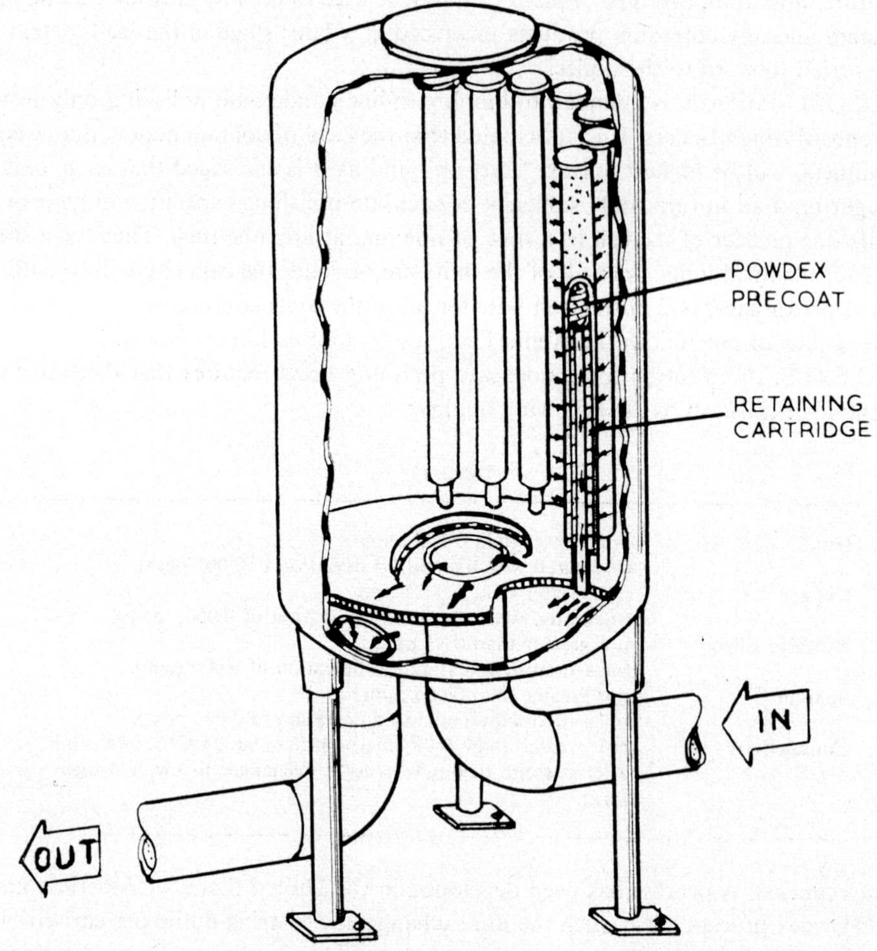

FIG. 3.6.3P. Powdex process cartridge unit sectional view (Reproduced by kind permission of the Graver Water Conditioning Co.)

terminated to recoat the filter because of increased pressure drop caused by fouling, rather than by exhaustion of the resin's ion-exchange capacity. The combination of ion exchange and filtering in the one unit is also said to provide an efficient way of removing silica from water, including non-reactive silicon.

Magnetic filters. In addition to the ion-exchange processes used for condensate polishing, magnetic filters have been employed experimentally in the feed system of some C.E.G.B. stations. These filters will remove particulate ferro-magnetic material present in the feed water, and in practice it has been found that even some non-magnetic materials such as copper oxide are also removed, possibly due to physical entanglement.

3.6.4. **Practical Considerations**

(a) Life of Ion-exchange Resins

Although ion-exchange resins will usually give many years' service, eventually they deteriorate in performance and have to be replaced with new resins.

Some of the resin deteriorates through attrition, partly through purely mechanical damage, and partly because of expansion and contraction of the resin particles resulting from the differing ionic sizes of the ions being exchanged. The fine particles produced in this way cause an increase in the pressure drop across the resin bed and result in increased rinse times during regeneration. Eventually it is necessary to remove these "fines" by backwashing or screening, and to replace them with fresh resin.

The resins used in demineralisation plant are organic chemicals and as such are subject to oxidation by oxygen, chlorine and even high nitrate concentrations if present in the water. At elevated temperatures, oxidation can cause serious degradation of the resins particularly with anion-exchange resins, and this leads not only to a reduction in ion-exchange capacity but also to contamination of the water by the soluble resin degradation products which are formed. For this reason it is anticipated that the use of vacuum degassers to reduce the dissolved oxygen of the water to no more than 0·03 ppm, will lead to increased anion resin life. Unless the water is deaerated, anion-exchange resins and mixed beds should not be operated for prolonged periods at temperatures above 60°C, as might arise in the treatment of evaporator distillate, condensate polishing and drains recovery systems. The higher the degree of cross-linking in a resin, the better will be its resistance to oxidative degradation. Type I anion-exchange resins can be used at higher temperatures than Type II resins.

In addition to the loss of exchange capacity caused by oxidation the strong groups of anion-exchange resins tend in time to degenerate to weak groups. Thus:

$$\left[-\overset{|}{\underset{|}{C}}-N\begin{matrix}\nearrow R \\ -R_1 \\ \searrow R_2\end{matrix}\right]^+ [OH]^- \longrightarrow \left[-\overset{|}{\underset{|}{C}}-N\begin{matrix}\nearrow R \\ \\ \searrow R_1\end{matrix}\right] + R_2OH \qquad (39)$$

Quaternary ammonium group — Strongly basic resin; Tertiary group — Weakly basic resin

with the result that the resin's capacity for exchanging weak acids such as carbon dioxide and silica, diminishes.

The resin may also become irreversibly fouled with substances such as oil or organic matter if they are present in the raw water. These substances physically fill the resin pores and prevent ions reaching exchange sites within the resin beads, or they may be irreversibly absorbed at the exchange sites. This type of fouling results in:

(a) A reduction in the ion-exchange capacity, particularly where strongly basic resins are involved.
(b) An increase in the electrical conductivity of the final water.
(c) An increase in the silica content of the final water.
(d) A prolonged rinse period.
(e) A drop in the pH of the final water.
(f) An increase in the pressure drop across the unit affected.

Chemical fouling of resins can occur if sparingly soluble substances are formed during regeneration, the calcium sulphate precipitation referred to earlier being a typical example of this problem.

(b) Organic Fouling of Resins

Organic fouling, briefly referred to in the previous section, is a special problem since organic matter is present in the raw water supply in many parts of the country and is difficult to remove by chemical treatment. There is also some evidence to suggest that due to the greater re-use of water in some parts of the country, the problem of organic fouling is increasing. The organic matter is mostly of vegetable origin (dead leaves, etc.) and the quantity present in water varies seasonally, with a maximum concentration in late autumn and winter.

The organic material concerned has a high molecular weight, and consists of large polyvalent molecules of variable composition generally referred to as *humic acids*. Usually the humic acid is in true solution and is not precipitated by acid; in this case it is called *fulvic acid*.

Fulvic acid is preferentially adsorbed by the strong, rather than the weak, groups of anion-exchange resins. Because of its large molecular size relative to the pore size of normal gel-type resins, it is not easily removed during regeneration but tends to diffuse further into the resin beads during each regeneration.

Where a resin has become fouled and its exchange capacity has fallen, some improvement can usually be effected by treatment with a strong alkaline brine (sodium chloride) solution before regenerating. Where analysis of the water indicates that it has organic fouling potentialities, this "salt cleaning" or "brine washing" is best carried out at regular intervals (possibly every 25 regenerations) before the degree of fouling has reached serious proportions. Sodium hypochlorite, which was sometimes used in the past instead of salt, should not be used since it will cause oxidative degradation of the resin.

Treatment of "organic fouling" water. The problem of fouling is gradually being overcome and at some stations where chemical pretreatment of the raw water is employed, up to 80% of the organic matter may be removed.

Another solution is to place an additional exchange unit, using an anion resin in the chloride form, to precede the main demineralisation plant. This unit, often called an *organic trap* or *scavenger unit* serves to remove the organic matter before it can reach the main plant. The quantity of anion resin employed in the organic trap is small in comparison with that in the main demineralisation units and it is never allowed to run to exhaustion but is regenerated at frequent intervals, for example after 2500 bed volumes. Regeneration is carried out in two stages, the first employing 10% sodium chloride and 2% caustic soda at about 40°C, whilst approximately $1\frac{1}{2}$ bed volumes of 3·5% sodium chloride are used in the second stage. Initially after regeneration, all of the resin in the organic trap is in the chloride form. As water is passed through it, anion-exchange reactions take place, chloride being exchanged for sulphate, carbonate, nitrate, etc., and in a short time the resin will reach equilibrium with the raw water. Since fulvic acid ions are very strongly adsorbed by the resin, exchange will continue between the anions (whatever they are) present on the resin and the small quantity (usually less than 5 ppm) of fulvic acid in the water. This adsorption of fulvic acid would continue until a considerable proportion of the exchange sites were occupied by fulvic acid, but in practice the resin is always regenerated well before this stage is reached. An organic trap of this type may remove up to 90% of the organic matter present in the water and experience suggests that the remaining organic matter does not cause any serious fouling of the anion resin in the main plant.

Resin manufacturers have also tackled the problem in a different way and have produced a range of anion-exchange resins having a cellular structure, the size of the pores being many times greater than the pores in normal gel-type resins. These resins are called *macroporous resins* and like the normal gel-type resins, will readily adsorb organic matter. However, on regeneration the organic matter is more easily eluted from the resin, with the result that there is little or no progressive accumulation. If after a period of use some fouling does take place, the performance of the resin is readily restored by salt washing. Because of the cellular structure there is less ion-exchange resin per unit volume, and therefore this type of resin has a lower exchange capacity than the normal gel-type resins. For this reason the use of macroporous resins should be restricted to waters containing organic matter.

A more recent development in this field is the introduction of the Permutit *isoporous* resins. These are essentially gel-type resins manufactured by a different process which, it is claimed, results in the cross-links being more regularly distributed. Thus the *micropores* are all approximately the same size and are large enough to permit the diffusion of even large organic molecules. These resins are used in standard anion and mixed-bed units in place of the normal gel-type resins and their ion-exchange capacity is the same as the normal resins.

Thus in designing a water treatment plant to demineralise a water known to contain organic matter, there are several ways of dealing with the problem. In selecting the best treatment to adopt, the following approach has been suggested by The Permutit Company Ltd.:

$$\text{Let } A = \frac{\text{ppm oxygen demand}}{\text{ppm anions}}$$

where ppm oxygen demand = oxygen absorbed from permanganate (4 h at 27°C) as ppm oxygen.

Then if A is less than 0·004,	no special treatment is required and any suitable anion-exchange resin can be used.
If A is between 0·004 and 0·008,	isoporous resin can be satisfactorily employed in a mixed bed unit.
If A is between 0·008 and 0·015,	organic poisoning of mixed bed is likely; use either organic trap column or isoporous resin in preceding anion-exchange unit.
If A is greater than 0·015,	water pretreatment is essential, employing coagulation and filtration.

It seems likely that a similar approach would apply to the use of macroporous resins.

(c) Non-reactive Silicon

The word "silica" is used rather loosely in regard to water treatment in general. Although silica is properly silicon dioxide (SiO_2), which is nearly insoluble in water, the word is commonly used to include almost all compounds of silicon in water and, when in true solution, these are largely ionised—probably as silicates. In the present context, compounds of silicon, such as fine clay particles and similar material, which are not detected by the normal tests for "silica", will be termed "non-reactive silicon".

Although strongly basic anion-exchange resins will remove ionised silica from water, some waters contain silicon in an un-ionised form (non-reactive silicon), which may pass completely through a demineralisation plant and also escape detection by the tests normally employed. Under boiler operating conditions the non-reactive silicon may be converted into the reactive form and since in high-pressure plant it is necessary to control the total silica in the boiler water, it is important to check that the *total* silicon content of the demineralised make-up water is below the maximum level permitted. (This is discussed in Chapter 2.)

In continental Europe pretreatment of the raw water with a pre-coat filter is often used to remove non-reactive silicon and in some cases activated carbon is employed as the pre-coat material. The Powdex process, referred to previously, which employs finely divided mixed ion-exchange resins as the pre-coat material, has been used at a number of generating stations in the U.S.A. The current approach in the C.E.G.B. is to employ the Powdex process after the final mixed bed unit, if non-reactive silicon is known to be present in the raw water.

(d) Control Tests: Measurement of Resin Ion-exchange Capacity

The various causes of deterioration in the performance of ion-exchange resins have been discussed in Section 3.6.4(a). The indication that changes have occurred may be a reduction in the throughput between regenerations, a deterioration in the quality of the demineralised water, or extended rinse times, and when these changes reach a stage where they seriously affect the performance of the plant, remedial action has to be taken. A better approach is to continuously monitor the performance of the resins and this can be done by measuring both the total and the operational exchange capacity of the resins at regular intervals throughout the life of the resin.

The changes resulting from chemical degradation cause a loss of total exchange capacity, due to loss of strong and weak exchange groups; some strong groups are also down-graded to weak groups. These changes are normally very gradual except where the resin is exposed to severe conditions of temperature or oxidative environment, so that modern exchange resins normally have a useful life of many years.

When fouling of a resin occurs, a serious reduction in the operational exchange capacity will be noted, since exchange sites present in the resin are no longer available to take part in exchange reactions. If the measured operational exchange capacity of a resin increases as a result of cleaning, this is strong evidence that fouling is taking place. Where this is found, the cleaning and measurement of the exchange capacity should be continued until there is no further increase in the measured exchange capacity, thus ensuring that all of the material fouling the resin has been removed.

Full details of the recommended procedure for determining the total ion-exchange capacity of a resin and the operational ion-exchange capacity are to be found in the Central Electricity Research Laboratories Report No. 794. The operational capacity is almost always less than the total exchange capacity and its value will depend upon the conditions under which it is determined. For control purposes the operational capacity is determined under standard laboratory conditions but it should be noted that the value obtained may differ from that achieved in an operating plant.

The following notes summarise the methods given in C.E.R.L. Report No. 794.

Cation Resins

(i) *Total cation-exchange capacity*, C_T. This is determined by completely converting a known weight of the resin to the hydrogen form, by passing a considerable excess of 1 N nitric acid through it. The regenerated resin after rinsing to remove excess acid, is then allowed to stand for several hours in contact with decinormal (0·1 N) sodium hydroxide solution containing 10% (w/v) sodium chloride, and the reduction in sodium hydroxide concentration which results is determined by titration with standard acid solution. This reduction in hydroxide is equal to the amount of hydrogen ions released from the resin and hence the total exchange capacity of the resin.

$$\text{Then } C_T = \frac{[V_1 N_1 - 4 N_2 t_1]}{W_1[100-M]} \times 100 \text{ m-equiv/g of dry resin} \tag{40}$$

where V_1 = volume of 10% sodium chloride/0·1 N sodium hydroxide,
N_1 = normality of the decinormal sodium hydroxide solution,
N_2 = normality of the standard acid,
t_1 = titre, in ml,
W = weight of moist resin in hydrogen form, in g,
M = % moisture content of the resin.

(ii) *Strongly acidic exchange capacity*, C_S. This is determined by passing 1 litre of 4% (w/v) sodium sulphate solution through a known weight of the fully regenerated resin and titrating a 100 ml. aliquot of the resulting solution with standard sodium hydroxide solution.

$$\text{Then } C_S = \frac{1000\, N_3\, t_2}{W_2[100-M]} \text{ m-equiv/g of dry resin} \tag{41}$$

where N_3 = normality of the standard sodium hydroxide solution,
t_2 = titre, in ml,
W_2 = weight of moist resin taken, in g.

(iii) *Weakly acidic exchange capacity*, C_W. This is obtained by difference.

$$\text{Thus } C_W = C_T - C_S \text{ m-equiv/g of dry resin} \tag{42}$$

(iv) *Operational exchange capacity*, C_O. A sample of the resin completely converted to the sodium form, is regenerated to the hydrogen form with 1 M hydrochloric acid (700 ml acid/250 ml resin) at a flow rate of 30 ml/min. Then 0·01 N barium chloride solution is passed at constant rate through a measured volume of the regenerated resin contained in a column. The pH of the effluent from the column is then measured and the point determined at which the pH increases.

$$\text{Then } C_O = \frac{10V_1}{V} \text{ m-equiv/ml of moist resin} \tag{43}$$

where V_1 = volume of barium chloride solution passed through the column at the time of pH increase,
V = volume of moist resin.

Anion Resins

(i) *Weakly basic capacity* C_W. This is determined by completely converting the resin to the chloride form, passing 1250 ml of 1% (v/v) ammonia solution through a known weight of the resin in this form and collecting the first 1000 ml of effluent in one receiver and the remaining 250 ml fraction in a second receiver. The chloride content is then determined by titrating 100 ml aliquots of the fresh 1% (v/v) ammonia solution, and the first and second fractions in turn, against standard silver nitrate solution. The ammonia solution elutes the chloride from all weakly basic groups plus some of the strongly basic groups and the value of chloride content in the second fraction is used to estimate the amount of chloride derived from strong groups, which is contained in the first fraction.

$$\text{Then } C_W = \frac{1000N[t_1-t_2]}{W[100-M]} \text{ m-equiv/g of dry resin} \tag{44}$$

where N = normality of the silver nitrate solution,
t_1 = titre in ml, of the first fraction aliquot,
t_2 = titre in ml, of the second fraction aliquot,
t_3 = titre in ml, of the fresh 1% (v/v) ammonia solution,
W = weight of moist resin taken, in g,
M = % moisture content of the resin.

(ii) *Total basic capacity*, C_T. 1000 ml of 4% (w/v) sodium sulphate solution is then passed through the same resin and the effluent collected. The chloride content is then determined as before on a 100 ml aliquot of the effluent and on 100 ml of the fresh 4% (w/v) sodium sulphate solution.

$$\text{Then } C_T = \frac{1000N\left[t_1 - t_3 + \dfrac{t_2 - t_3}{4} + t_4 - t_5\right]}{W[100-M]}$$

$$= \frac{250N[4t_1 + t_2 - 5t_3 + 4t_4 - 4t_5]}{W[100-M]} \text{ m-equiv/g of dry resin} \qquad (45)$$

where t_4 = titre in ml, of the effluent after passing sodium sulphate,
t_5 = titre in ml, of the fresh 4% (w/v) sodium sulphate.

(iii) *Strongly basic capacity*, C_S. This is obtained by difference.

$$\text{Then } C_S = C_T - C_W \text{ m-equiv/g of dry resin} \qquad (46)$$

(iv) *Operational capacity*, C_O. A sample of the resin completely in the chloride form is regenerated to the hydroxide form with 2% sodium hydroxide solution, using regeneration conditions similar to those employed on the plant. 0·003 N sulphuric acid is then passed through a known volume of the regenerated resin, contained in a column, until a rapid decrease in the pH of the effluent occurs.

$$\text{Then } C_O = \frac{3V_1}{V} \text{ m-equiv/ml of moist resin}$$

where V_1 = volume of 0·003 N sulphuric acid passed through the column up to the point of pH decrease, ml,
V = volume of resin employed, in ml.

It will be noted that the total, strong and weak exchange capacities are reported as milli-equivalents per gram of dry resin, whereas the operational capacity is given as milli-equivalents per millilitre of moist resin.

3.6.5. **Recent Developments**

All the processes discussed in the previous sections dealing with demineralisation are either in use within the C.E.G.B. or have been used elsewhere. Before leaving the subject, it is useful to discuss some of the recent developments which are of interest to the C.E.G.B., and two of these have been selected.

(a) Permutit Double Cation Resin Unit

In Section 3.6.2 it was noted that demineralisation plant operating costs could be considerably reduced by employing both strong and weak resins and regenerating them in series. This arrangement gives the lowest operating cost but usually at the expense of capital cost, since two complete units are employed instead of one.

The Permutit Company Ltd. have recently introduced a process in this country, in which both the strong and weak cation-exchange resins are contained in a single unit. In operation, the weak resin forms a layer on top of the strong resin and water is passed downward through them as in a normal unit. As water passes through the weak resin, cations associated with weak acids (such as carbonic acid) are exchanged for hydrogen ions (H^+); this is followed by passage through the strong resin, when cations associated with strong acid radicals (sulphate, chloride, etc.) are exchanged for hydrogen ions. The net result is that each resin performs its normal function and complete cation exchange occurs by the time the water leaves the unit.

Regeneration of the unit is carried out in the reverse direction to that normally employed. After backwashing the regenerant flows upwards through the resin bed, first passing through the strong resin and then through the weak resin; this is followed by rinsing again in an upward direction. A modified internal arrangement is required to carry out these operations and at certain stages of the regeneration a small downward flow of water is maintained.

(b) Demineralisation of Water of High Dissolved Solids

As the T.D.S. of a water increases, a point is reached where it is uneconomic to employ demineralisation and either an alternative water supply having a lower T.D.S. or an alternative treatment, must be employed. In theory an equivalent of both acid and base is required to remove one equivalent of a neutral salt from water, but in practice even more is required since the regeneration efficiency of resins is always less than 100%. Where, in a water with a high T.D.S., a significant proportion of the dissolved salts is present as bicarbonates, demineralisation may still be an attractive proposition, since only acid is required for regeneration purposes, the carbon dioxide liberated by cation exchange being removed in a degassing tower.

For waters consisting mainly of neutral salts, the range of waters for which demineralisation can usefully be employed can only be extended by employing resins having a high regeneration efficiency and which can be regenerated by cheap readily-available acids and bases.

A new demineralisation process, which is claimed to be capable of extending the economic range of ion-exchange to waters having up to 3000 ppm, has been described by the Rohm and Haas Company in their publication Amber-Hi-Lites Nos. 89 and 91. The process employs the Amberlite resins IRA 68 (weakly basic anion exchange) and IRC 84 (weakly acidic cation) and depends for its success on the discovery that, unlike most other weakly basic anion resins IRA 68 is capable of forming a bicarbonate complex, thus:

$$\underset{\text{Free base resin}}{R-N}+H_2O+CO_2 \longrightarrow \underset{\text{Resin bicarbonate complex}}{[R-NH]HCO_3} \tag{47}$$

and this bicarbonate complex will exchange strong anions, such as chloride, for bicarbonate, thus:

$$[R-NH]HCO_3+NaCl \longrightarrow [R-NH]Cl+NaHCO_3 \tag{48}$$

resulting in the conversion of neutral salts to alkaline bicarbonate salts.

If after passage through an anion exchanger of this type, the water is passed through a weakly acidic cation exchanger [IRC 84], the bicarbonates will be converted to carbon dioxide and water.

$$NaHCO_3 + RH \longrightarrow RNa + H_2O + CO_2 \quad (49)$$

In the process described by Rohm and Haas, the water passes through the cation exchanger, and then through a third column containing IRA 68 in the free-base form, which adsorbs the carbon dioxide and yields pure water. The reaction occurring in this column is represented by equation (47) and hence the complete process is represented by equations (48), (49) and (47) in that order.

When the units are exhausted, the first anion resin unit will contain the resin in the resin/anion complex form as given by equation (48). It is then regenerated to the free-base form with any readily available alkali, and lime or ammonia solutions have been used.

$$2\,[R-NH]Cl + Ca(OH_2) \longrightarrow 2\,RN + CaCl_2 + 2\,H_2O \quad (50)$$

The cation unit is regenerated with an acid such as hydrochloric or sulphuric acid, but waste acids derived from other processes may also be used:

$$2\,RNa + H_2SO_4 \longrightarrow 2\,RH + Na_2SO_4 \quad (51)$$

The third unit containing IRA 68 will be in the bicarbonate form at the end of the cycle and for the next period of operation it is used as the leading unit, the water passing through the units in the reverse order to that which was used on the previous operational cycle.

Because the regeneration efficiency of the weak resins used is almost 100% and low cost regenerants can be employed, the process is economically attractive and provides an alternative to evaporation for the treatment of high T.D.S. waters.

A further development of this process dispenses with the third unit containing IRA 68 resin. The first unit contains IRA 68 in the free base form and carbon dioxide is fed directly into the raw water as it enters the unit. The reaction occurring is

$$R-N + NaCl + CO_2 + H_2O \longrightarrow [R-NH]Cl + NaHCO_3 \quad (52)$$

The result is the same as that occurring in equation (48). In place of the third unit containing IRA 68, a degassing tower is employed, the liberated carbon dioxide being recycled back to the first unit.

3.7. CHOICE OF MAKE-UP WATER TREATMENT PLANT

The various processes used in the preparation of boiler feed make-up water have been described. For the complete water treatment plant two alternative schemes are possible, one employing demineralisation as the main salt removal stage and the other, evaporation. The two schemes were illustrated diagrammatically as schemes (A) and (B) in Figure 3.3.

With scheme (A) a number of alternative treatments, prior to evaporation, are used depending on the quality and composition of the raw water, and Figures 3.7A and B are two examples. Also with both of the main schemes, where non-reactive silicon is likely to be a problem, an additional polishing treatment may be employed after the mixed bed.

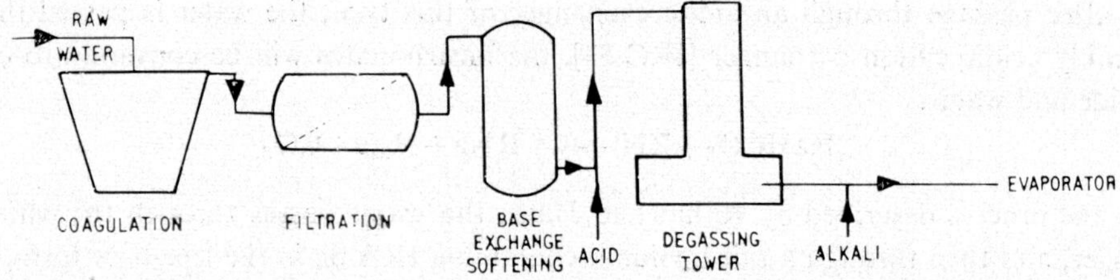

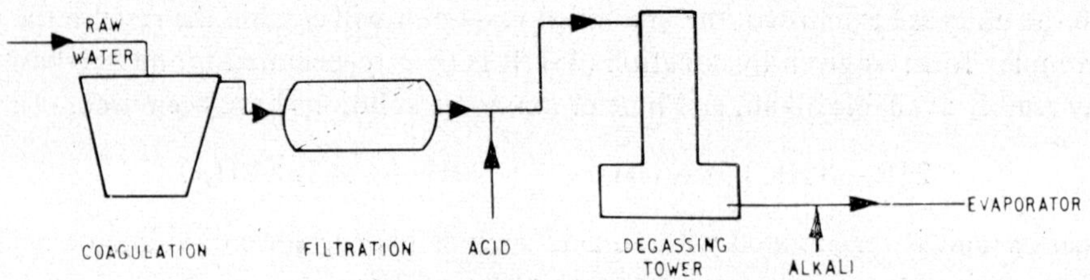

FIG. 3.7A, 3.7B. Make-up water treatment plant

Both schemes (A) and (B) are capable of producing water suitable for modern high-pressure boilers and the complete water treatment plant often comprises a combination of the two schemes as in Figure 3.6.3N.

It is worth remembering that evaporators require a supply of steam and that with bled steam evaporators, sufficient steam is available only when the turbine is operated at a sufficiently high load. Also the modern bled steam evaporator forms an integral part of the feed system, which means that it cannot be operated independently of the turbine.

Live steam central evaporators are not normally installed in modern high-pressure plant partly because of the practical difficulties associated with the very high pressures involved and partly because of their very high operating cost. Therefore, to provide make-up during periods of low load and for starting plant from cold, it is usually necessary to arrange for at least a part of the make-up to be produced by demineralisation. For a 100% evaporated make-up installation, an alternative supply of steam from auxiliary boilers and an alternative means of condensing the evaporator vapour would be required.

The complete water treatment plant must have a sufficiently large output to meet the peak demands for make-up. It is usual to refer to make-up as a percentage of the total boiler steam output for the station and for new high-pressure plant, make-up capacity of 3% is needed to provide for the peak demand, which arises largely during commissioning. When teething troubles have been overcome, make-up requirements should fall to about 1%. One way to provide the extra make-up for commissioning would be to use a transportable demineralisation plant, but this is now considered to be impracticable, partly because of the size of the plant that would be needed.

In a new station the units are usually planned to be commissioned successively so that, in theory, all of the surplus capacity of the water treatment plant is available to meet the commissioning needs of each unit in turn. Hence the overall percentage make-up required will be influenced to some extent by the number of units installed.

The 200–350 MW units now in service are mostly provided with a facility to reject condensate at the extraction pump discharge (condensate dumping), when the units are brought on to load, in order to ensure that water having high metal content is not fed to the boilers. The maximum rate of condensate dumping is usually about 10% of full-load flow, and the maximum quantity of water that can be rejected at each start-up, is limited by the availability of fully deaerated reserves to replace it as feedwater.

If a water recovery system (see Section 3.6.3d) is provided, the net demand for make-up will be reduced and may be as low as 0·5%. In this case, of course, the mixed-bed units will need to be sufficiently large to treat the water from the recovery system, as well as that from the water treatment plant.

The purification of condensate at start-up will be taken a stage further in some larger stations now under construction, where separate condensate polishing mixed-bed units will deal with the feed water at start-up, for up to 25% of full load feed flow. The water treated in this way is returned directly to the feed system and a single mixed bed polishing plant may be arranged to serve each boiler/turbine unit in turn, as the units are brought on load in sequence.

In considering the size of water treatment plant for a new station, the projected load factor of the station has to be taken into account. A low load factor implies two-shift operation and the more starts and stops involved, the higher will be the water requirements.

If after a study of the many aspects briefly referred to above, alternative water treatment plant schemes could satisfy the requirements, the choice between them must be made on economic grounds. The preferred scheme should show the lowest total cost, that is to say, capital plus operating cost. Evaporators and their associated plant have a higher capital cost than demineralisation plants of the same capacity, but against this the running cost for

TABLE 12

COMPARISON OF WATER TREATMENT COSTS

(WATER TREATMENT PLANT SUPPLYING 3% MAKE-UP TO A 1000 MW STATION)

	Demineralisation plant			Modern bled steam evaporators			Modern bled steam evaporators		
	Town supply water (T.D.S. 500 ppm)			Town supply water (T.D.S. 500 ppm)			Sea water		
Water treatment plant (percentage utilization)	20	50	80	20	50	80	20	50	80
Operating cost (*d.*/1000 gal) inc. cost of raw water	72	72	72	56	56	56	48	48	48
Capital cost (*d.*/1000 gal)	61	24	15	107	43	27	107	43	27
Total cost (*d.*/1000 gal)	133	96	87	163	99	83	155	91	75

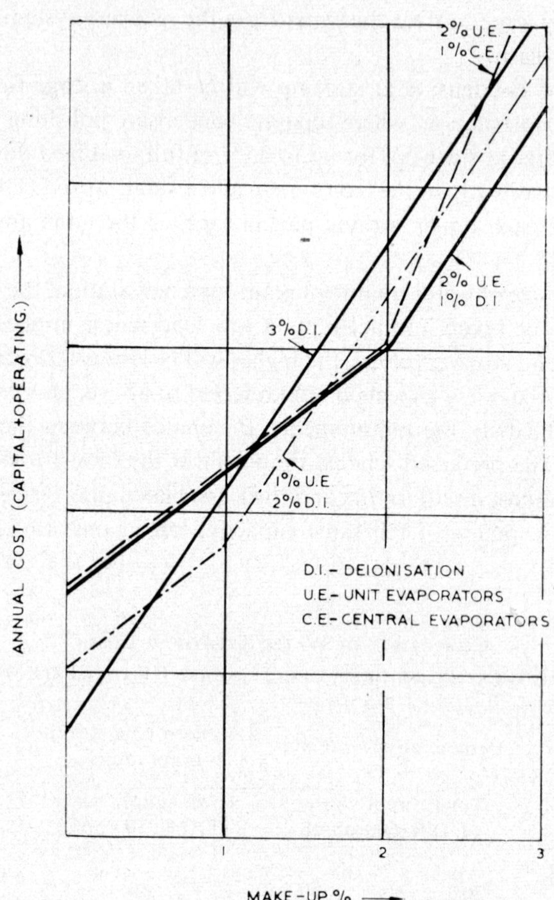

FIG. 3.7C. Comparison of costs for various make-up systems

demineralisation are often higher than for a modern bled steam evaporator. The total cost will be affected by the percentage utilisation of the water treatment plant and this is clearly demonstrated in Table 12, which shows a comparison of the approximate cost of demineralisation, versus evaporation. Usually the problem is much more complex and involves consideration of various combinations of demineralisation and evaporation. In such a combined scheme, it is assumed that the part of the treatment plant having the lowest operating cost, will be used to provide the major part of the make-up required. Figure 3.7C which is taken from a paper by F. J. R. Taylor, illustrates how the costs of make-up for a number of combinations of treatment plants are related to the quantity of make-up required.

3.8. COOLING WATER SYSTEMS

There are three main types of cooling water systems namely: direct or once-through, closed with cooling towers, and mixed systems. These have already been described in another volume. This volume is concerned with the chemical aspects which may be summarised under the following headings:

(a) Prevention of scale deposition on heat exchange surfaces.
(b) Prevention of organic growths and slimes (generally referred to as algae) on heat exchange surfaces.
(c) Prevention of mud deposition on heat exchange surfaces.
(d) Controlling marine growths on intake culverts (mussels and barnacles).
(e) Minimising corrosion throughout the cooling water system.

3.8.1. Scale Deposition on Heat Exchange Surfaces

If carbon dioxide is dissolved in water, the following reactions take place:

$$\underset{\text{(Dissolved)}}{CO_2+H_2O} \rightleftharpoons \underset{\text{Carbonic acid}}{H_2CO_3} \rightleftharpoons \underset{\text{Bicarbonate ion}}{H^+ + HCO_3^-} \rightleftharpoons \underset{\text{Carbonate ion}}{2H^+ + CO_3^{--}} \quad (53)$$

and

$$H_2O \rightleftharpoons H^- + OH^- \quad (54)$$

so that the solution obtained is an equilibrium mixture of all the ionic and molecular species CO_2 gas, CO_2 dissolved, H_2O, H^+, $(OH)^-$, $(HCO_3)^-$, $(CO_3)^{--}$. If the solution is now treated in such a way as to favour the production of one of these species, the equilibrium will be shifted in this direction.

Similarly a solution of a bicarbonate salt in water will contain equilibrium concentrations of all of the ionic and molecular species referred to above, in particular the solution will contain carbonate ions $[CO_3]^{--}$ and the free carbon dioxide present will exert a measurable vapour pressure. The free carbon dioxide present in a solution containing 60 ppm of calcium bicarbonate alkalinity (as $CaCO_3$) at 20°C, will exert a partial pressure equal to the normal partial pressure of carbon dioxide present in the atmosphere, so that such a solution would be in equilibrium with the atmosphere, if exposed to it. Solutions of higher concentration than this would tend to lose carbon dioxide to the atmosphere and solutions of lower concentration would tend to adsorb carbon dioxide from the atmosphere until equilibrium is again achieved. The reactions occurring would be:

$$2\,[HCO_3]^- \rightleftharpoons [CO_3]^{--} + H_2O + CO_2 \qquad (55)$$

the reaction from left to right occurring with high concentrations of bicarbonate alkalinity and the reverse reaction with low alkalinities.

It should be noted that although a solution of a bicarbonate salt in water will contain all the ionic and molecular species mentioned, if such a solution contains only sufficient free and combined carbon dioxide to form bicarbonate ions equivalent in concentration to the cations present, then on analysis, the solution will behave as if it contains only bicarbonate alkalinity since, in the process of titration with acid, the equilibrium will be upset. Similarly, pure water at 25°C containing no added carbon dioxide, will dissolve 12 ppm of calcium carbonate and analysis would show that the solution had a carbonate alkalinity of 12 ppm (as $CaCO_3$). In fact in this solution only about one third of the total carbon dioxide present will exist as carbonate ion and the principal ionic species will be the bicarbonate ion.

Once-through Cooling Water System

In a once-through system, the only changes in the cooling water composition which are of interest, are those which occur whilst the water is passing through the heat exchanger. In these circumstances there is practically no possibility of carbon dioxide being lost from the water but, even if this happens, the calcium carbonate may well exist as a supersaturated solution and for this reason, since the contact time is relatively short, carbonate scale deposition is seldom a problem. If, however, the cooling water is already saturated or supersaturated with calcium carbonate at the condenser inlet, the risk of scale deposition is increased to some extent. The degree of calcium carbonate saturation may be demonstrated by adding finely divided calcium carbonate to the water, stirring and then observing the pH. If the pH remains constant or falls the water is saturated or supersaturated with calcium carbonate; if it rises the calcium carbonate concentration is below saturation.

Langelier Index

A water which is unsaturated with respect to calcium carbonate is potentially corrosive so that to minimise the risk of both scaling and corrosion, the water should be just saturated. Langelier developed the following method of predicting the scaling or corrosive tendencies of a water, which requires a knowledge of the chemical composition of the water, the pH and the temperature:

The Langelier or Saturation Index

$$= pH_{(actual)} - pH_{(saturation)} \qquad (56)$$

$pH_{(saturation)}$ is defined as the pH the water would have, if it had the same total alkalinity but was just saturated with respect to calcium carbonate.

$$pH_{(saturation)} = [pK_2' - pK_s'] + p[Ca] + p[Alk] \qquad (57)$$

where p = the logarithm to the base 10, of the reciprocal of the adjoining symbol,
[Ca] = concentration of calcium as gram-equivalents/litre,
[Alk] = alkalinity as gram-equivalents/litre,
K_2' is the second dissociation constant of carbonic acid,
K_s' is the dissociation constant of calcium carbonate.

The quantity [$pK_2' - pK_s'$] varies with temperature and total ionic concentration and values for this quantity are given in Table 13 below, which is taken from *Industrial Water Treatment Practice* by Hamer, Jackson, and Thurston.

Note: The term "dissociation constant" arises from the basic theory of chemical equilibria, which cannot be discussed here. In the present context we are considering the dissociation of (a) the bicarbonate ion, HCO_3^-, and (b) calcium carbonate, $CaCO_3$, in water.

For (a) the equilibrium is: $HCO_3^- \rightleftharpoons H^+ + CO_3^{--}$

and the dissociation constant, $K_2' = \dfrac{(H^+)(CO_3^{--})}{(HCO_3^-)}$

For (b) the equilibrium is: $CaCO_3 \rightleftharpoons Ca^{++} + CO_3^{--}$

and the dissociation constant, $K_s' = \dfrac{(Ca^{++})(CO_3^{--})}{(CaCO_3)}$

The quantities in parentheses (), are approximately equal to the concentrations of the items within the brackets, at equilibrium.

TABLE 13

VALUES OF [$pK_2' - pK_s'$]

Dissolved solids content of the water		Values of [$pK_2' - pK_s'$] at a temperature of				
As ionic strength	As T.D.S. (ppm)	0°C	10°C	20°C	50°C	80°C
0·000	0	2·60	2·34	2·10	1·55	1·13
0·001	40	2·68	2·42	2·18	1·63	1·22
0·005	200	2·76	2·50	2·27	1·72	1·32
0·010	400	2·82	2·56	2·33	1·79	1·39
0·015	600	2·86	2·60	2·37	1·84	1·44
0·020	800	2·89	2·64	2·40	1·87	1·48

If the saturation index of a water is significantly positive, the water is potentially scale forming and, conversely, if it is negative the water is potentially corrosive. The method is not sufficiently precise to indicate the behaviour of a water having a sensibly zero saturation index, but since some degree of supersaturation can usually occur without scale deposition, it is better to ensure that the water has a small positive index at the highest temperature in the system (a value of about 0·4 has been suggested), rather than risk corrosion.

To achieve the required saturation index, treatment of the cooling water with either acid or alkali is employed, but can prove to be very costly.

It is implicit in the derivation of the Langelier Index, that the water condition approaches chemical equilibrium, but such a condition is rarely obtained in a cooling tower recirculating system. The principle would therefore appear to be more applicable to once-through systems, but, as has been noted earlier, condenser tube scaling seldom occurs in the latter case and chemical treatment would be economically precluded by the enormous volume of water to be treated. For these reasons the Langelier Index method of control is not commonly used in C.E.G.B. stations, although it appears to be applied more in the United States.

Recirculating Cooling Water Systems

In a recirculating cooling water system, any given increment of water added to the system as make-up will remain in the system for a considerable period of time, before being discharged from the system as purge water. The residence time, that is to say the average period of time during which any increment of make-up water remains in the system, may be sufficiently long for the composition of the water to change significantly, resulting in it becoming either corrosive or potentially scale forming.

If V_o = volume of water in the cooling water system (gal.)
C_o = cooling water flow rate (gal/h),
M_a = rate of make-up (gal/h),
P_u = rate of purge (gal/h),
E_v = rate of evaporation (gal/h).

Then
$$M_a = P_u + E_v \tag{58}$$

Also the average *residence time*, R_e will equal

$$V_o/M_a \text{ hours} \tag{59}$$

During this residence time, the same increment of make-up water will circulate N times around the system, where

$$N = \frac{C_o}{V_o} \times \frac{V_o}{M_a} = \frac{C_o}{M_a}$$

Thus both R_e and N are dependent on the rate of make-up, M_a, for any given system and hence the risk of a significant change occurring in the water composition, will be determined by the rate of make-up. The greater the rate of make-up the smaller will be the chance of change occurring. The once-through cooling water system is in fact merely a special case where C_o is equal to M_a and hence $N = 1$.

If the make-up water has a high temporary hardness, the cooling water will initially also have a high temporary hardness. As it cascades down the cooling towers on each circuit around the system, the warm cooling water will tend to lose carbon dioxide to the atmosphere (see equation (55)). If the contact time between water and air were sufficiently long, the loss of carbon dioxide would continue until the partial pressure of carbon dioxide above the water was equal to the partial pressure of carbon dioxide in the atmosphere. Thus at equilibrium, a water which initially contained bicarbonate alkalinity and free carbon dioxide, would contain both carbonate and bicarbonate alkalinity. The relationships between initial bicarbonate alkalinity and equilibrium carbonate and bicarbonate alkalinity are given in Figure 3.8.1.

In a cooling tower, however, equilibrium is never achieved on one pass, partly because the contact time is short and partly because the carbon dioxide content of the air within the tower shell is higher than in the atmosphere, due to the amount of carbon dioxide released from the water. Thus the amount of carbon dioxide lost from the water, whilst it remains in the system, will depend on the value of N, as will the concentration of carbonate ions produced. If however the make-up water contains free carbon dioxide, some of the carbonate ions will be converted to bicarbonate ions (the reverse of equation (55)), thus:

$$CO_2+[CO_3]^{--}+H_2O \rightleftharpoons 2\,[HCO_3]^- \quad (60)$$

resulting in a lower concentration of carbonate ions in the water fed to the condenser.

Where the carbonate ion concentration is sufficiently high, the solubility product of calcium carbonate may be exceeded and the water passing to the condenser is then potentially scale-forming. It is by no means certain that scale deposition will occur and in fact only a small proportion of C.E.G.B. cooling tower stations are troubled in this way. The reasons for this apparent immunity from scaling are often obscure and probably vary from station to station, with the result that different methods of treatment have been evolved to overcome scaling problems, where they have arisen.

Prevention of Scale Deposition

At some stations, chlorination of the cooling water to prevent slime growths has also successfully overcome an existing scale deposition problem, probably because the calcium carbonate does not easily crystallise out on surfaces maintained in a clean, smooth condition. At some of the older stations the cooling water is continuously treated with sodium hexametaphosphate (Calgon, Micromet) to maintain a small concentration (1 to 2 ppm) in the water at all times. The treatment is known as "threshold treatment" and apparently acts by interfering with the calcium carbonate crystallisation process. Metaphosphates are readily hydrolysed to orthophosphates by heating:

$$\underset{\text{Sodium metaphosphate}}{NaPO_3}+H_2O \longrightarrow \underset{\text{Sodium dihydrogen orthophosphate}}{NaH_2PO_4} \quad (61)$$

and this will form insoluble calcium and magnesium orthophosphates with calcium and magnesium salts present in the cooling water. For this reason the threshold treatment should only be applied in cooling systems where the required concentration of metaphosphate

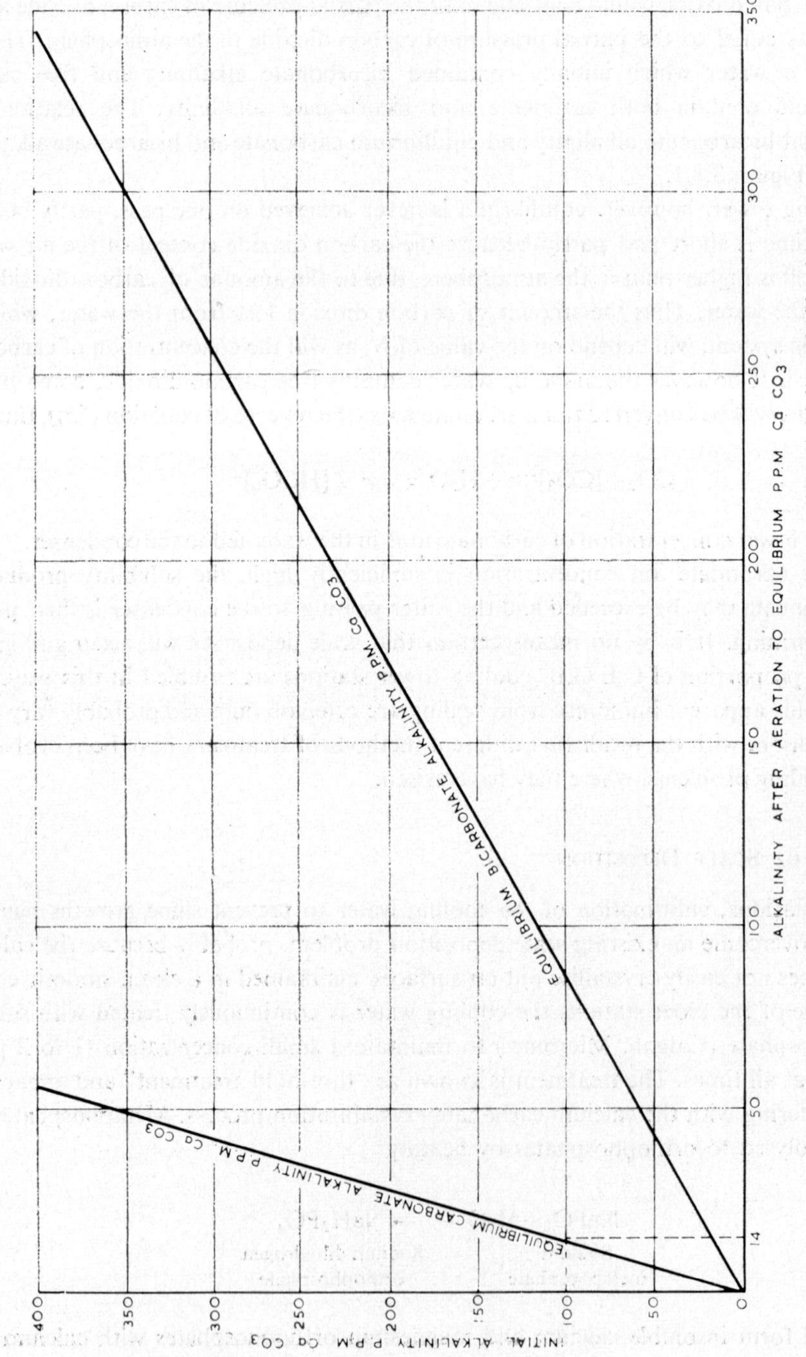

FIG. 3.8.1. Carbonate and bicarbonate alkalinity of a water containing initially only bicarbonate alkalinity, after bringing into equilibrium with the carbon dioxide content of the atmosphere

can be maintained with small additions of the metaphosphate; otherwise the deposition of insoluble phosphates on heat exchange surfaces may be a problem.

At a few stations, acid is added to the cooling water to reduce the carbonate alkalinity and so minimise the risk of calcium carbonate scaling:

$$[CO_3]^{--} + HCl \longrightarrow [HCO_3]^{-} + Cl^{-} \tag{62}$$

The rate of acid addition must be carefully controlled to ensure that the cooling water does not become corrosive and control is probably best effected by monitoring the pH of the cooling water. Acid treatment can be very expensive, particularly where the make-up water has a high bicarbonate alkalinity and a high rate of make-up is employed. For this reason sulphuric acid, which is much cheaper than hydrochloric acid, is often used but care must be taken to ensure that the concentration of sulphate ions in the water does not become high enough to risk calcium sulphate scale deposition. The solubility of calcium sulphate is about 2000 ppm at ambient temperature and it exhibits retrograde solubility.

In some cooling tower systems where the water is not maintained in a sterile condition (for example, where little or no chlorination is practised), acid is sometimes produced in the water as a result of bacterial oxidation of ammonia to nitric acid:

$$NH_3 + 2\,O_2 \longrightarrow HNO_3 + H_2O \tag{63}$$

If the ammonia concentration in the water is high, sufficient acid may be produced in this way to prevent scale deposition.

It is also interesting to note that chlorination, used to control slime growths, also results in the production of acid and this may be significant where a high level of chlorination is practised:

$$Cl_2 + H_2O \longrightarrow 2\,HCl + O \tag{64}$$

As mentioned earlier, the likelihood of calcium carbonate scale occurring is related to the hardness and the bicarbonate alkalinity of the make-up water. Where supplies of make-up water are limited, water softening is sometimes employed to reduce the bicarbonate alkalinity and the hardness.

In addition to these changes in the chemical composition, the loss of water by evaporation from the cooling towers, which may amount to about 1% of the cooling water flow, results in the salts present in the water being concentrated. The degree of concentration occurring is equal to M_a/P_u but since some salts are lost by scale and sludge deposition in the system, the measured concentration ratio, where:

$$\text{concentration ratio} = \frac{\text{concentration of a constituent in the cooling water}}{\text{concentration of the same constituent in the make-up}} \tag{65}$$

will be less than M_a/P_u by the amount of that constituent lost from the water in the ways suggested. The best estimate of the degree of concentration occurring, is obtained from the ratio of sodium in the cooling water to that in the make-up, by direct measurement.

Although calcium carbonate is the more usual constituent of condenser scales, if the cooling water is allowed to concentrate excessively, the solubility of calcium sulphate may be exceeded and scaling from this condition will then occur. This is a particular problem where

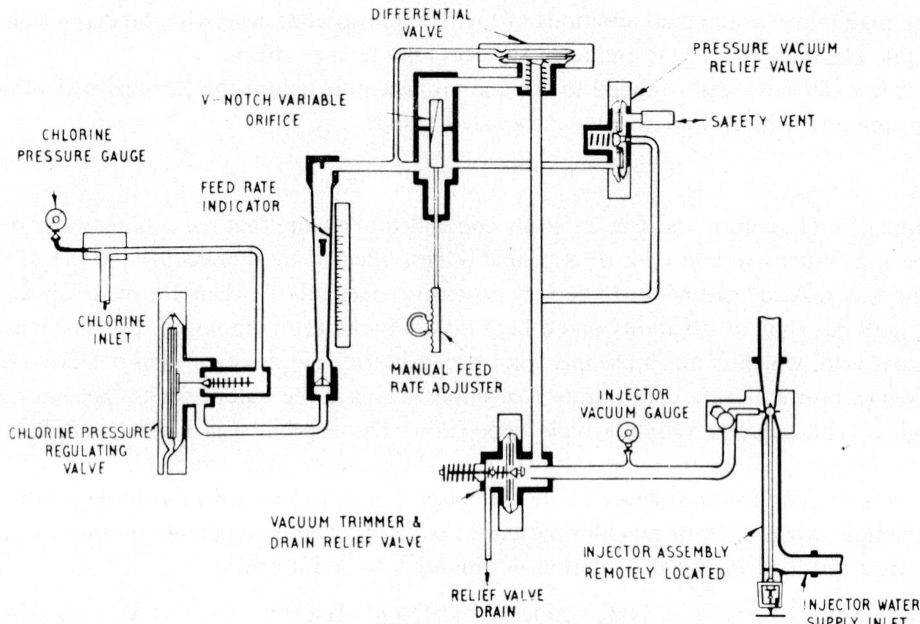

FIG. 3.8.2. V-notch chlorinator flow diagram

the make-up water already has a high sulphate concentration or where acid treatment with sulphuric acid is used.

If scaling does occur, heat transfer is impaired and the condenser or heat exchanger performance suffers. If the scale is largely calcium carbonate (20% or more) it can usually be removed with dilute inhibited acid, but calcium sulphate is not easily removed by chemical means and scaling due to this chemical should be prevented by every means possible. Further details of methods of chemical cleaning are given in Chapter 4.

3.8.2. Control of Organic Growth

Micro-organisms present in the cooling water tend to grow and multiply on the warm heat exchange surfaces, producing an organic film which reduces heat transfer. These micro-organisms can be killed and fouling prevented by use of disinfectants such as chlorinated phenols, by poisons such as copper salts, and by chlorine, which is also a powerful disinfectant.

Most power stations use chlorination for slime control, and liquified chlorine gas is purchased in steel drums. It is an expensive chemical but fortunately it is only necessary to maintain a small concentration in the cooling water as it passes through the heat exchanger, for a few minutes every few hours, in order to prevent slime growth. The chlorine is added to the cooling water by means of a chlorinator, which is a metering instrument capable of varying the rate of addition within certain limits. One type of chlorinator made by Wallace & Tiernan Ltd., is illustrated in Figure 3.8.2.

Chlorine is also a powerful oxidising agent which will react with other constituents in the water, and therefore for maximum economy the chlorine should be added to the water just before it enters the heat exchanger. Most inland stations practise intermittent chlorination, in which each of the main condensers and auxiliaries is chlorinated in turn for a period of about 10 min and the whole cycle is repeated every 4 h. The amount of chlorine added must be sufficient to ensure that there is a small chlorine residual in the water leaving the heat exchanger. The quantity of chlorine which it is necessary to add in order to achieve a detectable residual concentration after a stated time interval, is known as the *chlorine demand* of a water. It is determined by adding known amounts of chlorine to the water, and measuring the chlorine residual concentration after a time interval equivalent to the time taken for the cooling water to travel from the point of chlorine injection to the heat exchanger.

When chlorine gas is dissolved in pure water, the resulting solution will contain free chlorine, hypochlorous acid (HOCl) and hydrochloric acid, all in equilibrium and dissociating to form ions, thus:

$$\underset{\text{Chlorine}}{Cl_2 + H_2O} \rightleftharpoons \underset{\text{Hydrogen ions}}{2\,H^+} + \underset{\text{Hypochlorite ions}}{(OCl)^-} + \underset{\text{Chloride ions}}{Cl^-} \tag{66}$$

If alkaline salts are present in the water, hypochlorites and chlorides are formed.

If ammonia is present, chlorine will react with it to form a mixture of chloramines, known as monochloramine [NH_2Cl], dichloramine [$NHCl_2$], and nitrogen trichloride [NCl_3]. Thus:

$$NH_3 + Cl_2 \longrightarrow NH_2Cl + HCl \tag{67}$$

$$NH_2Cl + Cl_2 \longrightarrow NHCl_2 + HCl \tag{68}$$

$$NHCl_2 + Cl_2 \longrightarrow NCl_3 + HCl \tag{69}$$

If a sufficient excess of chlorine is added, ammonia is completely oxidised to nitrogen, a process known as *break point chlorination*, thus:

$$2\,NH_3 + 3\,Cl_2 \longrightarrow N_2 + 6\,HCl \tag{70}$$

Since these are competing reactions, the composition of the water after reaction between chlorine and ammonia will depend on the relative initial concentrations of the two reactants and the speed of addition of chlorine. If further chlorine is added to a water already containing monochloramine, some dichloramine will be formed but since monochloramine and dichloramine react together thus:

$$NH_2Cl + NHCl_2 \longrightarrow N_2 + 3\,HCl \tag{71}$$

monochloramine and hypochlorites are the main products found, when water containing ammonia is chlorinated. These chlorine–ammonia reactions must be taken into account when determining the chlorine demand of a water and on the practical scale, where a cooling water contains ammonia, it is more economical to add the chlorine as a dilute solution.

To ensure that the correct concentration of chlorine is being added to the cooling water, the water is analysed periodically for chlorine content. The test normally used involves adding a small quantity of acid orthotolidine reagent to a sample of the water. Free chlorine, hypochlorite chlorine and chloramines react with the reagent to produce a yellow coloration, the intensity of which increases with the level of concentration of total active chlorine present. The reaction with free chlorine and hypochlorite chlorine is almost instantaneous but chloramines react only relatively slowly and this forms the basis of a simple test, used to distinguish between the various forms of active chlorine present. A more sophisticated test, used to analyse a water for all forms of active chlorine present, was developed by Palin, but will not be described here.

3.8.3. Maintaining Condenser Cleanliness

If the cooling water contains mud in suspension, the mud will tend to deposit throughout the cooling system including the heat exchange surfaces, where it will reduce heat transfer. Where mud deposits form, the metal surface under the deposit may suffer corrosion because of electrochemical action, due to "differential aeration", the area under the deposit being depleted in oxygen relative to the surrounding metal exposed to the cooling water (see Chapter 2). Corrosion may also occur if the mud contains sulphide formed by bacterial action on the sulphates present in water which contains little or no oxygen.

It is essential to maintain the metal surfaces in a clean condition either by preventing the mud from depositing or by employing some regular method of cleaning. At one station, condenser tubes are kept clean by adding a chemical, of the type known as *polyelectrolytes*, to the cooling water at a concentration about 1 ppm for periods ranging from 10 min. to 1 h per shift. The way in which small traces of these chemicals act, is not fully known although it is suggested that these act partly by modifying the physical characteristics of mud in suspension in the cooling water, so that it no longer adheres to the metal surfaces, and partly by resuspending any mud already deposited out on the metal surfaces. It is further understood that some of these chemicals have corrosion inhibition properties.

Other stations carry out periodic cleaning during plant outages using small brushes, or *bullets* as they are called, which are forced through the tubes by water or compressed air.

Cleaning on load can be carried out by the *Taprogge process* in which sponge-rubber balls (approximately 10% larger than the tube's internal diameter), are carried through the tubes by the cooling water. The balls are fed into the inlet water box, and are recovered on screens placed in the outlet from the condenser and recycled back to the inlet. Figure 3.8.3 illustrates the principles of the process.

Experience with this process has shown that it can maintain condensers sufficiently clean to maintain condenser performance and hence it is often economically attractive, but the evidence concerning corrosion prevention is less convincing. At one station which has had the Taprogge process in service since shortly after commissioning, condenser tube corrosion has occurred, whilst at a second station, which had a severe corrosion problem, it is thought that the rate of corrosion has fallen since the Taprogge process was installed.

Plant-scale experiments are also in hand to investigate the effectiveness of the process in preventing calcium carbonate deposits on the condenser tubes when scale-forming water is used. These experiments have relevance to the design of future cooling tower stations, where relatively high concentration of the cooling water may be unavoidable.

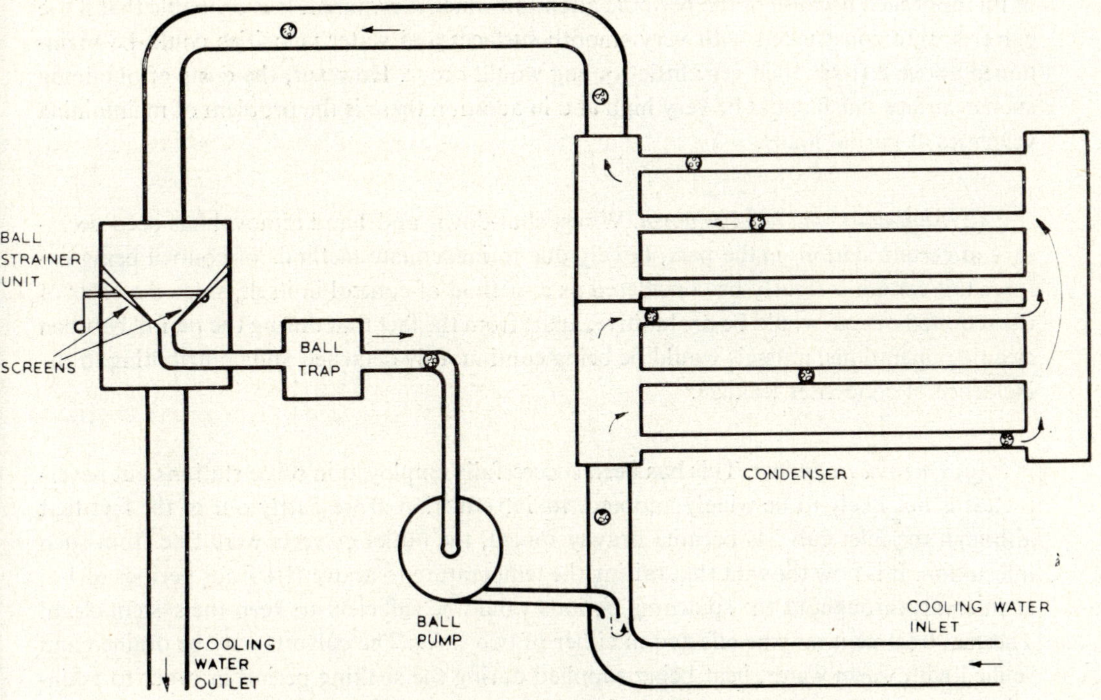

FIG. 3.8.3. The Taprogge system of on-load condenser tube cleaning

3.8.4. Control of Marine Growth

At coastal stations, marine organisms such as barnacles, oysters and mussels colonise the surfaces of the cooling water culverts. If allowed to develop in the intake culverts they eventually die and the shells are carried forward with the cooling water into the condenser. When this stage is reached the mussel shells are usually too large to pass through the condenser tubes and become wedged in the tube mouths, partially blocking the tubes. The turbulence thus set up causes rapid erosion/corrosion of the metal and in a relatively short time the tube is punctured, with consequent leakage of cooling water into the condensate.

Several methods for controlling marine fouling have either been proposed or used and these include the following:

(a) Surface treatment or design considerations.

(b) Shut-down and hand removal.

(c) Thermal treatment.

(d) Chemical treatment.

Each will be considered individually:

(a) *Surface treatment and design.* Surface treatments include the use of toxic paints and the use of copper oxide concrete, from which copper slowly leaks out into the water. Anti-fouling paints can minimise mussel fouling for two or more years, after which re-treatment will be necessary, thus necessitating plant shut-down. Limited use has been made of this approach because of the periodic attention which is required. It is probable that if the culverts were constructed with very smooth surfaces and water velocities could be maintained above 8 ft/sec then very little fouling would occur. However, the costs of obtaining such a surface finish could be very high and in addition there is the problem of maintaining velocities at partial loads.

(b) *Shut-down and hand removal.* Whilst shut-down and hand removal has been necessary at certain stations in the past, largely due to inadequate methods of control being enforced, it cannot seriously be considered as a method of control in itself, since the costs of the required outage would be prohibitive, apart from the fact that during the period between cleaning operations, mussels would be being continuously detached and contributing to the incidence of condenser leakage.

(c) *Thermal treatment.* This has been successfully employed in some stations but nevertheless is not likely to be widely adopted. Its introduction arose partly out of the fact that although the inlet culverts became heavily fouled, the outlet culverts were free from such infestation. It is now thought that raising the temperature to above 104°F for periods of 1 h, at intervals throughout the spawning periods would be sufficient to keep the system clean. Thermal treatment may be effected in either of two ways. The culverts may be drained and refilled with warm water, heat being supplied during the soaking period in order to maintain temperature, or alternatively, the inlet culverts can be flushed in reverse flow by warm water taken from the condenser outlet. To incorporate this design feature requires heavy capital outlay and in addition, further costs would be incurred at times of treatment since it would probably be necessary to restrict output from the station. Attendant difficulties arise from the effect of the increased temperature on the concrete culverts and there is a reported case of concrete breaking away from the wall when thermal treatment was applied at a temperature of 110°F.

(d) *Chemical treatment.* The only chemical treatment which has been successfully applied is chlorination, and fouling which has occurred whilst employing this treatment can usually be attributed to under-treatment in an attempt to effect some measure of economy. If chlorination were applied continuously there is no doubt that fouling would not occur; this of course would be extremely expensive and it is necessary to determine a programme of chlorination which makes optimum use of the chlorine whilst still preventing fouling.

The chlorination programmes employed vary considerably from station to station but in essence there are two basic approaches. In the first scheme, continuous chlorination is carried out for periods from 8–14 weeks at 1–2 ppm chlorine; this would probably be carried out twice during the year, for example May–June and August–November. By employing such a process, mussels would be killed before they had grown to a sufficient size to block the condenser tubes. In the second scheme, continuous injection is applied throughout the spawning season, normay April–November, at a lower injection rate, 0·6–0·8 ppm chlorine. This process, unlike the first one, is not designed to kill the mussels but rather to provide an environment in which they will not settle. In applying any chlorination schedule it should be borne in mind that in the presence of chlorinated water the mussels close their shells and can survive in this condition for many days without feeding. In order to obtain maximum efficiency from the chlorination schedule, some stations have employed methods to determine the spawning periods, which include examination of the adult mussels and the use of settlement frames to detect newly formed mussel "spat". Other useful information is obtained by observing the life cycle of the mussel and noting the sea temperature, since spawning usually occurs with an optimum temperature of 10–12°C, although this may vary. At one station, an experimental plant has been installed to produce the required chlorine by the electrolysis of the sea water. The chemical reaction is represented as follows:

$$2NaCl + 2H_2O \longrightarrow 2NaOH + H_2 + Cl_2 \qquad (72)$$

and the hydrogen obtained as a by-product may be used for the alternator cooling.

Experimental work has been carried out at some stations, utilising sound energy of ultrasonic frequencies as a method of controlling marine fouling, but has not proved effective.

With any method of controlling mussel fouling, it is important to ensure that adult mussels do not enter the inlet culverts. To achieve this the cooling water screens must be maintained in good condition.

3.8.5. Corrosion in Cooling Water Systems

In the construction of a cooling water system, various metals and metal alloys are used. The condenser tubes and other heat exchanger tubes in the system are normally made of a copper alloy; the condenser tube-plates are usually of naval brass, whilst condenser water boxes, pipework, valve bodies, etc., are of cast iron or mild steel. Valve and pump internals are often stainless steel or phosphor-bronze. The metals usually give good service providing the cooling water is not saline or heavily polluted, since there is insufficient electrolyte to support electrolytic action between the dissimilar metals.

(a) Graphitic Corrosion of Cast Iron

With oxygenated estuarine or sea water, severe electrolytic corrosion of iron components can occur, especially where these are in contact with more cathodic materials and thus cast iron condenser water boxes are particularly vulnerable. As cast iron corrodes, the iron content is taken into solution and forms corrosion products, leaving the graphite intact and still retaining the original shape. This is known as *graphitic corrosion*, the resultant porous mass has no mechanical strength and may escape detection unless the moist surface is carefully probed with a pointed tool.

Three methods are used to alleviate electrolytic corrosion in cooling water systems, these are:

(i) Paints and other protective coatings.
(ii) Cathodic protection using sacrificial anodes.
(iii) Cathodic protection using an impressed current.

Paints and other protective coatings. Although, in the past, painting has been used to afford some measure of protection to condenser water boxes, maintenance costs were high because of the limited life of paint films in a sea water environment and because the adequate surface preparation necessary for painting to be successful, is very difficult to achieve. There was also the danger that severe localised corrosion might occur at the site of any defects in the paint film. Painting can of course only be used in accessible areas of the system, with the result that other areas are not protected. A few power stations still use painting in conjunction with cathodic protection and it is claimed that this improves the efficiency of the process.

Natural or synthetic rubber coatings, if properly applied, can successfully be used to protect condenser water boxes. They have a long life and little maintenance is required but the initial cost of application is high. In a few stations, rather than apply protective coatings, the condenser water boxes have been constructed from reinforced fibre glass.

Where water boxes are unprotected, the corroding iron offers some protection to the copper alloy condenser tubes by a cathodic protection mechanism (discussed later in this section). Conversely, it follows that where coatings are successful in protecting the waterbox, the condenser tubes are no longer protected and corrosion, particularly of the tube ends, may occur. It is essential therefore, where coatings are used, to extend the coating a few inches down the condenser tubes. Alternatively plastic inserts may be placed in the tube ends.

Cathodic protection. In order to appreciate how cathodic protection works, the mechanism of electrolytic corrosion must be understood. This is discussed more fully in Chapter 2.

If two different metals are electrically connected and immersed in an electrolyte, current will flow from one metal to the other. The metal at which the current enters the electrolyte is known as the *anode* and the other metal as the *cathode*. This simple circuit is shown in Figure 3.8.5A (a). Current flows from one metal to the other because a potential difference exists between them, the magnitude of this potential being different for various metal combinations. It is possible to arrange metals and metal alloys in a table such that each metal is more electronegative than the metal above it and conversely more electropositive than the one below it, under standard conditions. Such a table is known as the *electromotive series* and in Chapter 2 the relationship between some of the more familiar metals and metal alloys, is given.

In cooling water systems, the iron components will be the anodes of an electrochemical cell such as is shown in Figure 3.8.5A (a) whilst the copper alloys will form the cathodes. If a third metal or electrode, which is more electronegative than iron, is added to the system and electrically connected to the iron and copper, the new electrode will corrode in preference to the iron. Zinc, magnesium and aluminium anodes have been used for this purpose and are referred to as *sacrificial anodes*. They are often used to protect small items of equipment but for the large equipment they have several disadvantages, in particular the large number of bulky anodes required and their limited life.

If instead of a more electronegative metal, an external d.c. potential is applied between the thrid electrode and the original two electrodes, in such a way that the third electrode is

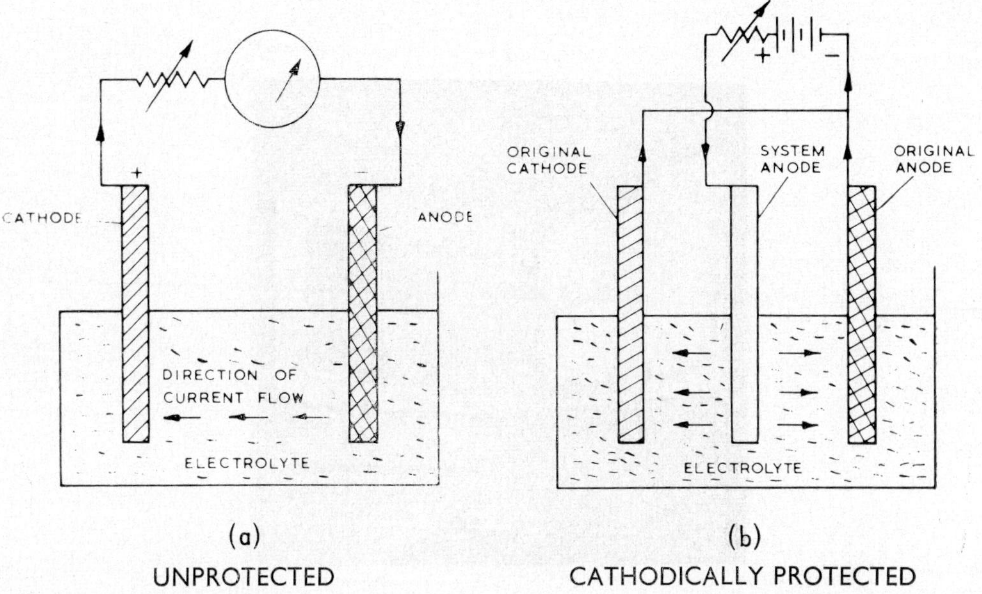

FIG. 3.8.5A. Galvanic cell

FIG. 3.8.5B. Cathodic protection anode

Rod type anode installed through condenser door

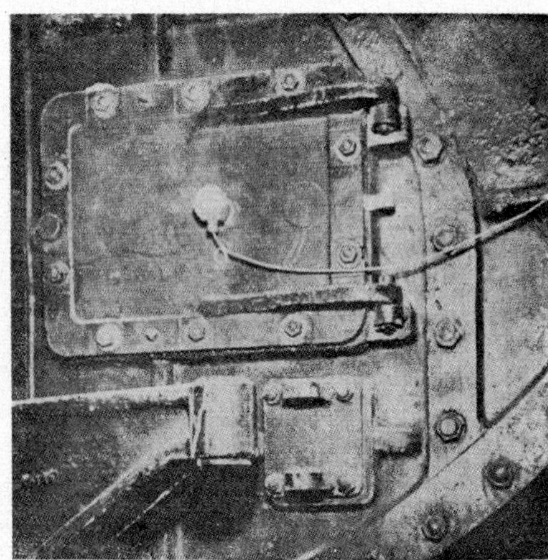

9 and 17 — External connection to anode

R3 — Reference test point

FIG. 3.8.5C. Cathodic protection anodes

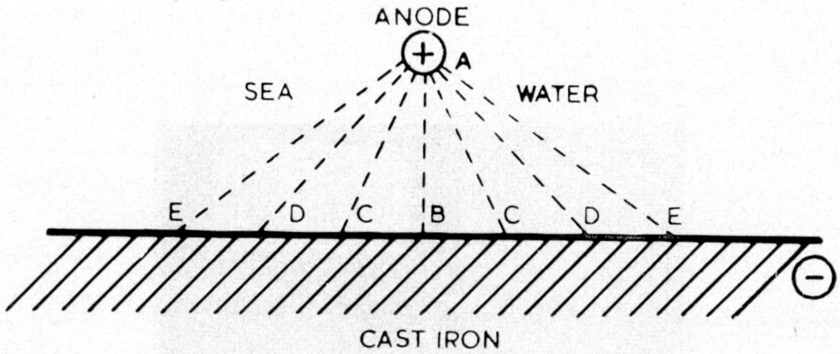

FIG. 3.8.5D. Protection afforded by a single anode

now the anode to the whole system, then the original anode will no longer corrode. This arrangement is shown in Figure 3.8.5A (b). This use of an applied potential to prevent electrolytic corrosion is known as *impressed current cathodic protection.* In power stations, *platinised titanium* is now used for the anode material and it will pass the required current without itself being corroded or taken into solution, so that in theory it should have a very long life. The two types of anodes in use in the Board's power stations are shown in Figures 3.8.5B and C. The rod or cantilever type needs to be correctly positioned and a number of electrodes are required to ensure that all parts of the system are protected. The continuous anode which is a more recent development, is clipped to the surface being protected. It is more efficient electrically and fewer electrodes are required; it is also much less liable to mechanical damage.

Referring again to Figure 3.8.5A, it is seen that a negative potential is applied to the metal to be protected. Obviously the applied potential must be sufficiently high to suppress the corrosion and the only sure way of achieving this is to measure the metal's potential with respect to its environment. The metal being protected and its cooling water environment, form half of an electrolytic cell or a *half-cell.* Its potential is measured by coupling it, that is to say completing the electrolytic cell, with a half-cell having a known e.m.f., called a *standard* half-cell. The half-cells normally used are copper/copper sulphate and silver/silver chloride. The generally accepted criterion for protecting cast iron is that its potential should be suppressed to −850 mV with respect to the copper/copper sulphate half-cell and −800 mV with respect to silver/silver chloride. To enable the potential to be measured from the outside of the equipment, a porous reference point is used in which the reference half-cell is incorporated. On a small installation, manual checking and adjustment can be carried out but recently, automatic control equipment has been developed by the South-Eastern Region, Research and Development Department. Figures 3.8.5 E and F show some typical half-cells.

In principle, cathodic protection provides the complete answer to electrolytic corrosion and it can be successful if applied and controlled correctly, as may be seen from Fig. 3.8.5G, which shows two water boxes of one condenser, both previously painted, one having cathodic protection and the other unprotected.

Consider a system in which a single rod anode is adjacent to a flat plate of cast iron as in Figure 3.8.5D:

Corrosion meter with $Cu/CuSO_4$ half-cell

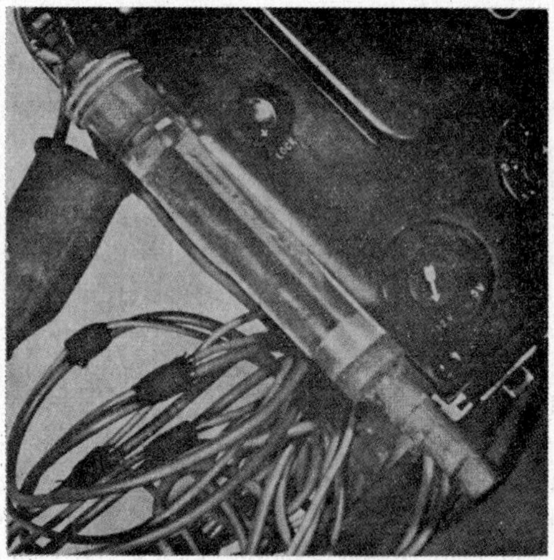

Close-up of half-cell

FIG. 3.8.5E. Cathodic protection—standard half-cell

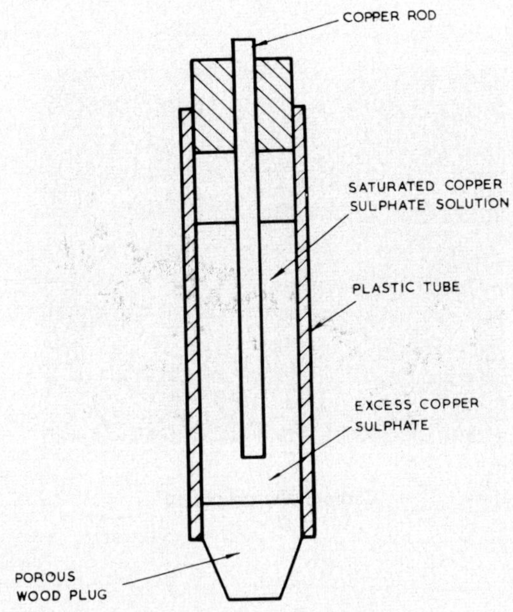

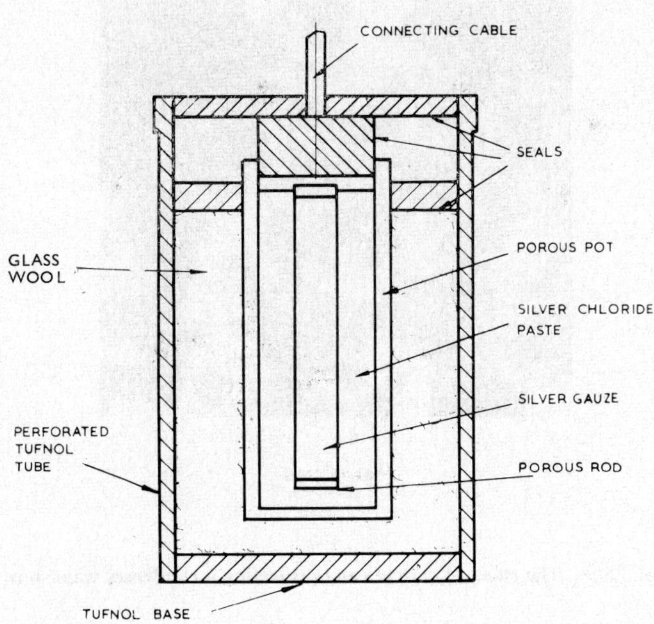

FIG. 3.8.5F. Silver/silver chloride half-cell and copper/copper sulphate half-cell

Cathodically protected

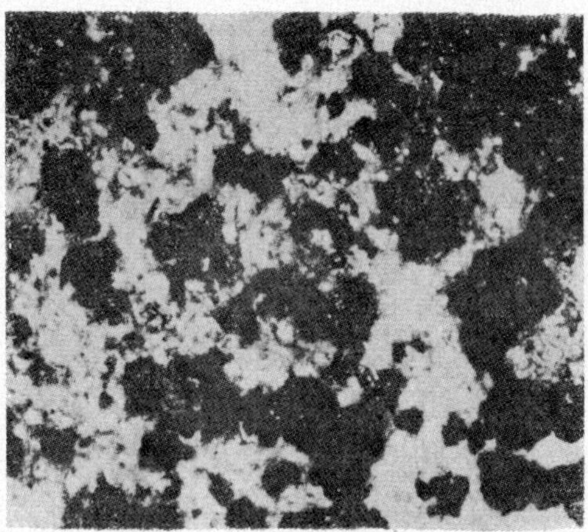

Unprotected

FIG. 3.8.5G. The effect of cathodically protecting a condenser water box

Adopting the same convention as in Figure 3.8.5A current flows from the anode, *A*, to points such as *B*, *C*, *D*, *E*, on the cast iron. The electrical resistance will be least for path *AB* and will increase progressively from path *AB* to *AE*, etc. Hence the electrical potential of the cast iron will be depressed most at *B* and will be depressed less and less at points more remote from *B*. At some distance from *B* a position will be reached where the potential will not be depressed sufficiently for the cast iron to be protected. Therefore, in an actual cathodic protection installation sufficient electrodes must be provided and their position carefully chosen to ensure that all parts of the system are protected. It also follows that sufficient monitoring reference points must be provided and their positions chosen with equal care, if adequate control is to be maintained.

It is of course possible to ensure that all parts of the system are protected, using fewer anodes, by increasing the anode potentials but this will result in some areas being grossly over-protected. Over-protection is, however, wasteful since electrical consumption is increased, and it also results in heavy scale deposition particularly around the anodes.

Two other factors which must be taken into account in controlling a cathodic protection installation, are cooling water flow rate and cooling water composition. Experience has shown that when the associated plant is taken off-load and the cooling water flow stopped, chlorine is generated in the stagnant water unless the anode potential is reduced to approximately half that of the normal operating level. Chlorine produced in this way can cause severe corrosion of cast iron and loss of platinum from platinised titanium anodes. Seasonal and tidal variations in cooling water composition and in particular changes in dissolved oxygen content, also affect the degree of protection achieved and should be taken into account, although this is taken care of to a large extent if automatic control equipment is used.

(b) Condenser Tube Corrosion

The copper alloys used for condenser tubes, when exposed to cooling water containing dissolved oxygen, undergo slight surface oxidation resulting in the formation of a very thin oxide film. The film is very adherent and the oxidation continues until the film has thickened to such an extent that further oxidation is stifled. If this protective film is mechanically damaged it rapidly reforms and the metal is protected from corrosion as long as it is present.

If the condenser tube is exposed to aggressive water before the film has had a chance to form, or if the film is damaged in the presence of aggressive water, then corrosion can occur. The degree of corrosion will depend on the conditions of service and the film-forming characteristics of the tube alloy.

Iron, present in either the tube metal or in the cooling water, assists in the formation of a tough adherent protective film. If the water boxes are unprotected these may provide the iron required and it follows that where they are protected an alternative supply of iron may be necessary. This may be obtained from scrap iron placed in the cooling water culverts or by dosing the cooling water with iron salts, such as ferrous sulphate. Ferrous sulphate treatment has been shown to be particularly effective with aluminium brass tubes but is not very effective with cupro-nickel alloys.

The types of attack which result in the breakdown of the protective film include:

(i) Corrosion/erosion.

(ii) General wastage.
(iii) Concentration cell corrosion (deposit attack).

Corrosion/erosion is almost invariably associated with sea or estuarine water and with velocities in excess of 5 ft/sec. It is caused by entrained gases or silt present in the rapidly moving water.

General wastage may be the result of abrasion or some form of chemical or electrochemical action causing continuous damage to the protective film. This may arise when the water contains no oxygen, that is, it is anaerobic and under these conditions free sulphide is often present, usually as a result of bacterial action.

TABLE 14
CONDENSER TUBE ALLOYS

Alloy	Composition (% by weight)	Relative price*	Properties and notes
70/30 Brass	Cu 70 Zn 29·96 As 0·04	100	Arsenic employed to inhibit dezincification
Admiralty Brass	Cu 70 Sn 1 Zn 28·96 As 0·04	102	Good general performance particularly in fresh unpolluted water applications.
Aluminium Brass	Cu 76 Al 2 Zn 21·96 As 0·04	108	Rapid healing film, resistant to impingement attack, used for less severe conditions than cupro-nickel alloys—Used in clean sea water.
9/10 Cupro-nickel	Cu 87 or 88 Ni 10 Fe 1 or 2 Mn 1	148	Not so resistant to abrasive attack as the 70/30 cupro-nickel alloys—more resistant than brasses to deposit attack.
70/30 Cupro-nickel (low iron and manganese)	Cu 68 Ni 30 Fe 1 Mn 1	156	Resistant to impingement and deposit attack.
70/30 Cupro-nickel (high iron and manganese)	Cu 66 Ni 30 Fe 2 Mn 2	156	Greater resistance to impingement attack than the low iron and manganese alloy but no greater resistance to deposit attack.
Stainless steel (type 316)	Cr 18 Ni 12 Mo 2.5	188	Performance not yet established but its use reduces the amount of copper alloys in the feed system.
Titanium	Pure	459	Very resistant to abrasive and corrosive attack—has no scrap value.

*Based on 1955 prices.

Concentration cell corrosion (deposit attack) was described earlier in the section dealing with condenser cleanliness and it occurs because the metal surface under a deposit is depleted in oxygen relative to adjacent areas of metal exposed to the cooling water. This aspect is discussed more fully in Chapter 2.

The composition and properties of the most common condenser tube alloys are given in Table 14.

The alloys most resistant to corrosion are also the most expensive and because of their good film-forming characteristics, their thermal properties are often worse than for the brasses. These points must be borne in mind when selecting condenser tube alloys. For unpolluted fresh water, brass or admiralty brass is usually chosen and both of these alloys contain a trace of arsenic to prevent dezincification.

3.8.6. **Dry Cooling Towers**

The limited availability of water, even for stations employing a recirculating cooling water system, imposes a serious restriction on the siting of large power stations. To overcome this problem, air-cooled radiators have been developed, which are suitable for cooling large volumes of water and they are generally referred to as "dry" cooling towers.

(a) Description

The first dry cooling tower system in the C.E.G.B. was commissioned at Rugeley in 1961 and is associated with a 120 MW unit (1500 lb/in^2, 538°C). The tower consists of a reinforced concrete shell mounted on concrete stilts 50 ft high. The radiators or heat exchangers are situated around the base of the tower between the stilts. These consist of aluminium tubes (99·5% aluminium) surrounded by transverse fins.

The cooling system is shown diagrammatically in Figures 3.8.6A and B. Steam from the turbine exhausts into a jet condenser into which is sprayed the cooled water returned from the cooling tower. The cooling water plus condensate is then extracted from the condenser; between 1% and 2% passes on to the boiler as feed water in the normal way and the remainder returns to the tower to be recooled.

(b) Chemical Aspects

In the air-cooled radiators, the aluminium surface in contact with water totals some 350,000 ft^2. The rest of the system is mainly composed of iron and steel, some of which is lined with a protective coating of an epoxy-resin paint to minimise contamination of the water with iron.

This is in some ways an unfortunate combination of metals since under conditions favourable for the protection of iron (namely high pH), aluminium is likely to corrode, and vice versa.

Experience at Rugeley has shown that as long as the protective paint on the steel work remains sound, a pH of about 7 in the tower circuit is satisfactory, the soluble aluminium and iron contents in the water being 0·015 ppm aluminium and 0·02–0·03 ppm iron. How-

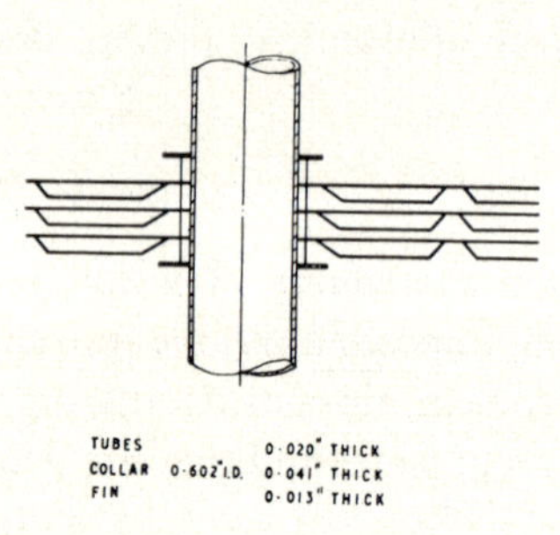

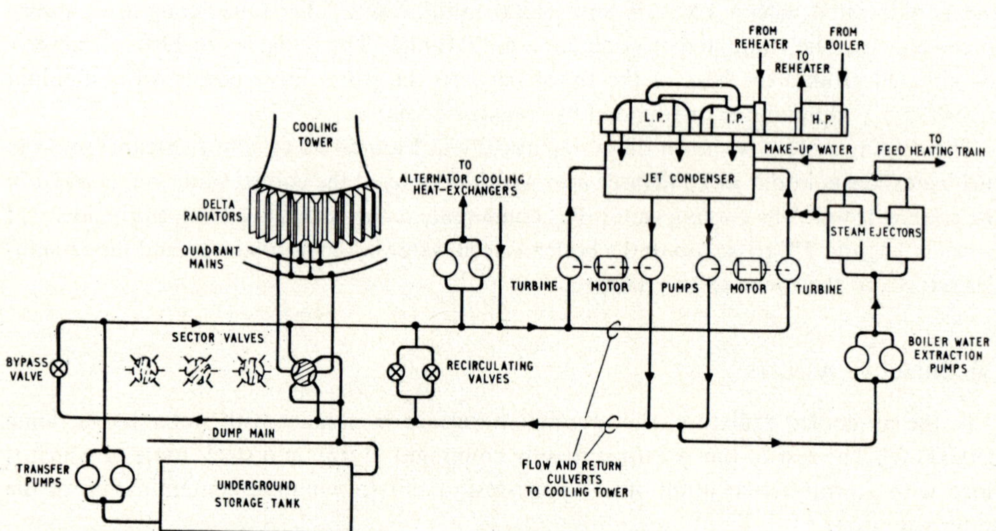

FIG. 3.8.6A, 3.8.6B. Dry cooling tower system at Rugeley

ever, when the paint failed, the soluble iron content rose to 0·06 ppm and it was necessary to control it by raising the pH. Morpholine was used to achieve this and the pH was raised to 8·8 with a subsequent reduction in soluble iron pick-up to 0·02 ppm (total iron below 0·04 ppm), and a slight increase in the aluminium content to 0·025 ppm. These conditions have so far presented no problem in the associated feed train and boiler, the aluminium content of the boiler water being maintained in the range 2–4 ppm with little boiler blowdown.

The most serious problem that has occurred has been external corrosion of the aluminium tubes due to pollutants in the atmosphere, and a protective coating has been applied externally to combat this.

3.8.7. **Protection of Cooling Tower Timber**

The cooling towers used in large recirculating cooling water systems are normally of the natural draught type. They are concrete structures and their hyperbolic shape produces a chimney effect and promotes an upward flow of air through the packing inside the tower.

Wood is the commonly used packing material, although it is being replaced by corrugated asbestos-cement to a certain extent in new plant. Although the latter is more expensive than wood it is much less likely to need replacement than timber, even when this is impregnated.

The air passing upwards from the cooling packing will contain entrained water droplets which would precipitate in the neighbourhood of the cooling tower and create a nuisance. To prevent this an additional packing known as a *mist eliminator* is placed a few feet above the main cooling packing. These mist eliminators are also constructed of wood or asbestos.

Wooden packings are subject to damage due to a number of causes and the following have been recognised:

(a) Mechanical damage—particularly resulting from ice formation in winter, and (rarely) due to the weight of hard scale formation.
(b) Attack by salt crystallisation.
(c) Attack by alkali—occurs usually where cooling water is base-exchange softened.
(d) Attack by chlorine—mainly a problem when the chlorine concentration is greater than 1 ppm.
(e) Degradation near iron fixings.
(f) Attack by dry and wet rot *(Basidiomycete)*—this is mainly a problem in cooling towers which are not in constant use and in those timber sections that are not continuously wetted when the tower is in use.

Some twenty years ago premature failure of tower packings, after only a few years in service, was reported both in this country and in the United States of America (see Fig. 3.8.7A). Following investigations conducted at the Forest Products Research Association, in 1949, species of microfungi, different from those responsible for dry and wet rot, were discovered in packing timber. In 1950 it was confirmed that these microfungi could cause surface decomposition of timber and this form of attack was given the name *soft rot*. About a dozen different fungi have been identified in samples of affected timber, including several

FIG. 3.8.7A. Collapse of cooling tower timber after soft rot attack

FIG. 3.8.7B. Soft rot attack on timber

FIG. 3.8.7C. Appearance of timber after soft rot attack

species of *Ascomycetes*, and a fungus called *Chaetomium globosum* was proved to be present in over 75% of the samples examined.

Soft rot differs from wet and dry rot both in the fungi responsible and in its effect on the attacked timber. With wet and dry rot, the attack occurs in depth through the thickness of the timber or may occur in the middle of a piece of timber, leaving the outer layers intact. Soft rot attacks and destroys the cellulose of the cell walls in the outer layers of the timber (see Fig. 3.8.7B.), and leaves the surface soft when still wet, and cork-like with shrinkage cracks when dry. An advanced case of soft rot is illustrated in Figure 3.8.7C. The attack occurs progressively inwards and the damaged surface layer may be washed off by the flow of water. In time the thickness of the timber may be reduced to such an extent that it fails due to insufficient mechanical strength.

Arising from the investigations referred to earlier, timber preservatives and methods of timber impregnation have been developed which it is confidently expected will protect the timber for 25 to 30 years, and this is the planned operational life of the associated plant. The preservatives are solutions of copper chromate, with or without arsenic and the treatment results in the preservative becoming fixed in a water-insoluble form, in the timber, with an average net dry preservative salt retention of 1·25 lb/ft^3 of timber. The preservatives and the methods of application are given in C.E.G.B. specification number TC/3001/1.

REFERENCES

Industrial Water Treatment Practice, by P. Hamer, J. Jackson and E. F. Thurston.
Water Treatment for Industrial and Other Uses, by E. Nordell.
Ion Exchange Resins. The British Drug Houses Ltd.
Water Treatment for Public and Industrial Supplies. The Patterson Engineering Company Ltd.
The Examination of Power Station Make-up Water for Different Forms of 'Non-reactive' Silicon. Central Electricity Research Laboratories Report No. RD/L/N/79/65.
The preparation of water for large power stations, by F. J. R. Taylor, *Proc. Soc. Water Treatment and Examination*, Vol. 14, 1965.
Amber-Hi-Lites. Rohm and Haas Company, Philadelphia.
Electrochemistry, Principles and Applications by E. C. Potter.

EXERCISES

1. In considering three alternative sources of raw water, namely river, deep chalk well and local public supply, to be used to prepare boiler feed make-up water, what factors would influence your choice of supply for a new generating station?

2. If water of quality 'A' in Table 1 were to be used for preparing boiler feed make-up water, what treatr ment would you recommend for boilers operating at (a) 600 lb/in^2 and (b) 2350 lb/in^2? Give reasons for you- recommendations.

3. (a) What factors influence effective precipitation and settlement in lime or lime/soda softeners?

(b) Describe the function of coagulants in water softening and suggest how the optimum dosage of coagulant may be determined.

(c) Calculate the charge of lime and soda required to soften 1000 gal of water of quality 'I' in Table 1, to give a final alkalinity of 20 ppm hydroxide and 40 ppm carbonate (both as $CaCO_3$), assuming that the lime is 95% pure and the soda is 99% pure.

4. (a) What are the principal causes of deterioration in performance of demineralisation plant, and in what way do they affect plant performance?

(b) What consideration should be given to the chemical constituents of waste effluents from demineralisation plants, for discharge into sewers or rivers?

5. Show diagrammatically the arrangements of water treatment plant for boiler feed make-up, based on demineralisation, which would have the lowest operation cost when treating water of (a) quality 'C' and (b) quality 'L', in Table 1. Explain the reasons for your choice of the various types of ion-exchange material to be used.

6. What problems would you anticipate if water of quality 'M' in Table 1, were to be used as the source of boiler feed make-up water? Suggest possible ways to overcome the problems.

7. (a) Where intermittent chlorination is used to control slime growth in CW systems, how is the dose-rate determined and what precautions must be taken when the cooling water is returned direct to a fresh-water river or canal? What chemical testing is required?

(b) Summarize the principal factors which influence the alkalinity and hardness of the water in a closed CW system with cooling towers, receiving make-up from a fresh-water river. Indicate the chemical reactions involved, where appropriate.

8. What condenser tube material would you select for (a) fresh-water river, (b) polluted estuary water, (c) sea water, and what treatment or precaution should be applied in each case to minimize corrosion of the condenser?

CHAPTER 4

PLANT CLEANING AND INSPECTION

4.1. INTRODUCTION

It is evident that the functioning of major parts of generating plant, for example the boiler, the feed system and auxiliaries, depends very largely on the transfer of heat between two fluids which are separated by a metal wall. The thermal conductivity of metals is substantially greater than most other substances, and hence the deposition of non-metallic substances (including metal oxides) on the heat transfer surface inevitably reduces the efficiency of the heat-exchange process. Such deposition may not only detract from operational efficiency—it may cause corrosion, as discussed in Chapter 2, or overheating of boiler tubes, occasionally to the point of failure.

The internal cleaning and protection of new plant prior to commissioning has assumed much importance and considerable complexity during the last twelve years or so. The scale of the processes employed is reflected in the cost which, for a 2000 MW station, is in excess of £200,000. Most significant among the reasons for such preparation for service is the fact that the risk of destructive boiler corrosion occurring in service is greatly increased by the presence of debris in boiler tubes; this and other reasons will be discussed in the sections which follow.

Having prepared the working surfaces of the plant, it is necessary to maintain them in good condition if loss of efficiency or availability is to be avoided. Since some degree of fouling cannot be avoided during operational life, need arises occasionally to clean the boiler, various components of the feed system, and other auxiliaries, after various periods of service. These "operational" cleaning processes usually involve chemical treatments which are different from those used for pre-commissioning cleaning, and will therefore be described separately.

The need to inspect and test plant scientifically is obviously related to the matters already mentioned, but it also arises in order to detect metallurgical defects (see Chapter 5) and more general evidence of the effects of service on the condition of the plant. This chapter is concerned with inspection and the cleaning of metal surfaces normally in contact with steam or water; the cleaning of the gas-side surfaces of boilers is considered in Chapter 1.

4.2. PRE-COMMISSIONING CLEANING OF STEAM/WATER CIRCUITS

During the last decade there has been a considerable change in the Board's approach to the pre-commissioning cleaning of plant, so that the boiler surfaces are now more effectively cleaned and conditioned, and also most of the feed system surfaces are cleaned and protected, at works and/or after erection on site. Steam pipes, condenser shells and tanks are also cleaned, either as part of the main cleaning circuits, or separately.

In addition to the development of cleaning processes to remove foreign matter from these surfaces, efforts have also been made to establish clean site conditions so as to minimise ingress of dirt into the plant during erection and subsequent maintenance.

4.2.1. Reasons for Pre-commissioning Cleaning

Pre-commissioning cleaning is a time-consuming process, several weeks being required for a complete cleaning programme to be carried out on a modern unit. In view of this, and the high cost of the work, it is worthwhile reconsidering why it is necessary at all.

The reasons may be summarised as:

(i) To enable the plant to be commissioned and operated without incurring physical or corrosive damage.

(ii) To eliminate deposits which may not only harm the plant by obstructing heat transfer or by restricting the flow of steam or water, but may also adversely affect operational efficiency.

(iii) To ensure that the Board's chemical standards for feed and boiler water are achieved as soon as possible after commissioning.

Oil, grease and protective coatings (other than those which may be applied to give permanent protection) must be removed at an early stage in the cleaning programme, mainly because they would interfere with subsequent cleaning processes.

The miscellaneous debris usually present in the boiler at the end of the construction stages, such as rust, mill-scale, cement, lagging, welding material, etc., must be removed physically and by chemical cleaning. For many of these contaminants it will not be possible to completely dissolve them by the cleaning process, but it is possible to attack the material which is bonding them to the metal surface and so enable them to be removed by a flushing process, or at least to be moved to an accessible part of the plant, where they can be manually removed. Failure to remove this type of material will lead, in the early days of commissioning, to blockage of strainers and valves, drain lines, and other parts having small clearances. Mechanical damage may also be caused by the impingement of debris carried by high-velocity steam or water, particularly in turbines or pumps.

Apart from the physical aspects of debris which has to be removed from the plant, there is a need to remove debris for important chemical reasons. For example, silica, calcium, magnesium and aluminium are present in sand, cement, P.F. dust and lagging, and if allowed to remain in the system, could lead to the formation of scales, similar to those associated with cooling water ingress.

In addition to the more general considerations there are two major operating problems which in themselves constitute principal reasons for plant cleanliness requirements. These are (a) the incidence of boiler tube internal corrosion, and (b) the problem of silica control during the early days of operation, and these will now be discussed in more detail.

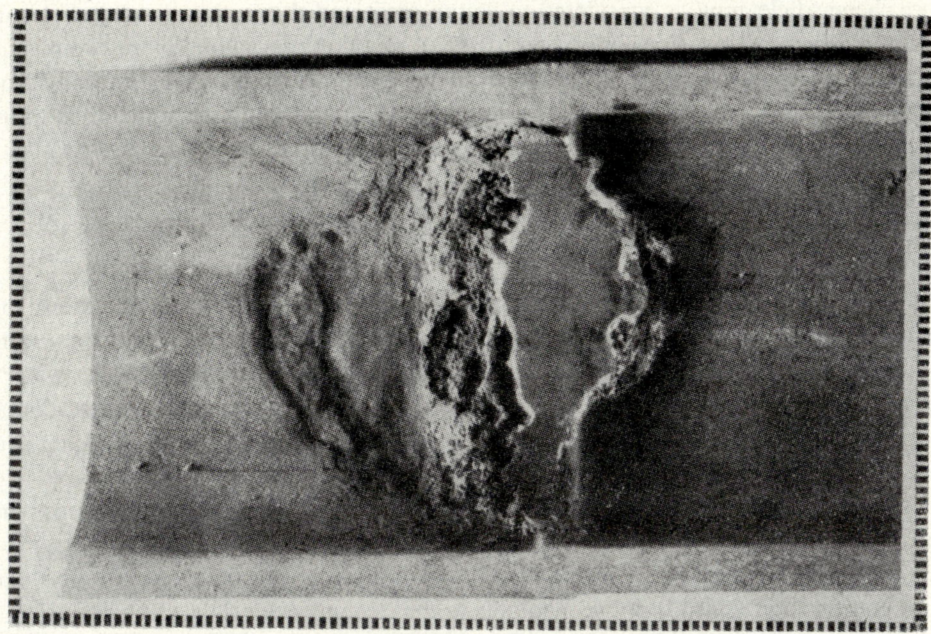

FIG. 4.2.1. Boiler tube corrosion at a weld protrusion

(a) BOILER CORROSION

As was discussed in some detail in Chapter 2, the deposition of solid matter in steam-generating boiler tubes is probably an important contributory cause of "on load" corrosion. It can be envisaged that the deposits may cause raised metal temperature because of their poor thermal conductivity, and that they may interfere with the flow of fluid in the boiler tube. The latter effect may also be brought about by a protrusion of weld metal inside the tube; this cannot be removed by the chemical cleaning process but it is relevant here since such a protrusion may well be a site for deposits to accumulate. Figure 4.2.1. shows deposit and corrosion at a weld protrusion. Tube welding techniques have now been developed which should largely prevent the production of internal protrusions.

The presence of mill-scale is conducive to corrosion in service, as is discussed in Section 4.2.2, and it must therefore be removed. The only way this can be achieved in a boiler within a reasonable period of time, is by an acid-cleaning process.

A final procedure ("passivation") is applied, designed to minimize the susceptibility of the steel surfaces to corrode during the interval between cleaning and operation, and to encourage the development, in subsequent service, of a uniform protective oxide film.

It has been argued that only physical cleaning to remove gross debris should be carried out before commissioning new plant, and that complete chemical cleaning should be delayed until the plant has been on load for a few weeks. Thus the large amount of metal oxides and other contaminants which are usually transported to the boiler during early operation, and delays between cleaning and commissioning, would be largely eliminated. These advantages are quite real, but the risk of initiating severe corrosion in the boiler whilst operating before chemical cleaning, is considered to be too great and this policy of delayed cleaning is not normally practised. Ideally perhaps, the full cleaning procedure before commissioning should be carried out and followed after a few months' operation, by chemically cleaning the boiler only, but the additional cost and plant outage would be difficult to justify unless there was clear evidence that the second clean was essential.

(b) Silica Control

Chapter 2 contains a discussion of the significance of silica in boiler water and shows that, because of the increasing degree of solubility of silica in steam of increasing temperature and pressure, it is essential in modern boilers to reduce the silica content of the boiler water to a very low level (0·3 ppm approximately) in order to ensure that the steam will not contain enough silica to deposit in the turbine. In a boiler containing say 250 tons of water, only about $2\frac{1}{2}$ oz silica need be dissolved to provide the concentration of 0·3 ppm.

Unfortunately, sources of silica abound in the environment of a power station; sand, cement, lagging material, refractories and fly-ash all contain substantial proportions of silica or silicates, and some of these materials are unavoidably introduced into new plant during construction. Silica is not very soluble in the cleaning chemicals which can safely be used in station plant, but the pre-commissioning cleaning processes do break down the bonding of siliceous material to metal surfaces, enabling the material to be washed out of the plant subsequently. Gross quantities of extraneous material should, of course, be removed manually before chemical cleaning commences.

When the plant is commissioned, the boiler should not be operated at a pressure greater than is commensurate with the silica concentration in the boiler water, as discussed in Chapter 2. It follows that if the cleaning procedure has left appreciable amounts of siliceous material in the boiler or feed system, (which will dissolve continuously but slowly in the boiler water), then the boiler must be operated at less than design pressure and with consequent restriction in output by the unit. When this condition occurs, recourse is taken to blow-down or drain and refill the boiler, until the silica content is reduced to the required amount, and this may involve high costs due to the losses of water, heat and power output. If the plant has been properly cleaned, the boiler water silica content should be reducible to 0·3 ppm within a week or so of first commissioning; in other circumstances it has been known to take many weeks, and may even require a shut-down to search for the source of silica.

4.2.2. **Development of Cleaning Processes**

Before the year 1950, approximately, the only pre-commissioning cleaning that was carried out consisted of an alkali boil-out of the boiler, using caustic soda and trisodium phosphate (sodium carbonate was sometimes used also) and water flushing of the feed system, together with a limited inspection of the plant to remove gross contaminants such as sacking and pieces of wood, which had entered during the construction stages. Under these circumstances the cleaning process was aimed at loosening some of the mill-scale, thus enabling it to be removed by blowdown, and also the removal of oil or grease used during construction. These processes were generally adequate for the plant designs of that period of time.

Mill-scale is a dense layer of magnetite produced on steel surfaces during rolling or fabrication at high temperatures, and, as discussed in Chapter 2, this layer is subject to discontinuities which, being anodic to the surrounding mill-scale, become corrosion pits in the presence of water. In these early years mill-scale could only be removed slowly, first by "weathering" (the exposure of plant components such as boiler tubes to the weather), and second by the gradual oxidation of mill-scale to ferric oxide during some years of normal service.

As working temperatures and pressures and also the physical size and complexity of the plant have increased, it has become less tolerant of foreign matter and operating difficulties causing loss of availability and restricted output have been directly attributed to the plant being commissioned in a "dirty" condition. This has led to the development of more elaborate cleaning and inspection methods, including the use of acids primarily to dissolve mill-scale and other metal oxides.

The early acid processes used inhibited hydrochloric acid (2% to 5%) as a static soaking technique in the boiler and economiser only, usually at or near ambient temperature. With the ensuing development of high-pressure designs (1500–2500 lb/in^2), it was realised that acid cleaning also had a marked effect in enabling the required low silica concentration in boiler water to be obtained reasonably soon after commissioning the plant.

Plant operators then realised that much of the advantage of chemically cleaning the boiler was being lost, since it became rapidly fouled by ingress of debris from the pre-boiler system. The first steps to improve this situation employed hot alkaline flushing of the feed system, the aim being to remove loose debris, particularly siliceous material which contributed to high silica levels in the boiler water. At a later stage of development this was followed by an acid cleaning procedure, usually employing citric acid. Considerable concern was expressed by manufacturers of the condenser, feed heaters and feed pumps regarding these developments, and quite often they insisted that some items of plant were by-passed in order to avoid possible damage by the acid.

During recent years, as turbines have become very costly to replace their output when shut down, progressively higher standards of steam purity have to be adopted. Steam contamination was evident in the early days of operation and a consequent development has been the application of cleaning processes both of a chemical nature (acid cleaning) and a physical nature (steam blowing) to the superheater, reheater and main steam and reheat pipes.

Developments eventually reached a stage where practically the whole of the unit, with the exception of the turbine, was chemically cleaned. The Borg-Siemens process, which has been used extensively in Europe for once-through boilers and also for some drum-type boilers, appeared to reach the ultimate in extent of chemical cleaning since it utilised the feed pumps for acid circulation (hydrochloric acid being employed) and the whole of the unit, with the exception of the turbine, was included in the circulation system. High temperatures and low acid strengths were used in this process and the feed pump was protected by injection of ammonia immediately before the pump to raise the pH above 5, injection of concentrated hydrochloric acid being carried out after the pump to depress the pH to about 3·5 again. The addition of citric acid was necessary to prevent iron hydroxide precipitation during the addition of ammonia. The Borg-Siemens process has been applied to only a few units operated by the Board, although complete unit cleans have been carried out by alternative techniques. On recently commissioned units, these have often included a hydrochloric acid clean of the boiler and economiser, followed by a "unit" clean with citric acid. Figure 4.2.2 shows the circuit for such a unit clean, excluding the steam side of the feed system.

When acid cleaning of the feed system was first introduced only the water side was included, but eventually cleaning was extended to the steam-side, thus allowing recovery of heater drains at a much earlier stage of commissioning than had previously been possible.

One objection to the cleaning of feed heaters was that the acid solution would be retained by capillary attraction in the interstices between expanded tube-ends and the tube-plate, and would not be removed by the normal rinsing procedure. However, this argument is largely defeated when the acid clean is preceded by an alkaline wash, since the alkali which first enters the interstices will neutralise any acid which subsequently enters, rendering it harmless.

Although the introduction of "unit cleaning" has substantially achieved its objectives it has, unfortunately, developed into a costly and time-consuming business. Furthermore, an immediate effect of acid-cleaning steel surfaces is to render them very sensitive to their subsequent environment: thus in moist air they will rust very quickly.

Current intentions for new plant are to obviate the need for acid cleaning the feed system, by cleaning the components at the manufacturer's works and treating and protecting the components so that their internal condition does not deteriorate either before or during construction on site. To be successful this needs to be coupled with "clean conditions" to be observed during assembly. Although this appears to be a sound approach there are obvious practical difficulties on site which are not easy to overcome and considerable inspection and control are necessary for it to be fully effective. Whilst this approach seems feasible for the condensate and feed systems it is obviously more difficult in the case of boiler plant. Significant progress has been made in the works cleaning of feed system components on plant to be commissioned in the next year or two and the current design specification for future plant calls for the entire condensate and feed system to be cleaned in the works, thus excluding the need for site chemical cleaning of these items immediately before commissioning.

In the early days of chemical cleaning of boilers and feed-heating plant, the processes used were mainly based upon recommendations by the contractors responsible for carrying out the cleaning, subject to agreement by Board personnel. Subsequently, as the technology of the cleaning processes developed, the Board took a more prominent part in formulating the specifications used.

As a result of research and accumulated experience, the Board has decided on the sequence of processes to be used and their essential technology for cleaning the various sections of C.E.G.B. plant. These requirements are stated in Generation Operation Memorandum No. 75 which is subject to revision as necessary, and the technical specifications in that document form the basis of "Uniform Specification M14A", which is issued to C.E.G.B. Project Groups for use when obtaining tenders for the chemical cleaning of new plant.

In Sections 4.2.3 to 4.2.15 which follow, various aspects of the several cleaning procedures

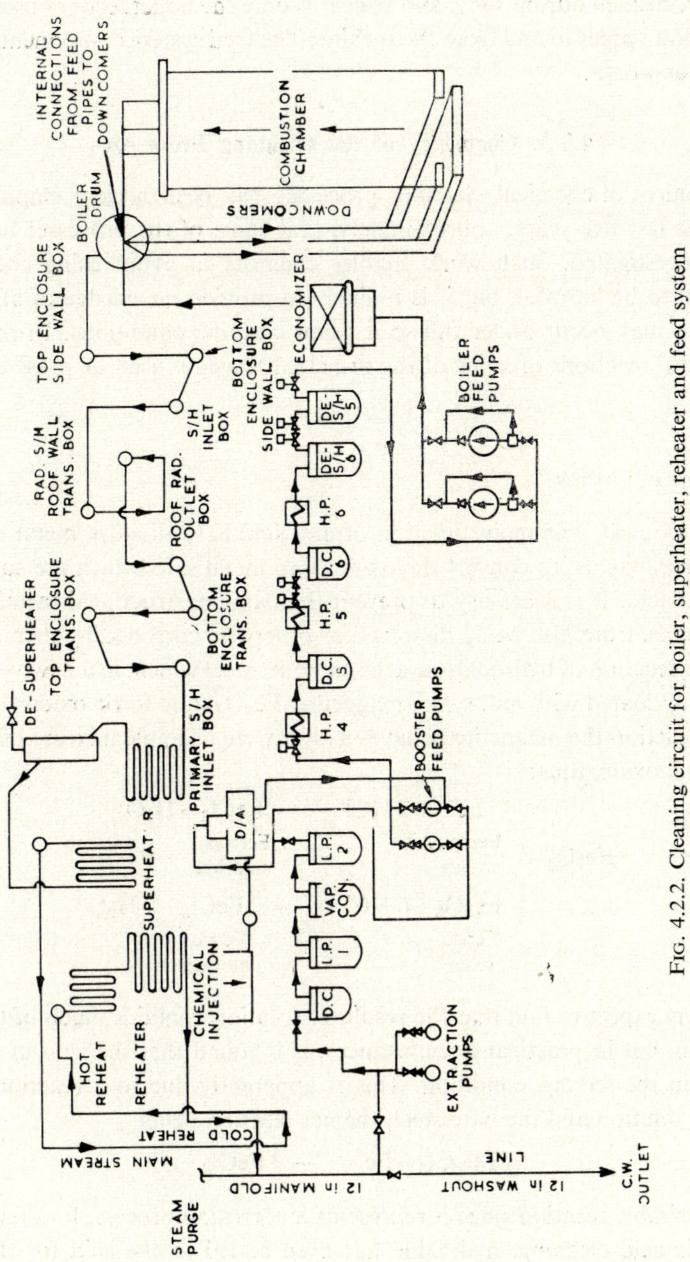

FIG. 4.2.2. Cleaning circuit for boiler, superheater, reheater and feed system

will be considered in some detail, after which the reader should refer to the Appendix which describes a fairly typical example of an actual cleaning operation. This example is No. 8 unit at Tilbury, cleaned during 1967, and concerns only the boiler, economiser, superheater, reheater and steam pipes to and from the turbine. The feed system components were cleaned and protected at works.

4.2.3. **Chemistry of the Cleaning Processes**

The development of chemical cleaning processes has been largely empirical and it is only during the last five years, approximately, that some of the processes used have been scientifically investigated. Such work enables chemists to avoid using chemicals which are thus shown to be harmful, but it is unlikely to provide knowledge of all the chemical reactions which may occur under full-scale plant cleaning conditions. In this section the relevant chemical reactions of some of the principal reagents used at present, will be discussed.

4.2.3.1. Action of Acids

Hydrochloric acid. The main function of any acid is to dissolve metal oxides and, in the case of mineral acids, to convert the oxides into metal salts which are soluble in water and the diluted acid. It is necessary to prevent the bare constructional metals, exposed by the cleaning fluids, from also being dissolved or otherwise corroded by them.

Consider the reaction of hydrochloric acid on boiler steel which, in a newly-erected boiler, is non-uniformly coated with mill-scale (magnetite, Fe_3O_4) and ferric oxide (Fe_2O_3), principally. In this reaction the magnetite behaves as if it were a simple mixture of ferrous oxide (FeO) and ferric oxide, thus:

$$Fe_3O_4 \begin{cases} \underset{\text{Ferrous oxide}}{FeO} + 2\,HCl \longrightarrow \underset{\text{Ferrous chloride}}{FeCl_2} + H_2O \\ \underset{\text{Ferric oxide}}{Fe_2O_3} + 6\,HCl \longrightarrow 2\,\underset{\text{Ferric chloride}}{FeCl_3} + 3\,H_2O \end{cases}$$

Thus one may expect to find that the resultant solution contains much of the iron in the ferric condition, but in practical circumstances, it is found that the iron in solution is almost entirely in the ferrous condition. This is apparently due to a reaction between the ferric chloride solution and the bare steel, the net reaction being

$$2\,FeCl_3 + Fe \longrightarrow 3\,FeCl_2$$

and is an undesirable reaction since it represents a corrosion process. In a few applications of hydrochloric acid cleaning, hydrazine has been added to the acid for the purpose of reducing ferric iron to the ferrous condition before it could react with the underlying metal but this procedure has not gained acceptance by C.E.G.B.

Citric acid. This substance is a crystalline organic acid having the chemical formula $HO{\cdot}C(CH_2{\cdot}COOH)_2{\cdot}COOH$ and since three of the hydrogen atoms are ionisable in aqueous solution, it is a tribasic (or tervalent) acid. When dissolved in water, citric acid is capable of dissolving the oxides of iron, very slowly in the case of mill-scale, producing the sparingly soluble citrates of iron. This is a similar reaction to that of hydrochloric acid and iron oxide, and is exemplified by the following reaction, in which citric acid is represented by "H_3 citrate", to simplify the formula.

$$\underset{\text{Ferric oxide}}{Fe_2O_3} + 2\,\underset{\text{Citric acid}}{H_3\ \text{citrate}} \longrightarrow \underset{\text{Ferric citrate}}{2\,\text{Fe citrate}} + 3\,H_2O$$

When ammonia is added to citric acid solution, sufficient to raise the pH value to 3·5 to 4·0 (approximating to ammonium dihydrogen citrate), the solubility and speed of solution of the oxides of iron are usefully increased. This is due to the formation of soluble iron "chelates", which are complex molecules. The chemical structure of these chelates is uncertain, but the reaction may be approximately represented thus:

$$\underset{\text{Magnetite}}{Fe_3O_4} + 3\,\underset{\text{Ammonium dihydrogen citrate}}{NH_4\cdot H_2\cdot\text{citrate}} \longrightarrow \underset{\text{Ferrous iron chelate}}{NH_4\cdot\text{Fe}\cdot\text{citrate}}$$

$$+\,2\,\underset{\text{Ferric iron chelate}}{NH_4\,(\text{Fe}\cdot\text{citrate}\cdot\text{OH})} + 2\,H_2O$$

Ethylene diamine tetra-acetic acid (EDTA) is another organic substance which has the property of forming water-soluble complex molecules with metals, including calcium, iron and copper, but its use by the C.E.G.B. is limited by its high cost.

Solvent capacity of cleaning agents. The extent to which a cleaning agent (usually an acid) can maintain iron in solution, and the effect of temperature change on this property, are important. It is evidently desirable to have a fairly high solubility and very necessary to avoid the deposition of iron compounds within the system. Table 1, below, shows the capacities of some cleaning solutions to hold iron in solution.

TABLE 1

CAPACITIES OF CLEANING AGENTS FOR DISSOLVING IRON OXIDES

Cleaning solution	Solution capacity (lb Fe_3O_4/gal solution)
5% Hydrochloric acid	0·33
3% Ammonium dihydrogen citrate	0·072
3% Phosphoric acid	0·008
19% Phosphoric acid	0·05
3% EDTA (ammonium salt)	0·025
6% EDTA (ammonium salt)	0·05
12% EDTA (sodium salt)	0·052

4.2.3.2. Inhibitors

It is necessary to prevent, as completely as possible, the solvent action of acid-cleaning solutions on bare metal without significant reduction of their main function of dissolving the oxides of iron. Inhibitors (sometimes called "restrainers") are substances added to the cleaning solutions to serve this purpose.

As was discussed in Chapter 2, the corrosion of metals involves electrochemical cells and in neutral or alkaline solutions it is not very difficult to introduce chemicals which will depress either the cathodic or the anodic reactions to a sufficient extent to reduce the corrosion process to a negligible level. In the case of acid solutions, this kind of inhibiting action is more difficult, but many organic substances have been found which will slow down the rate of attack by acids on metals, mainly it is thought, by forming an adsorbed film on the surface of the metal to be protected. Animal glue, formalin, and by-products of the tar and dyestuffs industries are examples of such inhibitors, though largely out-dated now.

Table 2

Effect of Inhibitors on Hydrochloric Acid and Citric Acid Corrosion of Mild Steel (Armour Hess Chemicals Ltd.)

5% Hydrochloric acid		5% Citric acid	
% Armohib 28	Corrosion rate lb/ft²/day	% Armohib 25	Corrosion rate lb/ft²/day
Nil	1·34	Nil	0·340
0·04	0·0074	0·01	0·034
0·06	0·0062	0·02	0·018
0·08	0·0047	0·03	0·010

In recent years the C.E.G.B. has considered certain proprietary inhibitors, named "Armohibs", to be the most effective, and "Armohib 25" and "Armohib 28" are usually specified for use with citric acid and hydrochloric acid, respectively. We know little of the chemical formulae of the Armohibs, but they are aliphatic compounds containing nitrogen, called "quaternary ammonium" compounds. (In this context, the word "aliphatic" indicates a chemical structure akin to that of paraffin wax.)

The effectiveness of the Armohibs is indicated in Table 2 which shows the results of tests published by the manufacturer. This table shows the amount of iron dissolved by 5% hydrochloric acid at 79°C from equal mild steel specimens, in the presence of various concentrations of Armohib 28, and similarly for 5% citric acid at 94°C, with Armohib 25.

The table shows that the inhibitors greatly reduce acid attack on the metal, but it also has to be borne in mind that increasing the temperature or the rate of flow of the solution reduces the effectiveness of inhibitors.

James (International Combustion Ltd.) has shown* that the corrosion rate of carbon steel

* Information taken from International Combustion Ltd.'s *The Corrosion Rate of Tube Material in Flowing Inhibited Acids* by P.C. James.

tubing containing 5% hydrochloric acid with 0·25% Armohib 28, at 82°C, varies with flow rate according to the formula

$$C.R. = 0{\cdot}12 d^{-1.25} \times f,$$

where

C.R. = corrosion rate, in mg/cm²/h,
d = internal diameter of tube, in in.,
f = flow rate of fluid, in gal/min.

The formula becomes $C.R. = 0{\cdot}0059 d^{-1.25} f$, if C.R. is required in lb/ft²/day.

Good correlation with the formula has been shown for tube bores in the range ¾ to 3 in., and flow rates from 2·5 to 75 gal/min, whilst extrapolation in tube bore to 12 in. is reported to be approximately correct.

TABLE 3

EFFECT OF INHIBITOR ON HYDROCHLORIC ACID CORROSION OF VARIOUS METALS (Armour Hess Chemicals Ltd.)

% Armohib 28	Corrosion rates, lb/ft²/day			
	Mild steel	Stainless steel (316)	Monel	Brass
Nil	1·34	0·096	0·00815	0·0020
0·04	0·0074	0·035	0·00185	0·0018
0·06	0·0062	0·039	0·00188	0·0016
0·08	0·0047	0·040	0·00185	0·0016

The effectiveness of inhibitors for reducing attack by acids on non-ferrous metals is substantially less than it is on steel. Nevertheless, for some non-ferrous metals, particularly those based on copper, attack by uninhibited acids is so much less than it is for mild steel or cast iron that the reduction in inhibitor efficiency with these non-ferrous metals is of little consequence. Table 3 shows these properties, according to Armour Hess Chemicals Ltd., in respect of 5% hydrochloric acid at 79°C with various concentrations of Armohib 28.

There is evidence that the reaction referred to earlier, namely the corrosion of steel by ferric ions, is not effectively inhibited, especially when chloride ions are also present. There is also some evidence that the presence of copper ions may adversely effect inhibitor efficiency.

4.2.3.3. EFFECT OF FLUORIDES

The addition of fluoride (either as ammonium bifluoride, NH_4HF_2, or sodium fluoride, NaF) to hydrochloric acid for pre-commissioning cleaning, has often been made on the assumption that it would assist in the solution of siliceous material. The solubility of silica in hydrofluoric acid is well known and is demonstrated by the use of this acid for etching glass.

Little experimental work appears to have been done to justify the addition of fluoride and Spillner of the V.G.B. has shown that the rate of solution of silica is extremely small. His results are from the exposure of particles of sand (700 particles per gram) to a mixture of 6·5% hydrochloric acid and 2·0% hydrofluoric acid, and show that at 25°C the rate of attack is 0·001 mm/h and even at 65°C it only reaches 0·01 mm/h. Since in plant cleaning the duration of a hydrochloric acid/fluoride stage is only about 6 h, this suggests that the solution of siliceous material is not practical during the acid cleaning process.

The use of fluoride may, however, be viewed from its effect on the solution of mill-scale, rather than as a "silica-solubiliser". Work by the Midlands R. and D. Department has included the effect of fluoride addition to a number of acid cleaning solutions, and has shown that, in all the acids tested, the presence of fluoride increased the rate of solution of mill-scale. Tests were carried out using pieces of mill-scale cemented to the inside of a length of plastics pipe, which was included in a rig through which the acids were circulated at 80°C, at various flow rates, with and without the addition of ammonium bifluoride.

The acids used in the test were:

(a) 5% hydrochloric acid,
(b) 5% sulphuric acid,
(c) 3% citric acid,
(d) 3% citric acid, ammoniated to pH 3·5,
(e) 3% sulphamic acid ($NH_2 \cdot SO_3H$),
(f) 3% disodium EDTA.

Figure 4.2.3.3 shows the curves obtained from all six solutions before and after fluoride addition and also the curve for ammonium bifluoride alone. Two general observations may be made from the figure:

(i) For all six solutions tested, the addition of 1% ammonium bifluoride increased the rate of dissolution of mill-scale.
(ii) The mineral acids (hydrochloric and sulphuric) were more effective than the organic acids in the absence of ammonium bifluoride, but with the exception of sulphamic acid, the reverse was true in the presence of the bifluoride.

It is of interest to consider the reasons for the second observation above, which probably lie in the degree of ionisation (or dissociation) which the various acids undergo in water. Reference to Section 2.1.1 of Chapter 2 should make it easier for the "non-chemist" to understand the following explanation.

Hydrochloric acid and sulphuric acid are termed "strong" acids because they are very largely ionised in water, producing many hydrogen ions, whereas the organic acids and hydrofluoric acid (contained in ammonium bifluoride) are ionised to a much smaller extent, producing fewer hydrogen ions, and are therefore called "weak" acids. This distinction is a principal reason for order of effectiveness indicated in Figure 4.2.3.3, for the various acids when tested without fluoride addition. Observation (i) indicates that the fluoride ion has a positive effect in assisting the solution of mill-scale by all of the acids, but when the ammonium bifluoride is added to the strong acids, the large concentration of hydrogen ions represses the ionisation of the hydrofluoric acid ($HF \rightleftharpoons H^+ + F^-$), and therefore much reduces the number of fluoride ions (F^-) which would otherwise be present. In the case of

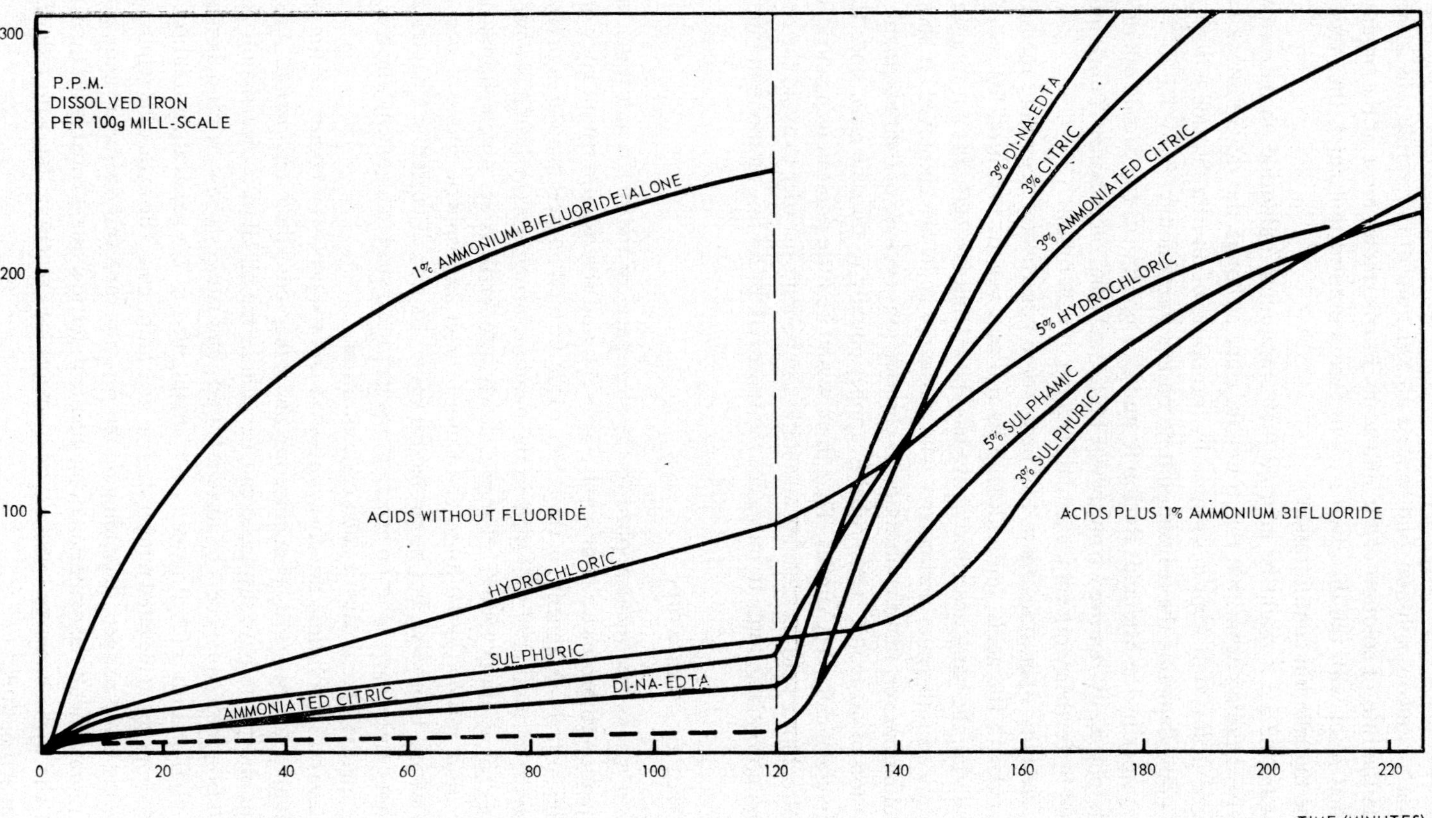

FIG. 4.2.3.3. Dissolution of mill-scale under dynamic conditions

ammonium bifluoride and the weak acids, there is no significant disparity in hydrogen ion concentration in aqueous solution, and hence a greater number of fluoride ions is available to accelerate the rate of solution of the mill-scale. Furthermore, most of the organic acid radicles tested react with the dissolved iron to form complex molecules, thus tending to maintain the fluoride ion concentration.

From Figure 4.2.3.3 it might be inferred that ammonium bifluoride on its own would be an effective cleaning agent, but this is not so, since although the rate of attack is high, the solubilised iron is only kept in solution by formation of an iron fluoride (either simple or complex) thus depleting the reagent, and becoming un-economic.

The above work by the Midlands Region R. and D. Department was carried out on pieces of mill-scale physically removed from newly fabricated mild steel boiler drums and is to be supported by further work on the actual cleaning of new tubes.

In plant cleaning operations there has been some controversy regarding the value of adding fluorides to the acids, and although its use may not be beneficial for solubilising siliceous material, the work described above has shown that it is certainly advantageous for the dissolution of mill-scale, particularly where organic acids are employed. It should be borne in mind, however, that the objection to chloride ions because of their association with stress-corrosion of austenitic steel probably applies equally to fluoride ions, since the two elements are chemically very similar. Thus there would be little point in using citric acid in place of hydrochloric acid for cleaning austenitic components in order to avoid the possibility of "stress corrosion", only to introduce the same risk by adding ammonium bifluoride.

4.2.3.4. Effect of Passivation

Passivation means treatment applied to the steel parts of the steam and water circuits with the intention to produce a continuous, coherent, and corrosion-resistant layer of magnetite on the surfaces. Magnetite is the stable oxide formed on steel in air at high temperatures (hence the formation of mill-scale in the manufacturing process), but it is unstable at temperatures below about 1500°C, changing slowly into ferric oxide. A comparable change takes place more rapidly in the presence of moist air at atmospheric temperatures, giving rise to rust which, of course, has no protective value. As was discussed in Chapter 2, Section 2.4.2, hot caustic soda solution generates magnetite on steel to an extent which depends on time, temperature, and caustic soda concentration.

Hence, passivating solutions are always alkaline and the process is conducted for as long, and at as high a temperature, as engineering and other considerations permit. Ammonia or trisodium phosphate have displaced caustic soda as the alkali of choice, mainly for fear that the solution may gain access to austenitic parts and cause stress-corrosion damage.

A substantial amount of hydrazine is also maintained in the passivating solution, some of which decomposes at elevated temperature to give ammonia. The hydrazine removes most or all of the dissolved oxygen, depending on temperature, and this assists the main purpose by making more difficult the oxidation of magnetite to ferric oxide. Probably the most useful reaction of the hydrazine is to reverse the last-mentioned reaction, that is to reduce ferric oxide to magnetite, thus:

$$6\,Fe_2O_3 + N_2H_4 \longrightarrow 4\,Fe_3O_4 + N_2 + 2\,H_2O$$

During steam-blowing, hydrazine is maintained in the boiler water and some hydrazine vapour passes into the superheater and reheater, together with ammonia produced by the thermal decomposition of much of the hydrazine. These effects most probably ensure that the superheater and reheater contain an alkaline and reducing atmosphere (and therefore are inimical to the formation of ferric oxide) during steam blowing, and it is found in the result that these parts are passivated to a useful degree. It is important to realise, however, that the very thin layer of magnetite produced by passivation of either the boiler, superheater or reheater, is soon oxidised in the presence of moist air; this layer must be protected from such an environment and gradually builds up to more substantial thickness and density during normal operation of the plant.

The essential comment to be made about passivation procedures is that a useful thickness of coherent magnetite is only produced at high temperature, in the time available. Thus in boilers, where passivation can be effected under pressure (600 lb/in^2, 253°C, or greater), a satisfactory coating of magnetite can be produced. In feed systems, where 90–95°C is usually the maximum temperature achieved, the magnetite produced is tenuous and probably has much less value as a protective coating.

4.2.4. Administration and Planning

A cleaning programme needs careful planning and supervision to ensure that it is carried out safely and efficiently, and of the many aspects which must be considered, the following are particularly important.

Consultation between all the parties concerned, including the cleaning contractor, is essential and a detailed working document is drawn up to cover the whole programme.

The design of the plant must be considered in order to establish the cleaning flow paths and pump requirements to give the necessary circulation rates, whilst the various metals and alloys in circuit must be taken into account so that the chemicals used will cause no damage to them.

For the purpose of applying the various processes to be used, it is normally necessary to divide the plant into a small number of circuits which can be isolated from each other. This enables each circuit to receive the chemical treatment appropriate to its materials of construction, to the condition of the internal surfaces, and to the pressure and temperature which each circuit can sustain. It also facilitates control of the processes, for example, control of fluid velocity in various parts of the system. Thus it is now common practice to separate the boiler from the feed system, and to provide for isolation of superheaters and reheaters when the remainder of the boiler is cleaned with hydrochloric acid. Similarly, the water side of the feed system is usually treated separately from the bled-steam side.

Means of heating water and chemical solutions, and conserving heat, have to planned, and this often necessitates the use of steam from auxiliary boilers. Large quantities of water, both fresh and demineralised, will be used and the water supply and availability of water treatment plant must be programmed accordingly. Several millions of gallons of water are required, and the maximum water storage capacity must be made available to meet the peak demands for water.

Similarly, considerable quantities of chemicals will be used, much of which are supplied in bulk liquid form, and adequate reserves need to be made available for use if required. This involves planning the required delivery service, usually by road tanker, and may be alleviated by the use of storage tanks.

A decision has to be made concerning the removal of drum internals, which are "tailored" and fitted before cleaning commences. It is desirable to remove them before cleaning the boiler and, having cleaned the various parts independently, to refit them in the drum after the boiler cleaning operations are completed. This ensures thorough cleaning and the avoidance of debris which would otherwise be likely to be trapped in the steam-purifying equipment. Nevertheless, there are two principal disadvantages; one is the valuable time taken to dismantle and reassemble the fittings, and the other is the attendant likelihood of damaging the cleaned surface and of reintroducing debris during reassembly. The trend at present is to leave drum internals *in situ* during boiler cleaning and, for final inspection, to remove only as much of the internals as is necessary to ensure that all debris in the drum can be removed.

Supervision of the whole cleaning operation must be maintained by engineers and chemists, on a shift basis, so that control tests can be made and appropriate action taken. A system of ready communication between operators on the plant must be devised in advance of the project. When the programme (which may take 3–6 weeks to complete) has been started, it should proceed continuously through its various stages to completion, since intervening delays can seriously detract from the desired result. Associated with this requirement, it is particularly desirable that the whole of the preparation of new plant for commissioning is programmed so that the chemical cleaning operations are following by the least possible delay before operating the plant.

4.2.5. **Criteria of Satisfactory Results**

Pre-commissioning cleaning programmes are now very complex and their justification has been discussed, but it is also desirable to consider the immediate aims and to define, as far as possible, the criteria of a satisfactory result. These criteria are of two kinds, of which the first (and of immediate consequence) are those to be found by thorough inspection before, during, and after the cleaning programme. Secondly, the effectiveness of the programme will be reflected, in an inverse sense, in the time taken in early plant operation to obtain the chemical conditions required by Generation Operation Memorandum No. 72; in particular the number of running hours on load required to reduce concentrations of silica in boiler water and to reduce iron plus copper in feed water, to satisfactory levels, should not exceed one week approximately.

The techniques of inspection will be described later (Section 4.5) and the final inspection should show that the internal surfaces of the plant are free of extraneous or undesirable substances. A list of such substances would be almost inexhaustible, but mill-scale, grease, metallic fragments and "dirt" are common examples. In the case of the boiler, the metal surfaces should finally be dry and uniformly coated with a grey or black adherent layer of magnetic oxide of iron. In this case also, because the passivation stage is carried out under pressure, the finished surfaces are fairly stable and protective, unless wetted.

The finished surfaces should also be free from pitting and a properly controlled chemical cleaning procedure will not cause such an effect. Nevertheless, areas of fine pitting are quite often observed at final inspection and are usually caused by corrosion resulting from inadequate protection of components during the period between manufacture and commissioning.

Debris of various kinds may remain in some of the inaccessible tubular components of the boiler, for example in superheater and reheater bends and in boiler tubes. A satisfactory condition in this respect is tested by the use of radiography for superheater and reheater bends and miniature cameras for boiler tubes, although only a proportion of the tubes can be economically tested in these ways. Examination of the steam-blow target plates indicates the overall effectiveness of the removal of small debris from the superheater and reheater.

4.2.6. Inspection before Cleaning

During construction of a modern station, substantial effort is made to achieve clean site conditions and to minimise ingress of debris into the steam/water circuits. Nevertheless, as Figure 4.2.6A shows, a pre-cleaning inspection is required to ensure that general debris has been removed as completely as possible, to extract obstructions such as sacks or tools and to assess the condition of the metal surfaces. Figure 4.2.6B illustrates the condition of a 350 MW boiler drum, before acid cleaning. Particular attention is necessary in checking that the contractor has not applied protective coatings other than those approved, since these may require special cleaning procedures. During this preliminary inspection use should be made of tube cameras to assess the initial conditions of the boiler tubes, and this may reveal (for example) excessive internal protusion of some butt welds which would require excision and rewelding.

4.2.7. Alkali Boil-out

It is usual to subject the boiler to an alkali boil-out carried out at a pressure substantially less than the normal working pressure, whilst the feed system is degreased by circulation of an alkaline solution at an elevated temperature, but at little more than atmospheric pressure.

The purpose in both instances is to remove any grease and oil from the surfaces, which could not only lead to difficulties in operation, but would certainly reduce the efficiency of the acid cleaning solutions by preventing them from coming into contact with the surfaces. Additionally the alkaline cleaning of the feed system may be necessary to remove temporary protective coatings, if these have been applied. A further function of the alkali boil-out is to loosen and remove some of the metal oxides which will be present and thus enable the acid stages to be carried out more efficiently, since the acid strength will not become depleted so quickly.

The concentration of chemicals used in an alkaline boil-out is not critical and a typical boiler clean may employ the following:

1000 ppm trisodium phosphate
300 ppm sodium hydroxide
0·05 ppm Lissapol C

The first two chemicals provide the necessary alkaline detergent action, for detaching loose scale and emulsifying any oils or greases, thus making them dispersible in the solution,

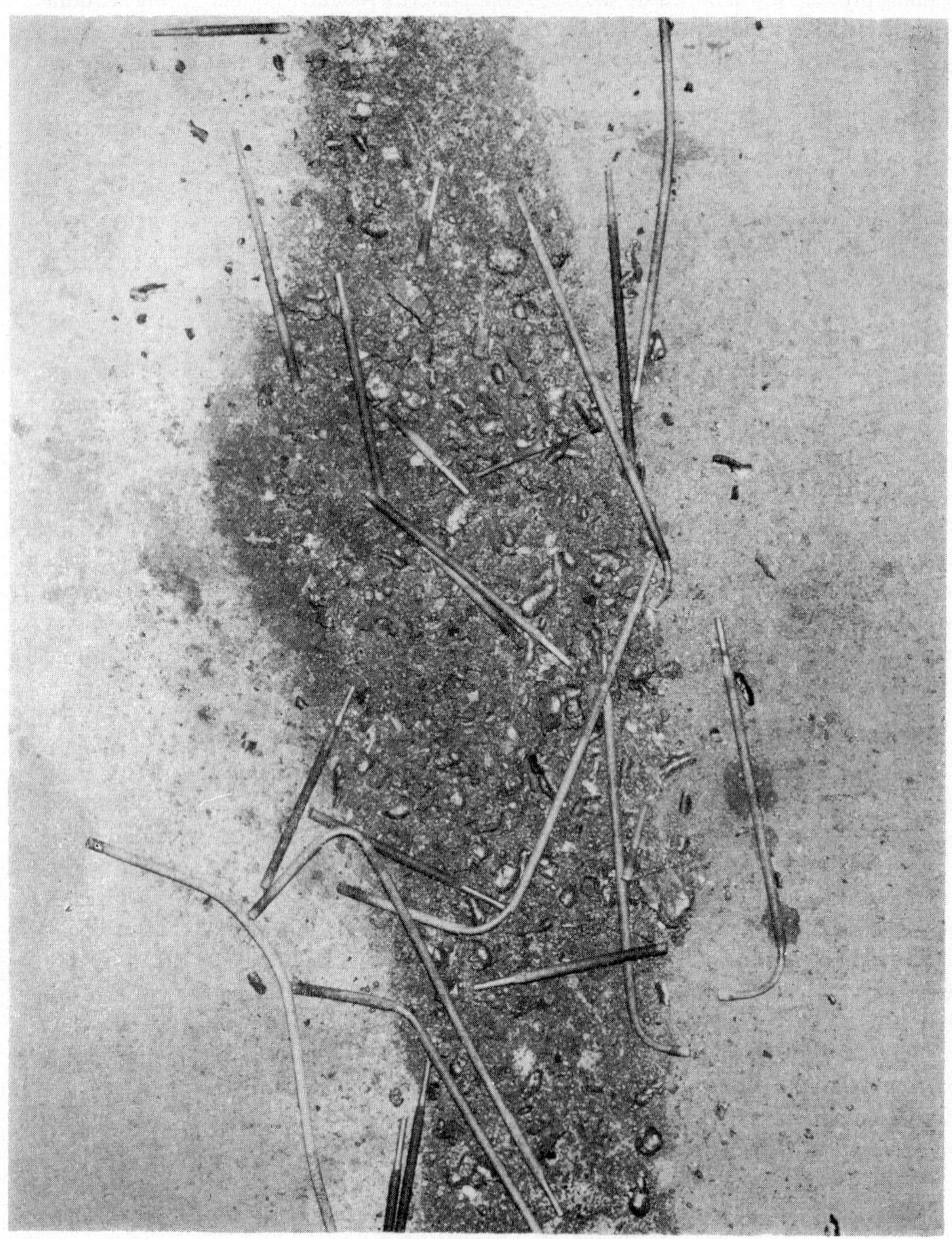

FIG. 4.2.6A. Debris from a feed heater flash box drain

FIG. 4.2.6B. Boiler drum condition before cleaning

whilst Lissapol C is a wetting agent which reduces the surface tension of the liquid and thus enables it to penetrate the material to be removed, more easily.

The pressure is slowly raised to approximately 600 lb/in^2 and maintained for at least 24 h.

Before draining, the water level is raised to the top of the drum and maintained there for 4 h. Throughout the process intermittent blowdown is carried out to remove suspended matter.

Precaution must be taken to prevent sodium hydroxide gaining access to austenitic steel parts, especially in the superheater and reheater. If prevention cannot be assured, it may be necessary to omit sodium hydroxide in order to avoid a risk of causing stress corrosion of austenitic parts.

Degreasing the feed system may be carried out by circulating a similar solution to that used in the boiler alkaline boil-out, at a temperature of 80–90°C for a similar period of time.

Extensive flushing of the plant is carried out before and after the degreasing, and serves to remove loose debris and residual cleaning chemicals.

4.2.8. **Acid Cleaning**

In order to remove the firmly attached scale and other debris which remain after the alkali boil-out, it is necessary to employ chemicals which are capable of taking it into solution. Some of the scale present is "mill-scale" and as the term implies it is formed during the manufacturing processes. At the high temperatures employed during these processes, the surface of carbon and low alloy steels oxidises to form a discontinuous coating of a complex mixture of iron oxides (mainly magnetite, Fe_3O_4), known as mill-scale. This scale is insoluble in water and alkalies and can only be dissolved by acids.

The chemical cleaning of power plants has involved the use of various acids each of which is accompanied by attendant advantages and disadvantages. The main considerations which lead to the selection of one particular acid are

(a) Efficiency and speed of dissolution of metal oxides.
(b) Availability and cost of reagent selected.
(c) The degree of safety with which it can be used, with respect to the personnel and plant involved.

Numerous acids have been used, including the inorganic (or mineral) acids: hydrochloric acid, sulphuric acid and phosphoric acid, and the organic acids: citric acid, sulphamic acid and ethylene-diamine tetra-acetic acid (EDTA). Although American literature makes reference to other acids such as formic acid and hydroxy-acetic acid, they do not appear to have provoked practical interest in this country.

Considering first the range of inorganic acids; although sulphuric acid and phosphoric acid were used occasionally in some of the first approaches made to acid cleaning of the Board's boiler plant, they are no longer seriously considered.

4.2.8.1. Phosphoric Acid

The use of phosphoric acid was largely based on the assumption that it would confer some degree of passivation on the newly cleaned surfaces by formation of a surface film of insoluble iron phosphate. In fact, phosphoric acid forms the basis of some commercial preparations which are used on a relatively small scale for cleaning rusted surfaces prior to painting. However, there is some risk that the insoluble iron phosphates would be rapidly removed from boiler surfaces under operating conditions and be carried over with the steam, leading to operating difficulties. Apart from this the high cost of phosphoric acid detracts from its selection as a primary cleaning agent.

4.2.8.2. Sulphuric Acid

In Great Britain sulphuric acid is the cheapest acid, it is effective and has been used occasionally for chemical cleaning. It has disadvantages; it is not volatile so that if the acid fills crevices and escapes subsequent neutralization it will not be removed by the later boiling stages and corrosion will probably occur in the crevice. Additionally, the strong acid (96% sulphuric acid, as received in bulk supply) liberates a large amount of heat when mixed with water, and this demands precautions when diluting the acid to the strength needed for cleaning. A further disadvantage of sulphuric acid is that the attack by dilute solutions on steel cannot be inhibited as effectively as for other mineral acids.

4.2.8.3. Hydrochloric Acid

Where an inorganic acid is used, the choice is almost invariably hydrochloric acid which is a comparatively cheap material. Unfortunately hydrochloric acid has an important disadvantage in that the chloride ion is known to promote cracking of austenitic steels in particular, by stress-corrosion. As the term "stress-corrosion" suggests, the resultant damage is caused by the combination of residual stresses within the metal, and the associated corrosive agent. Thus if the residual stress is sufficiently small, no harm is likely to be caused in subsequent service (which is usually the time when the damage is observed), as a result of cleaning with hydrochloric acid. Nevertheless, there is sufficient uncertainty about the circumstances which affect stress-corrosion, that the C.E.G.B. do not permit the use of hydrochloric acid in contact with austenitic steel parts of station plant. Figure 4.2.8.3 shows damage to a stainless steel pin, caused by hydrochloric acid, although in this case the alloy was found to be of poor quality.

4.2.8.4. Organic Acids

The organic acids are, in varying degree, less effective than hydrochloric acid in dissolving mill-scale, but they have the important advantage of being associated with a negligible risk of causing stress-corrosion in alloy steels. Equivalent concentrations of inorganic and organic acids in water show a marked difference in pH value, the organic acids being generally higher because they are only partially ionised. This property can be advantageous by

FIG. 4.2.8.3. Stainless steel pin damaged by hydrochloric acid

reducing the dependence necessarily placed on highly effective inhibitors to prevent corrosion of the cleaned metals, when using inorganic acids.

As was discussed more fully in Section 4.2.3.3, the addition of small amounts of ammonium bifluoride to most organic acid solutions, substantially increases their rates if dissolving mill-scale.

Citric acid. Citric acid is the organic acid commonly used in C.E.G.B. practice, although it requires to be used at a minimum temperature of 85°C to be effective on mill-scale, and it is expensive.

When citric acid was first used in chemical cleaning, it was employed as a 3% solution of the free acid and it was observed that insoluble phases sometimes appeared, which were identified as iron citrates. Investigation showed that the solubility of iron oxides in citric acid could be increased by adding some ammonia, and this led to the use of ammonium dihydrogen citrate which forms ferrous ammonium citrate, a complex salt of sufficiently high solubility to prevent precipitation in the cleaning solution. For this reason and because the acid alone is very slow in dissolving mill-scale, citric acid is now commonly ammoniated to a pH of 3·5–4·0, which corresponds to the formation of the ammonium dihydrogen citrate.

Obviously, there is a considerable range of acids which might be used for chemical cleaning, but it should be noted that in current C.E.G.B. practice the selection is usually restricted to hydrochloric acid and ammoniated citric acid.

4.2.8.5. Application of Acid Cleaning

Like most chemical reactions, the rate of attack by acid on iron oxides increases with temperature, but in practice, the maximum temperature is limited by the fact that the effectiveness of corrosion inhibitors decreases with increasing temperature. In general, a temperature of approximately 90°C is used for citric acid and approximately 75°C for hydrochloric acid.

The strength of the acid used might be expected to vary according to the amounts of oxides and other debris to be removed. However, the effectiveness of inhibition is also affected by the acid strength and this together with economic considerations has led to the general use of hydrochloric acid at an initial strength of 5% w/w and citric acid at 2 to 4% w/w. It is sometimes necessary to add more acid, when the strength becomes unduly depleted by the amount of debris present.

In the early development of acid cleaning, no positive circulation was employed, cleaning being carried out by static soaking. There are advantages and disadvantages to both the "soaking" and "circulation" techniques.

In the case of static soaking it can be argued that the acid adjacent to the surfaces becomes depleted in strength, leading to incomplete cleaning in heavily contaminated areas, and also that detached pieces of scale may fall to the bottom of the system and block drain lines, etc. Furthermore, it is not possible to obtain representative samples of the liquor in order to exercise chemical control of the process. On the other hand, static soaking ensures that the upper surfaces of drums and headers are flooded, since it is easier to maintain the level at the top of the drum with minimum risk of carrying the liquor into the superheater.

If pumped circulation is employed, then most of the disadvantages given above are overcome, but there is a risk that the inhibitor will become less effective unless the velocity is carefully controlled. In actual practice little difficulty has arisen from this effect, and precautions are usually taken to ensure that the flow rate does not exceed about 1 ft/sec. In a few cases, however, acid attack taking the form of longitudinal grooves has occurred in headers, etc., and has been attributed to ineffective inhibition as a result of excessive velocity. In practice, pumped circulation is always provided and may be interrupted by periods of static soaking.

The acid cleaning process must be carefully monitored throughout by sampling and testing, to ensure that the composition and temperature of the circulating fluid are correct, and that the precipitation of insoluble compounds is prevented.

The use of station pumps for the circulation of acids or other chemical cleaning solutions is not desirable, since the pumps may suffer as a result of chemical attack and/or impingement of the suspensed materials present during the process. Therefore, in present practice specially designed pumps are provided by the contractor. In a few cases, in the past, the use of station pumps has been permitted for acid circulation and whilst in at least one station this was not followed by any damage apparent during the subsequent examination of the pump, considerable care is required and the risk involved is now thought to be too great to be accepted. Figure 4.2.8.5A shows three contractors' pump assemblies, with some temporary pipework, and Figure 4.2.8.5B shows a temporary drum door and connection.

4.2.9. Draining Acid and Flushing

Various methods of acid draining and flushing have been employed. The acid may be drained off by displacement with air or nitrogen, the aim of the latter being to reduce access to the steel surfaces by oxygen from the atmosphere. Following this the system is usually rinsed with dilute citric acid, 0·1% to 0·5%, and in recent practice is ammoniated to pH 3·5–4·0 to complex the residual iron and thus prevent the precipitation of iron hydroxide as the pH increases during the flushing process. Alternatively, the acid may be displaced with a solution of dilute citric acid, but this is difficult to carry out in practice. Rinsing with dilute citric acid at pH 3·5–4·0, will impart a small degree of rust-resisting property to the steel surfaces.

In the case where the boiler has been cleaned with hydrochloric acid and is then to be included in a larger circuit including the superheater and reheater for a citric acid clean, it is important that as much of the hydrochloric acid as possible is flushed out. Failure to do this will result in chlorides entering the superheater and reheater, which if made of austenitic steel may suffer from stress-corrosion. To this end, flushing of the boiler should be done by repeated draining and refilling, and should include reversing the normal circulation pattern (if possible) and back-flushing the superheaters into the drum.

After draining the acid, flushing is carried out to remove any residual acid together with loose debris which remains undissolved. It is generally accepted that a minimum velocity of 3 ft/sec is required to ensure that settled debris is carried out of vertical tubes, and

Fig. 4.2.8.5A. Contractors' pumps and temporary pipework

FIG. 4.2.8.5B. Temporary drum door and connection

this is not easily achieved, especially with respect to pendant superheaters and reheaters. These components consist of multiple parallel loops of various lengths of tubing, and therefore the water flow will not be uniformly distributed, whilst any major obstruction by debris in a tube is likely to prevent any useful flow through that tube. Thus very high pumping rates are required for flushing, which is continued until the effluent is free from acid and suspended matter and it must be completed expeditiously, bearing in mind that the steel surfaces will soon rust on exposure to the air.

Demineralised water must be used for flushing superheaters and reheaters, and is normally used for flushing other parts also, although for these it may be expedient to substitute towns water for a proportion of the flushing water and to complete the process with demineralised water. The station's installed pumps may be used for flushing, providing there is no risk of debris or chemicals entering them.

4.2.10. **Passivation**

The more theoretical aspects of passivation have been discussed. The practical purposes are to produce at the end of the cleaning operations a surface which (a) will not rust quickly (that is, within a few weeks) when exposed to air, and (b) will provide a suitable basis for the more substantial magnetite film which normally develops during subsequent operational service—in the boiler, particularly.

Passivation of the boiler and associated components is usually achieved by filling to normal working level with demineralised water containing 300 ppm hydrazine (initially) and 50 ppm ammonia, and then raising pressure to 600–900 lb/in^2 for 24 h. The hydrazine is rapidly decomposed under these conditions and further additions are made to try to maintain a minimum concentration of 25 ppm during the process. The passivation procedure is terminated by reducing pressure to 100–150 lb/in^2 and draining the boiler at this pressure, by which means the internal surfaces are dried.

For the feed system it has been customary to attempt passivation by circulating a solution of similar composition to that given above, for 24 to 48 h at about 95°C, which is the upper temperature limit brought about by the necessity to restrict the pressure to near atmospheric. Passivated steel surfaces must be protected from moist air and water until the plant is commissioned. Unfortunately it is usually impossible to dry feed systems internally and passivation, as applied to these parts, is of rather doubtful value.

4.2.11. **Final Inspection**

Final inspection of the plant is difficult because of the limited access on a modern unit. Direct access is usually confined to boiler drums, headers, deaerator storage vessels, tanks, some large-bore pipes, and condenser steam spaces to a limited extent.

Unless special care is taken during the inspection and final closure of the plant, foreign matter will enter the system, but this can be largely prevented by the introduction of "clean conditions" working at all points of entry into the plant. In this system of working, personnel entering the plant are required to wear clean overalls and overshoes, and drum ends are enclosed by "penthouses" to act as changing rooms and to prevent entry of adventitious material. Where other parts of the plant are opened for inspection, they must not remain open for any longer time than is essential, making full use of temporary closures as necessary. (Fig. 4.2.11A shows the construction of a penthouse erected at a drum end.)

The various forms of thermal insulation which are applied after the erection of plant, provide a major source of siliceous contamination which is best denied access to plant internals by ensuring that all lagging work is completed and cleared up, before the cleaning programme commences. Completion of the lagging is in any case necessary to avoid excessive cooling during the circulation of acids.

Practically the entire amount of boiler tubing is, of course, inaccessible for direct inspection and internal examination is by the use of remotely operated tube cameras (described in Section 4.5.3.2.4), which are inserted in a proportion of the tubes. In some parts of the boiler it is necessary to cut the tubes for this purpose, the operation being sited at locations where the repair weld will not be subject to considerable heat transfer during operation. Photographs thus obtained in a boiler tube (a) before and (b) after acid cleaning, are reproduced in Figure 4.2.11B.

Boiler headers are examined by removing temporary pipe connections and nipples, and loose debris which may accumulate in headers can be removed through these points of access and drain connections.

FIG. 4.2.11A. Drum-end "penthouse" enclosure

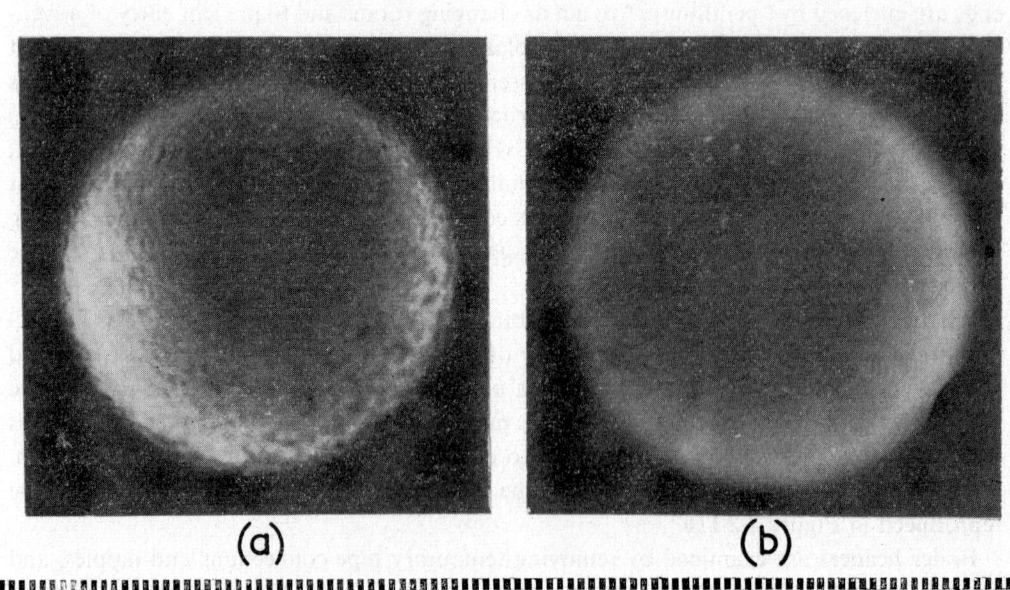

FIG. 4.2.11B (a), (b). Boiler tube camera photographs, before and after acid-cleaning

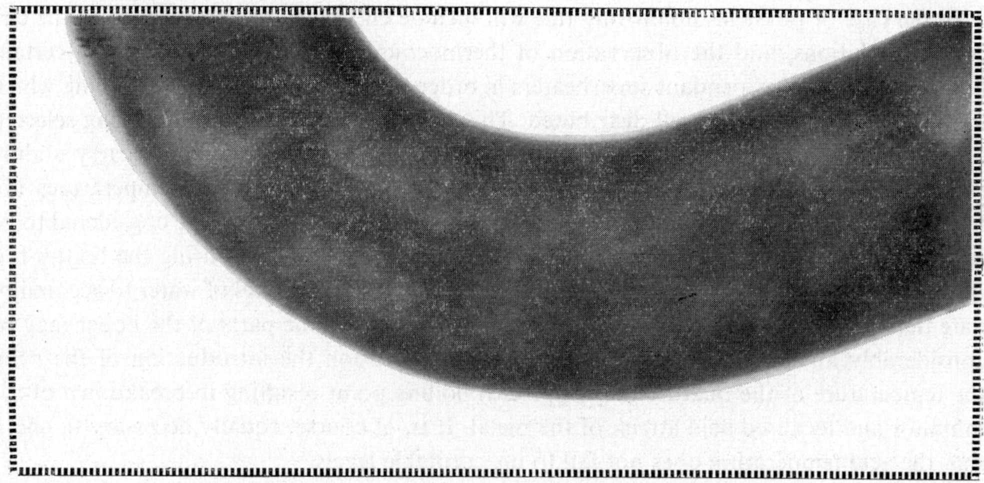

FIG. 4.2.11C. Radiograph showing "debris" in superheater bend

Superheater and reheater tubes which are suspected to have debris in bottom return bends, will have been identified by thermocouples or other means of assessing abnormal temperature during the flushing process. These bends, and a proportion of the remaining ones, should be subject to inspection by gamma-radiography, in order to confirm their internal condition. Figure 4.2.11C is a radiograph of a superheater bend found to contain half-a-dozen nuts and bolts.

Inspection of the feed system will be made at the deaerator, at temporary connections and such other points as may be agreed, having regard for the need to avoid breaking in to all-welded sections, as far as possible.

The criteria of satisfactory results from a cleaning programme have been discussed previously. During the inspection one looks particularly for a substantially uniform surface finish and for signs of corrosion pitting, which may indicate a loss of effective inhibition during acid treatment. There may also be areas (as at the top of a boiler drum) which have escaped cleaning due to air- or gas-locks, and in this event it is usual to manually reclean the areas concerned. Samples of deposits removed at various stages in the process are chemically examined, particularly to discover if they contain significant amounts of oil which can interfere with the action of the acids. Other forms of debris, including tools, wire and bolts, are often disclosed by radiography, whilst weld-protrusions, pitting and other surface imperfections in boiler tubes are visualised by tube cameras.

4.2.12. Control of the Cleaning Process

Close chemical control at each stage of the cleaning process, and inspection at appropriate intervals, are essential for a successful clean and this necessitates having experienced chemists available throughout the entire programme. If any stage is thus shown to have been inadequate it will be necessary to repeat it.

Control of the cleaning process may be considered as two kinds, namely, physical and chemical.

In the case of physical monitoring this will include checking the temperature of the circulating solutions, and the observation of thermocouples which may be fitted to certain items of plant such as pendant superheaters in order to demonstrate that the flushing which follows the acid stages, is well distributed. This is usually achieved by monitoring selected tubes which, from a knowledge of the plant design, are considered likely to carry unduly small flows. Thermocouples are also necessary to indicate that metal temperatures are not excessive when acid is added to the boiler. At first thought, this may be considered to be an unlikely occurrence, but it can arise in some boilers as a result of firing the boilers too fast when raising the water temperature, just before lowering the level of water to accommodate the required volume of acid. Thus the temperature of some parts of the boiler may be considerably higher than that of the remainder, and upon the introduction of the acid, the temperature of the mixture may approach boiling point resulting in breakdown of the inhibitor and localised acid attack of the metal. It is, of course, equally necessary to check that the acid temperature does not fall to unacceptable levels.

It is very desirable to be sure that the correct flow rates for acid injection and for flushing, are achieved. In recent cleaning operations this has been done by including flow meters in the appropriate flow paths.

Chemical control is more important to some stages than to others, the ones needing most control being the acid, flushing, passivation and effluent neutralisation stages. Sampling points should be available at various points in the circuit and may enable the amounts of materials dissolved in various sections to be determined, but their main value lies in enabling the chemist to ascertain the strengths of the solutions used in the various procedures, and to check the uniformity of a solution throughout the circuit.

During the preliminary flushing stages, analyses will be carried out to estimate the amount of debris that has been removed and also to indicate that most of the loose material has been removed. During the first alkali boil-out, analysis will confirm that the required quantities of chemicals are present, and the quantity of debris in suspension, together with siliceous material which is partially removed in this stage, will also be measured. Table 4 lists the amounts of iron oxide and silica, determined by analyses and weighing, removed at various stages in cleaning a 200 MW unit.

During the acid cleaning operation, the circulating solution will be analysed for acid strength and metal content (ferrous and ferric iron, copper, and nickel).

Frequent determination of the acid strength is necessary to ensure that the initial concentration is as specified, and that it does not become depleted to the point of being ineffective. In addition to these purposes in the case of citric acid, there is a risk that iron citrates, which have relatively low solubility, may become precipitated as sludge and this is said to be prevented if the concentration of free citric acid is maintained not less than 2%.

Regular analysis for metal content (particularly iron) will indicate the rate of dissolution of the metal oxides, and when the iron concentration becomes stable the process is considered to be complete, provided of course, that the acid is not spent. The determination of copper and nickel is more important in the case of post-commissioning cleaning which is discussed in Section 4.4.1, but analysis for these metals during a pre-commissioning clean serves to show if any attack on non-ferrous components (if present), has occurred.

TABLE 4

MATERIAL REMOVED DURING CLEANING OPERATIONS

	Iron oxide, lb	Silica, lb
Steam side system:		
Water flush	20	1
Alkali	35	18
Acid	1660	—
Final flushing	70	—
Boiler and feed system:		
Alkali	109	16
Hydrochloric acid	5020	—
Citric acid	3410	—
Loose debris	27	3
	10,351	38

Under normal circumstances the iron concentration increases rapidly as the acid is being introduced (a process which in itself may take 2–3 h), and within an hour of the total quantity of acid being introduced has usually reached a value stable enough to indicate that cleaning is complete. It is, however, usual for circulation to be continued for about six hours, but the process should not be prolonged unnecessarily because there is evidence that the action of inhibitors deteriorates with time.

Chemical analysis may indicate that some abnormality is occurring. For example, a confirmed fall in iron concentration may show that iron is being precipitated, although with good management this is unlikely to arise and, in the past, has usually been associated with the use of citric acid unaccompanied by ammonia. Similarly, a fall in copper concentration may indicate that plating or precipitation of metallic copper is occurring; this is likely only in postcommissioning cleaning and where non-ferrous components are exposed to the process.

When superheaters and reheaters are to be cleaned, usually by circulation of citric acid, it is essential to remove all air from the system in order to ensure adequate flow through the various parallel paths, thus "priming" the system. This may be done by "steaming out" before the chemicals are added, that is, by firing the boiler with all vents open until the air is discharged. Thereafter a small pressure must be maintained, or other means taken, to prevent air ingress until cleaning and flushing the superheater and reheater are completed.

During the subsequent flushing stages analysis will indicate when all the acid and suspended matter have been removed. This is of vital importance, where a complete circuit clean with citric acid follows a hydrochloric acid clean of the boiler evaporative surfaces only, in order to ensure that chlorides do not enter the austenitic components.

When passivating the boiler it is usual to include frequent and heavy blowdown in order to discharge as much suspended matter as possible. This of course adds to the loss of the passivating chemicals, normally hydrazine and ammonia, which are maintained by further injection based on analysis of the boiler water. It is usually required that a minimum reserve of hydrazine (25 ppm) is present during this stage.

To meet the needs for analytical control of the cleaning process, methods of chemical analysis have been developed and those currently recommended are quoted in G.O.M. No. 75.

Some additional control tests of a more qualitative nature are often useful. For example it is difficult and usually impracticable to measure inhibitor concentration, but the following simple test will indicate whether or not the acid is effectively inhibited. A degreased ball of steel wool is placed in a test sample of the cleaning solution; if the solution is not effectively inhibited, hydrogen gas is evolved and the steel wool floats, whereas it will sink in the case where the inhibitor is satisfactory. Another useful practice is to fasten removable specimens of mild steel within the circuit to be cleaned, the prior condition of the specimens being chosen either to indicate the effectiveness of the acid for removing mill-scale or deposit, or to see if any corrosion is caused, for example due to loss of inhibitor effectiveness during the process.

4.2.13. **Disposal of Effluents**

The discharge to waste of much of the effluent from flushing operations usually presents no difficulty since the water is inoccuous and does not pollute natural water courses. However, the used chemical solutions must be treated before they can be discharged to sewers, rivers or coastal sea, because of the harm that the solutions can cause to drains, sewers, sewage works, and the flora and fauna of natural waters. The protection of these items at risk is, of course, the business of various authorities such as river boards and local authorities, and it is necessary to comply with the conditions they may impose on those who discharge effluents into waters under their control.

In most cases it is only necessary to neutralise the effluent so that its pH value lies between pH 5 and pH 9; when it is discharged via the CW system it is advantageous to run a CW pump, by which means the effluent is greatly diluted as it is discharged.

If the chemical effluents are discharged first to a holding tank or sump, it is necessary to clear the tank of the effluent from each stage of the cleaning programme, before that of a succeeding stage is run into the tank. This is to avoid mixing of the undiluted effluents which, particularly those used in post-commissioning cleaning, can result in uncontrolled reactions which produce dangerously toxic gases.

4.2.14. **Purging of Superheater and Reheater**

This process, which is more succinctly called "steam blowing", has become an established practice to follow passivation for the purpose of physically removing any substances which remain deposited in the superheater and reheater and associated steam pipes to the turbine, after completion of the chemical cleaning operations. These substances are mainly metal fragments and metal oxides which have not been dissolved by the cleaning chemicals, and less dense insoluble matter which may possibly be transferred from the boiler to the superheater in the earlier procedures.

It is necessary to remove this debris to prevent harmful effects when the plant is put into service. These effects are (a) overheating of superheater or reheater tubes due to obstruction to steam flow; (b) gradual transfer of silica from siliceous debris, to the turbine; (c) mechanical damage to the turbine and its control valves by the impingement of debris carried by the steam. Finely divided debris in the steam entering the turbine will at least roughen the surface of the turbine blades, and thus reduce the turbine efficiency.

In order to accomplish the purposes of steam blowing, the procedure aims to pass steam through the superheater and reheater in a manner calculated to carry the debris with it more effectively than would be done by the steam flow in normal operation at maximum load on the unit. This leads to the need for steam blowing with greater kinetic energy than that of normal operation conditions, and the ratio of these two quantities (kinetic energy, blowing/kinetic energy, maximum load) is called the "disturbance factor" or "shifting factor", and should be more than unity. Since kinetic energy is proportional to the mathematical product of mass and the square of its velocity, it is not difficult to deduce, for a pipe of a given bore, the following relationships:

$$\text{disturbance factor} = \frac{D_m \cdot (F_b)^2}{D_b \cdot (F_m)^2} = \frac{V_b \cdot (F_b)^2}{V_m \cdot (F_m)^2}$$

where D_m = steam density at normal operating conditions,
D_b = steam density at steam blowing conditions,
V_m = steam specific volume at normal operating conditions,
V_b = steam specific volume at steam blowing conditions,
F_m = steam flow rate, maximum in normal operation,
F_b = steam flow rate, when steam blowing.

The concept of the disturbance factor cannot be applied to practice as precisely as the formulae may suggest, since the steam temperature and pressure are varying rapidly during the steam blow, and pipes and tube of several different bore sizes are contained in the circuit. It is of interest here to note that at the open end of the temporary pipework installed to discharge the steam finally, the velocity of the steam cannot exceed the velocity of sound in air (about 1100 ft/sec), and this will have a limiting effect on the steam flow in the parts to be cleaned, unless the discharge pipework length and diameter are optimised — an unduly long discharge pipe would have to be of uneconomically large bore. At the time of preparing this lesson the C.E.G.B. is reassessing various requirements for steam blowing and will issue a Generation Design Memorandum giving recommended circuits and parameters for the process.

There is at present some uncertainty about the interpretation of disturbance factor; most probably the steam blow is not sufficiently effective if the factor is less than 1·0, whilst values of 1·5 or more are desirable. It is to be expected that a firmer interpretation will be made with further experience in correlating disturbances factor with target-plate inspection (referred to below) and the effects in service of the turbine.

A typical steam-blowing procedure would be conducted in two stages, as follows. First the superheater and main steam pipes are blown through to the turbine main steam chest,

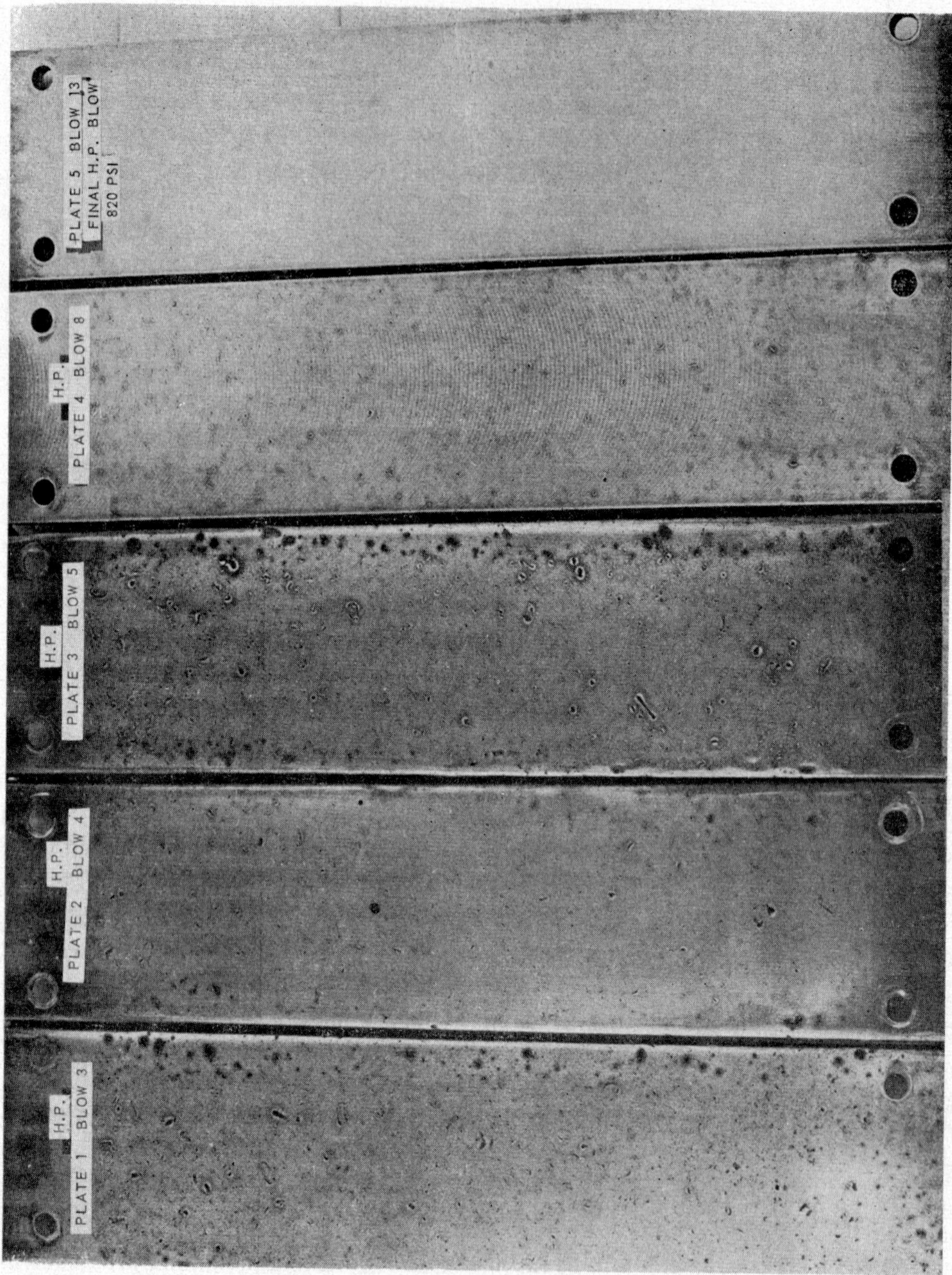

FIG. 4.2.14A. Steel target plates after steam blowing

to which is fitted temporary discharge pipework to atmosphere. Following this the circuit is rearranged, replacing the discharge connection at the steam chest with a temporary connection to the cold reheat pipes, and discharging to atmosphere from the turbine reheat intercept steam chest. Thus in the second stage, steam flows in turn through the superheater main steam pipes, cold reheat pipes, reheater and hot reheat pipes. A similar circuit is shown in Figure A.6 of the Appendix to this chapter.

The boiler is filled to normal working level with demineralised water containing only ammonia and hydrazine, maintained as closely as possible during the procedure, at 50 ppm and 25 ppm respectively. Pressure is raised to a value between 600 and 1000 lb/in², depending on the design of the plant, and the boiler stop valves (replaced, in recent practice, by temporary quick-acting valves) are opened as rapidly as possible. The pressure is allowed to fall to between 200 and 300 lb/in², when the stop valves are closed, and the period of pressure reduction and steam flow may range from 3 to 8 min.

During the release of pressure, the boiler is subjected to quite severe thermal shock and pressure change, and these must be controlled to avoid undue stresses developing, especially in the drum. For this purpose the fall in saturated steam temperature should be limited, for example in one particular boiler design, to 42°C.

This procedure of pressure raising and steam blowing is repeated numerous times for each of the two circuits, and the effect of each "blow" is observed on "target-plates" of polished mild steel or aluminium, which are securely fixed to intercept the discharged steam at short range from the point of discharge. A new plate is used for each blow, and the sequence of blows is terminated when the damage (craters and other indentations) produced on the target plate by solids in the steam, is judged by experienced observers to be minimal. Figure 4.2.14A illustrates typical effects on steel target-plates, and Figure 4.2.14B shows a closeup of an aluminium plate.

The total number of steam blowing excursions needed varies widely, being as few as 30 or as many as 200, but if the design of the procedure is successful the number should not much exceed 60.

The following summary of an actual steam blowing procedure is extracted from a commissioning report for a 500 MW unit.

A total of 34 steam blows were made. Thirteen were made through the superheater and main steam pipes via the boiler stop valves and main steam chest to atmosphere, and 21 further blows through the superheater and main steam pipes via the boiler stop valves and main steam chest and by temporary pipe-work to the cold reheat pipes and thence through the reheater, hot reheat pipework and reheat intercept steam chest to atmosphere.

In order to carry out the steam blowing safely and effectively the following conditions were specified for the boiler during the process:

(1) The drum level to be maintained during pressure raising periods.
(2) Priming to be avoided if possible.
(3) Maximum permissible change in saturation temperature to be 42°C.
(4) Time taken to open main boiler stop valves to be 70 sec.
(5) Maximum boiler pressure—900 lb/in².
(6) Maximum boiler temperature—427°C.

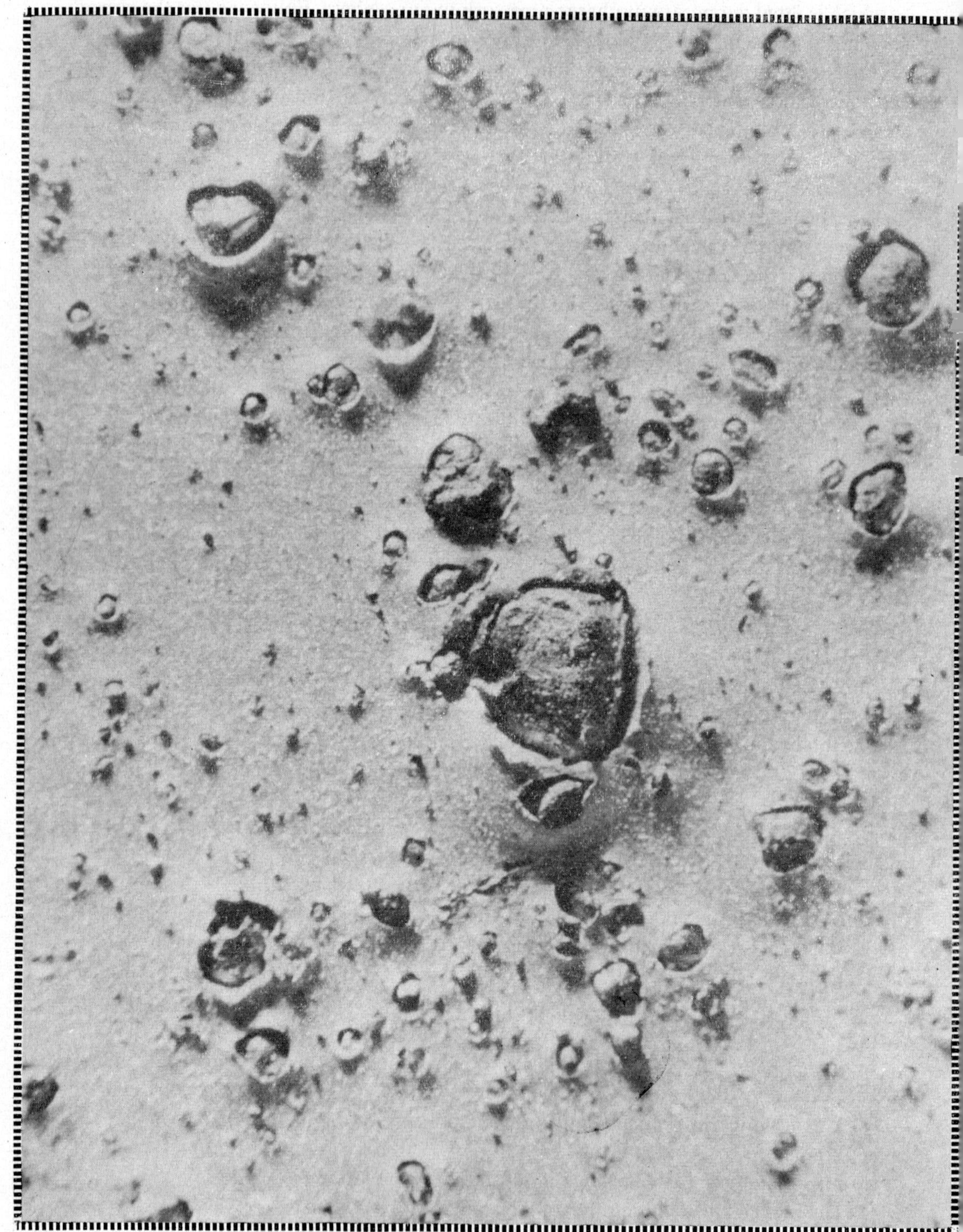

FIG. 4.2.14B. Aluminium target plate after steam blowing

The following calculations of disturbance factor were obtained during typical steam blows.

MAIN STEAM BLOW

(1) *At steam drum*

Design pressure $= 2650$ lb/in² ($V_m = 0{\cdot}117$)

Peak flow pressure during steam blow $= 770$ lb/in² ($V_b = 0{\cdot}581$)

Boiler MCR flow $= 3{\cdot}45 \times 10^6$ lb/hr $= F_m$

Peak flow during steam blow $= 1{\cdot}87 \times 10^6$ lb/hr $= F_b$

$$\text{disturbance factor} = \frac{V_b \cdot (F_b)^2}{V_m \cdot (F_m)^2} = \frac{0{\cdot}581 \times 1{\cdot}87^2}{0{\cdot}117 \times 3{\cdot}45^2} = 1{\cdot}45.$$

(2) At the flow nozzles situated in the main steam pipes a disturbance factor of 1·40 was similarly calculated.

REHEATER STEAM BLOW

(1) *Reheater inlet*

Design pressure $= 624$ lb/in² } $V_m = 0{\cdot}988$
Design temperature $= 365$°C }

Design flow $= 2{\cdot}52 \times 10^6$ lb/h $= F_m$

Peak flow pressure during steam blow $= 290$ lb/in² } $V_b = 2{\cdot}010$
Temperature during steam blow $= 320$°C }

Peak flow during steam blow $= 1{\cdot}89 \times 10^6$ lb/h $= F_b$

$$\text{disturbance factor} = \frac{2{\cdot}01 \times 1{\cdot}89^2}{0{\cdot}988 \times 2{\cdot}52^2} = 1{\cdot}15$$

(2) At the reheater outlet a disturbance factor of 0·95 was calculated.

4.2.15. **Safety**

Nearly all chemicals are harmful to people if ingested by them, and therefore require some care in handling. Particular care is needed in chemical cleaning operations, where some hundreds of tons of chemicals are used which, separately or together, may be caustic, toxic, corrosive, and/or potentially explosive.

The principal acids and alkalis used, namely hydrochloric acid, citric acid, caustic soda, ammonia and hydrazine, are supplied as strong aqueous solutions, and of these the first four are usually needed in sufficient quantities to warrant bulk delivery in tankers. Other reagents are used in comparatively small quantities and include trisodium phosphate, ammonium bifluoride, sodium nitrite and sodium bromate, which are obtained in crystalline form.

The chemicals should be securely stored away from sources of heat, and a water supply is necessary for hosing spillages. "Incompatible" chemicals, that is those which if mixed together would result in a hazardous reaction, should be stored separately. Acids and alkalis are in this category, as are bromates, nitrites and acids. In the dry state (or nearly dry) bromates and strong acids constitute an explosively powerful oxidising mixture and, for example, it is quite possible for a spilled solution containing citric acid and sodium bromate to dry out and be detonated by an accidental impact.

The normal handling precautions, such as the use of protective clothing and masks, the posting of warning notices and a good standard of personal hygiene, will obviate most risks. Nevertheless, it is also essential that temporary pipework to carry the chemicals, shall be pressure-tested beforehand, and if there are any flexible connections in the circuit, these must have isolating valves so that leaks or bursts in the flexibles can be dealt with immediately.

Perhaps the most insidious hazard of the acid-cleaning process is that associated with some inhibitors, Armohibs 25 and 28 in particular. At their recommended dilutions these inhibitors are virtually harmless, but there have been cases of localised over-concentration where as a result the inhibitor was probably present in deposits within the boiler. It is also possible that the concentrated inhibitor may be changed chemically into a more toxic substance during the cleaning process. Harmful effects have been occasionally experienced by some personnel who entered boiler drums immediately after the acid-cleaning phase; these effects may be delayed and consist mainly of strong irritation of the eyes and some loss of vision. In the cases which have been reported, no permanent harm has accrued.

This trouble can be prevented by ensuring that the correct amount of inhibitor is used and uniformly dispersed throughout the diluted acid, by the use of suitable protective clothing including an external air supply and hood for personnel directly at risk, and by handling procedures which will prevent the general spread of contamination by residues which are taken from the boiler.

4.2.16. **Works Cleaning and Protection of Plant**

The student will have observed from the matters already discussed that the chemical cleaning of plant after erection, has been developed to the point where practically all parts of the steam and water system (except the turbine itself) have been subjected to the process. Complete system cleaning is now a particularly onerous process and it has become necessary to reassess the economics and effectiveness of the work, especially when applied to 500 MW and 660 MW units.

The basic reasons for cleaning, as already given, are unchanged but it is apparent that acid-cleaning the feed and bled-steam systems is beset with problems, some of which have been mentioned. These problems are principally those associated with:

1. Weight-supporting in condensers, l.p. heaters, l.p. steam pipes.
2. Ineffective passivation.

3. Ineffective flushing.
4. Presence of austenitic and non-ferrous materials.
5. Limitations on temperature and pressure.

These difficulties have led the C.E.G.B. to consider to what extent they can be overcome by cleaning the components at works, obviating the need for any "drastic" treatment using acids, after erection on site. At present this policy is not applicable to boiler plant mainly because of the large number of site welds required and the attendant opportunities for the ingress of debris of all kinds. The policy is feasible for application to the feed system and is being increasingly applied to modern stations, Tilbury 'B' being the first to adopt it.

The methods of cleaning at works are to a specification approved by the Board and basically consist of acid-cleaning or shot- or grit-blasting. Immediately after cleaning, it is essential to apply some form of protection against rusting, pitting and contamination which would otherwise occur during the long periods of storage and erection. As will be seen, several kinds of protection may be used, but all must be removed just before the plant is commissioned, and are hence referred to as "temporary protection".

A further requirement which is essential for success after works cleaning, is that erection should be carried out so as to prevent the ingress of dirt or appreciable damage to the temporary protection. This can be done by adopting "clean conditions" during erection, and is discussed in some detail in the paragraphs to follow.

The available methods for providing temporary protection of plant can be divided into two main groups:

(a) protective coatings such as lacquers, paints and oil-based compounds,
(b) effective sealing by means of suitable air-tight closures, often supplemented by internal treatment with vapour-phase inhibitors or the replacement of enclosed air by nitrogen.

As the responsibility for various sections of the plant may lie with various contractors and sub-contractors, it is probable that a combination of some or all of these methods will be employed in practice.

4.2.16.1. Protective Coatings

(i) Temporary Coatings

There is a wide variety of materials available, ranging from products which give protection only in mildly adverse environments, to those which are suitable for the more arduous conditions of exposure, such as may be provided by our weather at all seasons. Apart from the degree of protection offered by any coating, an important property with regard to temporary protectives is ease of removal, and the nature of the processes required to effect removal. As one might anticipate, the coatings which give the highest degree of protection are also the most difficult to remove.

The following summary gives brief details of some of the coatings which have either been used or considered for temporary protection of the internal surfaces of the condensate and feed system.

"Serlac". Serlac was formulated by C.E.G.B. (S.E. Region) and has been used to coat the inside surfaces of feed pipework. It is a fluorescent lacquer based on maleinised oil, thinned with water and containing corrosion inhibitors. It can be applied by brush or spray and dries fairly quickly. Serlac coatings can be removed by cold dilute alkali, for example, 0·04% caustic soda, 0·15% trisodium phosphate or 2·3% sodium carbonate.

"Trilac 27". This is a polybutyral resin dissolved in trichlorethylene, which can be applied by spraying or dipping. This coating has been successfully applied for the protection of nuclear plant during erection. It requires special ventilating facilities for its application, because the solvent vapour has anaesthetic effects when inhaled in appreciable amounts. The coating can be removed by alkalis but a moderate temperature is required. A higher alkali strength than that for Serlac removal is also needed, and 0·5% caustic soda at 60°C is recommended.

"Tectyl" products. A range of these products is available, including Tectyl 511/M and Tectyl 506. These materials can be applied by dipping, spraying or brushing. Tectyl 511/M when applied leaves a thin oily film which has the property of displacing water, whilst Tectyl 506 gives a firm brown waxy film with a higher degree of protection, and it has been used to protect turbine blades and other machined parts.

The Tectyl products are best removed by petroleum solvents. It is also claimed that they can be removed by alkali washes and in the case of Tectyl 511/M, by lubricating oil and hot water, but there is doubt about these claims. In any case their use in power plant is undesirable if removal from one part of the plant by a process which does not completely dissolve or emulsify them, could lead to re-deposition in other sections of the plant. The oily nature of the products is a further deterrent to their use for the purposes we are considering here.

"Ensis" products. These products, which contain lanolin, have dewatering properties and provide a good superficial barrier to the atmosphere. The coating is said to be removable by warm alkaline solutions, though difficulties have sometimes been reported and have been attributed to ageing of the coating, leading to a hard film.

(ii) Semi-permanent Coatings

The protection of pre-cleaned modern condenser shells has received special consideration because their design precludes filling with the solutions required to remove temporary protective coatings. Mainly for this reason, "semi-permanent" coatings have been applied to some condensers, the intention being that no direct removal of the coating would be attempted and that it would be gradually dissolved or eroded away during some months or years of operational service. An important requirement is that the coating shall not fail in service by becoming detached in sufficiently coherent form as to cause any undesirable effect in passing through the system.

Two semi-permanent coatings which have been used are as follows.

Krustavit P6. This is a graphited bitumen. It has been found necessary to appply this in a fairly thick film for it to be effective, which is an undesirable feature bearing in mind that it will eventually be transported to the boiler.

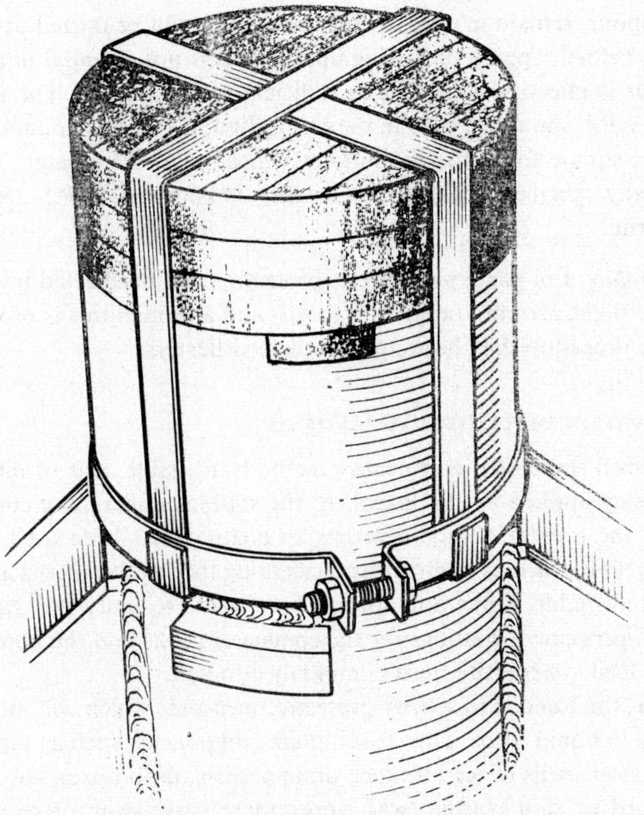

FIG. 4.2.16.2. End sealing of large bore unflanged pipe

C.E.G.B. 7. This product was initially developed as a rapid-drying primer for the protection of structural steel after shot-blasting, and it consists of resins and zinc chromate in a solvent base. Although relatively untried for the protection of condenser surfaces, it is now gaining acceptance for this particular application.

4.2.16.2. Closure of Cleaned Components: Vapour-phase Inhibitors

Certain plant items lend themselves to a procedure of thorough drying following cleaning, assembly and testing, after which they are hermetically sealed. This is carried out before the item leaves the manufacturers' works. Figure 4.2.16.2 shows the sealing of a large bore unflanged pipe. Various methods of sealing are used including the use of blank flanges, tight-fitting plastic closures, and taping. The obvious difficulties of these methods arise when the seals are broken in transit or during the site erection process. Further measures then have to be taken to ensure that the plant is maintained in a dry and clean condition.

Vapour-phase inhibitors (V.P.I.). In addition to the use of sealing closures, vapour-phase inhibitors have been used for additional protection. Vapour-phase inhibitors are powdered chemicals which slowly volatilise at ambient temperature. They do not react with moisture or oxygen but exert a corrosion-inhibiting effect on the metal surfaces with which they or their vapour, remain in contact. Ideally they should be dusted over all the surfaces to be protected before capping or boxing up, but this is not essential in an enclosed space since the vapour is effective if present in sufficient concentration. The manufacturers recommend that V.P.I. should be used in the proportion of 1 gram per cubic foot of enclosed air space, or per square foot of metal surface, whichever is the greater. The vapour-phase inhibitor normally specified in C.E.G.B. contracts is known as V.P.I. 260, which is cyclohexylamine nitrite.

Inert gas sealing. For some plant items, protection can be effected by filling with nitrogen gas under a slight pressure (approx. 5 lb/in^2) and although this is of somewhat limited application, the procedure has been applied to feed heaters.

4.2.16.3. Application of Protective Methods

Having described some of the protective methods available, it is of interest to consider how they are being applied by the Board. In the stations undergoing construction during the first half of the present decade a variety of coatings has been used, representing the different views held by various contractors concerning the relative merits and disadvantages of the products available. Some attempt has been made to unify this situation more recently, and semi-permanent coatings for the condenser shells and the application of Serlac or Trilac to the feed systems are most commonly adopted.

For the future, the trend is to specify protective measures which will minimise the use of coatings applied in liquid form. Thus feed system components such as pipework and valve bodies, and the steel shells of feed heaters, drain coolers, flash boxes, etc., will be required to be acid cleaned or shot blasted (with appropriate passivation or recovery of shot) at works. After hydraulic testing if necessary, the components will be dried and insufflated with V.P.I. 260 powder, following which they will be made pressure-tight with metal closures. The deaerator storage vessel will, of necessity, need to be treated in a similar manner only after erection on site, and even more stringent protective measures depending mainly on corrosion inhibitors and the exclusion of moisture and dirt, will be specified for mild steel feed heater tubes, which may have wall thickness no more than 0·010 in.

The successful application of these future procedures should obviate the need for cleaning the feed system after building it on site, except for flushing it with demineralised water. As yet, however, there is no practical experience on which to assess the effectiveness of the procedures, and the difficulties would seem to be mostly those associated with preventing the ingress of moisture and debris, during erection.

As was stated earlier, these developments directed towards the elimination of chemical cleaning of plant after erection on site cannot yet be applied to boiler components. Nevertheless, increasing attention is being given to these components which, in the past, have suffered mainly from rusting and pitting due to unsatisfactory tube end-sealing and unprotected site storage conditions.

To obviate some of these difficulties, future specifications for boiler tubes will require stringent protective measures to be applied in both the tube manufacturers' and boiler makers' works and also under site storage conditions. Thus, after hydraulic testing at the tube mill, ferritic tubes should be dried, insufflated with V.P.I. 260 and the ends capped, and the external surfaces should be coated with an etch-primer or other suitable paint. Austenitic tubes are required to be dried and capped but not otherwise protected. Similar precautions are required at the boiler makers' works, together with frequent inspections including internal inspection by an introscope, to ensure a satisfactory condition of the tubes.

On arrival at site all tubes should be inspected and any tubes with end caps missing should be returned to the works for retreatment. The tubes should be stored out of contact with the ground and under cover as far as possible. Periodic inspection of external coating should be carried out and where necessary it should be renewed. Internal inspection of the tubes should be carried out on a "sample basis" using an introscope and if any deterioration is observed, the whole batch should be examined. Where necessary tubes should be recleaned, re-inhibited and resealed, the aim being to keep the cap in position until it is necessarily removed for erection.

4.2.17. **Site Erection under Clean Conditions**

The previous section has described some of the steps which can be taken to avoid the necessity for extensive chemical cleaning on site. However, to be successful these measures need to be accompanied by application of "clean conditions" assembly at site. Whilst it has proved possible to obtain satisfactory works cleaning and protection of plant items during the storage period, it is much more difficult to achieve the erection of the plant without nullifying, to a large extent, the advantages thus obtained.

There are a number of reasons why clean conditions are not readily obtained during erection. Thus, if a contractor knows that the plant is going to receive a final chemical clean, he tends to relax his efforts to minimise the entry of dirt, etc. Furthermore, the precautions necessary to prevent the ingress of contaminants are time-consuming and tend to conflict with the erector's immediate interests; in some instances conditions less stringent than those required by the C.E.G.B. may have been adopted to enable the construction programme to be maintained. The human element also plays a major part in the success or otherwise of obtaining clean conditions; no matter what policy is determined, much reliance has to be placed on the man actually carrying out the erection.

In future contracts, clean conditions assembly will be specified in much more detail than hitherto. To attain the required degree of cleanliness it appears that several steps will need consideration, including:

(a) The education of the erection staff concerning the necessity for cleanliness.
(b) The provision by the Board of sufficient supervising staff to obtain adequate inspection and monitoring of erection procedures.
(c) Careful arrangement at the planning stage, of the site construction programme so as to avoid some of the more obvious sources of contamination. Thus considerable benefit would accrue if the main buildings were completed as far as possible before plant erection commenced: similar advantages would result from arranging to have civil work such as flooring completed at an early stage. Particular attention is required to prevent the possibility of dust getting into the steam/water circuits when lagging is being carried out, since this is a principal source of contamination by silica.

As an example of ways to achieve "clean" construction, individual feed heater components can be effectively sealed whilst in storage, such sealing being effective up to the time the item is integrated into the feed system. Thereafter protection is more difficult and the current trend is to maintain a non-corrosive atmosphere by circulating dry air or by displacing the air with nitrogen and closing the system progressively. An additional complication arises when the plant is subjected to hydraulic testing, since a choice must then be made between trying to dry the plant by warm air circulation, and storage in chemically treated water.

The types of contamination which clean conditions assembly aims to prevent, can be divided into three categories:

(i) Corrosion products arising from surface rusting.
(ii) Fine particulate matter entering the system in the form of dust, such as P.F. dust, lagging material, sand and cement.
(iii) Gross contaminants, usually material left in the plant during the erection process, due to carelessness and less charitable causes.

Concerning contamination by dust, the steps taken to prevent its entry into plant vary considerably from one section of the plant to another. Justifiably, most attention is given to preventing the entry of dust into the turbine, its control and lubricating systems, and electrical components. To maintain cleanliness of these parts during erection, a "clean conditions area" is usually established around the machine. The area is fenced off and access is limited to those personnel required for erection; it is kept as clean as possible by vacuum-cleaning plant. Personnel working in the area are supplied with protective clothing, including regularly-laundered overalls and over-shoes.

The presence of gross contamination is all too common and has been the cause of considerable outage of plant during the early days of operation. In a 2350 lb/in^2 boiler, frequent tube failures during the commissioning period were found to be due to a bundle of welding rods which had been pushed into a tube bore thereby preventing water flow. A similar experience occurred on a 350 MW unit of which the boiler suffered 41 tube failures during the first three months of service, due to sand used for tube-bending having been left in some of the tubes. Depending on the nature of the contamination and its location, some of the gross debris may be removed during the steam-blowing procedure, but it is evident that much trouble would be avoided by better control of the sources of contamination. Progress of this kind is being made, albeit slowly.

4.3. PRE-COMMISSIONING PREPARATION OF TURBINE OIL SYSTEMS

4.3.1. Introduction

More care is now taken in the preparation for service of turbine oil systems than in earlier decades, for a number of reasons including the increasing complexity and duty of the systems, the penalties of failure to lubricate or control the turbine, commissioning delays, and an increased general awareness of the advantages of eliminating all extraneous matter from the plant internals.

The oil or other operating fluid in the turbine control system may be quite separate from that in the lubricating system, which is normally associated with the seal oil and jacking oil systems, and it may be convenient to treat each of the four systems separately when preparing the assembled plant for service.

The total preparation for service consists mainly of cleaning the components at works followed by suitable protective measures, erection on site under clean conditions, and thoroughly flushing the systems with oil before commissioning.

4.3.2. **Purposes**

As will be readily appreciated, the first requirement is to finally achieve an absence of all forms of debris in the oil system, since harm may otherwise be caused by restriction of oil flow, by scoring or abrading bearing surfaces, by interfering with the control mechanisms, and by damaging hydrogen seals.

Some rotating parts have journals made of 3% chromium steel and this alloy is particularly susceptible to fine debris, suffering extensive damage by scoring.

A further important reason to exclude debris is that, when dispersed in the oil charge, it may adversely affect the oil stability since some metals or metal oxides present in the debris have a catalytic effect and accelerate oxidation of the oil. Finely divided debris suspended in the oil may also lead to foaming and consequential problems in service.

4.3.3. **Works Cleaning and Protection**

Present-day practice is to clean at works practically the entire oil system. Pipework is chemically cleaned and large components such as oil tanks and oil-cooler shells, are shot-blasted.

Temporary or permanent coatings are applied to the cleaned surfaces, the temporary coatings being formulated for ease of removal by the eventual circulation of flushing oil. In past years the paints used for permanent coating often proved to be unsuitable, being softened and removed by the oil in service. Suitable materials are now available, such as the polyester resin coatings, and these may be applied to oil tanks, bearing pedestals and oil wells, and oil return pipes.

The oil return pipes are particularly liable to corrode in service, since there are extensive lengths installed at a small angle to the horizontal, which may normally run half-full of oil with water vapour in the upper half. In some modern stations this situation is resolved by fabricating these pipes in stainless steel.

The components are, of course, suitably sealed for transport and storage before erection which is carried out as a "clean conditions" procedure, the principles of which have been described already.

4.3.4. **Flushing**

Even when all of the preparatory precautions have been taken, the newly erected oil system is likely to contain some dust, grease and small debris, and if temporary coatings have been used, they have to be removed. Final cleaning is therefore effected by circulating a flushing oil charge, the grade of which is either equal to the operating charge, or of slightly lower viscosity.

The usual procedure is to temporarily divide the system into a number of small circuits, through each of which the oil is circulated at higher velocity than in normal operation, for periods of time up to 24 h and at temperatures between 80° and 95°C. The oil may then be passed through a purifier to a reserve tank, returned to the main tank after debris or sediment is removed from same, and circulated through each circuit again for 6 to 9 h.

Certain parts of the system which might be damaged by the transfer of debris during flushing, such as some components of the control gear, bearings and journals, are isolated from the flushing circuits and separately cleaned by hand.

4.4. PLANT CLEANING DURING SERVICE LIFE

Most of the remainder of this lesson is concerned with maintaining clean surfaces in the steam/water circuits and the turbine oil system. The cleaning processes thereby required, referred to as "post-commissioning cleaning" or "post-operation cleaning", are often different from the processes used prior to commissioning new plant, mainly because the chemical nature of the material to be cleaned off is usually different from that in new plant.

As stated in the introduction to the lesson, heat-exchange surfaces must be kept clean to maintain maximum efficiency in operation and in boiler tubes there is the vital additional purpose of avoiding the corrosion hazard associated with deposited matter.

The technology of cleaning the gas-side surfaces of boilers and auxiliary parts will not be discussed here since the subject arises in Chapter 1 and other volumes.

4.4.1. **Boilers and Economisers**

4.4.1.1. Incidence and Nature of Fouling

If a modern boiler needs cleaning internally, it is usually because "on-load" corrosion has occurred or is feared likely to develop because of deposits, particularly in the steam-generating areas. The technical aspects of this relationship have been discussed previously (Section 4.2.1.1), and more fully in Chapter 2. Superheaters and reheaters are not susceptible to on-load corrosion and they rarely need cleaning for any other reason.

The economisers of modern boilers should seldom need cleaning; if they do, or in older plant, the need arises normally because of an accumulation of rust nodules which are the products of corrosion caused mainly by excessive amounts of oxygen dissolved in the feed water. The remaining discussion is therefore concerned only with cleaning the boiler in relation to boiler corrosion. Where acid cleaning is resorted to the economiser is often included, partly because it may be inconvenient to exclude it.

In the boiler, after periods of service varying from about 5000 to 40,000 h, there may be considerable accumulation of deposited matter and/or boiler corrosion products in the generator tubes. The deposited matter usually consists largely of magnetite and ferric oxide with widely varying proportions of copper oxide and metallic copper, and small amounts of nickel and zinc compounds, all depending on the materials of construction of the feed system, from which much of these deposits originate. They may also contain calcium and magnesium compounds, including calcium phosphate where phosphate treatment is practised, resulting from condenser leakage.

The products of on-load corrosion may take the form of scabs or crusts, composed mainly of magnetite, which tend to adhere to the corroding area producing them but are readily detached, and may also take the form of very hard, dense plugs of magnetite which are usually firmly bonded to the corroding area. These corrosion products may also contain small proportions of the constituents of the deposited matter.

When the welding of steel parts is to be carried out, for example for the replacement of corroded tubes, there is an additional reason to remove copper deposits before welding. This is because copper is likely to penetrate the heat-affected grain boundaries and the resultant weld is prone to failure by cracking.

4.4.1.2. Deciding when to Clean the Boiler

There are no generally accepted rules which will always lead to a clear answer to this question nor, if it is accepted that the regular acid cleaning of boilers is necessary, is there common agreement regarding the optimum frequency of such cleaning.

Within the C.E.G.B., cleaning as a routine is not generally practised and most commonly a boiler or group of boilers in a station will be the subject of individual consideration each time the question of cleaning arises. The factors to be considered in reaching a decision are as follows:

(a) Plant history and the results of regular inspection: this would include the record of tube failures due to corrosion, and the occurrence of metal oxide debris distributed within the boiler—sometimes observed to be increasing in quantity as the load factor diminishes.

(b) Results of non-destructive tests and perhaps tube camera surveys, applied to the boiler tubes.

(c) Chemical aspects, for example, the appearance of a marked degree of "hide-out" of boiler water salts would support the view that cleaning was necessary, as would an operating history of persistent condenser leakage.

(d) Other aspects, such as boiler design, current developments in cleaning technology, and the nature and amount (so far as may be known) of debris to be removed.

Of course, chemical cleaning will do nothing towards the repair of wasted or embrittled metal, and where the wastage is severe or embrittlement (hydrogen damage) is known to be present, it is necessary to cut out and renew the affected tubes, before chemical cleaning is effected. Present-day procedures have limited efficacy in dissolving "operational" debris of the kinds described; they will remove all protective films of magnetite and most of the dispersed metal oxides deposited within the boiler, but they are not very effective for the removal of massive "plugs" of magnetite, which must be dealt with by cutting out the affected section of tube.

Many American power companies acid-clean boilers as part of routine maintenance, arguing that the procedure can be integrated into the maintenance programme and carried out with speed and efficiency, and that as a result freedom from corrosion caused by deposits, can be assured. One group of the French electricity supply industry acid-cleans boilers at intervals so calculated that the integrated quantities of iron, copper and nickel fed to a boiler between cleaning operations do not exceed a specified weight per unit area of water tube surface. Such policies have some proponents in Great Britain but are not generally adopted here, for various reasons which include a more conservative outlook, objection on the grounds of cost, time taken, and/or ineffectiveness of the procedure, and the possibility that the procedure will remove a well-developed protective oxide film, only to replace it with one which may be inferior.

4.4.1.3. Methods of Cleaning

The main differences between pre- and post-commissioning cleaning arise because:

(a) There is no need to have a preliminary boil-out to remove grease and oil, unless it is known that some mischance has introduced quantities of oil into the unit.
(b) There is a possibility that some parts of the boiler tubes have suffered hydrogen damage due to corrosion in service. If these parts are cleaned with hydrochloric acid, then the surface will be attacked and the underlying area made slightly porous, although this effect does not appear to arise if citric acid is used. This phenomenon is called "subsurface cavitation", and it can in theory provide a site where boiler chemicals will concentrate when the plant is in service, thus causing hideout and, probably, corrosion.
(c) The debris to be removed will usually contain copper and possibly nickel and zinc.
(d) The drum furniture will normally be left in place during the clean.

Post-commissioning cleaning otherwise is very similar to pre-commissioning cleaning and the precautions necessary are the same.

Because hydrogen damage cannot be detected non-destructively, cleaning is usually done using an organic acid, and citric acid is the most widely used. A preliminary stage under oxidising conditions to remove non-ferrous metals (principally copper), is followed by a stage designed to remove oxides of iron, and this is followed by a second process for the removal of copper. Finally there is a passivation stage.

The non-ferrous metal removal process is usually based on the oxidation of nickel and copper to their oxides, which are then dissolved in the form of stable complexes. As oxidising agents it is customary to use sodium bromate or sodium nitrite or, less commonly, ammonium persulphate or hydrogen peroxide. The complexing agent is practically always ammonium citrate which is formed *in situ* by reaction between ammonia and citric acid. The conditions for the efficient operation of the process are quite critical, the temperature and pH in particular needing careful adjustment to 50°C and pH 10, respectively.

The iron removal stage which follows, uses citric acid with sufficient ammonia to give a pH 3·5–4·0. As was shown earlier, this reagent is rendered much more efficient by the addition of sodium fluoride or ammonium bifluoride, and the addition of one of these is advisable unless austenitic steel components are involved.

Following the iron removal stage, the final cleaning stage is a process designed to remove any residual copper which may have been protected from the reagent used in the first stage, by magnetite. Depending upon the amount of copper which is judged to be present, this can be done either by repeating the first stage with fresh reagents, or by adding an oxidant and suitable pH adjustment to the reagent used in the iron removal stage.

The reagents and procedure for the post-commissioning cleaning of boilers are summarised in Generation Operation Memorandum No. 75.

The requirements for the proper disposal of waste effluents and for safety, as discussed in Sections 4.2.13, and 4.2.15, apply equally to any cleaning operations conducted after commissioning. In particular, the hazard of toxicity associated with inhibitors demands the same precautions and, in this connection, it is probably unnecessary to use an inhibitor during the copper removal process, since it is carried out at a relatively high pH value.

4.4.2. **Turbine Oil Systems**

Using modern additive-containing turbine oils and with discriminate use of the installed oil purifying equipment, turbine oil systems seldom require cleaning during service life. Such cleaning occasionally becomes necessary if the oil has produced much sludge, or if the system has become heavily contaminated with circulating water (particularly salt water), or, very rarely, if substantial amounts of rust develop in the system.

Oil sludge is a resinous substance formed by the polymerisation of oxidation products, and its formation is normally prevented by correct management of the oil in service. The sludge is not readily soluble in most solvents, nor easily dispersed by detergents, so where sludge deposits occur it is usually necessary to dismantle the system and clean manually. Solvents such as trichlorethylene are sometimes used successfully to remove mixtures of sludge and other residues which may have accumulated in oil coolers; this process is similar to that described later in relation to feed heaters.

It is common experience that some water finds its way into turbine oil in service; it usually separates from the oil satisfactorily and is removed at drain points or in the purifying equipment. Excessive accumulation of water may lead to the production of fairly stable emulsions from which separation is slow, and a particular hazard with mixtures of oil and sea water is that where there are fine clearances, corrosion may occur and restrict oil flow or cause seizure of moving parts. Extensive contamination with cooling water may necessitate cleaning the oil system after the cause of the trouble has been eliminated. The cleaning is best done by emptying the system, flushing with warm condensate or demineralised water, and drying by blowing through with warm, clean air. The water should not be passed through bearings or hydraulic mechanisms, and these must be dismantled and cleaned by hand. It is preferable not to use alkalis or detergents in this cleaning process, because they tend to encourage emulsion formation and consequential difficulties.

The development of extensive rusting in an oil system is symptomatic of a persistent excess of water, perhaps with depletion of the rust inhibitor in the oil, and/or a long period of idleness of the turbine affected. The condition is sufficiently rare to warrant only an outline of the cleaning procedure required, which is to drain the system, isolate bearings, control gear and other vulnerable components, degrease with solvents, acid-clean, flush, and dry. In one example where such cleaning was necessary, it was found to be simplest to dismantle the pipework for acid cleaning, and to clean the remainder of the system manually.

4.4.3. Feed Heaters

When feed heaters become fouled it is normally detected by their consequent loss of performance, and the cause is usually contamination of the feed water by oil; occasionally the steam side of the heater is also found to be affected. The fouling material may thus be largely oil in low-temperature heaters, or particularly in h.p. heaters, it may be oil degraded by the temperature and possibly bonded with small amounts of finely divided oxides of iron and perhaps copper. Less commonly, the loss of performance is caused only by adherent deposits of metal oxides, but the fouling of h.p. heaters associated with oil is usually the most troublesome and difficult to deal with.

There are two principal methods of removing oily deposits and these are treatments with either aqueous detergents or with organic solvents. Where possible, it is best to obtain samples of the deposit and to experiment to determine which is more effective. The circulation of detergents, hot or cold, can usually be done by station maintenance staff but the use of solvents will generally call for contractors.

A wide variety of detergents is available. The commercial ones are usually alkaline and are basically soda ash or sodium silicate to which may be added organic materials. The organic detergents fall into three classes, namely cationic, anionic and non-ionic, and the anionic ones are most frequently used. These materials are large hydrocarbon molecules into which an anionic group, generally a sulphate or sulphonate, is introduced. Another class is based on cresylic acid. All of these materials enable oil-bonded deposits to be broken down to emulsions which can be removed by flushing.

The solvent-based cleaners are commonly trichlorethylene, which dissolves oil. The most effective method of using these is to expose the part to be cleaned to the vapour, which condenses on the metal and enables the oil to be continuously extracted with fresh solvent. The contaminated solvent drains to an electrically heated boiler where the oil is retained and from which the solvent distils, to be condensed again on the exposed parts.

Detergents are generally preferable to solvents because the latter remove the oil but leave a residue of oxides which may be difficult to remove, whereas detergents are capable of removing both oil and deposit in one operation.

4.4.4. Condensers

The cleaning of condensers may be undertaken for one or more of the following reasons:

(a) Steam-side fouling by oil.
(b) Steam-side fouling by iron oxides.
(c) CW-side fouling by scale, deposits, algae or slime.

All of these impede heat transfer across the tubes and a significant decline of heat transfer will be shown by measuring the condenser performance.

It is sometimes difficult to assess visually whether a decline in performance is due to the fouling of the steam-side or the cooling water-side, but this distinction can be readily made using a simple heat transfer testing rig, if it is possible to remove a tube for this purpose. With heavily fouled surfaces there is, of course, little difficulty.

4.4.4.1. Steam-side Cleaning

(a) *Oil fouling*. This is usually removed using trichlorethylene vapour degreasing of which a brief description has been given above.

(b) *Iron oxide fouling*. Some operators have observed deposits of iron oxide on the steam side and have used chemical cleaning processes to remove them; in principle these usually consist of soaking in cold dilute hydrochloric acid with a suitable inhibitor. Care should be taken to prevent vapour from the acid reaching the l.p. turbine blading, for example by means of impervious sheeting fixed below the turbine exhausts. The acid soaking is followed by thorough rinsing which is continued until the conductivity of the effluent has reached a satisfactory value. The process is not widely favoured because of the risk of leaving some acid in the packings of the tube ends, and because it is difficult to satisfactorily inhibit mineral acids from attacking some brasses and cupro-nickel alloys, especially at dissimilar metal junctions.

4.4.4.2. Water-side Cleaning

The methods of preventing or removing fouling caused by algae or slimes are normally those of chlorination and/or "bulleting", but these will not be considered further since they have been discussed in Chapter 3.

With regard to scale and precipitated deposits in condenser tubes these are nearly always confined to stations having recirculating cooling systems, in which concentration of the cooling water occurs. Hard compact scale is generally the product of a slow separation of insoluble matter and tends towards a softer, more granular form of deposit when the environment gives rise to comparatively rapid precipitation. In either case the fouling material usually consists mainly of calcium carbonate together with a variable proportion of magnesium hydroxide and perhaps small amounts of the phosphates of these metals. There is rarely any significant amount of calcium sulphate present.

Such material is readily attacked by cold dilute acids, forming more soluble salts, and it is customary to descale condensers by circulating inhibited hydrochloric acid, about 2% strength initially, until the acid concentration ceases to fall appreciably. Should it fall below 1%, the concentration should be restored to this figure by adding more acid. Recently, citric acid has been used in place of hydrochloric acid in order to reduce corrosive attack on the metals in the system. The process is completed by draining the acid and thoroughly flushing by means of a CW pump.

4.4.5. **Evaporators**

Most evaporators are fed with softened fresh water (see Chapter 3), which contains some bicarbonate hardness, 1 or 2 ppm at least, and in due course scale consisting mainly of calcium carbonate will build up on the evaporating surfaces, causing loss of performance.

There is a small but increasing number of sea water evaporators in coastal and estuarine stations, where the evaporator feed water is usually treated to remove practically all of the alkalinity without appreciably reducing the total hardness. Unless these evaporators are suitably designed and carefully operated, they develop very troublesome scales containing a large proportion of calcium sulphate, which are not appreciably soluble in acids or alkalis.

Since the scale in fresh-water evaporators is normally of a similar nature to that which may be found inside condenser tubes, a similar procedure for cleaning can be used. Inhibited hydrochloric acid is again normally used but because of the smaller size of an evaporator, pumped circulation is usually unnecessary and the technique of static soaking, sometimes agitated by compressed air, is adopted. After flushing the evaporator, it should be recommissioned with the distillate diverted to waste until the distillate quality, as determined by pH value and conductivity, is normal.

It is possible to clean evaporators on load, using the complexing agent EDTA and citric acid, in the following way. Blowdown is shut off and EDTA solution is injected into the evaporator feed water. As reaction with the scale proceeds, the pH value of the liquor rises and has to be controlled within the range pH 7·0–8·5 by injecting citric acid solution. The cleaning process may take several days during which the distillate can be taken into service, provided that carry-over is prevented by blowing down as necessary. The spent EDTA solution, which is expensive, can be regenerated externally by treatment with a mineral acid. Evaporator cleaning with EDTA is rarely practised within the C.E.G.B., mainly because rigorous chemical control must be maintained throughout the process. There is little or no experience of the process as applied to calcium sulphate scale.

The well established technique of "cracking" to remove scale from evaporator tubes which are steam-heated internally, is often of use in such plant, but this design of evaporator is not normally adopted in modern stations. This technique is to drain the water from the evaporator, then to pass steam through the tubes or coils, and then to admit the feed water. The disparity in thermal expansion coefficients and conductivities, between the metal tube and adhering scale, should cause the scale to spall and loosen from the tube surface.

The cleaning of sea water evaporators containing calcium sulphate scale is often a severe problem, and if the scale contains less than about 20% of calcium carbonate and/or magnesium hydroxide, chemical cleaning methods are likely to be ineffective. The "cracking" procedure may be helpful where it can be applied, but in many situations where calcium sulphate scale occurs it has to be cleaned off manually and mechanically—a tedious process which is likely to do considerable damage to the underlying metal. As the reader will have inferred, the prevention of this kind of scale is much more profitable than the effort required to remove it.

4.4.6. **Turbine Blading**

The fouling of turbine blades may occur as a result of inadequate control of impurities in the steam supply; these impurities may be due to the solubility of silica and caustic soda in high-pressure steam (see Chapter 2), or to carry-over of liquid from the boiler drum. In supercritical boilers the solubility in the working fluid of all materials carried by the feed water, is sufficient to cause deposition in the turbine, but supercritical plant is not considered in the following account.

When silica deposits occur on turbine blades in the absence of appreciable amounts of other impurities in the steam, the deposit is usually hard and insoluble in all feasible solvents, and contains more than about 75% of silica as SiO_2. In 500 MW machines and with 0·03 to

FIG. 4.4.6. Deposit on turbine l.p. rotor.

0·08 ppm silica in the steam supply, it is to be expected (theoretically) that silica would be deposited on a few rows within the l.p. cylinders, encroaching into the i.p. cylinder blading at higher concentrations in the steam. Figure 4.4.6 is a photograph of the l.p. rotor of a modern turbine, showing moderate blade deposits consisting of ferric oxide and silica.

Turbine blade fouling caused by boiler water carry-over is spread more widely than silica and tends to deposit within the h.p. and i.p. cylinders as much as in the l.p. counterparts. These deposits and any caused by caustic soda solubilised in the steam, may also contain silica but are generally soluble in water, so their presence in the turbine may be deduced from the transient development of non-volatile alkalinity and conductivity in the turbine condensate, when restarting a machine that has been allowed to cool after a period of base-load running. This effect is caused, of course, by solution of the deposits in the condensed steam formed within the turbine during start-up.

Turbine blade cleaning. Deposits which contain a high proportion of silica are not usually amenable to chemical cleaning processes, and in practice a sample of the deposit would be tested chemically to confirm this. If the deposited silica is small in amount and limited to one or two rows of blades, it may be practicable to remove it with hand tools. Otherwise it will be necessary to lift the affected rotor and subject the blades to grit-blasting—usually done by a specialist contractor who should be aware of the need to cause the least possible harm to the surface finish and contours of the blades.

When turbine blade deposits which are largely water soluble are seen, they are often visibly moist because of the deliquescent nature of some of the constituents. These deposits are normally removed by the process of "blade washing", which is effected when the machine is cold, by turning it on the barring gear whilst passing steam at low pressure and temperature through the turbine. The condensate is frequently tested for conductivity (and analysed if it is required to determine how much material has been removed), and is put to waste until cleaning is judged to be complete.

It will be apparent that a degree of blade-washing is unavoidably applied every time a turbine is started up from cold, for which reason machines which are subject to two-shift operation and/or week-end outage, seldom have blade-fouling problems.

4.5. INSPECTION OF PLANT

4.5.1. Introduction

Inspection of power plant is carried out for several reasons. Thus inspections are carried out by the chemical and mechanical staff in the event of breakdown, and also as a routine for many items which are known to be subject to deterioration either by mechanical damage or chemical processes; inspections are also carried out for statutory and insurance purposes.

It is not intended in this chapter to deal with all these inspection requirements but rather to discuss the principles of those which are normally carried out by the chemical staff, with some reference to the work carried out by metallurgical and non-destructive testing teams.

The examination of *in situ* evidence of the condition of plant is a job for specialists, and in addition to the engineering aspect there is likely to be evidence of significance only to the chemist, the metallurgist, or the biologist. Since the engineer's assessment usually first requires the plant to be cleaned free of any debris, scale or corrosion products that may be found on "opening up", it is essential that the chemist (in particular) should be given prior warning of an impending examination and that he should have the opportunity to examine the plant before any other person. Quite often valuable diagnostic evidence is lost because other personnel, with doubtless good intention, have disturbed or even cleaned up the area of interest.

Since it is not always possible to have plant out of service solely for the purpose of inspection, the opportunity to examine plant is generally taken whenever it is opened up for some other purpose. Inspections may be carried out to determine a course of either corrective action or preventive action. In the case of breakdown of plant, for example failure of a boiler tube, detailed examination of the plant is made to try to determine the cause and to enable corrective action to be implemented; another important purpose is to determine whether an extension of the trouble is imminent and to take appropriate preventive action. The institution of planned, regular inspection programmes together with knowledge of the plant items which are susceptible to the various kinds of damage, can significantly improve the availability of the plant.

Most stations have their idiosyncrasies with regard to the variety and situation of the problems which they must control. These cannot be defined in advance of experience, but some of them are likely to require inspection and other attention more frequently than at planned overhauls. Problems of this kind often include gas-side fouling of superheaters, economisers and air heaters, fouling and/or corrosion of condenser tubes, and scaling of evaporators and their preheaters.

Breakdown of plant, if it is not of entirely mechanical or electrical nature, is likely to require inspection by the chemist by methods selected according to the circumstances, but generally in accordance with the principles outlined in this section.

The special requirements for inspecting new plant before service have been dealt with in the earlier parts of this chapter, and it is convenient to divide the remainder of this section into two parts, the first being concerned with the general principles of inspection of various parts of the plant, and the second with inspection techniques, including instrumental methods.

4.5.2. **Inspection at Planned Outage or Overhaul**

The major overhaul periods provide a suitable opportunity for extensive examination. This should include the following:

(a) Complete examination of all boiler gas passes from the combustion chamber to the chimney. The main aspect of the examination is to determine the extent of any corrosion or excessive fouling which may be present. Attention is paid to the appearance and colour of deposits (which may occur in distinct layers) apart from the extent of the deposition, and samples are removed for chemical analysis. Micrometering of selected tubes in the boilers (particularly superheater and reheater tubes) at each major overhaul period also enables a check to be made on the rate at which corrosion or erosion is occurring. Air heaters and associated ducting, especially in oil-fired boilers, are examined to assess the extent to which wastage by corrosion may have occurred, and to consider what method of cleaning may be needed. On a modern boiler, extensive scaffolding is required in order to reach every part of the boiler which requires examination. This is both time-consuming and costly, and in one station considerable savings have been made by employing cameras with telephoto lenses. By using these through various inspection ports available on the boiler, detailed examination can sometimes be made without entering the plant. A typical photograph is shown in Fig ure 4.5.2A.

(b) The steam-water circuit will be examined as far as possible for evidence of corrosion, such as "radiussed" pieces of metal oxide scale in boiler water headers, and for more general deposition of iron oxides. Since viewing access is limited it may be necessary to employ special techniques to confirm the condition of suspected corrosion areas. These include the multi-shot still camera and closed circuit television cameras which will be described later. The results of such camera surveys provide a useful guide on which to base a decision regarding the necessity for subsequent acid cleaning of the boiler.

Some dismantling of the plant may also be necessary, for example, if marked deterioration of feed heater performance has been encountered, then it may be necessary to lift the heater body to determine the cause and the best method of cleaning, where necessary.

On the CW plant it is necessary to examine the condensers and auxiliary heat exchangers for tube end and tube plate erosion, for mud, corrosion products and other debris in the tubes, and for evidence of graphitic corrosion of water boxes in stations using salt water. This evidence is described later, and is most likely to be found in close proximity to the tube plate. Corrosion within condenser tubes is not amenable to visual detection, and it is sound policy for stations using salt water or polluted water to program for regular (annual or biannual) testing of about 10% of the tubes in each condenser, by means of the "Probolog" or "Introview" type of instrument (see Section 4.5.3.2.1). This should be done until experience shows that the corrosion risk is negligible.

It is also necessary to examine annually the internal surfaces of various tanks, such as those associated with R.F.W., water treatment plant, drains and deaerators, in order to detect and remedy the presence of corrosion or oil, or to maintain protective coatings in sound condition.

(c) In addition to the boiler gas-side and steam/water circuits, examinations may be made of those items of the cooling water system not normally accessible except at periods of major overhaul, for example CW culverts, etc. Items such as auxiliary coolers (turbine oil coolers, distilled water coolers, etc.) should be inspected, though this may have been done at other times, since stand-by plant is normally available. In sea-water-cooled stations it is usually necessary to adopt measures such as chlorination to control marine fouling, particularly by mussels, whilst cathodic protection may be installed to prevent graphitic corrosion. Examination of the cooling water system should serve to show that effective control of these factors is being obtained.

FIG. 4.5.2A. Photograph by telephoto lens of furnace wall tubes

(d) Regarding the turbine itself, the parts of the plant normally requiring inspection by a chemist include turbine blading if accessible, where siliceous and other deposits may have formed, and also sections of turbine oil system, where corrosion or sludge deposition may constitute potential problems.

The aim during an annual overhaul programme should be to inspect as much plant as is possible, having regard to lack of access at other times. Thus the opportunity may be taken to inspect the internal lining of chimneys particularly on oil-fired boilers, where in some stations periodic cleaning by either washing or vacuum cleaning is carried out as a safeguard against smut emission. In such cases internal inspection of chimneys may be carried out by steeplejacks, but in one station the use of closed circuit television has been employed successfully. The television camera was mounted on a rig which was hauled to the top of the chimney and lowered down in stages. "Panning" and "tilting" and also the changing of lenses from wide angle to telephoto were done by remote control. A complete survey of the chimney interior was carried out in approximately three hours, using wide angle and telephoto lenses. Figure 4.5.2B indicates the picture produced by the telephoto lens.

4.5.3. **Inspection Techniques**

4.5.3.1. Visual Examination: Sampling and Examination of Deposits

To the trained eye, much can be learned from a visual examination of the plant and the structure and location of deposits or corrosion products. In some types of corrosion, visual examination can often suggest the cause and mechanism of attack, whilst the colour of deposits and the presence or absence of a laminar structure provide useful information. A photographic record, either in colour or monochrome, is useful for subsequent reference or comparison.

Typical examples of such corrosion diagnosis are the horseshoe shaped pits associated with corrosion/erosion, or impingement attack, which sometimes are found on the cooling water side of condenser tubes using estuarine or sea water for cooling. This is commonly caused by partial obstruction by a piece of wood or mussel shell, and in such cases the closed loops of the "horseshoes" are orientated towards the inlet ends of the tubes. Similarly graphitic corrosion of cast iron components in the cooling water system where sea water is used, can be diagnosed by physical examination. The outer surface of the corroded area consists of nodules of iron oxide corrosion products, which are brown in colour. On removing these nodules, the surface underneath is often found to have retained its original shape but to be black in colour; however, it is soft when wet and is easily cut with a pocket knife.

With some deposits the colour can often be a guide to their chemical constitution, for example the presence of copper metal particles can sometimes be seen in boiler water-side deposits. The use of a magnet will quickly establish the nature of a deposit with a high metallic iron or magnetic iron oxide content. Similarly, visual examination of boiler gas side deposits will give valuable clues as to the possible type of deposit which is present, particularly if the deposits are made up of more than one layer, for example in the case of alkali matrix deposits where there is a white inner layer of alkali metal sulphates, on top of which is found the brown/grey deposits of normal ash content. The publication *Cleaning*

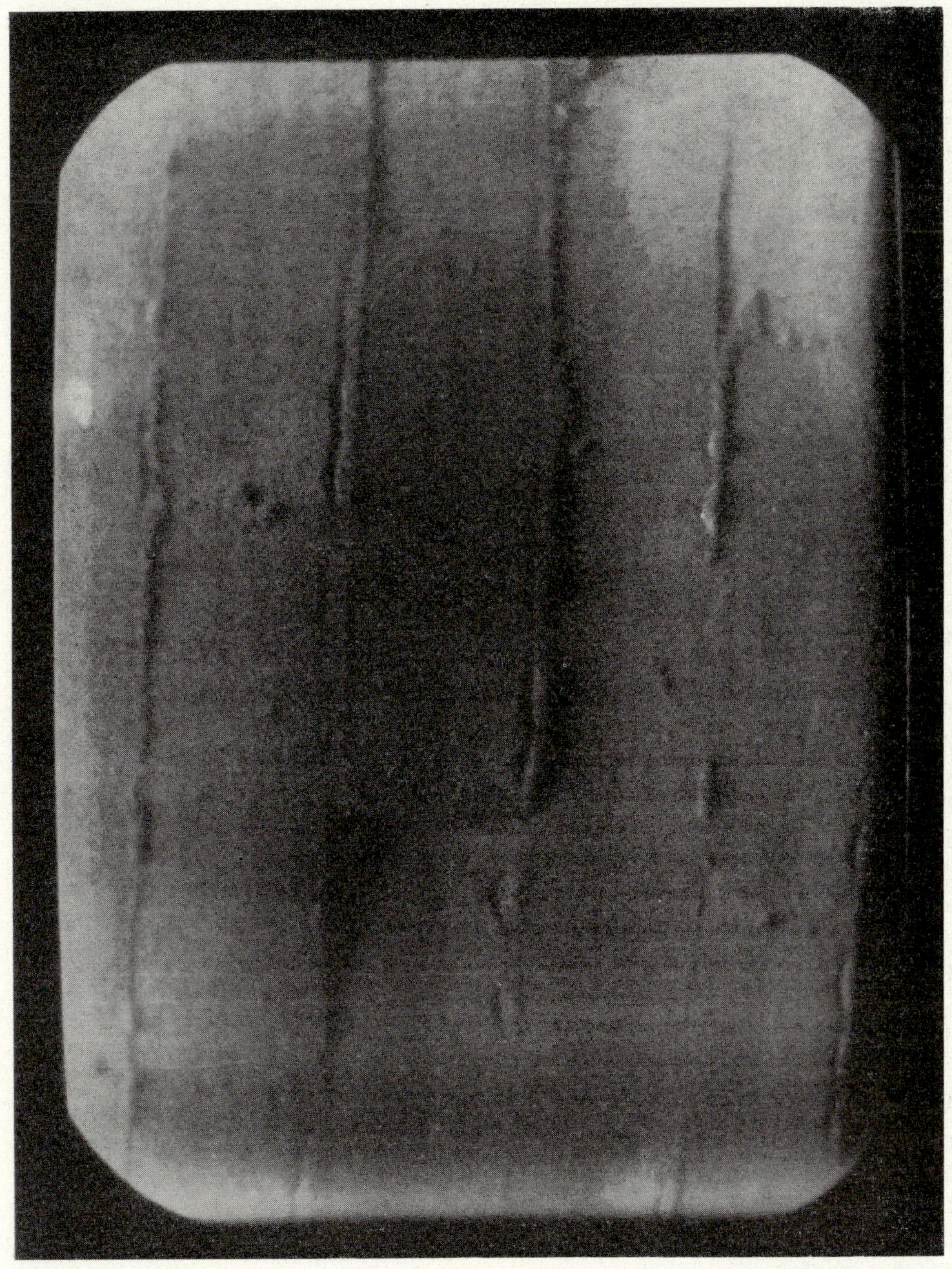

FIG. 4.5.2B. Picture by closed circuit television of chimney interior

of Modern Watertube Boiler Plant by the Boiler Availability Committee, contains useful information on this subject as well as that of its title, and there is an excellent monograph, produced by C.E.G.B. entitled *The Identification of Atmospheric Dust by use of the Microscope*, which shows the appearance of all the normal constituents of chimney emissions and other sources of dust.

Corrosion pitting of steel may present a variety of visual forms, which with experience can be characterised to some extent. Thus, pitting caused by dissolved oxygen is commonly in the form of isolated rather shallow oval pits overlaid with brown crusts or "tubercles", whilst on-load corrosion usually shows the overlapping of many such pits and has other characteristics which have been described more fully in Chapter 2. Pitting due to acidic condensates (sometimes found, for example, in non-draining superheaters) occurs most often as irregular shaped, smaller and steeper-sided pits than are caused by dissolved oxygen; the very small steep-sided pitting, with smooth contours, occasionally seen on moving parts such as journal bearings, are commonly caused by electrical discharges at the areas of pitting. It is often very useful to make a record of the size and shape of small areas of corrosion, in order to demonstrate the subsequent progress or suppression of the trouble, and this can be done by making lead or "plastic" casts of the areas, after cleaning them, as well as by a photographic record.

In addition to the visual examination, the chemist also takes samples of any deposits for chemical analysis, the results of which are complementary to the examination *in situ;* when considered together these should provide valuable diagnostic information.

A station chemist should, after a period of time, build-up knowledge of the characteristics of the plant in his care and thus be able to form a picture of "normality", based on inspection, operational experience, and chemical analysis. When a significant divergence from "normal" occurs he then starts investigation to find the cause. For example, the analysis of deposits found in a section of the plant which is normally free from such deposits, may well indicate that the source of deposited material is some other section of the plant, to which the investigation will then be extended. Thus, although variable quantities of copper and nickel are nearly always present in boiler water-side deposits, the presence of excessive quantities would lead to investigations to determine from which section of the feed system they originated. This, of course, would be supported by the routine analysis of the feed water for these metals.

4.5.3.2. Instrumental Methods

A number of instrumental techniques have been specifically developed to enable plant inspection to be carried out more effectively and quickly: in some cases they also enable the rate of wastage or deterioration to be monitored by their use as routine. Many of these instruments come clearly within the discipline of the metallurgist or the non-destructive testing engineer and therefore are dealt with in Chapter 5; however, brief mention will be made of some of them in this chapter.

4.5.3.2.1. *Condenser tube examination.* The very stringent limits imposed on the boiler water chloride concentration in high-pressure boilers, particularly those operating on the "zero solids/volatile alkali" treatment, has necessitated greater attention to condenser tube leakage problems.

For many years now it has been customary to subject condenser tubes thought to be suffering corrosion or other kinds of wastage, to an internal examination by equipment which can detect the location and extent of any thinning. Such equipment includes the "Probolog" and the "Introview", which use a probe that is drawn through the tube and induces eddy currents in the region of the tube which surrounds the probe. This technique is used to establish the general condition of the tubes at times of annual maintenance or other similar suitable outage.

Almost invariably there are times when condenser leakage occurs during operation and necessitates taking the plant out of commission to locate and remedy the leakage. Until recently the established method of locating condenser leaks has been to fill the steam side of the condenser with water containing fluorescein; the cooling water side of the tube plate is then scanned with ultraviolet light and the source of leakage is indicated by the visible fluorescence near to it. However, this technique is not sensitive enough for modern plant and leaks undetectable by this method can make it very difficult to maintain the required boiler water condition. No single method of locating condenser leaks has emerged as the accepted alternative to the fluorescein method, and this remains as a considerable problem for high-pressure plant.

Other methods which have been tried but are not in general use as yet, make use of ultrasonic equipment and foam blanketing. Both require the condenser under test to be drained down on the cooling water side and to be under vacuum. The ultrasonic device utilises a sensitive microphone and amplifier capable of picking up noise in the 35–45 kc/s range, this being approximately the frequency of the sound made by air as it is pulled into the vacuum space through a minute hole. In the foam blanketing method, the cooling water side of the tube plate is covered with a stable foam and the point of leakage is indicated by a disappearance of the foam where it is sucked into the leaking tube. The effectiveness of the method depends on the type of foam used and the method of application (among other factors), and a report by the South-Western Region R. and D. department (No. RD/SW/M. 94), gives practical details and suitable materials.

4.5.3.2.2. *Metal sorters.* In recent years there have been a number of tube failures, particularly in superheaters, which have been due to the inclusion of tube lengths fabricated from incorrect material. Mild steel has occasionally been accidentally introduced instead of low chromium alloys and this has resulted in early creep failure.

A number of instrumental techniques ("metal sorters") have been developed to differentiate between the materials involved, such as $2\frac{1}{2}\%$ chromium/1% molybdenum, 1% chromium/$\frac{1}{2}\%$ molybdenum and mild steel, and the instruments generally work on one of the following principles:

(a) magnetic, including eddy current meters,

(b) thermo-electric,

(c) spectrographic.

One of the problems is that the most reliable types of equipment are rather bulky and thus are difficult to use in sections of the boiler which are not readily accessible. In addition to employing these techniques which enable a complete survey of an entire superheater to be carried out fairly quickly, recourse sometimes has to be made to chemical analysis of filings which have been carefully removed from the suspect tube so as to avoid contamination. These chemical methods of analysis are, of necessity, slower than the *in situ* instrumental techniques but are of value not only in areas where access is difficult, but also as confirmatory back-up in cases where a clear-cut indication is not obtained from the equipment.

4.5.3.2.3. *Metal flaw detectors.* A number of techniques are available for detecting metal flaws, such as cracks, slag inclusions, and laminations. These include dye penetrants, magnetic crack detectors, ultrasonic equipment and gamma-ray methods using radioactive isotopes, whilst X-ray methods are also employed as a method of inspection during the manufacturing stages. All of these are discussed in Chapter 5.

4.5.3.2.4. *Boiler tube inspection.* Internal boiler tube corrosion has led to the need for techniques which will enable the internal condition of tubes to be determined during maintenance periods, so that tubes undergoing corrosion may be removed before failure occurs. Such inspection methods may also be used during construction for examining the internal profiles of welds, thereby ensuring that a high standard is obtained. Several techniques may be used including ultrasonic probes, tube cameras and endoscopes.

Ultrasonics. Ultrasonic examination has been used as a method of non-destructive testing for many years and two ways in which it may be applied to boiler tube examination, may be mentioned. In the first, ultrasonic equipment may be used to determine the tube wall thickness, and hence the loss of metal due to corrosion. The probe is applied to the tube exterior which must be thoroughly cleaned by shot blasting or grinding, before testing. The technique is rather slow and needs an experienced operator to interpret the results with confidence. It has been successfully used on many boilers and is particularly valuable where it is not possible to examine the tube by internal inspection devices. More recently a "bracelet" of ultrasonic probes which is clipped around the diameter of the tube, has been developed for the specific examination of welds during the construction period.

Internal inspection cameras. Two types of equipment are available, these being the multi-shot still camera and the closed circuit television camera.

The multi-shot still camera. The multi-shot still camera was developed by the South-Western Regional R. and D. Department in conjunction with Inertia Switch Limited. It has been progressively improved over the last few years and successive developments have included a reduction in size to enable it to be used in small bore tubes and to negotiate bends of small radius. The camera with which most experience has been obtained is illustrated in Figure 4.5.3.2.4A. This camera is $1\frac{5}{8}$ in. in diameter and approximately 6 in. long and is capable of traversing clear bends of 12-in. radius, in tubes of 2-in. internal diameter. It carries its own lighting source and several adaptations can be made to allow "close-up" photography of any required area. It uses 16 mm film preloaded in interchangeable cassettes, each of which will give 500 pictures and is sufficient to inspect 200 ft of boiler tube. The

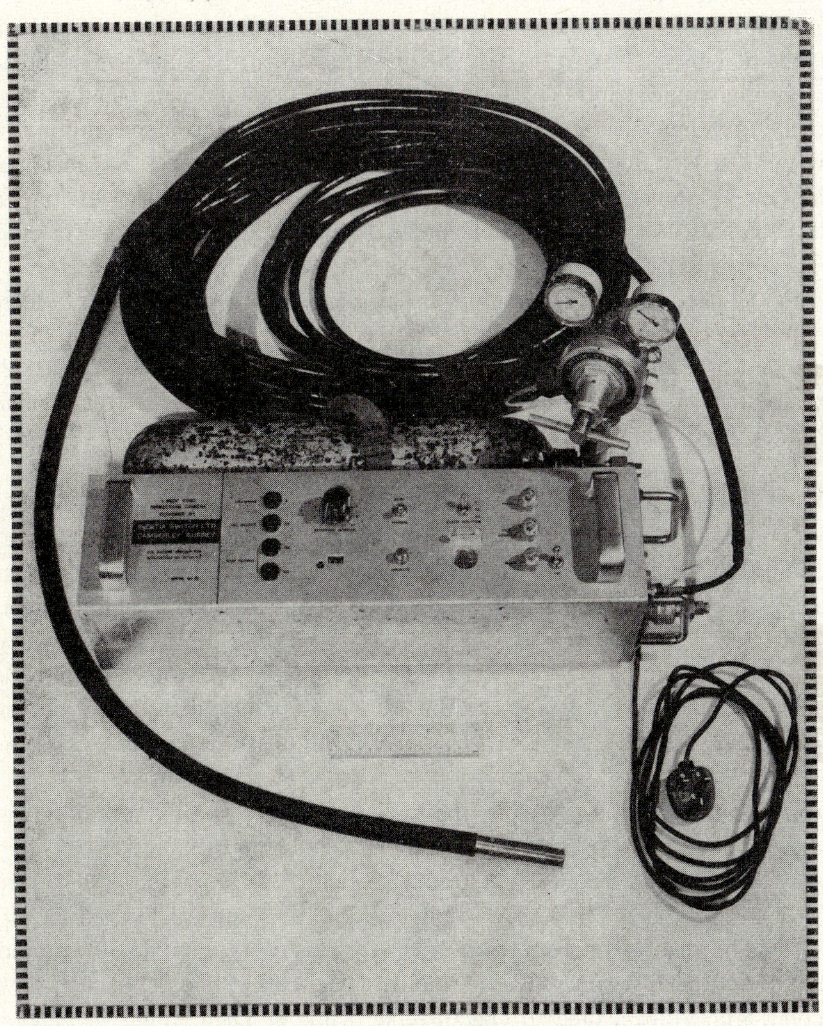

FIG. 4.5.3.2.4A. Boiler tube "still" camera

camera lens has a focal length of 6·5 mm, giving a good depth of field with a viewing angle of 50°, and exposure time can be varied from 0·25 to 15 sec. Attached to the camera is a flexible cable of 100 ft or more with a colour-coded marker along its length so that the position of the camera in the tube can always be determined. At the other end of the 100 ft cable is the control box, which is connected to the battery power supply.

The normal method of operation is either to lower the camera down the tube and take photographs at intervals, or to lower the camera down to the maximum depth at which photographs are required and take them as it is retrieved. Pictures are usually taken at 4-in. intervals giving a slight overlap of each one. The angle of the camera lighting is designed to throw tube surface defects into contrast with the clean surface, by causing shadows. Depending on a number of factors, such as the time available and whether corrosion is expected to be present mainly in specific areas, the procedure may be concentrated in the suspect areas.

Generally the film used is a normal 16-mm black and white negative film, which may be viewed after processing, by projection onto a screen. Where defects are observed, prints may be made for record purposes and for future assessment. Typical photographs of defects taken by the camera are shown in Fig. 4.5.3.2.4B. Experimental work has also been done with "positive" films and also colour films, but results to date have been disappointing.

A recent development of the multi-shot still camera is a model which is 0·9 in. in diameter and 3-in. long, and is capable of negotiating bends of 12-in. radius in 1-in. bore tubes. It uses a split 16-mm film held in a cassette in the form of a flexible tail, and the film transport mechanism is actuated pneumatically by nitrogen obtained from a small cylinder. When using this camera it is necessary to take photographs every 2in. to obtain full coverage; the cassette holds enough film for 100 flashlight photographs to be taken, covering approximately 15 ft of boiler tube: operation of this camera is similar to the larger one and it has proved particularly useful in some of the smaller bore tubes used in nuclear stations.

Closed circuit TV cameras. Whilst the multi-shot still camera is a major step in tube inspection techniques, it is still time-consuming largely due to the necessity for film processing. Also when surveying a boiler, much time is usually lost in photographing tubes which are free from defects. It is obviously desirable to see the results of the inspection without delay and thus be able to concentrate on defective areas. With these points in mind a miniaturised closed circuit TV camera was developed which is $1\frac{1}{2}$ in. in diameter, 5 in. long and is based on a $\frac{1}{2}$-in. Vidicon tube; it can traverse bends of 12-in. radius in tubes of $2\frac{1}{4}$-in. diameter. A wide-angle lens and integral axial light source are included. The camera head, which weighs $1\frac{1}{2}$ lb, has a 20-ft control cable tail; by attaching an extension cable, satisfactory pictures can be obtained on the monitor screen, even when this is mounted up to 400 ft away from the camera itself. One of the attachments available is a remotely controlled rotating mirror mounted in front of the camera head, which allows the field of view to be adjusted as required.

Both the multi-shot still camera and the TV camera have been widely used for surveying the internal tube surfaces of boilers. Probably the best method of using these techniques is to rapidly survey a boiler using the TV camera and then give a more detailed examination

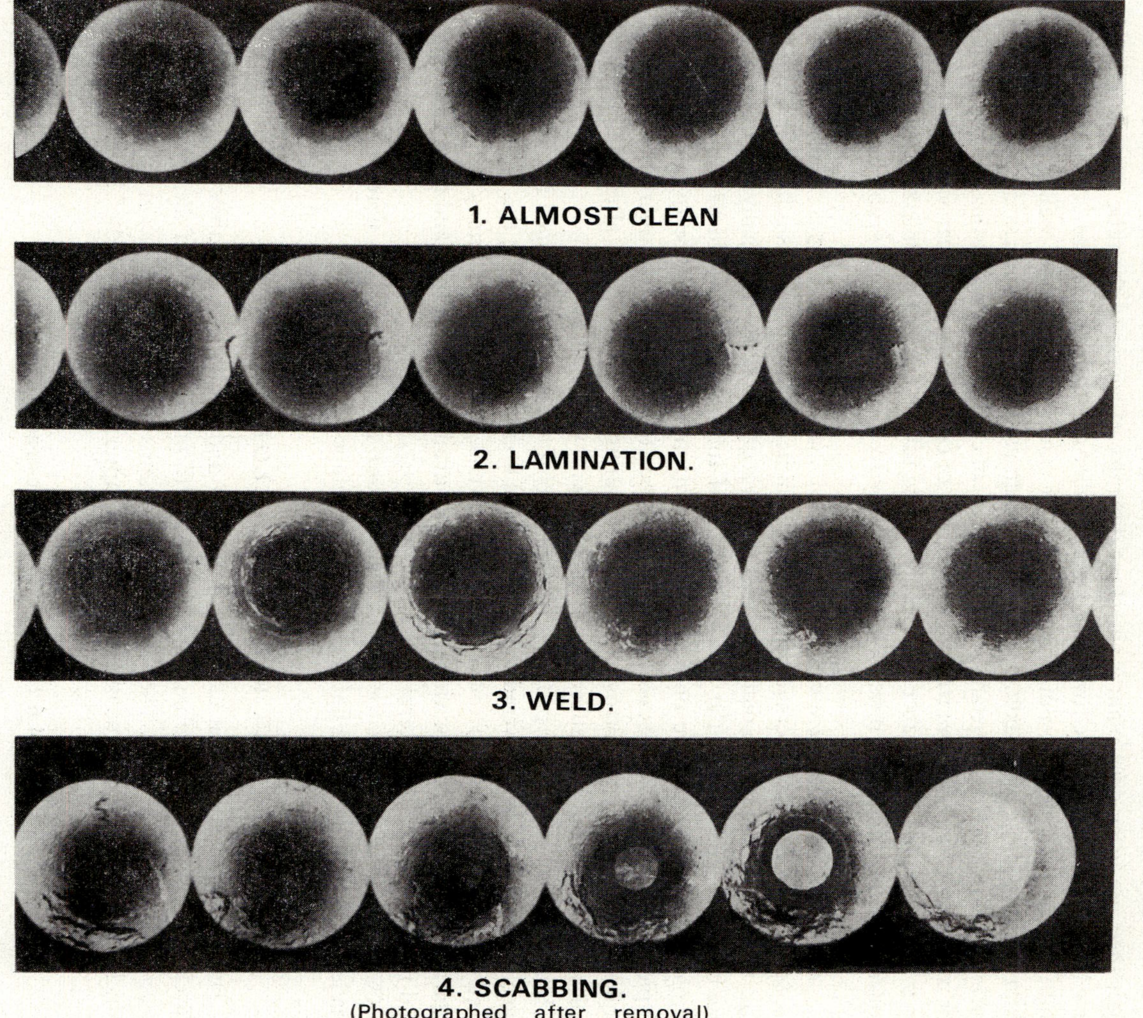

FIG. 4.5.3.2.4B. Defects shown in boiler tubes by $1\frac{3}{4}$-in. tube camera

of specific areas where defects have been observed, using the photographic camera, which gives better definition and picture quality.

The practical difficulties with the use of tube cameras have been associated in the past with insufficient robustness to withstand the severe conditions to which they are subjected, such as dragging them round a number of bends. One important factor which tends to limit their use in modern plant is the sheer difficulty of access to the tube interior. Thus, new boilers are not equipped with the large number of header caps used on older boilers, and furthermore it is often not possible to lower cameras down water wall tubes directly from the drum since intermediate headers are often used. However, access is not so limited during the construction stage and the cameras are of considerable use at this stage to monitor weld quality and freedom from blockage. It is of course possible to carry out inspection on a boiler during maintenance outages by cutting tubes at some convenient section in the boiler "dead space" where rewelding is least likely to cause subsequent problems.

Endoscopes. Increasing use is being made of quartz rod endoscopes, which may be used for the short-range internal inspection of plant items to which access is limited. A small hole approximately $\frac{3}{8}$ in. in diameter is all that is required and occasionally this technique has been used to examine the internal surface of tubes where it would be impossible to employ other methods, the only alternative being to remove the tube from the boiler. Typical of such situations would be the sharp bends formed by burner quarl tubes. In one boiler which has suffered corrosion, tubes were examined by endoscopes after drilling small holes in them, which were subsequently filled in with weld material.

Essentially an endoscope probe consists of a solid quartz rod, contained in a steel tube, which transmits a high-powered light beam to the area to be examined; this eliminates the need to carry a bulb at the front of the lens at the end of the probe. Running parallel to the quartz rod is a telescope for viewing. Probes are available for forward and lateral viewing and range from 5 in. to several feet in length.

Of possible development for inspection purposes, a "rope" of glass fibres can also be made to transmit light, and has the further advantage of being flexible.

4.6. RECENT DEVELOPMENTS

The technologies of plant cleaning, protection and inspection are subject to continual development and it is therefore scarcely possible to produce a survey of the subject which is up-to-date when printed. The following paragraphs summarise significant developments which were recent at the time of finalising this chapter, that is, at the end of 1967.

4.6.1. Protection of Austenitic Parts

Because of the failure by cracking in service of some austenitic superheater or reheater tubes at a few stations, more attention has been given to exclude chemical conditions during pre-commissioning cleaning which could possibly cause such failures. The chemicals most suspect are, of course, chlorides (or other halides) and caustic soda.

In order to eliminate a risk of carrying chloride-laden foam into superheaters whilst cleaning the boiler with hydrochloric acid, wetting agents (for example, Lissapol) are now required by G.O.M. 75 to be excluded from the cleaning solution. Similarly, citric acid and other solutions shall not, when used in austenitic superheaters and reheaters, contain more than 2 ppm of chloride.

It is possible during the preliminary alkali boil-out, passivation, and steam blowing that small amounts of the boiler solutions may be carried over into the superheater. Since this risk cannot be eliminated, it is now considered necessary to omit caustic soda and wetting agents from all of these procedures. Chlorides must also be controlled, particularly during steam blowing.

4.6.2. Boiler Plant Purging with Nitrogen

Arising mainly from the "critical path" time (and therefore cost) required by steam purging procedures, the Southern Project Group of the C.E.G.B. has considered alternative procedures. Purging by the rapid release of air or nitrogen compressed in the boiler to 800 to 1000 lb/in^2, was found to be feasible, and disturbance factors at least as good as when using steam should be obtainable. The major items of equipment necessary are a compressor, quick-acting valves, and manifolds. It is proposed that the boiler water circuit and those associated with the superheater and reheater, should be separately purged at a stage between the initial alkali boil-out and the acid clean of the boiler, thus removing debris which would otherwise have to be dissolved or flushed away.

The advantages claimed for this procedure are, principally, time-saving, improved efficiency of subsequent cleaning, the elimination of risks associated with thermal shock, and absence of a final purging operation. The elimination of thermal shock could be regarded as a disadvantage in so far that rapid changes in temperature may bring about the disruption of firmly bonded debris, weld spatter, etc.

This form of purging, using nitrogen, was introduced in the pre-commissioning cleaning of a 350 MW boiler at Tilbury in 1968.

4.6.3. Safety

The C.E.G.B. has approved a Code of Practice on safety precautions to be used in chemical cleaning operations, for issue as an appendix to Generation Operation Memorandum No. 75. This "code" is essential reading for anyone with responsibility for the cleaning of C.E.G.B. plant.

4.6.4. Sub-surface Cavitation

This phenomenon, discussed previously in relation to the post-commissioning acid cleaning of boilers, has been investigated by the former Midlands Region R. and D. Department. It has been found that the effect of producing subsurface porosity on hydrogen-damaged steel, previously ascribed only to hydrochloric acid, can also be produced by others acids including citric acid. There is, therefore, no reason now for the exclusion of hydrochloric acid when cleaning corroded boilers (except where austenitic parts are involved), but any boiler tubes which have suffered hydrogen damage must first be replaced, so far as it is possible to identify them.

4.6.5. **Passivation**

Some further research in the Midlands Region indicates that the passivation procedure used in feed systems which have been acid cleaned (25 ppm of hydrazine for 24–28 h at 90° to 95°C) has little or no practical value in producing a surface condition which will resist atmospheric corrosion. Accordingly, this particular procedure will be deleted from G.O.M. 75, but the procedure for passivating boiler surfaces under pressure is effective and remains an essential requirement.

An important fact thus confirmed is that the protective oxide on mild steel, however it may be produced, is rapidly damaged by the combination of moisture and air.

APPENDIX

RECORD OF THE CLEANING OF THE BOILER SYSTEM OF A 350 MW UNIT

1. Plant Details

Maximum continuous rating of boiler	2,350,000 lb steam/h
Turbine stop valve conditions	2300 lb/in^2, 566°C, with single-stage rheating to 566°C
Materials of construction	
Boiler tubes	Mild steel
Primary superheater	Mild steel
Front wall radiant superheater	1% Cr, $\frac{1}{2}$% Mo
Radiant platen superheater	1% Cr, $\frac{1}{2}$% Mo and $2\frac{1}{4}$% Cr, 1% Mo
Secondary convection superheater	$2\frac{1}{4}$% Cr, 1% Mo; 1% Cr, $\frac{1}{2}$% Mo and 18% Cr, 11% Ni
Horizontal reheater	Mild steel and 1% Cr, $\frac{1}{2}$% Mo
Pendant reheater	$2\frac{1}{4}$% Cr, 1% Mo; 18% Cr, 8% Ni and 1% Cr, $\frac{1}{2}$% Mo

Plant capacities	
Boiler and economiser to N.W.L.	78,000 gal
Boiler and economiser to top of drum	82,000 gal
Boiler, economiser, superheater, reheater and steam lines	182,000 gal

2. Summary of Cleaning Procedure

2.1. *Stage I. Initial Alkali Boil-out* (Boiler and economiser)

2.2. *Stage II. Hydrochloric Acid Clean* (Boiler and economiser)

(a) Circulation and soak using 5% hydrochloric acid.
(b) Citric acid rinse and water flush.
(c) Store with ammonia and hydrazine.

2.3. *Stage III. Priming of Boiler Unit* (Boiler, economiser, superheater, reheater and steam pipes)
(a) Steam out superheater and reheater.
(b) Priming of superheater and reheater.

2.4. *Stage IV. Citric Acid Clean* (Boiler, economiser, superheater, reheater and steam pipes)
(a) Citric acid circulation with ammonia to pH 3·5–4·0.
(b) Reheater flush.
(c) Superheater and boiler flush.

2.5. *Stage V. Final Alkali Boil-out* (Boiler and economiser)
Passivation process.

2.6. *Stage VI. Steam Purging of Superheater, Reheater and Steam Pipes*

3. SCHEDULE OF CHEMICALS USED DURING CLEAN

Stage I. Initial Alkali Boil-out

Trisodium phosphate anhydrous	10½ cwt
Lissapol C	3½ cwt

Stage II. Hydrochloric Acid Clean

Hydrochloric acid (sp. gr. 1·16)	61 tons
Armohib 28	820 lb
Lissapol N	420 lb
Caustic soda liquor (46%)	43 tons
Citric acid—anhydrous	15 cwt
Ammonia (sp. gr. 0·880)	82 gal
Hydrazine 35%	235 lb

Stage IV. Citric Acid Clean

Citric acid liquor (46%)	53 tons
Armohib 25	550 lb
0·880 ammonia to raise pH to 3·5	1824 gal
Additional ammonia to raise pH to 5·0	2290 gal
Ammonia for storage	200 gal
Hydrazine, 35%	235 lb

Stage V. Final Alkali Boil-out

Ammonia (sp. gr. 0·880)	78 gal
Hydrazine, 35%	670 lb
Additional hydrazine to maintain 25 ppm	2850 lb

Stage VI. Steam Purge

Hydrazine, 35%	4730 lb

4. Schedule of Water used

Stage I. Initial alkali boil-out—200,000 gal
Stage II. Hydrochloric acid clean—660,000 gal
Stage III. Steaming out and priming circuit—100,000 gal
Stage IV. Citric acid clean and flushing—1,140,000 gal
Stage V. Final alkali boil-out—240,000 gal
(No record available for Stage VI.)
Total Quantity used in Stages I to V—2,340,000 gal

PROCEDURE

2.1. Stage I. Initial Alkali Boil-out

The decision was made to leave the drum internals in position, due to the time required for refitting afterwards.

The boiler was boiled out using 1500 ppm trisodium phosphate and 300 ppm wetting agent (Lissapol C). The pressure was raised to 600 lb/in^2 and maintained at this pressure for 24 h. The level was raised to the top of the drum gauge glass for 4 h during this period and a routine blowdown from all points was given every hour. After 24 h the pressure was lowered to 100 lb/in^2, the boiler part drained and refilled with hot water from the deaerator. This was repeated to enable the drum temperature differentials to be maintained within the required limits and the boiler was finally drained at 50 lb/in^2.

2.2. Stage II. Hydrochloric Acid Clean

(a) Circulation and Soak with 5% Hydrochloric Acid

Temporary pipework was erected to enable a circulation system to be established. Large vents were fitted to the drum to allow adequate venting during filling and draining. Restrictor plates were fitted to all the downcomers within the drum to divert the flow of acid preferentially through the wall tubes. Accumulations of loose debris within the drum were also removed at this stage.

The following procedure was adopted to ensure that the hot 5% hydrochloric acid solution was not chilled and to ensure adequate distribution of the acid and inhibitor through the boiler. The boiler and economiser were filled with water and the temperature was raised to 80°C using the oil burners. Circulation was maintained through the boiler and economiser to equalise the temperatures within the unit. Meanwhile the deaerator was filled with water and heated to 88°C. Backflushing of the superheater was carried out to ensure that the superheater was filled with water to form a hydraulic seal against acid carry-over. The boiler was drained with the exception of the cavity and rear enclosures and economiser. It was necessary to leave these full of hot water, which in addition to the hot water in the deaerator, was used to refill the boiler with 5% w/w hydrochloric acid and 0·1% w/w inhibitor (Armohib 28). The level was raised to the top of the drum using a temporary gauge glass and maintained at this level throughout the hydrochloric acid clean.

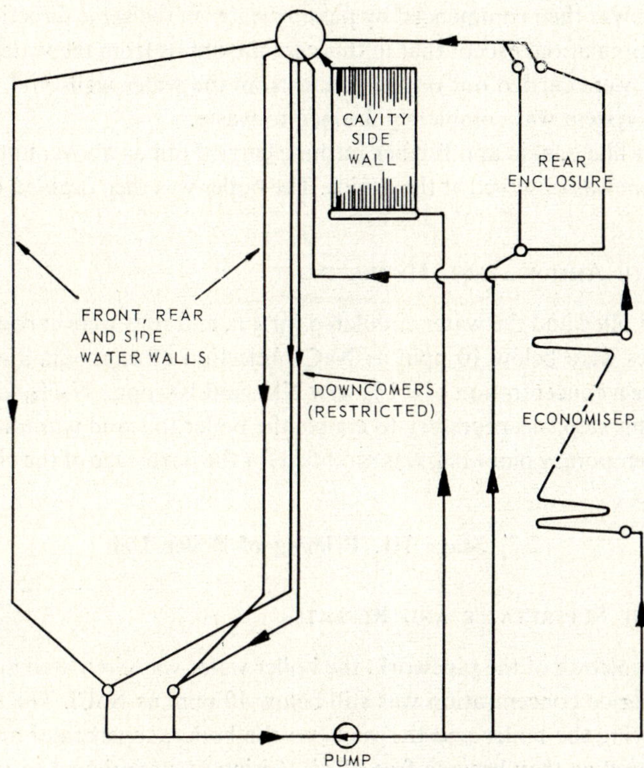

FIG. A.1. Circulation during hydrochloric acid cleaning

The circulation pattern was up through the economiser and cavity and rear enclosure headers into the drum and down through the water wall tubes (see Fig.A.1). Circulation periods of 30 min were interrupted by static soaking for 1 h.

Chemical tests to determine acidity, ferric and ferrous ions were carried out during the circulation periods, and the temperature was recorded. From these tests, the progress of the clean was followed and when equilibrium conditions were considered to be established, the boiler was ready for draining. Drainage was direct to the CW system, with a CW pump in operation, but to conform to the River Authority requirements the pH of the effluent was controlled between 5 and 9 by the addition of caustic soda liquor at the point of entry to the CW system. The hydrochloric acid cleaning period lasted for 12 h.

(b) 0·1% w/w CITRIC ACID RINSE AND WATER FLUSH

To minimise rusting following the hydrochloric acid treatment, the boiler was filled to the top of the drum with 0·1% w/w citric acid. The drum level was then lowered to normal working level and the superheater back-flushed with deionised water. This operation was done to ensure that the superheater remained full of deionised water and that any small amount of hydrochloric acid which may have entered the superheater previously, was flushed back into the drum.

Boiler flushing was then commenced by passing water in the same direction as in the hydrochloric acid circulation, except that in this case the drains from the water walls were put to waste. Checks were carried out on the drains from the water walls and when these were visibly clean the system was completely drained to waste.

The boiler was filled again and further flushing carried out as above, until a conductivity below 250 rmo/cm was observed at the drains. The boiler was then drained completely.

(c) STORAGE WITH AMMONIA AND HYDRAZINE

The boiler was filled and the water circulated for $\frac{1}{2}$ h, and tests were carried out to confirm that the chlorides were below 10 ppm as NaCl. Injection of ammonia and hydrazine was then made to give a concentration of 0·1% w/w NH_3 and 100 ppm N_2H_4, followed by circulation for 2 h. This step was necessary to enable the boiler to stand without excessive corrosion, whilst the temporary pipework was modified for the next stage of the clean.

2.3. **Stage III. Priming of Boiler Unit**

(a) STEAM-OUT OF SUPERHEATER AND REHEATER

Following completion of the pipework, the boiler water was circulated and tested to confirm that the chloride concentration was still below 10 ppm as NaCl. The temperature was then raised by firing the boiler and the vents on the boiler, superheater and reheater were progressively closed as they became free of air. Firing was continued to maintain a drum pressure of 20–30 lb/in^2 and a good steam blow at the blowdown vessel.

(b) PRIME SUPERHEATER AND REHEATER

Firing was stopped and immediately the superheater and reheater were pumped full of demineralised water and the whole boiler system kept at 50 lb/in^2 drum pressure.

2.4. **Stage IV. Citric Acid Clean**

(a) CIRCULATION OF AMMONIATED CITRIC ACID

Whilst maintaining a pressure of 50 lb/in^2 on the system the boiler, economiser, superheater and reheater were circulated, using sufficient pumps to maintain a reasonable velocity (see Fig.A.2). A further check was carried out to confirm that the chloride concentration was still below 10 ppm as NaCl.

Firing was restarted and the temperature of circulation was raised to 96°C, when firing was stopped and the boiler dampers closed to minimise heat loss within the system.

The tankers containing the liquid citric acid were arranged for discharge and checks were carried out to ensure that the citric acid was sufficiently low in chloride concentration.

Following confirmation that the temperature of the circulating system was remaining steady, injection of citric acid and inhibitor (0·03% Armohib 25) commenced until a concen-

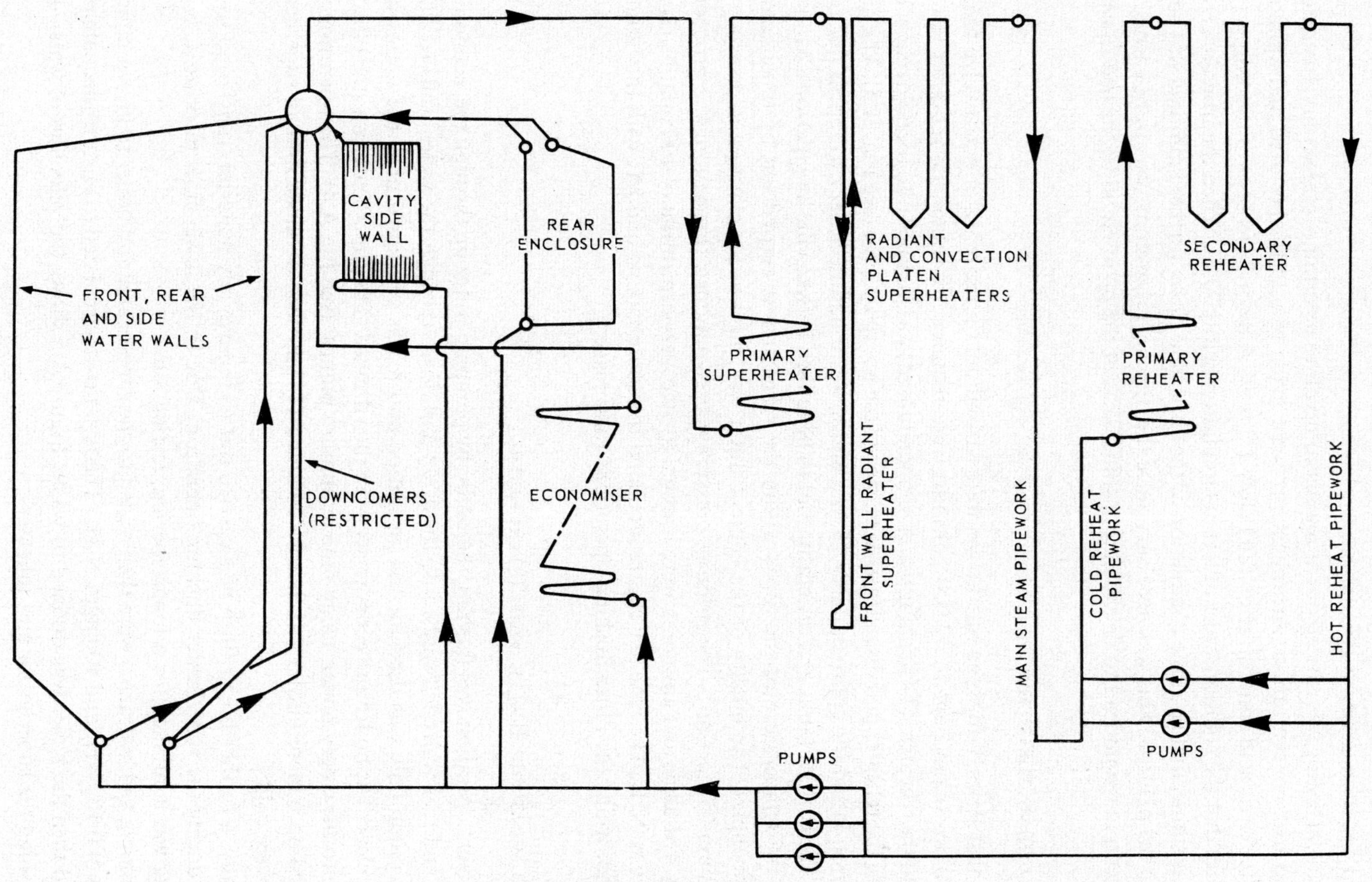

Fig. A.2. Circulation during citric acid cleaning

tration of 3·0% citric acid was achieved. Ammonia was then added to raise the pH to between 3·5 and 4·0, with constant circulation.

Chemical testing was carried out at regular intervals to determine total acidity, free acidity, pH and iron concentration. Arrangements were made to add further acid and ammonia, if the free citric acid concentration fell below 2% w/w. A record was kept of the temperature to ensure that the solution was not left in circulation below 70°C. When the iron concentration was stable, that is, at the completion of the citric acid clean, further ammonia was added to raise the pH to 5. This was necessary since the flushing programme was to be carried out in stages and would lead to delays before some parts of the plant could be flushed to waste.

(b) Flushing the Reheater

The first series of flushing operations were carried out on the reheater and twelve temporary pumps (equivalent to 12,000 gal/min) were required to obtain the required velocity through the reheater. The flushing water (demineralised) was pumped into the cold reheat line, through the reheater and discharged to waste via the hot reheat line (see Fig. A.3). The flushing period lasted for approximately 15 min and it was found necessary to repeat this twice more. The R.F.W. tanks and deaerator had to be refilled between flushing periods. Monitoring of the temperature of the tubes within the reheater was carried out by touch and by thermocouples. The quality of the effluent water was checked for clarity and conductivity (maximum 25 rmo/cm). The priming pressure was maintained throughout the procedure. Finally hydrazine and ammonia were injected to give a concentration of 100 ppm N_2H_4 and 0·1% w/w NH_3. The reheater was recirculated to ensure mixing and left with this solution to stand, whilst the other flushing operations took place.

(c) Flushing the Boiler and Superheater

The next operation was flushing the boiler by pumping water up through the economiser cavity and rear wall enclosures, into the drum and down through the water walls to waste.

One pump only was used for this operation and it was continued until the drains were visibly clear. The boiler and economiser were then drained. The superheater was then flushed back into the empty boiler, using nine temporary pumps (see Fig.A.4). The operation was repeated once more until a satisfactory conductivity has been obtained and all superheater loops were cold.

Finally, a further boiler flush was carried out to remove any material transferred to the boiler during the superheater flushing operation. Due to the lower flow rates used, boiler flushing was continued for a longer period, lasting 3 h.

Ammonia and hydrazine were then added to the superheater and boiler to give a concentration of 0·1% NH_3 and 100 ppm N_2H_4. The system was circulated to ensure adequate mixing and then the boiler was drained to waste. Figure A.5 shows the velocities of water used for flushing the superheater and reheater.

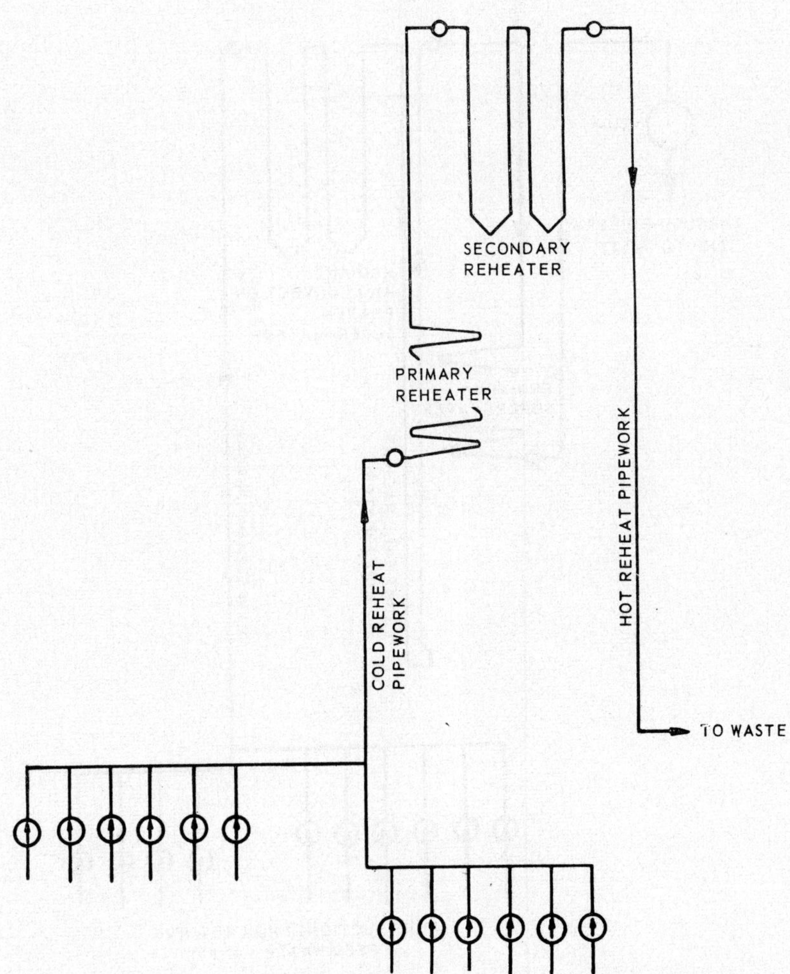

FIG. A.3. Circulation during reheater flushing

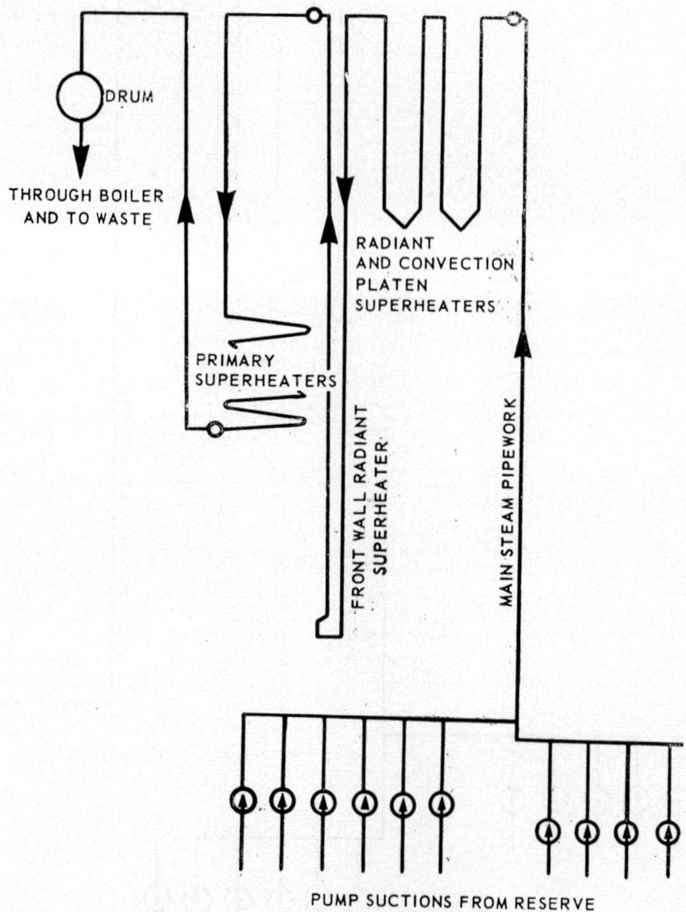

FIG. A.4. Circulation during superheater flushing

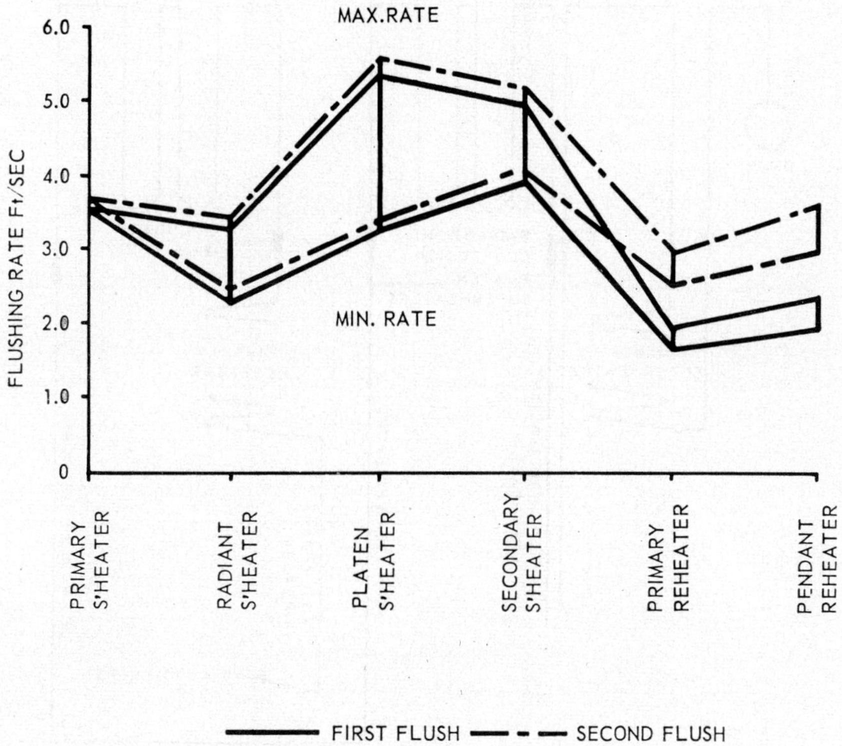

FIG. A.5. Water velocities during flushing

2.5. Stage V. Final Alkali Boil-out (Passivation)

The temporary pipework was removed from the boiler, although other temporary pipework associated with the superheater and reheater was left in position. The drum doors were opened, ensuring that clean working conditions were maintained by slightly pressuring the drum with filtered air at one end and an airlock at the other end. Samples of deposit were removed from the drum by personnel using remote breathing apparatus and protective clothing. These samples were analysed to check that the inhibitor content was negligible, and clearance was then given for entry to remove the restrictor plates from the downcomers.

The drum doors were then closed and the boiler filled with demineralised water to normal working level. Hydrazine and ammonia equal to 300 ppm N_2H_4 and 0·05% w/w NH_3 were then injected. The boiler pressure was raised to 600 lb/in^2 and the passivation process continued for 48 h at this pressure. Injections of hydrazine were made at regular intervals to maintain a hydrazine concentration in excess of 25 ppm. Blowdown was carried out from all points to remove suspended matter which developed during this stage.

At the end of the alkali boil-out, the boiler was drained hot under pressure and a further internal inspection carried out. Again particular care was taken to ensure clean working conditions were maintained. The boiler was found to have been satisfactorily cleaned and reasonable passivation had been obtained. Some small quantities of loose deposit still re-

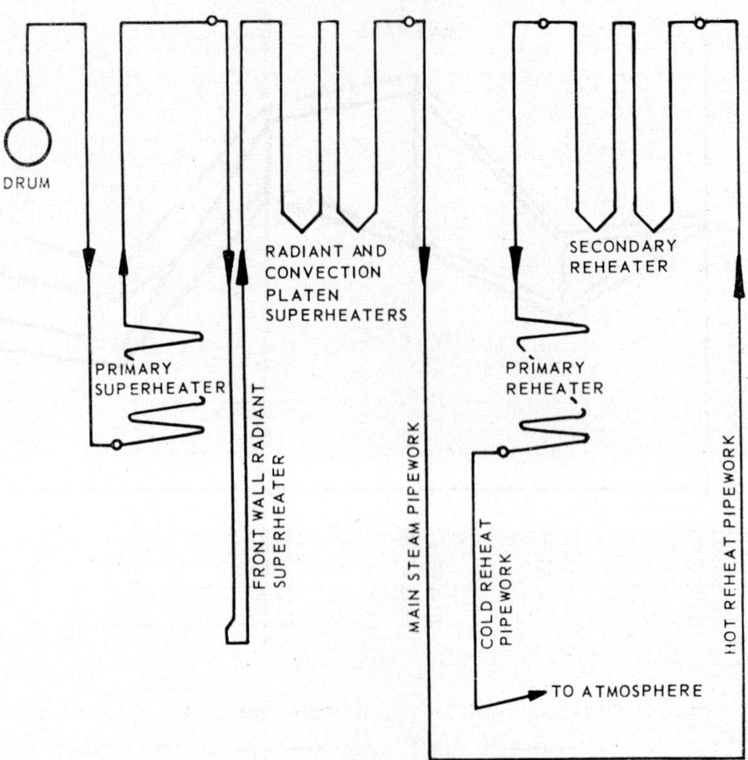

FIG. A.6. Flow path for purging the complete circuit

mained in the drum and headers; these deposits were removed by vacuum extraction and flushing some headers, but with care to prevent any water passing into other parts of the boiler.

2.6. Stage VI. Steam Purging of the Superheater, Reheater and Steam Lines

Steam purging was carried out at pressures up to 2200 using steel target plates to determine the number of blows necessary. In order to reduce the noise nuisance to the local communities, a silencer was used. Throughout the purging procedure, the boiler water was maintained practically free from non-volatile constituents, in order to ensure that if any carry-over occurred it would have no harmful consequences in the superheater or reheater.

The procedure used was first to purge the superheater and main steam lines thence discharging to atmosphere. When this section was clean, the pipework was rearranged to enable purging from the superheater, main steam lines and hot reheat lines to atmosphere. Finally, a further rearrangement allowed the whole of the steam circuit—superheater, main steam lines, hot reheat lines, reheater and cold reheat lines to be purged (see Fig.A.6). A total of 71 steam blows were required to purge the system to a satisfactory state of cleanliness, as judged by target plate examination.

Disturbance factors were calculated and results approximating to 1·5 and 1·1 were found for the superheater and reheater respectively, with fluctuations about these figures for individual blows.

Following completion of the steam purge, an exhaustive examination was carried out on the superheater and reheater pendant loops, using gamma-ray and X-ray techniques. A total of 1786 radiographs were made, representing 81% of all loops. As a result of the examination, two tubes were cut and pieces of welding rod were found in each tube.

REFERENCES

Because of the continuing development of the subject, there are no published textbooks which can be recommended to the reader. Of the following list of papers, those issued internally by the C.E.G.B. were about to be reissued in revised form at the time this chapter went to print, and should be read by the reader.

1. *Generation Operation Memorandum No. 75*, 2nd revision, including safety requirements*
2. *Uniform Specification No. M. 14A* Generation Design Branch.*
3. Fundamentals of acid-cleaning of boilers, by F. Spillner, *Mitt. V.G.B.*, vol. 90, June 1964 (C.E.Translation No. 3855).
4. A new look at pre-service cleaning, by J.M. Brown and K.B. Hellstrom, *Engineering and Boiler House Review*, April 1967.

EXERCISES

At least *seven* of the following questions should be attempted, and in some of them the student will be expected to make use of information from other chapters in this volume.

1. What are the principal factors to be considered when planning a programme for the pre-commissioning chemical cleaning of a unit boiler and feed system. including austenitic components in the superheater and reheater?

(It is *not* necessary to prepare a timetable, nor to give more than the briefest statement of the chemical processes to be used.)

2. Discuss the methods commonly used for removing (a) small amounts of mineral oil, (b) mill-scale, (c) rust, (d) siliceous materials, and (e) small metallic fragments, from newly-erected plant which has not been pre-cleaned. Include, where appropriate, an explanation of how the various methods serve their purposes.

3. What are the principal undesirable effects which should be considered when acids are to be used in plant cleaning operations? Comment on the effectiveness of various procedures used to prevent these undesirable effects.

4. Make recommendations for the protection during storage and erection of the following works-cleaned components of a unit turbine and feed system, and give reasons for your proposals:

(a) turbine rotors and cylinders;
(b) condenser shell interior surfaces;
(c) bled steam pipes of large diameter;
(d) deaerator storage vessel;
(e) feed heaters tubed in mild steel.

5. What are the causes of loss of performance in (i) feed heaters, and (ii) evaporators, and how is the deterioration detected? Give brief descriptions of suitable procedures for restoring performance in each of these items of plant.

6. Write notes on the following subjects:

(a) The association of internal boiler corrosion with: debris in the boiler, dissolved oxygen in feed water, condenser leakage, and the pattern of operation.

(b) The methods of assessing the internal condition of a boiler which, during two years approximately, has experienced a few generator tube failures due to on-load corrosion.

7. If you were the chemist in a station using a moderately polluted estuary for cooling water, where in the CW system would you look for evidence of damage by corrosion, and by what means would you make a thorough examination of the CW side of the condenser?

* Issued internally by C.E.G.B.

8. State briefly the principal safety precautions to be observed when planning and carrying out the following operations:

(a) cleaning with acids;

(b) disposal of waste chemical liquors;

(c) inspection of plant after chemical cleaning;

(d) inspection of plant at maintenance periods.

CHAPTER 5

METALLURGY AND WELDING

5.1. INTRODUCTION

This chapter will deal with the materials used in fossil-fuelled power stations; the properties of materials used in nuclear power reactors are discussed in Volume 8.

Metals may be defined as substances that are good conductors of heat and electricity, and are generally malleable and ductile. They occur naturally in ores in the form of chemical compounds such as sulphides or oxides. With the exception of the "noble" metals such as platinum and gold, metallic materials tend to react chemically with their environment and revert to compound forms, a process which can be regarded as corrosion.

Although a detailed discussion of the extraction of metals from their ores, their subsequent refinement and the principles of modification of properties by alloying are beyond the scope of this lesson, the essential background to the relevant ferrous materials is outlined in the following sections.

Many non-ferrous materials are used in power station construction, for example, copper-based alloys for condenser tubing, copper for alternator windings, tin-based alloys for bearings, and aluminium for busbars, but the majority of components are made from iron-based materials and it is with these ferrous materials that this chapter is largely concerned.

We shall examine the principles underlying the behaviour of ferrous materials and go on to consider the assessment criteria used for their selection, the requirements of fabrication into components, and problems arising from their operating environment.

5.2. FERROUS METALLURGY

Ferrous materials can be broadly classified into three groups:

(a) Pure iron (which has no industrial significance for our purposes).

(b) Cast iron.

(c) Steels.

Steel, which consists essentially of iron alloyed with small amounts of carbon and other elements, will be the major subject of the following account, although some reference will be made to cast iron.

5.2.1. **Production of Iron and Steel**

The first process in the conversion of iron ore to cast iron or steel is carried out in a blast furnace, where iron ore, coke, limestone and air react together at high temperatures to produce an impure form of iron called "pig iron". Pig iron can then be remelted in a Cupola furnace to produce "cast iron".

Steel is produced by either "hot metal" or "cold metal" melting practices. In the former, molten pig iron from a blast furnace is transferred to either open hearth or converter-type furnaces, and the conversion to steel involves the use of oxygen to improve the production rate. In cold metal melting, steel is produced from a furnace charge compounded of pig iron and scrap in varying proportions. In this case electric arc melting furnaces may be used as an alternative to the open hearth, but it is common to make oxygen additions to either type of furnace.

These processes are based essentially on the removal of excessive amounts of carbon, silicon and manganese, and impurities such as sulphur and phosphorus by chemical reactions under certain types of slags, whereby the impurities are transferred from the metal to the slag. This is then followed by the addition of controlled quantities of carbon, silicon, manganese and aluminium to produce the required composition of steel. Other alloying elements are added in the case of alloy steels.

The removal of impurities depends upon the nature of the slag used. An acid slag which contains a high proportion of siliceous materials, is suitable for the removal of carbon, silicon and manganese, and a basic slag which contains a high proportion of lime, is suitable for the removal of sulphur and phosphorus in addition to carbon, manganese and silicon. Both slags depend on oxidation reactions which are achieved by introducing oxygen to the molten steel, in the form of gaseous oxygen or iron oxide.

Because acid and basic materials react together chemically it is necessary for a process using acid slags to be carried out in a furnace lined with acid refractories (for example, silica) and for a process using basic slags in a furnace lined with basic refractories (for example, dolomite). This leads to the terminology of acid and basic steels.

After completion of the steelmaking operation, the molten metal is poured into either a refractory mould for the production of shaped castings, or into a cast iron mould and allowed to solidify for the production of ingots for subsequent fabrication into forgings, billets, bars, sheets or piping.

Steel in the molten state can contain in solution relatively large quantities of gases, particularly oxygen and hydrogen, but the solubility diminishes with falling temperature, and gases released during cooling may be entrapped in the solidifying steel, giving rise to extensive porosity. All cast ingots contain a small proportion of cavities but these have little significance and are welded up by a pressure-welding mechanism during rolling or forging. It is necessary, however, to remove most of these gases whilst the steel is still molten. Oxygen can be removed quite easily by adding elements such as manganese, silicon or aluminium which, because of their high affinity for oxygen, react with it to form non-metallic oxides which rise into the slag. Steels which have had most of their dissolved oxygen removed are called "fully killed steels".

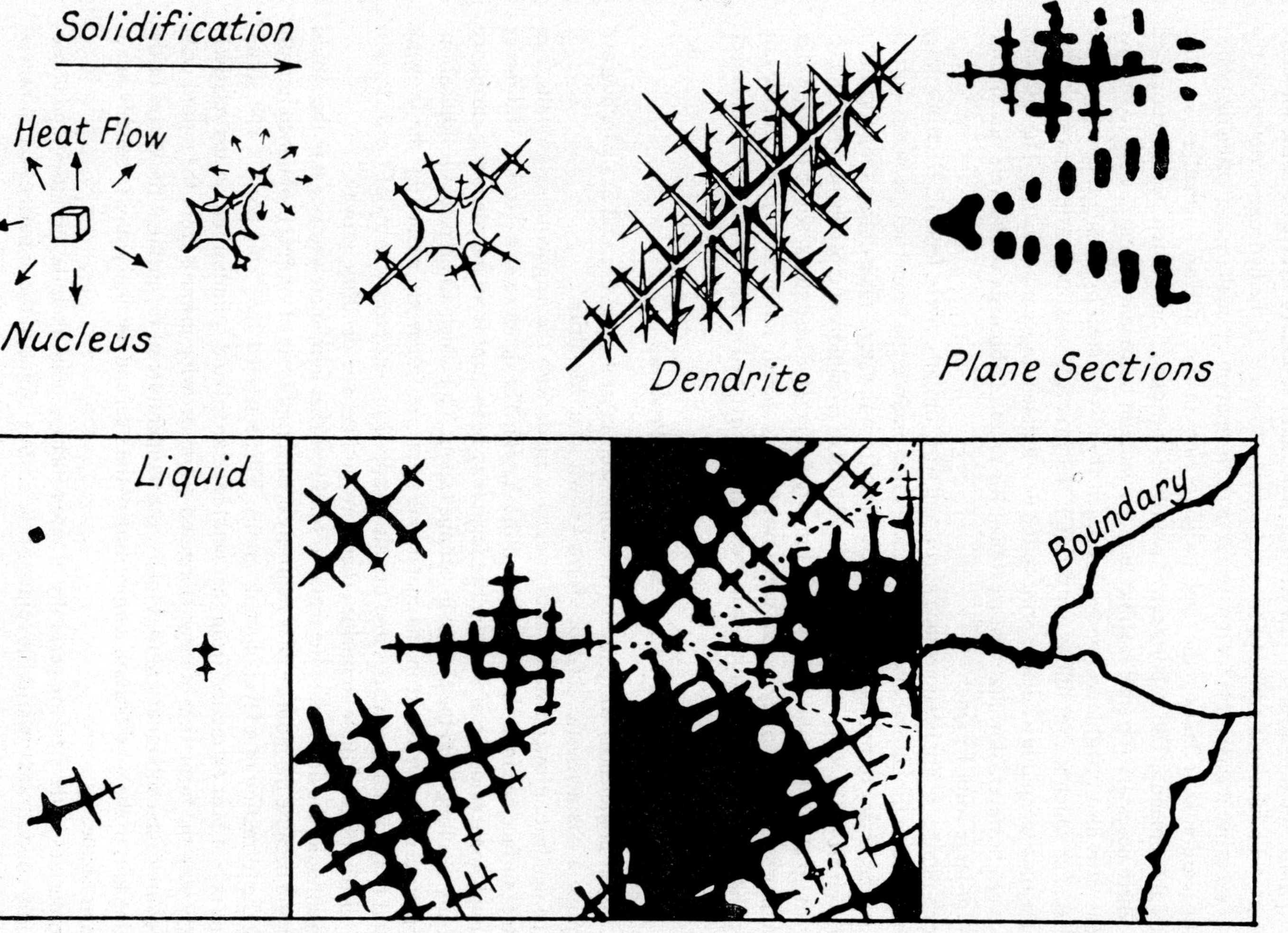

FIG. 5.2.2A. Solidification of metals—crystal growth (From *Metallurgy for Engineers*, Rollason, Edward Arnold Ltd.)

Hydrogen on the other hand cannot be removed easily from molten steel and it is necessary to limit contamination by this element. The main source of hydrogen is the moisture content of the materials added to the steel-making furnace. If the hydrogen content of the steel is excessive it is possible for it to result in the formation of small fissures often described as hair-line cracks or flakes in the steel. Large forgings in alloy steel are particularly sensitive to this phenomenon. The importance of removing hydrogen from molten steel is recognised and where necessary the steel is subjected to a vacuum treatment during casting to remove the gas. It is also possible to remove hydrogen from solid steel by raising the temperature to a level at which it can diffuse out through the structure without causing cracking. This was the only technique available prior to the development of vacuum casting. The diffusion process is, however, slow and heat treatment for times of about six weeks were not uncommon for large rotor forgings.

For a number of applications such as rod and wire for nails, rivets, fencing, electrode cores and thin sheets for automobile bodies (where good engineering properties are not essential), it is not necessary to remove the dissolved gases from the steel, since during the fabrication operation from ingot to wire or sheet, the extensive reduction in cross section is sufficient to weld up all the blow-holes present in the original ingot. These steels, which are known as "rimming steels", generally have a very low carbon content, and hence are very ductile and amenable to the extensive reduction in cross-section involved in fabrication to wire and sheet.

5.2.2. **Solidification Processes**

A metal commences to freeze by forming seed crystals (see Fig. 5.2.2A) which grow, adopting a characteristic "fir-tree" form known as a "dendrite".

The dendrites grow outwards until contact is made with the neighbouring growths; this contact surface becomes the boundary of the crystal. The dendrite arms thicken and merge although some evidence of the dendrite form may be evident from interdendritic shrinkage porosity resulting from the change in volume (about 3%) which accompanies solidification. In practice impurities exist in the metal and these are frequently pushed by the growing crystal to the grain boundaries and are also trapped in between the arms of the dendrites. The distribution of these impurities therefore outlines the dendritic growth.

In a mould, the first part of the metal to solidify is that immediately adjacent to the mould wall and because the mould wall is relatively cool, a rapid rate of cooling is obtained initially with the formation of a layer of small grains. This is termed the "chill zone". The subsequent rate of heat extraction from the metal is retarded and solidification occurs relatively slowly with the formation of large elongated grains growing inwards from the mould wall. These are called columnar grains. Finally, the temperature of the metal at the centre of the casting falls to the solidification temperature of the steel and the final grains to solidify are of regular shape.

Impurities in the steel are generally of lower melting point than the steel itself and consequently concentrate towards the centre of the casting. Some of the impurities are, however, trapped between the growing columnar grains and solidify *in situ*. These impure areas are referred to as "segregates". During solidification, the steel is contracting and special precautions are necessary to avoid the formation of a contraction cavity, termed "pipe", at the top

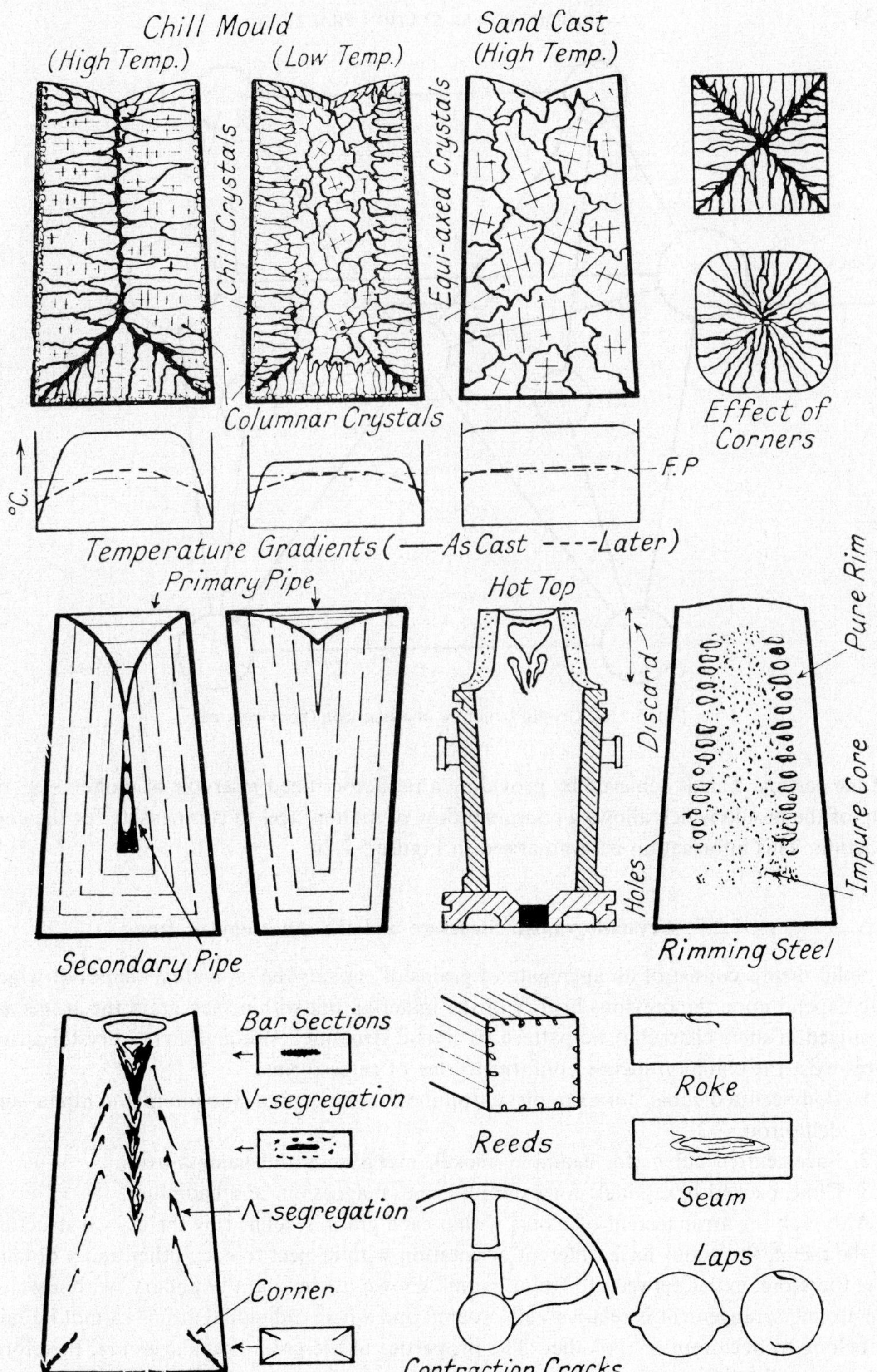

FIG. 5.2.2B. Structure, contraction effects and segregation in ingots (From *Metallurgy for Engineers*, Rollason, Edward Arnold Ltd.)

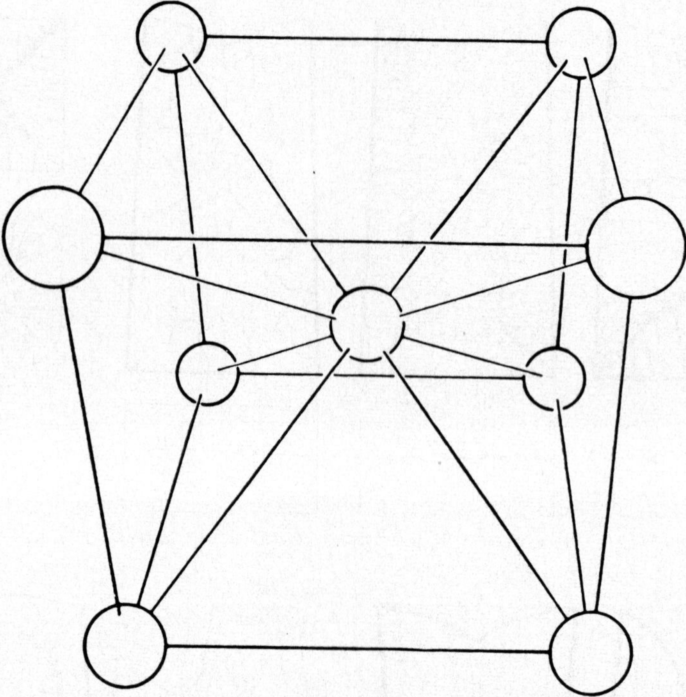

FIG. 5.2.3A. Crystal structure of alpha-iron (body-centred)

of the casting. This is achieved by providing a refractory lined reservoir of molten steel on top of the mould which allows a continual flow of molten steel to compensate for the contraction. This information is summarised in Figure 5.2.2B.

5.2.3. **Crystallographic Structure and the Allotropy of Iron**

Solid metals consist of an aggregate of grains of crystals the individual shapes of which will depend upon the previous history of the material, but within each grain the atoms are arranged in some characteristic pattern or crystal structure. Although many crystal structures exist the common metals conform to one of three forms:

1. Body-centred cubic: for example, chromium, molybdenum, vanadium and alpha- and delta-iron.
2. Face-centred cubic: for example, nickel, manganese and gamma-iron.
3. Close packed hexagonal: for example, zinc, magnesium and cadmium.

Although the arrangement of atoms within each grain is defined by the crystal structure of the metal, the grains have different orientation with respect to each other and a boundary, therefore, exists between touching grains known as the grain boundary, within which the atomic arrangement is relatively disordered and where individual atoms cannot be said to belong to one grain or the other. The properties of the grain boundaries are, therefore, fundamentally different from the grains and this has important consequences on the behaviour of materials.

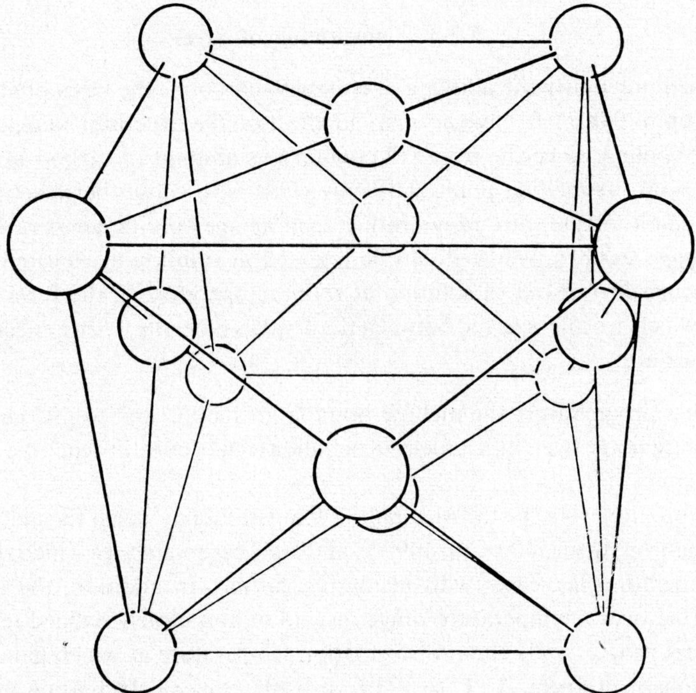

FIG. 5.2.3B. Crystal structure of gamma-iron (face-centred)

Some elements are able to exist in more than one crystallographic form. Each form is termed an "allotrope" or "allotropic modification". Iron itself exists in two forms of the cubic pattern, namely, body-centred cubic (bcc) (Fig. 5.2.3A) and face-centred cubic (fcc) (Fig. 5.2.3B). The bcc form exists between room temperature and 910°C, and between 1400°C and the melting point at 1539°C. The lower temperature form is known as "alpha"-iron and the higher temperature form as "delta"-iron. An aggregate of grains of this crystallographic structure is termed "ferrite", the alpha form being "alpha-ferrite" and the delta form "delta-ferrite". The face-centred cubic form exists between 910°C and 1400°C and is referred to as "gamma-iron". An aggregate of grains of this crystallographic structure is termed "austenite". Under the microscope both ferrite and austenite appear similar.

It is not surprising to find that these two differing crystalline structures exhibit different physical properties and to illustrate two of the more important ones, austenite has a coefficient of thermal expansion approximately $1\frac{1}{2}$ times that of ferrite and has a lower thermal conductivity.

If a piece of iron is heated, its crystallographic structure changes from alpha-iron to gamma-iron at 910°C, from gamma-iron to delta-iron at 1400°C, and it melts at 1539°C, the reverse occurring on cooling. One further change does occur at a temperature of 769°C, when alpha-iron loses its magnetism; above this temperature iron is non-magnetic. The changes in crystal structure which occur at 910°C and 1400°C are accompanied by changes in volume, and on heating, a contraction occurs at 910°C and an expansion at 1400°C. The reverse occurs on cooling through these temperatures.

5.2.4. **Constitution of Steel**

As mentioned previously the alloying of iron with carbon is the basis of all steel making, Additions of up to 0·008%C have no marked effect on the structural changes which occur on heating or cooling, since the iron will contain this amount of carbon in solution at all temperatures up to its melting point. The only effect is to cause the structural changes to occur over a small temperature range rather than at one specific temperature. Increasing the carbon above 0·008%, whilst it can be retained in solution at elevated temperatures, results in it being thrown out of solution at room temperature in the form of particles of iron carbide, which are interspersed between the crystals of ferrite. Other effects of increasing carbon contents are:

1. To depress progressively the melting point from 1539°C and to introduce a widening temperature range over which melting or solidifying occurs, instead of a clearly defined melting point.
2. To raise progressively the transformation temperature at which the delta-iron changes to gamma-iron, from 1400°C to 1492°C at 0·10% carbon, above which the transformation temperature decreases with increasing carbon. In addition, the transformation then occurs over a temperature range instead of at a clearly defined temperature.
3. To depress progressively the transformation temperature at which gamma-iron changes to alpha-iron, from 910°C to 723°C at 0·80% carbon. Increasing the carbon content above 0·80% raises the transformation temperature to 1130°C at 2·0% carbon. The transformation also occurs over a temperature range rather than at a clearly defined temperature.

All these changes in structure of necessity take time to occur, and whilst those occurring at high temperature occur quite rapidly, it will be appreciated that those occurring at low temperatures are very sluggish and long periods of time are necessary to allow them to take place.

If we now construct a graphic diagram, plotting the temperatures of the structural changes against carbon content, with pure iron at the left-hand edge, a diagram of the form shown in Figure 5.2.4A is obtained. This represents equilibrium conditions and is known as the iron–carbon (or more exactly, as applied to steels, the iron–iron carbide) equilibrium diagram. The point *J* on the diagram represents a complex solidification mechanism called a peritectic reaction, but this is of little significance to the final properties of steels and will not be discussed here. The point *C* represents another complex solidification mechanism called a eutectic reaction which will be referred to briefly later in discussing cast iron. The point *S* refers to a similar mechanism occurring in this instance in the solid state and is called a eutectoid reaction.

The region bounded by points *N*, *H* and *A* represents the solid solution of carbon in delta iron, the region *N–J–E–S–G* represents the solid solution of carbon in gamma-iron, termed austenite, and the region *G–P–O* represents the solid solution of carbon in alpha-iron. These illustrate the variation in solubility of carbon in iron at all temperatures up to melting. The point *S* denotes the minimum temperature and the carbon content at which austenite can exist and below this temperature any austenite is immediately transformed to a constituent

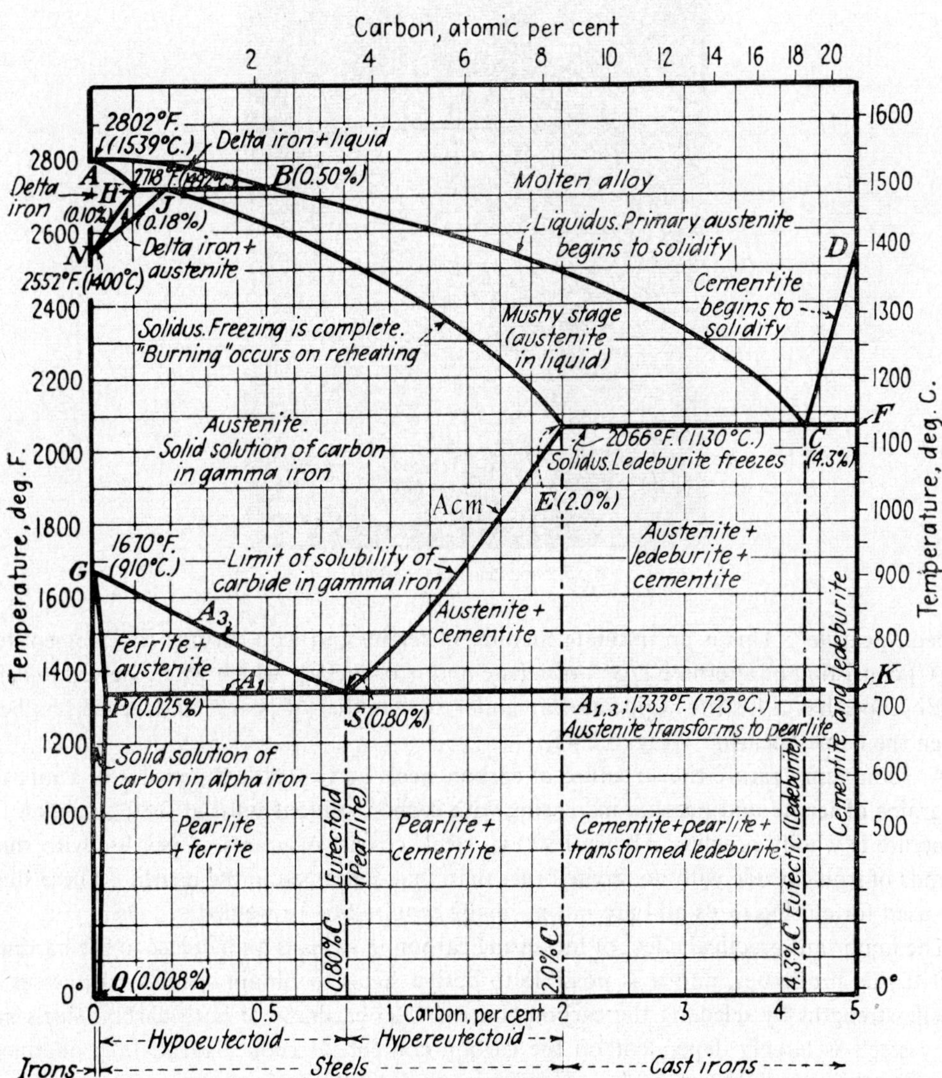

FIG. 5.2.4A. The iron/iron carbide equilibrium diagram (Reproduced by kind permission of the American Welding Society.)

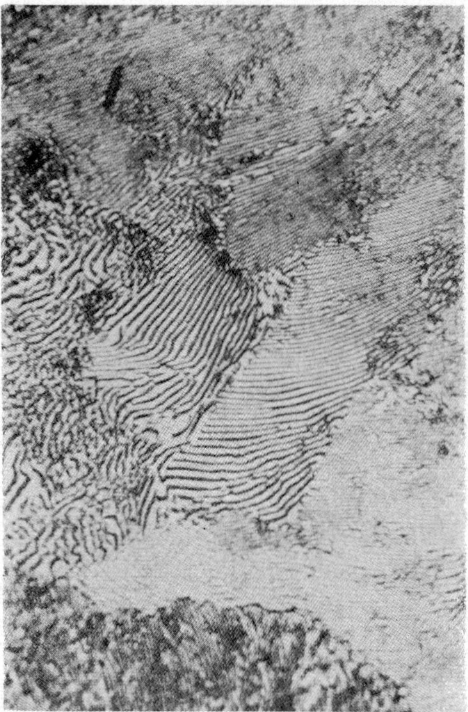

FIG. 5.2.4B. Pearlite (×1000)

called "pearlite". This is an intimate mixture of ferrite and iron carbide (or "cementite", Fe_3C) consisting of alternate layers of ferrite and iron carbide which when viewed under a microscope, has a lustrous appearance similar to "mother-of-pearl" and hence has been given the name "pearlite" (Fig. 5.2.4B).

At room temperature the structure of carbon steels up to 0·80%C consists of a mixture of grains of ferrite and pearlite increasing with carbon content until at 0·80% carbon the structure is wholly pearlite. Above 0·80%C, steels consist of grains of pearlite with small islands of iron carbide with no ferrite other than that contained in the pearlite. These steels are used for cutting tools and are not normally required to be welded.

The important practical effect of increasing carbon in steels is an increase in the hardness and tensile properties, and it is possible to design steels to obtain specific hardnesses or tensile strengths by selecting the carbon content. The hardness of both carbon steels and alloy steels is largely dependent on the carbon content. Carbon also has a pronounced effect on the weldability of steels, and it is always desirable to keep the carbon content as low as possible in steels which are required to be welded.

5.2.5. **Heat Treatment of Steel**

Heating a eutectoid steel, that is one containing 0·80% carbon at room temperature and composed exclusively of pearlite, causes a transformation in its structure as the temperature passes through the point *S* on the diagram, at 723°C. Figure 5.2.5 is a simplified form

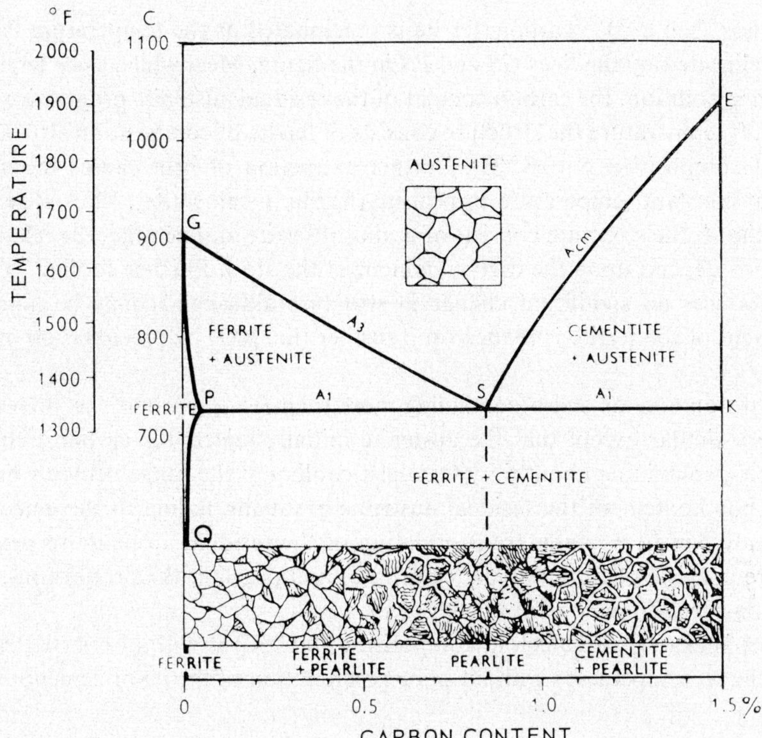

FIG. 5.2.5. "Steel" portion of iron/iron carbide diagram (Reproduced by kind permission of Sandvik U.K. Ltd.)

of the "steel" portion of the iron–iron carbide equilibrium diagram, and illustrates the appearance of the ferrite/pearlite, pearlite, pearlite/cementite structures, when viewed through a low-power optical microscope, as the carbon content varies from 0 to 1·5%. This transformation consists of a disassociation of the pearlite, its constituent parts (cementite and ferrite) going into solution to form a new stable phase called "austenite". Under practical conditions of heating, the formation of austenite requires a certain length of time and takes place, in the case of plain carbon steels, within a temperature range of 5° to 10°C.

Similarly with steels containing more or less than 0·80% carbon, which are designated hyper- or hypo-eutectoid steels respectively, the transformation into austenite occurs when the temperature passes the line *PSK* on the diagram. The transformation temperature is commonly referred to as the A_1 point. If the temperature rises beyond this point the excess ferrite or cementite, as the case may be, dissolves in the austenite. Lines *GS* and *SE* in the diagram mark the completion of the transformation according to the carbon content of the steel and are known briefly as either the A_3 and the A_{cm} points, respectively. Beyond these critical temperatures, ferrite and cementite are completely in solution and the steel has a purely austenitic structure. It should be noted that whereas practical requirements of steel generally call for a fine grain structure, the grain size of austenite increases with the temperature and duration of the temperature.

If a plain carbon steel is slowly cooled from above its critical temperature (A_1, Fig. 5.2.5), that is, from the austenite phase region, the transformations are reversed. In steels containing less than 0·80% carbon, ferrite is precipitated as the temperature falls through the range delineated by the lines *GS* and *PS* in the figure. Meanwhile, since ferrite contains little carbon in solution, the carbon content of the residual austenite progressively increases until at the A_1 temperature the structure consists of ferrite of composition *P* (0·025%C) and austenite of composition *S* (0·8%C). Further extraction of heat causes the austenite to transform at constant temperature to pearlite (ferrite + cementite). Thus at temperatures just below the A_1 the structure consists of grains of ferrite and pearlite, the relative proportions of which depend upon the carbon content of the steel. Further cooling to room temperature produces no significant change in structure although it may be noted that the carbon content of the ferrite is reduced still further (line *PQ*) by precipitation of more carbide.

The transformation of steels containing more than 0·8% carbon (i.e. hyper-eutectoid steels) is very similar except that the austenite initially rejects the carbon rich-phase, cementite (Fe_3C, containing about 6·8%C) whilst cooling in the range between lines *SE* and *SK*, the carbon content of the residual austenite gradually falling to the eutectoid value. Further removal of heat causes transformation of the residual austenite to pearlite. Thus the structure consists of iron carbide (cementite) and pearlite, the proportions, again, depending upon the carbon content of the steel.

If the steel is exactly of eutectoid composition (0·8%C) it will, of course, transform to pearlite at the A_1 temperature without prior precipitation of ferrite or cementite.

5.2.5.1. Effects of Cooling Rate

The events described above occur if the steel is cooled slowly under equilibrium conditions (hence the term "equilibrium diagram"). However, the attainment of equilibrium conditions is governed by the ability of the various atoms present to move, that is, to diffuse, and arrange themselves in the position required by the crystal lattices of the new phases. This diffusion is dependent in turn upon the temperature and the time available for the necessary movements to be accomplished. The higher the temperature and longer the time the easier is the attainment of equilibrium. In plain carbon steels the structures produced by annealing (heating followed by slow cooling) are reasonably well represented by the equilibrium diagram. At faster rates of cooling, however, equilibrium is not approached and significant differences in structure arise. Indeed if the rate of cooling is sufficiently rapid the transformations which depend upon diffusion may be suppressed and other modes of transformation occur. It is this ability to change the structure by varying the rate of cooling, which forms the basis of the heat treatment of steel.

The first effect of a higher rate of cooling than the equilibrium rate is to lower the temperature at which transformation occurs, and an understanding of the different structures which may be produced, can be gained by considering the processes which would occur if the steel were cooled instantaneously to a temperature below the equilibrium transformation temperature, and then held at constant temperature until the transformation was complete. This can be simulated experimentally by rapidly quenching specimens of the steel from the austenite phase region to the transformation temperature and holding them isothermally to determine the initiation, progress and completion of the transformation. The

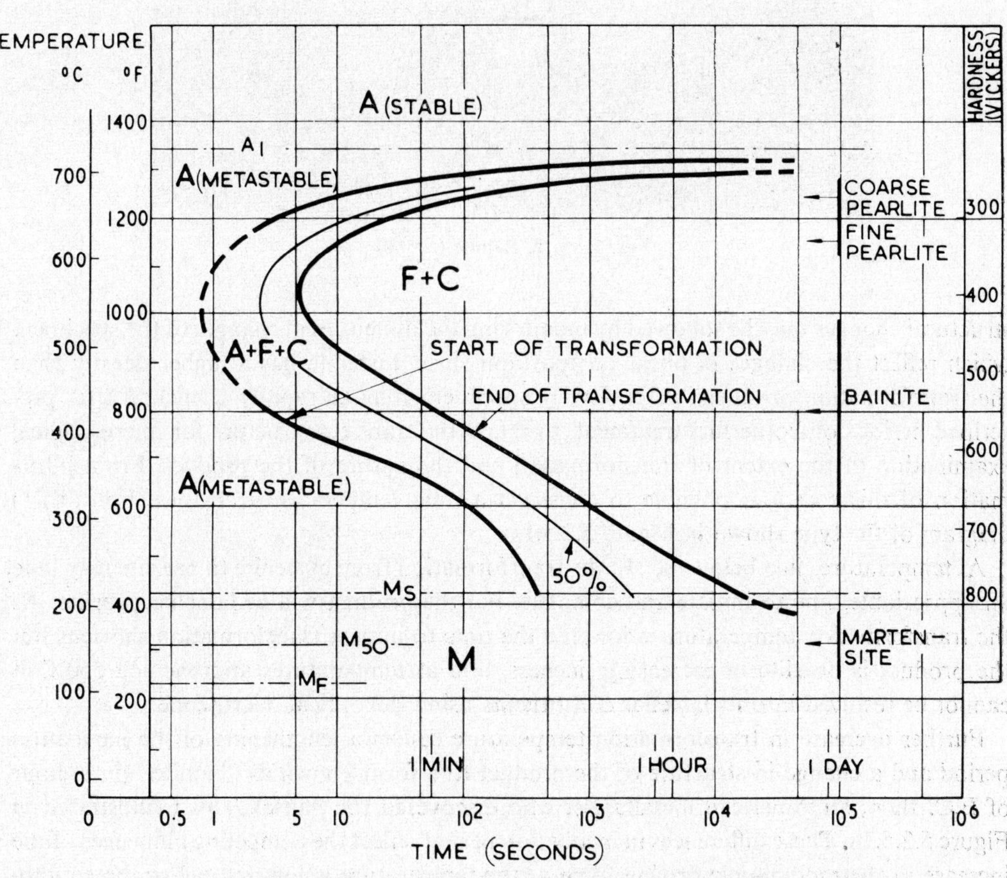

FIG. 5.2.5.1A. Typical isothermal transformation diagram of a steel (TTT diagram)

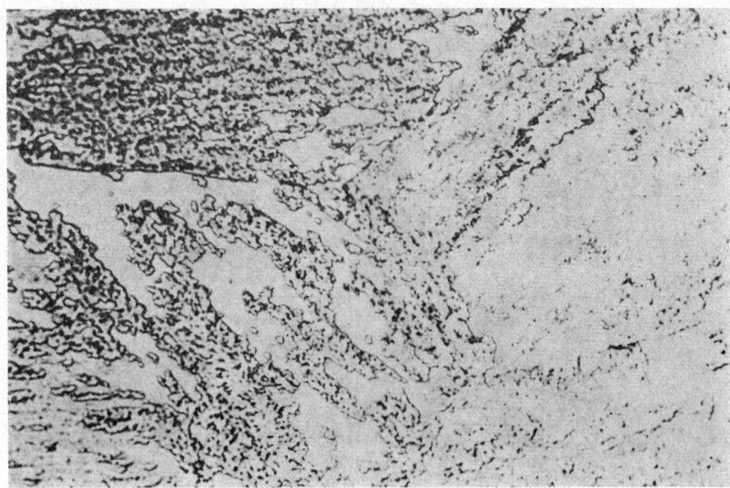

FIG. 5.2.5.1B. Bainite (×500)

structural changes may be followed by monitoring the dimensional changes of the specimen, which reflect the changes in phase composition since austenite has a higher density than the transformation products. Alternatively specimens may be rapidly quenched after prescribed periods of isothermal treatment, to retain the transient structure for microscopical examination of the extent of transformation and the nature of the product. From information of this sort it is possible to construct a time–temperature–transformation (TTT) diagram of the type shown in Figure 5.2.5.1A.

At temperatures just below the A_1, the transformation from austenite to pearlite may take an appreciable time to initiate and complete, but the product will be lamellar pearlite. As the transformation temperature is lowered the time to initiate transformation shortens but the product is pearlite of increasing fineness, and at temperatures approaching 550°C it cannot be resolved into its lamellar constituents using the optical microscope.

Further decrease in transformation temperature causes a lengthening of the incubation period and a change in structure of the product to a form known as "bainite" (in honour of E.C. Bain, an American metallurgist who discovered the phase). This is illustrated in Figure 5.2.5.1B. These differences in incubation period reflect the competing influences of the increase in thermodynamic driving force as the temperature is lowered below the equilibrium value (A_1) and the increasing difficulty of diffusion of atoms to form the new phases.

If the temperature is lowered sufficiently, the diffusion-controlled nucleation and growth modes of transformation are suppressed completely and the austenite transforms by a diffusionless process in which the crystal lattice effectively shears to a new crystallographic configuration known as "martensite" (Fig. 5.2.5.1C). This phase has a tetragonal crystal structure and contains carbon in supersaturated solid solution. The formation of martensite, unlike pearlite and bainite, is dependent not on time but on temperature. Thus there is a temperature at which the first traces of martensite will form, known as the martensite start temperature (M_S). Further transformation will only occur if the temperature is lowered and will be complete at some temperature known as the "martensite finish temperature"

FIG. 5.2.5.1C. Martensite (×500)

(M_F). The M_S temperature is between 150° and 350°C and the M_F temperature about 100°C lower. Under the optical microscope martensite has a characteristic acicular appearance.

The above account has considered only the transformation of austenite below the critical temperature, and in all but steels of precisely eutectoid composition this will be preceded and accompanied by precipitation of ferrite (pro-eutectoid ferrite) in hypoeutectoid steels or cementite in the case of hyper-eutectoid steels. Lines indicating the formation of these phases can also be depicted on the TTT diagram.

In practice the transformation of a steel on cooling occurs normally under continuous cooling conditions rather than isothermally, but similar processes are involved.

In the martensitic condition the steel is hard and brittle as shown by the comparison of steels of different carbon content in the slowly cooled (annealed) and rapidly cooled (quenched) conditions, shown in the following table.

	Carbon content, %					
	0·1	0·3	0·5	0·7	0·9	1·2
	Hardness, VPN					
Slow-cooled	109	121	170	229	248	285
Water-quenched	390	640	810	920	920	920

Steels are rarely used in the hard but brittle martensitic condition, but high strength coupled with improved ductility can be obtained by tempering, i.e. reheating the martensitic structure at an appropriate temperature below the critical temperature (A_1). This procedure of hardening followed by tempering is the basis of the heat treatment of steels.

Martensite, as pointed out previously, is a non-equilibrium structure and the reheating or tempering treatment will tend to restore the material to its equilibrium phase constitution

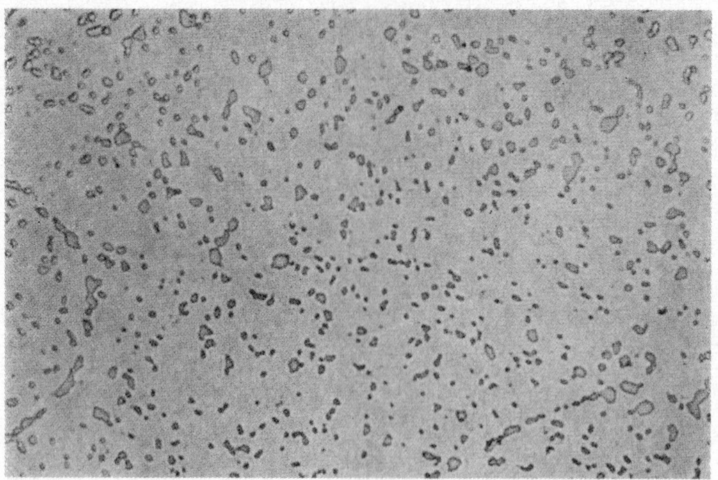

FIG. 5.2.5.1D. Spheroidised pearlite (×500)

of ferrite + carbide. Thus if a steel is tempered at progressively higher temperature the first effect is to relieve the high internal stresses which characterise the martensite structure. The carbon in supersaturated solid solution then commences to precipitate as carbide, in this case in the form of discrete particles rather than the lamellar form associated with its formation as pearlite. As the temperature of tempering is raised and the time increased, the carbides grow in quantity and size and the matrix becomes depleted in carbon. These changes are accompanied by a reduction in hardness and strength, and by an increase in ductility. If tempering is continued the structure ultimately consists of approximately spherical particles of carbide in a relatively soft matrix of ferrite. By judicious choice of tempering conditions the desired mechanical properties may be obtained.

The lamellar form of carbide in pearlite may also be spheroidised by long-term heating at temperatures approaching the lower critical temperature (the A_1), that is, 500° to 720°C. In this case the driving force for spheroidisation is the reduction of interfacial energy between the carbide phase and the matrix—analogous to the spherical shape of a soap bubble, which is caused by surface tension effects—but the rate of approach to the equilibrium condition is determined by the diffusion of atoms which, in turn, depends on the temperature and duration of heating. Given a knowledge of the original structure, it is possible to assess the thermal history of a specimen from the degree of spheroidisation which has occurred. In this way it is possible to demonstrate that a high-temperature component, such as a superheater tube, has been overheated and this is useful in the investigation of plant failures. Figure 5.2.5.1D shows the appearance of spheroidised pearlite, under the microscope.

5.2.5.2. EFFECTS OF ALLOYING

A prerequisite to the hardening of steels is that martensite should be formed on cooling, but this can only be achieved if the rate of cooling is great enough to suppress the formation of pearlite or bainite and in plain carbon steels this can be achieved by quenching relatively small specimens in water. Larger specimens, however, cannot be cooled sufficiently rapidly through the whole section because of the limited rate at which heat can be extracted through the surface of the piece. However, this difficulty can be overcome by alloying the steels to modify their transformation characteristics. For example, some alloying additions such as nickel and molybdenum reduce the rate of diffusion so that pearlite formation is suppressed in favour of the martensitic transformation; thus it is as if the curves on the TTT diagram are displaced to the right, i.e. to longer times. It is also possible to change the M_S as M_F temperature by alloying.

Thus a wide selection of mechanical properties may be obtained by selection of carbon content, alloying additions, cooling rate through the austenite transformation, and conditions of tempering.

The following glossary summarises the terms used to describe the various heat treatments used to condition steels:

Annealing. Heating into the austenite region and slow cooling in a furnace or in a container filled with a thermal insulating material such as vermiculite.

Normalising. Heating into the austenite region and cooling in still air.

Quenching. Heating into the austenite region and immersing in circulating oil or water.

Tempering. Reheating after normalising or quenching, to a temperature below the change point (about 723°C), and cooling in air, oil or water.

Subcritical annealing. Similar to tempering but generally longer times are involved.

The effects of alloying elements are numerous and a list of a few of the more important effects is given below.

1. To alter the transformation temperatures and times.
2. To modify the room temperature and elevated temperature strengths of given structures by (a) stiffening the crystals and (b) introducing complex precipitates which tend to harden the steel.
3. To modify the type of oxide film formed on the surface of the steel and thereby affect its corrosion resistance.

Alloying elements can be broadly classified into two groups.

(a) *Austenite stabilisers*, which have the effect of extending the temperature range over which austenite is formed. Such elements are carbon, manganese, nickel, copper and cobalt.

(b) *Ferrite stabilisers*, which have the effect of extending the temperature over which both alpha and delta ferrite are formed, which consequently reduces the temperature range over which austenite is formed. Such elements are silicon, chromium, molybdenum, tungsten, titanium and niobium.

Some of these elements, for example, chromium, molybdenum and vanadium, also form carbides which replace or modify the iron carbide in the structure.

Additions of the austenite-stabilising elements reduce the temperature at which the austenite to ferrite change occurs and will consequently facilitate the formation of martensite or bainite, with slower rates of cooling than are necessary with plain carbon steels. This also means that for a given cooling rate, larger cross-sections can be fully hardened uniformly throughout their section.

Chromium, although in itself a ferrite and carbide former, has a side effect of making the structural changes very sluggish, which makes the suppression of the austenite to ferrite change in heat treatment easier to achieve. It is, therefore, extensively used in steels to be hardened and tempered. A further important property of chromium, particularly marked when present in quantities above about 5%, is to improve resistance to corrosion and oxidation.

Resistance to corrosion and oxidation of steel depends on the film of oxide formed on its surface and in carbon and many low-alloy steels, this oxide film offers little or no resistance to atmospheric corrosion. At elevated temperatures these steels have good resistance to oxidation in air or flue gas up to 575°C (1065°F) but above this the rate of oxidation increases rapidly. The presence of chromium, however, in excess of about 5% promotes the formation of a more protective oxide film and although 5% is insufficient to obtain useful resistance to atmospheric and aqueous corrosion, it is enough to improve the oxidation resistance up to about 600°C. Further increasing the chromium content produces a more resistant oxide film and at 13%, satisfactory resistance to mild corrosive media, such as wet steam, is achieved. Applications of this type of steel are steam turbine blades, propeller and pump shafts, impellers and water turbine runners. Increasing the chromium content above 13% produces improved resistance to more corrosive media and at 28% chromium, satisfactory oxidation resistance at 1100°C can be obtained.

Steels with chromium contents of up to 14%, provided the carbon content is above 0·12%, can still be hardened and tempered. With chromium greater then 14%, however, because of its strong ferrite forming tendencies, the alpha- and delta-ferrite regions join together at the expense of austenite, with the result that such steels are wholly ferritic up to their melting point and cannot be hardened by heat treatment. Such steels tend to have a very coarse crystal structure and are then relatively brittle at room temperature and difficult to fabricate. They are not, however, brittle at elevated temperature, and find application in furnace parts such as grate bars, clinker dams, and dampers, generally in the form of castings.

In order to be able to utilise the good corrosion-resisting properties of these high chromium steels and at the same time attain satisfactory engineering properties, it is necessary to re-establish the austenite region. This can be done by adding nickel and with an 18% chromium steel the addition of about 2% of nickel does this and produces a steel which can be hardened and tempered. This is the well-known S80 steel (En 57) which is widely used for pump shafts in the marine field. Maintaining a chromium level of 18%, the addition of increasing amounts of nickel extends the re-established austenite region until at 8% nickel the temperature of the change from austenite to ferrite is depressed below room temperature, and the structure at room temperature consists of grains of austenite. These steels are termed austenitic and include the well known 18/8 stainless steel.

Since the austenite–ferrite change, on which hardening and tempering are dependent, is suppressed below room temperature, these austenitic steels are similar to the high chromium ferritic steels in that they cannot be hardened by normal heat treatment. They are different from the ferritic high chromium steels in that they are extremely ductile, and ideally suited for deep pressings and similar applications. In addition, since they are austenitic they are non-magnetic and they have a high coefficient of thermal expansion and a low thermal conductivity.

Although these steels are not hardenable and have relatively low tensile strength at room temperature, they do have good elevated temperature tensile properties which, when combined with their good corrosion resistance, suit them to applications demanding this combination. These include superheater tubing and steam piping where the metal temperatures are in excess of 550°C, gas turbine components, and numerous types of pressure vessels employed in the chemical and allied industries.

The bulk of the austenitic steels produced are of the chromium and nickel type and other elements such as titanium, niobium, molybdenum, copper, tungsten, cobalt and aluminium, may be added to impart special properties. A common example found in modern power stations is the superheater tube and steam pipe austenitic steel containing 18% chromium, 10 to 12% nickel and $2\frac{1}{2}$% molybdenum.

5.2.6. **Cast Irons**

Cast irons, like steels, are alloys of iron and carbon, but they contain larger proportions of carbon, silicon, sulphur and phosphorus. A typical composition of cast iron is:

Total carbon	Silicon	Manganese	Sulphur	Phosphorus
3·5%	2·0%	0·8%	0·1%	0·35%

Reference to the iron/iron carbide equilibrium diagram (Fig. 5.2.4A) shows that cast irons occupy the region between approximately 2% and 4% carbon. The carbon in cast iron can exist at room temperature as iron carbide, or as graphite which is the more stable form. Irons containing carbon as graphite are soft, easily machinable and are called "grey irons". Irons with carbon present as iron carbide are extremely hard, difficult to machine, and are called "white" or "chilled" irons. Irons with fairly equal proportions of graphite and iron carbide have intermediate hardnesses and are called "mottled" irons. The factors mainly influencing the form of the carbon are the rate of cooling during solidification and the chemical composition. Rapid rates of cooling prevent the formation of graphite and result in a high proportion of iron carbide and hard white irons. Conversely, slow rates of cooling promote graphite formation and result in soft grey irons. This effect of cooling rate on the structure of cast iron is reflected in the precautions necessary for the satisfactory welding of the material. White irons can be regarded as unweldable. Grey iron can be welded but it is imperative to avoid the formation of a zone of white iron adjacent to the weld deposit as a result of the chilling effect of the parent metal. It is essential, therefore, that grey iron be preheated to retard the cooling rate in the weld metal and adjacent parent metal.

The rate of cooling of a casting is largely dependent on the mass of the casting and consequently castings of a relatively thin cross-section tend to be "white" and castings of large cross-sections are always "grey". Castings with varying sections often contain a gradation of structure from grey to white.

With regard to chemical composition, elements such as manganese, nickel, copper, carbon and silicon promote graphite formation whereas chromium and other carbide formers promote carbide formation. Sulphur and phosphorus are present in larger quantities than in steels, because the processes involved in the manufacture of cast iron do not include refining operations to remove these elements. Grey iron is extensively used in engineering because of five notable characteristics, these being:

(a) Cheapness.
(b) Low melting point and high fluidity making it suitable for castings of intricate shape.
(c) Relatively good erosion- and corrosion-resistance.
(d) High damping capacity, with respect to vibration.
(e) Relatively good mechanical properties under compressive loading.

The graphite in grey irons exists in the form of flakes which act as stress-raisers under tensile loading and consequently grey irons have relatively low tensile strength and ductility.

White irons, because of their extreme hardness and brittleness are generally confined to applications demanding high abrasion resistance and where they can be fully supported by less brittle materials.

Small castings which tend to be white because of their small cross-sections can, however, be heat-treated to convert the iron carbide into graphite which results in a soft, machinable iron suitable for general engineering applications. This process, which is similar to the annealing of steel, is termed "malleablising" and the final product is "malleable iron". The graphite in malleable iron is in the form of regular particles rather than flakes and as such has relatively little embrittling effect under tensile loading. The room temperature tensile properties of these irons approach those of mild steel.

Deliberate alloying additions are made to cast irons, in the same way as to steels, to enhance their mechanical properties and resistance to corrosion, abrasion and high temperature.

5.2.7. **Deformation and Fracture**

All engineering structures are required to sustain a load in one form or another, and a useful measure of a material for a given engineering purpose is its tensile strength. The design stresses imposed are generally within the elastic limit of the material, as defined by the tensile test, but a structure may be subject to local stresses above the yield point at certain locations due to discontinuities (such as stress raisers within the structure) and therefore must be capable of deformation without fracture. This requires that the material should have an adequate level of ductility in addition to strength.

In a metal the properties of individual grains are direction-dependent according to the particular crystal form, but since they are generally randomly oriented and small in comparison with conventional material section sizes, the directional effect is usually eliminated. It is thus entirely feasible for the engineer to regard a piece of iron or steel as having uniform properties in all directions, for design purposes. Certain exceptions to this generalisation include grain-oriented silicon-iron sheet for transformer cores, where a special effort is made to produce material in which the grains have the same crystalline directions, in order to reduce electrical losses. It is important to consider the generally harmful effect of segregations of non-metallic inclusions, which are always present to some extent in commercial steels and which, if not properly controlled, may appear as laminations and strings of inclusions in plate and tube. Such inclusions impart directional mechanical properties by introducing weaker and/or more brittle material, thus for example, great care is taken in the working of important forgings such as those for turbine rotors, so that any non-metallics are oriented in the least harmful direction.

Deformation of crystalline materials such as metals involves both "elastic" and "plastic" strain. The elastic portion is not permanent and disappears when the load is removed, but plastic strain involves the movement of atoms within individual crystals along certain planes, by a shearing process known as "slip". It has been found in practice that metal crystals are much weaker than would be expected from theoretical predictions, based on considerations of the inter-atomic forces which must be overcome to produce slip. This has now been explained by the concept of line imperfections within the crystal lattice, known as "dislocations", which are able to move under relatively low stresses through the crystal lattice, producing slip of one part of the crystal with respect to the remainder. Modern methods of microscopy using high magnifications and resolutions, now permit the observation of some of these phenomena, confirming the theories previously formulated.

If a specimen is subjected to an increasing uniaxial stress as in a normal tensile test, it initially deforms in an elastic manner until a point is reached (the elastic limit or yield point), where the metal ceases to behave elastically and suffers permanent or plastic deformation (see Fig. 5.3.1A(a)). If the stress continues to increase, the metal will extend a certain amount and then fracture, producing a crack which is usually trans-crystalline at ambient temperatures. Under creep (high-temperature) conditions and relatively low stress values, movement is generally concentrated at the grain boundaries usually resulting in an intergranular type of failure.

As described previously, the plastic deformation of a metal occurs by a shear process but if the stress system imposed on a specimen is such that shear cannot occur readily, the material may fail by rupture of the inter-atomic bonds without significant plastic deformation. These conditions may obtain when multi-axial stresses are present. Thus, if three equal but mutually perpendicular tensile stresses are applied to a specimen, it can only deform elastically until the stresses reach a value at which the specimen is torn apart. Such triaxial stress conditions occur in practice and must be considered in evaluating materials (see Section 5.3.3).

Triaxial stress conditions may be obtained by notching a specimen and this situation more readily initiates a crack by increasing the ratio of tensile to shear stresses. Important factors in this class of test are the stress situation (notch severity and section size), speed of deformation, and temperature. The fracture is usually composed of three zones, an initial small ductile (fibrous fracture) zone where some plastic deformation occurred, a second brittle (cleavage fracture—across certain crystal planes) zone where no plastic deformation occurred and possibly a third zone of fibrous fracture where plastic deformation recurs. The extent of these zones in conjunction with the energy required to break the specimen, gives some indication of the toughness of the material. This property varies considerably with temperature, and in some plain carbon and low alloy steels the change from tough to brittle behaviour is often within the ambient temperature range.

5.3. CRITERIA FOR THE ASSESSMENT OF MATERIALS

Metals to be used in power plant are not necessarily very different in character from those used in other fields of engineering but two requirements, resistance to stress at temperature and ability to obtain the requisite mechanical properties in large masses, are of increasing importance.

It is useful to indicate the metals that are used throughout the plant, and these are listed in Table 1 for boilers, Table 2 for turbines and Table 3 for piping.

Before discussing these materials in more detail it is necessary to outline the assessment criteria that are commonly used to select them initially. These may be summarised as:

1. Room temperature mechanical properties.
2. Physical properties.
3. Temperature-dependent mechanical properties.
4. Weldability.
5. Corrosion resistance.

5.3.1. Mechanical Properties at Ambient Temperature

The most important qualities to be assessed are strength and ductility. The former is indicated either by the stress required in a standard tensile test to break the metal (ultimate tensile stress, U.T.S.), or the stress required to produce a specific amount of plastic deformation (proof stress). These two criteria are shown on the stress–elongation curves (Fig. 5.3.1A).

TABLE 1

STEELS FOR LAND BOILERS

Steel type	Chemical composition									Application
	C	Mn	Si	Cr	Mo	V	Nb	Ti	Ni	
Carbon	0·15	0·5	0·3	—	—	—	—	—	—	Boiler tubing
Cr, Mn, Mo, V	0·15	1·5	0·3	0·7	0·3	0·1	—	—	—	Boiler drums: by-pass vessels
0·5 Mo	0·1	0·5	0·3	—	0·5	—	—	—	—	Superheater tubing
1 Cr, ½Mo	0·1	0·5	0·3	1·0	0·5	—	—	—	—	Superheater tubing, headers
2¼ Cr, 1 Mo	0·1	0·5	0·3	2·25	1·0	—	—	—	—	Superheater tubing, headers
9 Cr, 1 Mo	0·1	0·5	0·7	9·0	1·0	—	—	—	—	Superheater & reheater tubing
18 Cr, Ni, Nb (AISI 347)	0·1	1·5	0·3	18	—	—	1	—	12	Superheater tubing
18 Cr, Ni, Ti (AISI 321)	0·1	1·5	0·3	18	—	—	—	0·5	12	Superheater tubing
18 Cr, Ni, Mo (AISI 316)	0·1	1·5	0·3	18	2·5	—	—	—	13	Superheater tubing
Esshete 1250	0·1	6·0	0·5	16	1	0·25	1	—	10	Superheater tubing

TABLE 2

STEELS FOR TURBINES

Steel type	Chemical composition										Application
	C	Mn	Si	Cr	Mo	V	Nb	Ni	W	Co	
1 Cr, Mo	0·4	0·6	0·2	1·2	0·6	—	—	—	—	—	Bolts up to 480°C (900°F)
$\frac{1}{2}$ Cr, $\frac{1}{2}$ Mo, V	0·1	0·5	0·2	0·4	0·6	0·3	—	—	—	—	Turbine loop piping
Cr, Mo, V	0·25	0·5	0·2	1·0	1·0	0·3	—	—	—	—	Turbine rotors
Cr, Mo, V	0·15	0·5	0·2	1·0	1·0	0·3	—	—	—	—	Turbine casings (castings)
Cr, Mo, V	0·2	0·6	0·2	1·0	1·0	$0{\cdot}7_5$	—	—	—	—	Bolts up to 565°C (1050°F)
$2\frac{1}{4}$ Cr, Mo	0·15	0·5	0·3	2·2	1·0	—	—	—	—	—	Turbine valves
Ni, Cr, Mo, V	0·25	0·5	0·2	0·3	0·5	0·1	—	2·8	—	—	Turbine rotors
3 Cr, Mo, V, W	0·25	0·3	0·4	2·7	0·5	0·3	—	—	0·5	—	Turbine casings (castings)
12 Cr	0·08	0·4	0·3	12·5	—	—	—	—	—	—	Turbine blading
12 Cr, Mo	0·1	0·6	0·3	12·5	$0{\cdot}7_5$	—	—	—	—	—	Turbine blading
12 Cr, Mo, V	0·1	0·6	0·3	12·0	0·6	0·2	—	—	—	—	Turbine blading
12 Cr, Mo, V, Nb	0·15	0·5	0·4	12·0	1·0	0·3	0·3	—	—	—	Turbine blading
Stellite 6	1·0	—	—	26·0	—	—	—	—	5·0	68·0	Turbine blade erosion shields
High-speed steel	0·7	—	—	6·0	—	—	—	—	19·0	—	Turbine blade erosion shields
AISI 316	0·05	1·5	0·5	17·0	2·5	—	—	11	—	—	Steam chests
AISI 347	0·07	1·5	0·5	18·0	—	—	1·0	12	—	—	Turbine casings

TABLE 3

STEELS FOR STEAM PIPING

Steel type	Chemical composition							
	C	Mn	Si	Cr	Mo	V	Nb	Ni
0·5 Mo	0·1	0·5	0·3	—	0·5	—	—	—
$\frac{1}{2}$ Cr, $\frac{1}{2}$ Mo	0·1	0·5	0·3	1·0	0·5	—	—	—
$\frac{1}{4}$ Cr, $\frac{1}{2}$ Mo, $\frac{1}{4}$ V	0·1	0·5	0·3	0·5	0·5	0·25	—	—
$2\frac{1}{4}$ Cr, 1 Mo	0·1	0·5	0·3	2·25	1·0	—	—	—
AISI 347	0·07	1·5	0·5	18	—	—	1·0	12
AISI 316	0·05	1·5	0·5	17	2·5	—	—	11
Esshete 1250	0·1	6·0	0·5	16	1·0	0·25	1·0	10

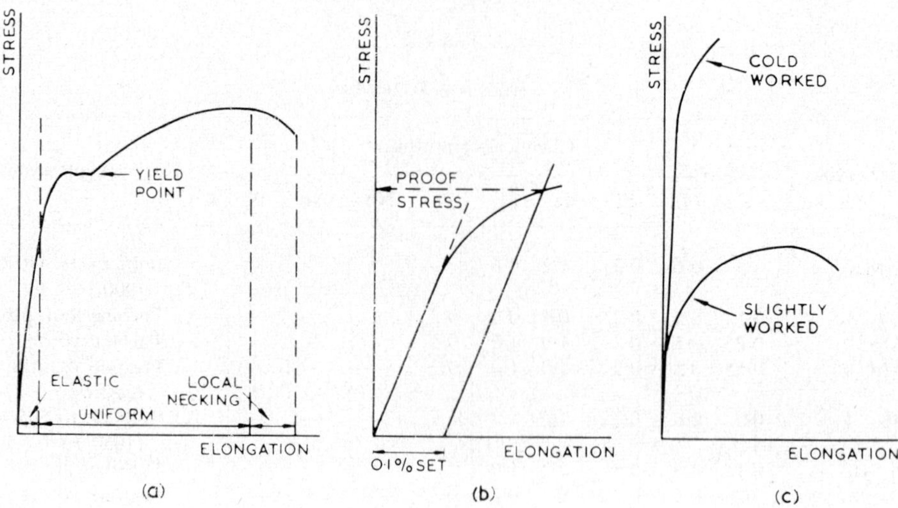

FIG. 5.3.1A. Stress–elongation curves for mild steel (From *Metallurgy for Engineers*, Rollason, Edward Arnold Ltd.)

The well-defined yield of mild steel is shown in (a) and the method of assessing proof stress, corresponding to 0·1% extension, in a steel not showing a sharp yield, is shown in (b). Design stresses for pressure parts not subject to creep are based upon arbitrarily selected fractions of one or another of these mechanical properties. The hardening effect of cold deformation is shown in (c), in the figure.

Ductility is measured by noting the elongation undergone by a test specimen during straining to fracture. It may also be gauged by measuring the reduction of area at the point of fracture. If a metal exhibits low values of elongation and/or reduction of area, it is unlikely to be able to withstand stress concentrations or shock loading without premature failure.

The most common cause of failure of engineering components is fatigue under cyclic stresses. A tensile test gives an indication of how a steel will behave under static tensile loading, but it gives only an approximate indication of how a steel will behave under fluctuating stresses. Ferrous metals differ from many other metals in that they exhibit an endurance limit, which means that there is a maximum level of fluctuating stress which can be tolerated indefinitely. In most steels this stress is approximately 50% of the ultimate tensile strength and the endurance limit is defined as the stress which can be endured for ten million reversals of stress. The method of presentation of fatigue data as "S–N" curves, is shown in Figure 5.3.1B where the range of the cyclic stress (for a given mean stress), is plotted against the logarithm of the number of cycles. A definite endurance limit is only obtained in tests in air at temperatures below about 300°C. Tests at higher temperatures or in a corrosive environment, do not display an endurance limit, and the type of curve obtained under such conditions is also shown in the figure.

The presence of stress concentrations in the form of notches, holes, sharp changes in sections, or re-entrant angles, have a pronounced effect on the behaviour of a component under fluctuating stress conditions and it cannot be emphasised too strongly that these effects must be carefully considered in the design stage.

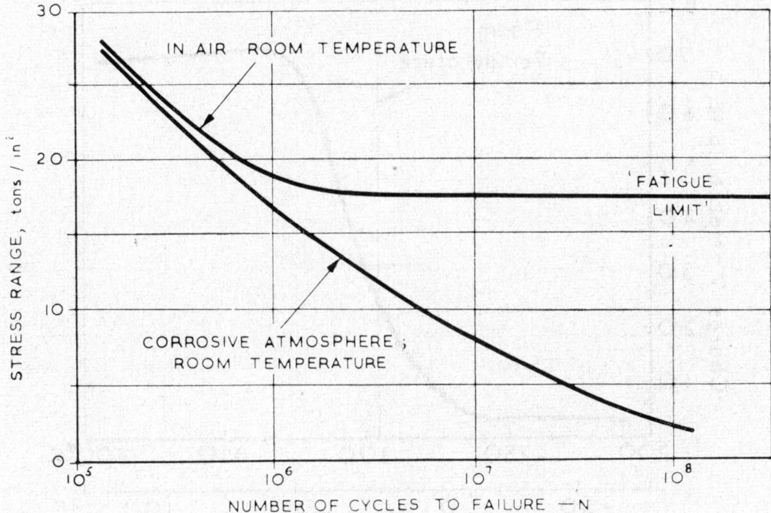

FIG. 5.3.1B. Typical S–N curves

5.3.2. Physical Properties

When a metal is heated over a range of temperatures it expands. The rate of this expansion varies with the type of metal, being higher for austenitic than for ferritic steels. Similarly the ability of a metal to conduct heat is a function of its crystal structure and associated chemical composition. Thus austenitic steel has a lower thermal conductivity than ferritic steel. The density of iron is 7·87 g/cc, and this is modified very slightly by alloying additions but not sufficiently to be of technological interest.

Typical physical properties of ferritic and austenitic steels are summarised below:

	Coefficient of thermal expansion 20°–600°C (per °C)	Thermal conductivity at 600°C (C.G.S. units)
Ferritic steel	$14{\cdot}3 \times 10^{-6}$	0·080
Austenitic steel	$18{\cdot}5 \times 10^{-6}$	0·050

In situations where metal is heated and cooled rapidly, ferritic steels have intrinsic advantages over austenitic steels. Thus in the example of heating a thick section of metal under restraint, a greater build-up of strain can be expected in the case of austenitic steels due to their higher rate of thermal expansion and lower coefficient of thermal expansion and lower coefficient of thermal conductivity. Furthermore the lower elastic limit of austenitic steels enhances the probability of some of the strain being plastic. When it is necessary to join ferritic and austenitic steels these differences in physical properties must be allowed for, if the initiation of cracking is to be avoided.

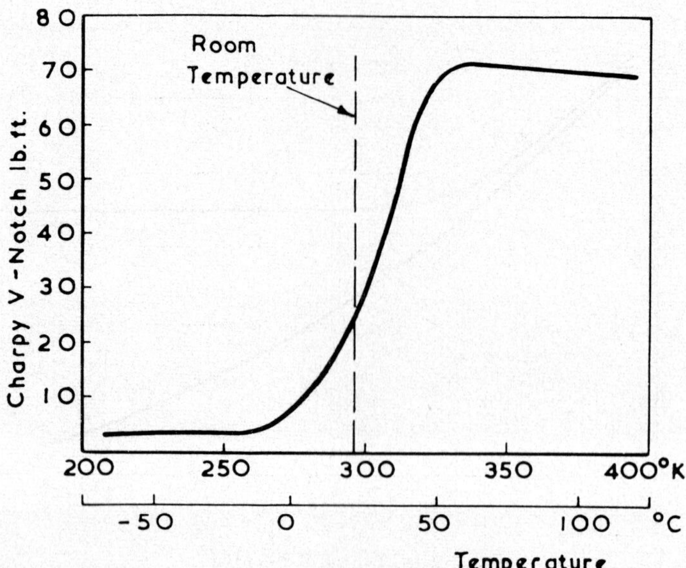

FIG. 5.3.3A. Typical "Charpy" V-notch values for a semi-killed low carbon steel (Reproduced from J. G. Tweedale, *Mechanical Properties of Metals*, Allen & Unwin)

Other physical properties of steels bear a direct relationship to the crystal lattice structure and these may have a strong influence on the efficacy of methods of non-destructive testing. Magnetic methods are not applicable to the austenitic steels and the coarse grain size of austenitic stainless steel weld metal increases attenuation of ultrasonic signals, thus largely invalidating ultrasonic inspection techniques. Depending on the type of crystal structure and chemical composition of the metal, there are variations in electrical properties such as permeability, but these are not very significant in relation to surface magnetic inspection where it can be applied. Nevertheless, such variations may be important in more refined techniques of examination (for example in metal sorting), where a small deviation in homogeneity or heat-treatment may markedly influence a specific electrical or magnetic parameter which can be readily monitored.

5.3.3. Temperature-dependent Mechanical Properties

TOUGHNESS AND BRITTLE FRACTURE

A test often called for in material specifications is the notch impact test, where a standard specimen containing a notch to induce triaxial stress conditions, is broken by subjection to an impact of known energy. The two tests in common use are the Izod (specimen stressed as a cantilever) and the Charpy (specimen stressed as a beam). This test is usually regarded as confirming the "toughness" of the product, and as some measure of the quality of a batch of material.

In practice it is found that the toughness of a steel is dependent on temperature, as shown in Figure 5.3.3A, where the energy absorbed in the Charpy test is plotted against the temperature of testing. At temperatures well above the value termed the "transition temperature",

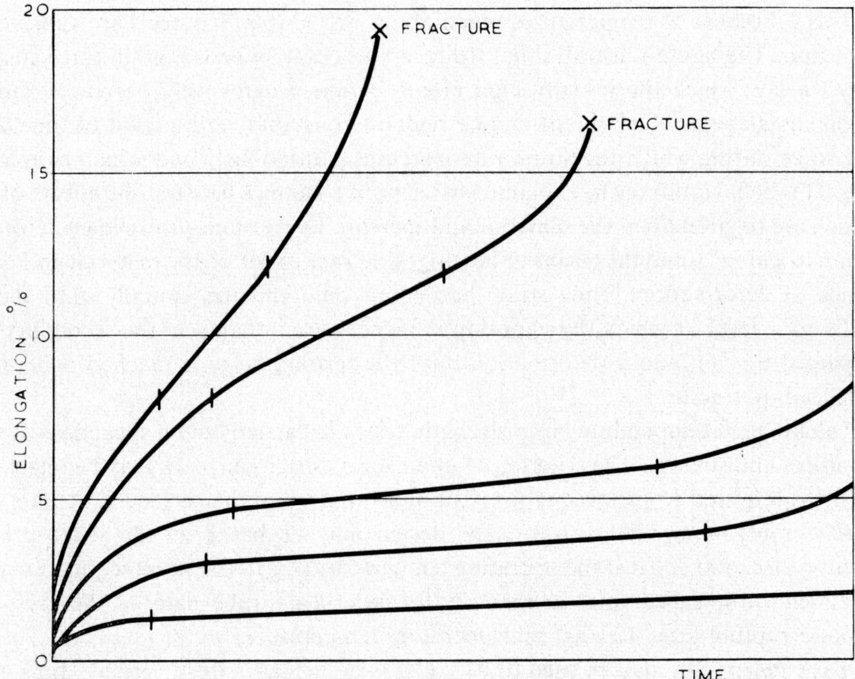

FIG. 5.3.3B. Typical family of creep curves

the material behaves in a tough manner and the fracture exhibits a fibrous appearance; below the transition temperature the material behaves in a brittle manner and the fracture is characterised by bright facets (often described incorrectly as "crystalline"), which arise because the fracture propagates by cleavage in certain preferred crystal planes. Brittle fracture propagates rapidly and, in engineering service, may result in catastrophic failure of a vessel or component. In the transition temperature region, the fracture is partially fibrous, partially brittle, and there is a transition temperature zone over which the behaviour changes from one mode to the other. The transition temperature itself may be defined in a number of ways, for example (in the Charpy test) the temperature at which the fracture is 50% fibrous, or the temperature at which the energy absorbed is 20 lb.ft.

In commercial steels the transition temperature may be greater than the ambient temperature and this illustrates the need to evaluate the brittle fracture characteristics of such materials. The size of a component may also affect the mode of failure, large components being more susceptible to brittle fracture than smaller ones, and this factor must also be taken into account. Thus the matter of toughness under fracture conditions is of much importance, and is the subject of considerable research effort.

CREEP

When a metal is stressed at a sufficiently high temperature it will continue to deform with time, although at a relatively low rate. This process, known as "creep", is illustrated graphically in Figure 5.3.3B and is of major importance in selecting metals for service in power plant. This is because at temperatures much above 350°C (660°F), steels are subject to this phenomenon. The figure illustrates that the creep process can be viewed in three stages: the primary stage in which the initially high rate of strain progressively decreases with time, the secondary stage in which the strain rate remains constant over a period of time, and the tertiary stage during which the strain rate increases continuously and which culminates in rupture. This behaviour may be explained in terms of a balance between the effects of straining, which are to strengthen the material and increase its resistance to further deformation ("strain-hardening"), and the effects of heating which are to soften the material and decrease resistance to deformation. Thus strain-hardening predominates initially until the strain rate falls to a level at which the opposing influences are in balance (the secondary stage), accompanied by continuous deformation until the tertiary stage is reached, where loss of strength leads to fracture.

To evaluate high-temperature creep strength, tests are carried out on specimens at various temperatures and stresses. The criterion of interest for design purposes may be related either to the stress required to produce a given amount of strain in a specified time interval (for example, for high-temperature rotors the design may be based on the stress producing 0·1% strain after 100,000 h at the operating temperature), or to the stress required to rupture the specimen in a specified time (for example, for superheater tube material, the stress required to cause rupture after 100,000 h at operating temperature).

The basic creep data may be used to derive curves such as "stress" versus "time to rupture" at various temperatures or "stress" versus "time to give various amounts of plastic strain", which are of direct use to designers. A typical design stress criterion is 60% of the stress that will, on average, lead to failure in 100,000 h at a specific temperature. Such a criterion measures the resistance to deformation exhibited by a metal and is useful in determining its strength relative to competing metals. It does not provide a complete assessment of the probable service performance of the metal in power plant. In practical engineering situations the behaviour may depend equally upon the capacity to deform before initiating a crack, and this feature will be affected by component geometry and the manner of loading during plant operation.

Since modern power stations operate with steam cycles of either 538°C (1000°F), 565°C (1050°F) or 600°C (1112°F), it will be evident that the majority of the boiler components and most of the turbine rotating parts will come within the time/temperature category of behaviour. A detailed discussion of the deformation process and the complex features of the metal that govern it, are beyond the scope of this lesson. Readers wishing to understand the phenomenon and its role in material selection for power plant, should consult the appended literature recommended for further reading.

Fatigue

Metals undergoing high-temperature service may also be subject to fatigue. This process is one in which failure may arise after exposure to very many cycles of alternating stress, with or without the superimposition of a mean stress. Although the commonest cause of metal failure in general engineering, it is comparatively rare in power plant. This is due to the fact that it proves possible to design away from high levels of alternating stress, and that the predominant failure mechanism at high temperatures is creep and not fatigue.

In power plant, it is possible to encounter situations that are classified as "thermal fatigue". In these the frequency of straining is given by the number of stops and starts endured during the full life of the plant (say 5000 to 10,000). The level of strain is enhanced by the creation of thermal gradients during operation, and/or by geometric strain concentration. This problem is believed to involve both creep and fatigue processes: the relevant properties of a metal are still a matter of research investigation.

5.3.4. Weldability

A modern boiler, producing 3,450,000 lb of steam per hour to supply a 500 MW turbine, contains about 90,000 pressure part welds. These are mainly butt welds between tubes in the boiler furnace, superheater and reheater. Where tubes are joined to larger cylinders, as in the junction of superheater tubes to headers or feed heater tubing to tube plate, fillet welds are required. It is therefore vital that metals chosen for their creep resistance and/or corrosion resistance, should be readily weldable.

We shall discuss the more detailed aspects of welding metallurgy when dealing with the problems of fabrication. The assessment of weldability has, in the past, relied upon a variety of *ad hoc* testing procedures in which welds were subjected to various types of mechanical restraint or where completed weldments were deformed, often in some kind of bend test, to measure their ductility.

Where welds are involved in high-temperature applications it becomes important to know the manner by which the welding method has distributed internal strain throughout the weldment and to appreciate the way this might modify the metal properties. It is sometimes noted, for example, that the zone of the parent metal immediately adjacent to the weld metal (the heat-affected zone) may achieve a low-ductility condition. This may result in eventual crack initiation during service from what has been a sound weld—in an inspection sense—upon completion of fabrication. The detection of such conditions requires more sophisticated mechanical property tests aiming to simulate the welding conditions, and tests to assess the time-dependent behaviour of prestrained metal. These approaches underline current researches in this quite complex subject.

5.3.5. Corrosion Resistance

If mechanical properties are the principal guide to metal selection for power plant, resistance to corrosion is perhaps the property least easy to anticipate. The corrosion resistance of a metal is either intrinsic as in a noble metal like gold or platinum, or is achieved by the formation of a protective oxide film.

Assessment of corrosion resistance in power plant environments is a formidable problem. Earlier lessons in this chapter have dealt with gas-side corrosion, "on-load" boiler corrosion and graphitic corrosion of cast iron. Metallurgically speaking, tests have been made on large boilers to assess the gas-side corrosion resistance of steels against various oil and coal combustion products. Experiments are being made in the use of low alloy chromium-molybdenum steels to combat "on-load" corrosion, whilst alloyed cast iron has been shown to be slightly better than ordinary cast iron in preventing graphitic corrosion.

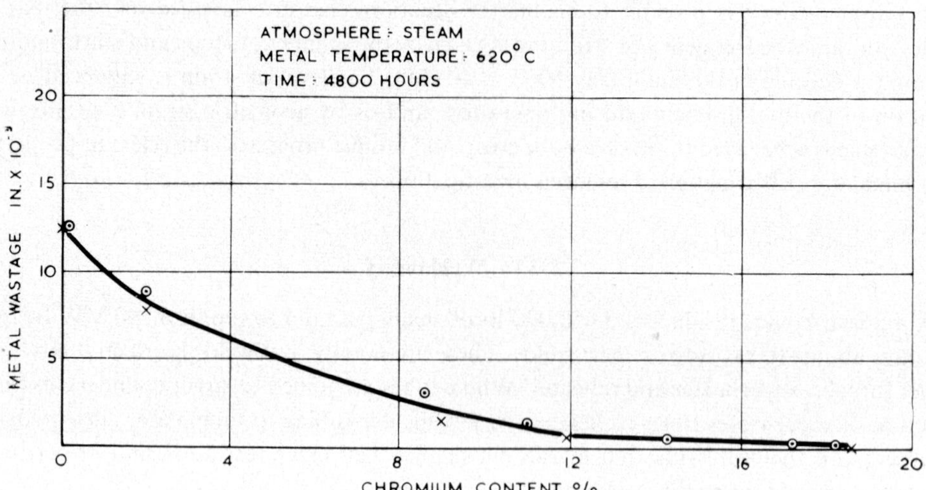

FIG. 5.3.5. Corrosion of steel in steam, related to chromium content

These assessments often turn out to be specific to the situation examined and not of general application. The behaviour of oxide films on metal is sensitive to environment in ways that cannot yet be stated quantitatively. Thus in the case of gas-side corrosion, the transport of metal outwards through the oxide film is occurring at the same time as ingress of oxygen or sulphur inwards from the gaseous atmosphere. In the example of "on-load" corrosion it can be expected that the rate of heat transfer across the metal and its oxide will also affect the corrosion process. In other cases, such as graphitic corrosion of cast iron, *in situ* electrochemical effects within the metal structure play an important role.

The above comments refer to the breakdown of the protective film: the normal situation is that the metal is protected by its oxide. In steels there are three elements known to be beneficial in conferring resistance to oxidation. These are chromium, aluminium and silicon and they appear to act by forming oxides (Cr_2O_3, Al_2O_3 and SiO_2), particularly able to withstand breakdown against corrosive gases. Unfortunately both aluminium and silicon tend to detract from the mechanical properties of steel so that chromium remains as the most useful of these elements in this respect. Table 4 lists the upper temperature limit for various steels as a function of their chromium content, assuming the absence of known aggressive atmospheres.

Figure 5.3.5 is of interest in showing one example where laboratory scale experiments and plant trials have given a similar result. The environment, dry superheated steam, would not be expected to introduce unusual effects upon the oxide films and the overriding importance of chromium content is clearly shown.

5.4. FABRICATION

Fabrication at works and plant construction at site require the production of material in the various shapes required, for example, pipe, tube and sections. This involves forming the material, such as by the bending of tube and plate, and the joining of parts and items, almost

TABLE 4

EFFECT OF CHROMIUM CONTENT ON HIGH-TEMPERATURE OXIDATION

Chromium content %	Typical steel	Upper temperature limit, °C
2	$2\frac{1}{4}$ Cr, 1 Mo	580
9	9 Cr, Mo	630
12	12 Cr	750
18	18 Cr, 8 Ni	800
21	21 Cr	900
23	23 Cr, Ni, W	1050
29	29 Cr	1125

invariably by fusion welding. Inspection and testing is carried out on the basic material, on weld joints and on the completed component.

5.4.1. Material Production and Forming

A great deal of the information on mechanical and physical properties of materials has been obtained by primary producers, from material in intermediate stages of production, for example, blooms, bar and small cast test pieces. It has been assumed that these properties will be representative of final items such as pipe, tube and major castings. This may be true for many ferritic materials when the final form receives the same completion heat treatment, for example, anneal, normalise, or normalise and temper, for most of the low-alloy creep-resisting steels. However, this is not always the situation and it has been shown that the method of tube production in one particular austenitic creep-resisting steel, may have some influence upon the structure of the material and subsequently its properties at high temperature. Data is now being obtained on an international basis for material in its final production form, and this procedure should ultimately extend the confidence placed by designers in the steels available.

The effect of cold work upon the subsequent performance of a steel, particularly when it is to be subject to high-temperature conditions, is not fully understood. Tubular components for superheaters have given rise to some concern in the past. Due to their relatively small cross section, e.g. 2 in. o.d.$\times\frac{3}{8}$ in. wall thickness, it is economical to bend the tubes cold and to weld them into complete elements without subsequent heat treatment. Up to 15% cold deformation has been accepted before a post-bending heat treatment was required. In this condition 0·5% Mo superheater tubing has been particularly subject to failure at the bends in service. This steel is now known to be susceptible to cracking because of its inability in this condition to deform very slightly in service, thus dissipating peak local stresses. It is no longer used in boiler construction for the C.E.G.B. (see ref. 2).

The effect of cold work on several different grades of steel used for superheaters is now under investigation in an attempt to ascertain the necessity or otherwise for introducing further heat treatment cycles into production schedules.

5.4.2. **Welding**

Welding has now become an indispensable tool in industry for joining components or producing complicated fabrications. The recognition of welding as a technology in its own right has been slow but development was hastened when it was recognised that certain fabrications were possible only by welded construction. The subject has grown during this century from the bare wire electric arc welding of non-critical parts, to the present time when all aspects of the process are under critical examination in an attempt to provide a weld joint potentially equal in all respects to the parent material.

Welding processes were exploited by engineers during a period when mild steel and plain carbon steels constituted the bulk of structural and engineering plant. Only with the necessity to join the more complex materials for high duty service has adequate consideration been given to the properties of weld metal and the effects of welding on the parent material.

In fusion welding, the surfaces to be joined are themselves brought to the melting point. An explanation of the various processes in common use using ferrous metals as our example, may prove useful. An essential requirement is the provision of a controlled source of heat and Figure 5.4.2 is useful in illustrating the sources available in terms of power and power intensity, and it is possible to trace the evolution of processes by the diagram. The blacksmiths forge weld used low intensity and medium power. The intensity was next increased quite substantially by the introduction of gas welding, then by an order of magnitude by conventional arc and more recently by the development of electron beam and ruby laser systems. Processes with which we are closely concerned at the present time are outlined below.

5.4.2.1. Gas Welding (Oxy-acetylene)

Acetylene is burned in a high efficiency torch in an atmosphere of oxygen and the heat of the flame is used to melt the parent plate—and a filler wire if so required (Fig. 5.4.2.1). The maximum temperature at the tip of the flame is approximately 3000°C. This process is still widely used for joining relatively thin sections, e.g. tubes (boiler and economiser), particularly for site butt welds and for sheet metal.

5.4.2.2. Arc Welding

This is the most widely used process in industry. It can be subdivided as follows, depending on the nature of the electrode and of the shielding medium used.

(a) *Arc welding with a coated electrode.* This is the classic process which has evolved from the early bare wire welding. The electrode consists of a steel wire covered by a flux coating which contains the necessary ingredients to provide a shielding gas to avoid contamination and stabilise the arc, to provide slag cover and to control the composition of the weld metal and its quality (Fig. 5.4.2.2A). The process has been used for almost all thicknesses and types of material although it becomes less economic for thicknesses much in excess of 1 in.

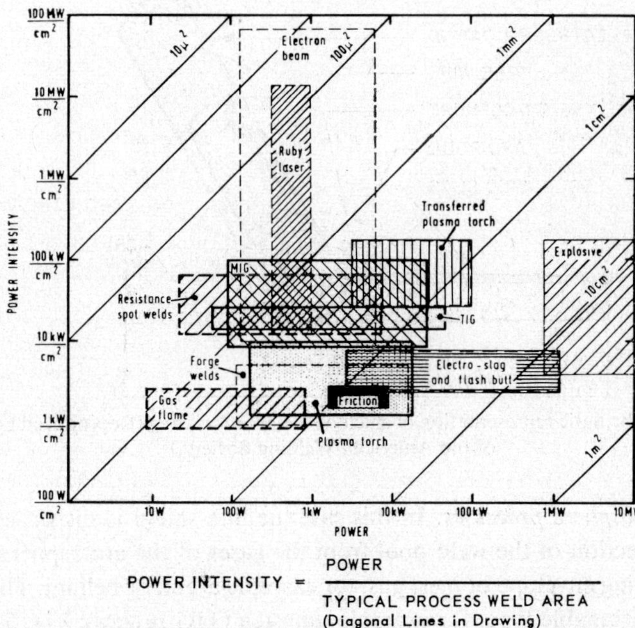

FIG. 5.4.2. Welding processes—power and power intensity (Reproduced by kind permission of the Welding Institute)

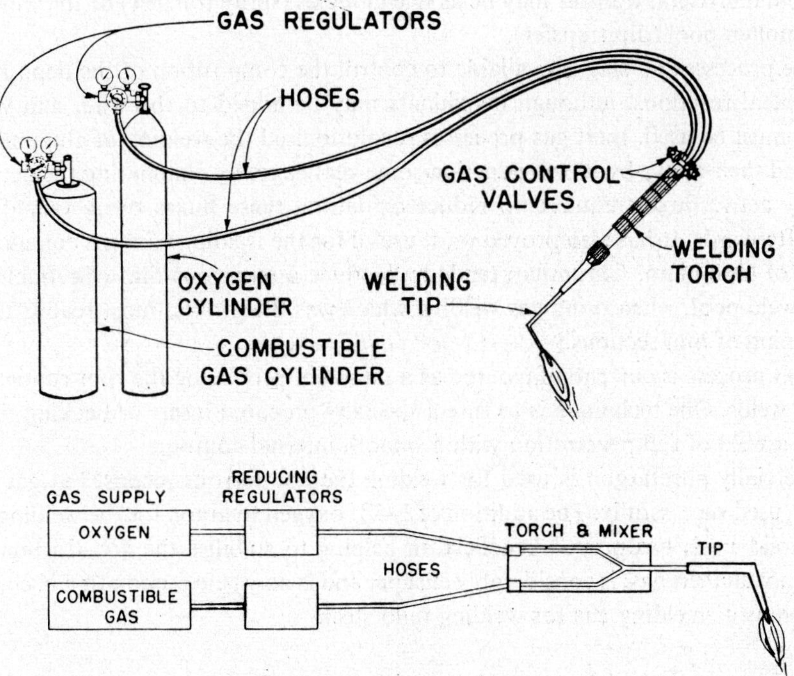

FIG. 5.4.2.1. Basic gas welding equipment (Reproduced by kind permission of the American Welding Society.)

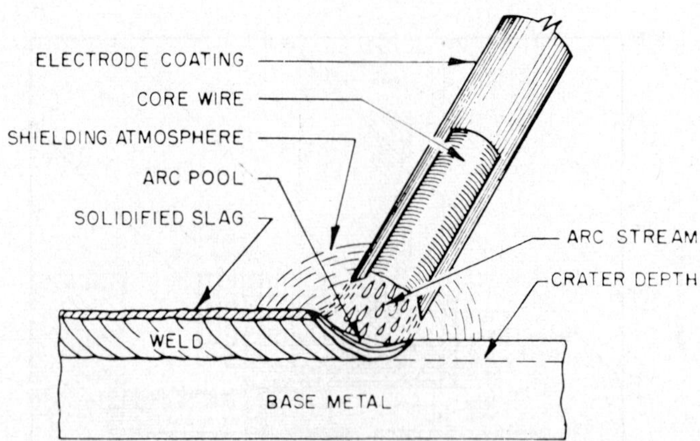

Fig. 5.4.2.2A. Schematic representation of shielded metal arc welding (Reproduced by kind permission of the American Welding Society.)

(b) *Inert atmosphere processes.* In this case the flux shield is dispensed with, and arc stability and protection of the weld pool from the gases of the atmosphere are effected by the use of a shielding envelope of inert gas, for example argon or helium. The electrode may be of the non-consumable type, for example tungsten (TIG process, Fig. 5.4.2.2B), or consumable in which a fine filler wire of the type required for the weld is fed automatically into the weld pool (MIG process, Fig. 5.4.2.2C). In the MIG process the electrical system is arranged to maintain the required arc length constant and therefore the process is inherently semi-automatic. Metal transfer may be as fine globules (spray transfer) or the wire may dip into the molten pool (dip transfer).

In these processes no slag is available to control the composition of the deposited metal by slag–metal reactions, although deoxidants may be added to the wire, and very clean materials must be used. Inert gas processes revolutionised the welding of aluminium, magnesium and their alloys by its introduction some years ago, by eliminating the necessity for the highly active fluxes required to reduce oxidation, these fluxes being very difficult to remove afterwards. It has also proved most useful for the welding of steels containing more than 5% of chromium. Chromium tends to produce a tenacious film of refractory oxide over the weld pool, when using gas welding which was the process most readily applicable to the joining of thin sections.

The TIG process is currently favoured as a means of producing the root run in pipe and tube butt welds. One technique is to fuse a specially prepared insert or backing ring which results in a weld of full penetration with a smooth internal contour.

Commercially pure argon is used for welding the non-ferrous metals, but gas mixtures have been used successfully. The addition of 1–5% oxygen to argon for the welding of stainless and mild steels has proved beneficial in helping to stabilise the arc. Carbon dioxide, although not an inert gas, is considerably cheaper and is now being used often in conjunction with argon as a shielding gas for welding mild steel.

(c) *Submerged arc welding (e.g. Unionmelt process).* The arc is struck between a continuously fed electrode and the workpiece under cover of a mineral flux powder. Usually the

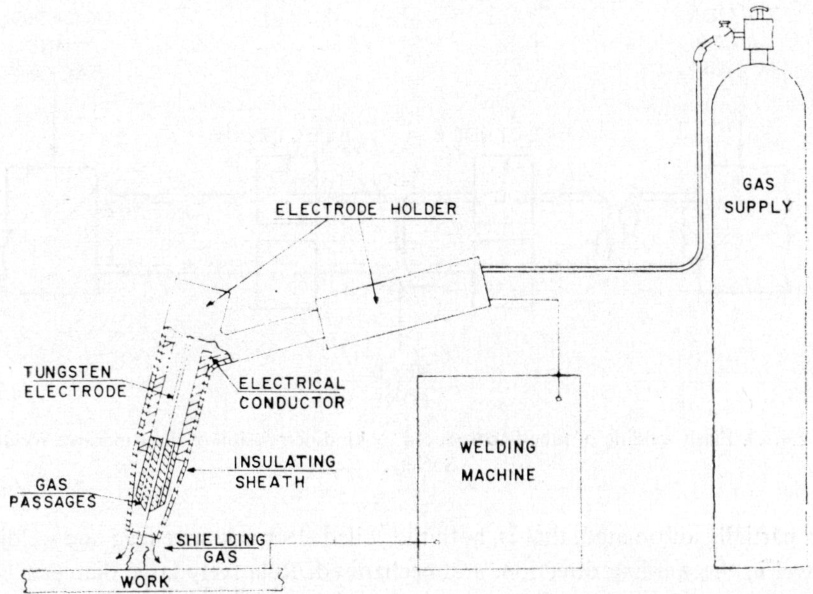

FIG. 5.4.2.2B. Schematic diagram of inert-gas metal-arc nonconsumable (TIG) process (Reproduced by kind permission of the American Welding Society.)

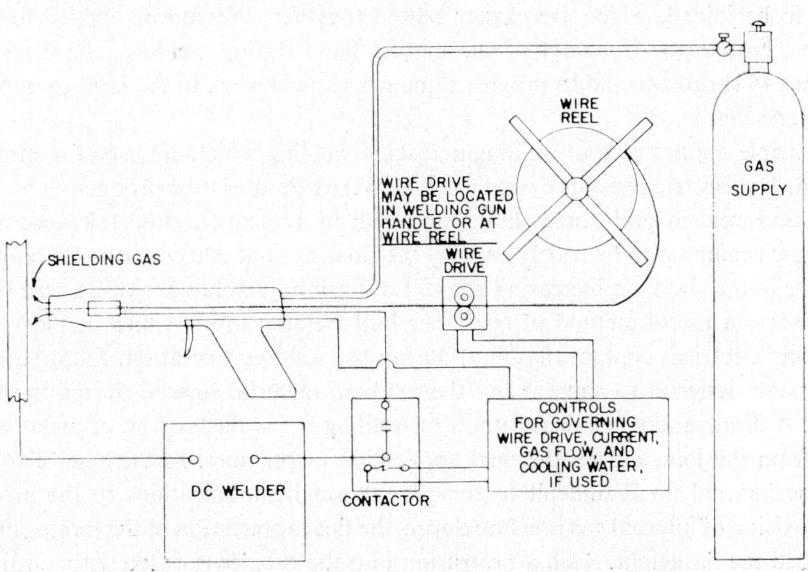

FIG. 5.4.2.2C. Schematic diagram of inert-gas metal-arc consumable (MIG) process (Reproduced by kind permission of the American Welding Society.)

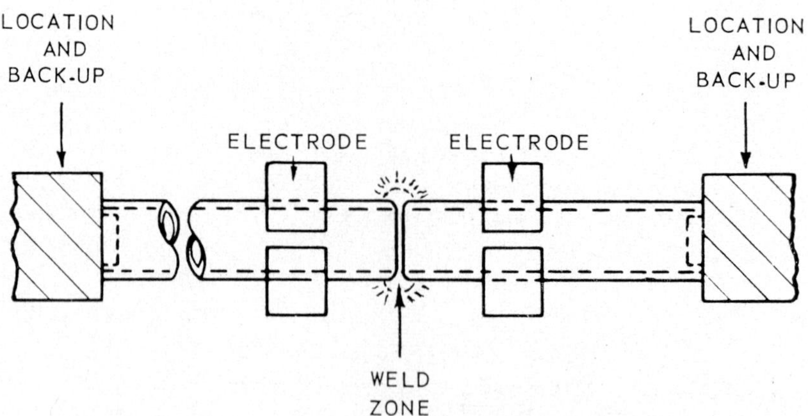

FIG. 5.4.2.3. Flash welding of tube (Reproduced by kind permission of the American Welding Society.)

method is partially automated, that is, both the coiled electrode wire feed and welding head which travel in the welding direction, are mechanised. Relatively large diameter electrode wire is possible and in the case of automatic welding, high currents may be used with a corresponding high deposition rate and good penetration. This method is most useful where heavy sections such as boiler drums must be joined, since high deposition rate permits heavy weld runs to be made with a reduction in welding time.

5.4.2.3. Electric Resistance Welding

In this process the heat required for welding is developed by passing a high current across the parts to be joined, which are closely butted together. Pressure is applied to rupture oxide films, ensure good contact of the mating faces during welding, eliminate central porosity due to shrinkage and to provide some mechanical work to the joint to improve its general properties.

This principle applies to spot welding and seam welding which are used for sheet metal fabrication. The electric resistance welding (E.R.W.) method of tube production by making a longitudinal weld in preformed sheet, can result in a cheaper, close tolerance material which is now beginning to be widely used by the Board's contractors for boiler tubing. For joining tube in the shop, an increasing amount of flash butt welding is being used (see Fig. 5.4.2.3). This is a useful method of resistance butt welding tubes, which includes making and breaking electrical contact (flashing) during the welding operation, followed by final heavy pressure designed to squeeze out the oxidised material formed during the flashing operation. A disadvantage of this method of welding is the flash or fin of metal which is extruded from the joint during the final application of pressure. In the case of tube butt welding the internal fin is difficult to remove, although modifications to the process to include provision of internal gas pressure during the final application of the forging pressure, have reduced the fin height. Such a protrusion into the tube bore is likely to obstruct the flow of water in service, and boiler corrosion has been observed to occur at internal potrusions associated with butt welds.

5.4.2.4. Electroslag Welding

This process is particularly suitable for the welding of very thick sections. Heat is generated by passing an electric current via the electrode(s) through molten slag to the workpiece. Welding is arranged to take place in the vertically upwards direction and the parent metal preparation is a square butt joint. This process is similar to submerged arc welding in that electrode wire and powdered flux are fed into the welding pool but in this case there is a more positive reaction between the active flux and the molten metal. Figure 5.4.2.4 illustrates this form of welding. It should be noted that several electrodes are often used and that these may be designed to traverse across the weld in addition to being fed into the pool, by a welding head, as in the submerged arc process. Water-cooled copper moulds close each end of the molten pool of weld metal and slag, and the entire assembly of welding head and moulds is moved vertically upwards to keep pace with the deposition of weld metal. The quality of the weld metal is high, but one disadvantage is the large grain size which is produced, and it is usual to normalise these weldments in order to refine the grain and improve mechanical properties in the weld joint. This procedure is somewhat unsatisfactory due to the high temperatures (about 900°C) required for normalising, and efforts are being made to reduce the grain size in the as-welded condition without recourse to this high-temperature treatment.

5.4.2.5. New Welding Processes

Electron beam welding and laser welding. These processes depend upon the energy of a focused beam of electrons and a coherent monochromatic beam of light, respectively. They are presently restricted to special applications but introduce a new order of power intensity into welding (approximately 10^7 W/cm^2). This can result in deep penetration (electron beam) and very precise location (electron beam and laser) of the weld. Disadvantages are that the electron beam techniques must be carried out in a vacuum chamber—although out-of-vacuum methods are being developed—and although the laser does not require a vacuum, it was first generated by a solid state system which can produce only pulses of energy and is thus of relatively low total power. Recent developments using a gas laser system have shown that continuous energy emission is possible, and an early application of this system is likely to be precision cutting.

Applications to date for electron beam welding have been confined mainly to the welding of highly reactive and refractory metals, whereas laser beam welding has been used for the very small welds required in electronic components.

Plasma processes. These processes depend upon a modification to the tungsten inert gas (TIG) process, although other than the purely inert gases may be used. The arc may be struck between the electrode and the nozzle (known as plasma jet process) or between electrode and workpiece (transferred arc, known as plasma arc process) and the welding nozzle of the torch provides a constriction through which the arc passes on its path from the tungsten electrode to the workpiece. This constriction results in a higher arc temperature with improved directional stability of the arc, the heat intensity of which is less dependent

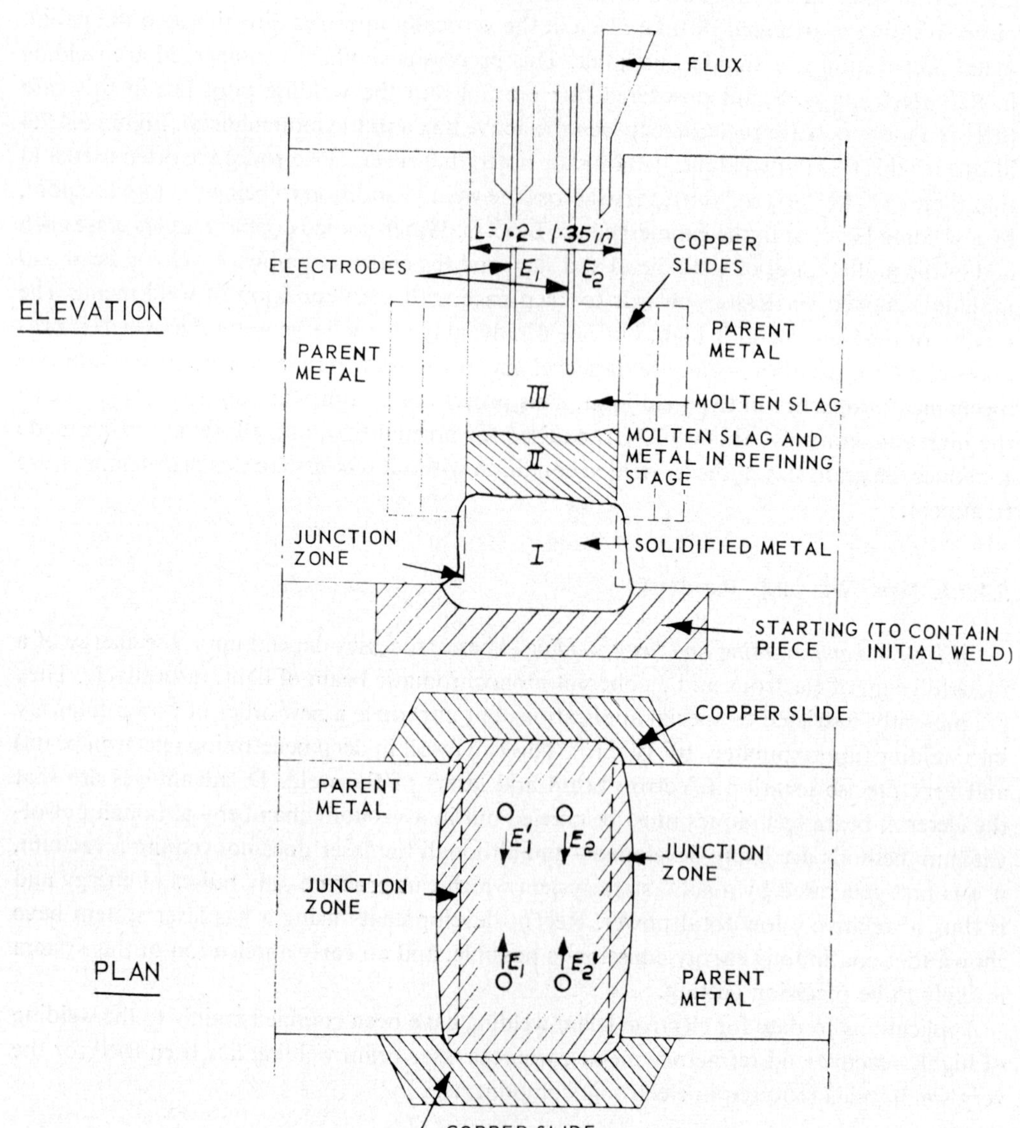

FIG. 5.4.2.4. Fundamentals of electroslag welding (From *The Metallurgy of Welding*, D. Seferian. Chapman & Hall Ltd.)

upon arc length than with a conventional arc, due to its cylindrical shape. Although plasma is present in all welding arcs, this particular technique is referred to as a plasma process and it may be used for welding, cutting and surfacing operations. The welding application of plasma arc is still at a very early stage and cutting applications are more fully developed. The process is fundamentally faster than conventional TIG and advantage is likely to be taken of this by varying degrees of mechanisation.

Friction welding. Heat is generated by the friction caused by the rotation of one item against the stationary component to which it is to be joined. When bonding occurs the rotating part is braked, the other released, and forging pressure is applied. Liquation is minimised since fusion at the forging surfaces produces a reduction in frictional force, so providing a self-compensating temperature control. The temperature gradient across the joint is shallow which reduces the cooling rate at the weld joint and so provides a suitable situation for welding hardenable steels. Friction welding is thus potentially a very attractive system making high quality forge welds without a cast structure, which is applicable to difficult materials and dissimilar metal joints, and is particularly suitable for the welding of tube. The process was first developed in the U.S.S.R. and although machines have been constructed in this country, work to date has been mainly experimental. A disadvantage with existing machines is that only material of restricted cross section (up to 1 in^2) may be joined, the main difficulty being not the attainment of higher speeds or pressures, but the provision of sufficient braking power.

5.4.2.6. Automatic Welding

An important factor retarding the general acceptance of welded components is that the welding process has always depended heavily upon the human element. The complete or partial automation of existing processes and/or the introduction of new processes where the approved sequence of operations may be reproduced exactly and almost indefinitely, offers a solution to operator defects such as lack of penetration and fusion, overpenetration and inadequate removal of slag between runs. This ideal situation is still remote but the first stage toward automation, that is welding by a machine, is already being applied to welds which are of relatively simple geometry. A good example of this is the electric resistance welding of tubes from strip by making a longitudinal seam. Butt welds between tubes are now also made by machine in the shop whenever possible, this being economical in relation to the number of joints involved. Submerged arc and electroslag welding of boiler drums and the flash butt welding of tube are also examples of machine welding. The suitability of the argon arc processes for machine welding has already been suggested, with particular reference to the MIG method. The refinement of control mechanisms for all welding processes amenable to machine operation, may be expected to make steady and possibly rapid progress as the full advantages of complete automation are recognised in industry.

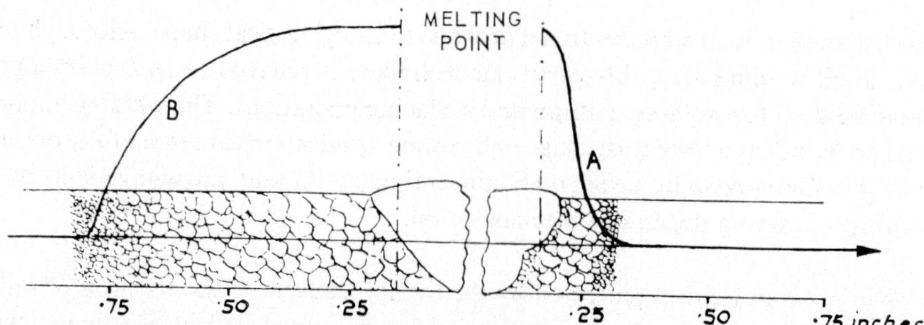

FIG. 5.4.3. Influence of the temperature cycle in welding on the parent metal grain size (curve B—oxy-acetylene welding; curve A—arc welding) (From *The Metallurgy of Welding*, D. Seferian Chapman & Hall Ltd.)

5.4.3. Weldability and Defects in Welds

Fusion welding is essentially a process of introducing cast metal between the surfaces to be joined. A small quantity of the parent metal is taken into solution and the weld metal will be to some extent a mixture of the base material and that introduced by the welding process. There is a close analogy between welding and the furnace melting and refining of metals, since both processes involve essentially the same principles and constituents. That portion of the parent metal next to the weld will have its temperature raised to just below its melting point and there will be a temperature gradient from this location through the base material down to ambient temperature. Figure 5.4.3 illustrates the various zones present in a typical weld joint. It should be noted that the width of the heat-affected zone of the parent plate, and in certain cases the amount of grain coarsening will depend upon the time at high temperature; this will be generally greater with low power intensity processes such as gas welding than with the higher intensity processes typified by electric arc welding. Electroslag welding is an exception since the total heat input is high due to the large quantity of metal deposited and the speed of deposition.

The total heat input controls to some extent the cooling rate of the weld, by heating the surrounding base material to a high temperature; thus, in general, the cooling rate of a gas weld is slower than that of an arc weld in material of given dimensions. A gas weld may therefore be expected to have a coarser grain size and a structure consistent with a lower rate of cooling, than an arc weld. Cooling rate has a profound influence on the final structure and therefore on mechanical properties, and may be critical in a transformable alloy steel (see Section 5.2.5). Coarse grain size is one reason why site gas welds in mild steel boiler tubes are often flame-normalised after welding. Normalising involves reheating to a high temperature (about 900°C), with transformation to the austenite form which permits recrystallisation during cooling, to produce a more desirable grain size. As previously stated this also applies to electroslag welds if a fine grain structure is required.

A weld is subject to two classes of defect:

(a) Those due to faulty technique on the part of the operator such as lack of fusion, lack of penetration, overpenetration, slag lines. These are basically the fault of the welder or choice of process and will not be considered in detail in this lesson, although this does not imply that they are unimportant.

(b) Those resulting from fundamental difficulties in a welding operation such as cracking and porosity. Although faulty technique on the part of the operator can cause these defects, their incidence may be inherent in the process. Prevention of this class of defects requires action based on a sound knowledge of the reactions involved during the fusion and cooling of the materials concerned. Porosity and cracking in welds will now be considered in some detail.

5.4.3.1. Porosity

Porosity is caused by the entrapment of gas bubbles by the freezing dendrites during the cooling of the weld pool, and therefore will occur essentially only in the weld metal.

Gas may be evolved from the melt for two reasons and the first of these is a chemical reaction, such as between ferrous oxide and carbon, which liberates carbon monoxide ($FeO + C \rightleftharpoons Fe + CO$). Porosity from this cause is usually effectively prevented by the presence of efficient deoxidisers such as silicon and manganese in the filler wire or coating of an electrode, which combine preferentially with the oxygen to give compounds which enter the slag.

The second reason is the supersaturation of gases, for example, hydrogen and nitrogen, in the melt. This was typical of early bare wire welds in which nitrogen was picked up from the atmosphere. Porosity may also arise in the atomic hydrogen process (little used now) in which considerable quantities of hydrogen may be absorbed by the weld pool if an incorrect technique is employed. A high sulphur content can induce porosity, probably due to the delaying effect which low-melting compounds may have on the nucleation of gas bubbles, by coating the sharp profiles of the freezing dendrites on which the bubbles more readily form.

5.4.3.2. Cracking in Weld Metal

Cracking is the most serious defect in a weld since it can lead to premature failure, particularly in a dynamically loaded component.

Cracking may occur at two stages during the cooling of the weld metal:

1. At a high temperature just below the freezing point, and known as "hot cracking".
2. At room temperature or a little above, and known as "cold cracking".

(i) Hot cracking (carbon and low-alloy steels). Hot cracking is influenced by the sulphur and carbon content of mild steel weld metal. Sulphur tends to form a low melting point compound with iron, namely iron sulphide (FeS), which can form a low melting eutectic of Fe/FeS and segregate to form a network at the grain boundaries of the steel, remaining liquid after the metal has frozen. There is, therefore, no cohesion between the grains and the weld metal may tear apart under contractional stresses. This difficulty is overcome by ensuring that sufficient manganese is present. If the manganese : sulphur ratio is appreciably greater than 10, then manganese sulphide is formed in preference to iron sulphide and, since the former solidifies at a temperature slightly above that of the steel, the inclusions appear as discrete spheres and as such are relatively harmless. The sulphur content of wire for electrodes is limited to 0·04% max. and that of the parent steel to 0·05 or 0·06% max.

Generally, values are considerably below this, especially in the core wire. High carbon content in weld metal also tends to promote hot cracking and should be kept below 0·15%. These remarks apply equally to the low-alloy steel weld metals and, for example, in the presence of nickel coupled with a high sulphur content, a Ni/NiS eutectic of even lower melting point than that of Fe/FeS may be formed. Hot cracking usually occurs only in the weld metal itself but an example of parent metal hot cracking is given in Section 5.4.3.6 and 5.4.5.2.

(ii) Cold cracking (carbon and low-alloy steels). This occurs at a relatively low temperature, considerably below that at which the steel solidifies and in ferritic transformable steels is due to stresses arising from the following factors, usually in association with a susceptible microstructure typical of which is martensite.

1. Volume changes accompanying transformation from austenite to ferrite/pearlite/bainite/martensite.
2. Cooling rate—dependent upon size and temperature of weldment and heat input of welding process.
3. Hydrogen—introduced by the electrode or its coating as chemically combined water, damp coatings or rusty and wet workpieces.

The transformation product is related to the alloy content of a steel and to its cooling rate. The higher these are, the lower will be the transformation temperature and the harder and more brittle the product formed. Mild steel, being unalloyed, has a high transformation temperature so that the residual stresses arising from transformation are low: in consequence it is not normally susceptible to this form of cracking.

The solubility of hydrogen in austenite is significant: in ferrite, however, it is virtually insoluble. It follows that some hydrogen in the diffusible atomic form may be trapped within the weld following transformation from the austenitic to the ferritic state. The atomic hydrogen may diffuse to microscopic voids in the lattice where it can form pockets of molecular hydrogen gas at high pressure. The diffusion rate is dependent upon temperature and the solvent metal, being greater at elevated temperatures. The resulting internal stress may be high enough to initiate cracking either for this reason alone or in association with other factors such as cooling rate and alloy content.

5.4.3.3. Cracking in the Heat-affected Zone

The above remarks have been concerned with the weld metal itself but apply equally to the heat-affected zone. This is the zone of parent metal which has had its structure altered, but not completely melted by the heat of the welding process, and into which hydrogen may diffuse from the weld metal. When welding low-alloy steels the parent material is often of higher carbon and alloy content than the weld metal, and the heat-affected zone is usually the more prone to cracking. An important contributing factor is that, due to the lower transformation temperature associated with the higher alloy content, this zone remains austenitic after the weld metal has transformed to the ferritic condition and is thus very receptive to hydrogen diffusing from the weld metal.

Therefore, before a steel is welded, a preliminary assessment of the situation must include consideration of carbon and alloy content, the welding process or processes (low hydrogen or otherwise), which can be used and if the component can be economically preheated. Modifications to the steel composition, welding process or temperature at which welding is carried out, either individually or collectively, may be required to ensure the production of a crack-free weld. In carbon steels and low-alloy steels, preheat is valuable in maintaining elevated temperatures for a longer period and, more important, in reducing the cooling rate, which encourage transformation to occur at a higher temperature. Thus the stresses associated with transformation are reduced, and the transformation product may itself be modified to a form less susceptible to cracking. Diffusion of hydrogen through steel is more rapid at higher temperatures and since the time at elevated temperature is increased by preheating, a greater proportion of the hydrogen introduced by welding may be removed.

5.4.3.4. Low-hydrogen Welding Processes

The main features of these processes are in producing a weld metal of:

1. Low hydrogen content, and in same cases:
2. Low sulphur content due to desulphurising action, if a basic slag is formed. Not all low hydrogen coated electrodes have an entirely basic coating.
3. Low total gas content (low oxygen, nitrogen, etc.).

The above well-known features are now self-explanatory. Low hydrogen content reduces a major stress raising component, permitting relatively high cooling and transformation stresses in the weld and heat-affected zone to be realised before cracking occurs. Low sulphur and total gas content reduce the probability of hot cracking and porosity in the weld metal, and, in fact, high sulphur (free cutting) steels may often be welded without visible cracking or porosity, although this is unwise if the weld is to be stressed.

The hydrogen present in the stick electrode coating is mainly present in the form of water, since the silicate binders which are a common feature of these coatings are hygroscopic. For these reasons it is necessary to bake low hydrogen electrodes prior to using them and until recently, temperatures up to 100°C were recommended for this purpose, but the present view is that this temperature should be 300°C.

5.4.3.5. Weldability Tests (With Specific Reference to Cold Cracking)

A large number of laboratory tests to determine the cold crack susceptibility of various weld metal/parent plate combinations has been devised and found useful application in practice but it is perhaps sufficient to describe only one of these, namely the simple fillet weld tests known as Controlled Thermal Severity (C.T.S.) Tests devised by the Welding Institute.

The test correlates the electrode type and size (hydrogen content and cooling rate) and steel thickness and quality (cooling and transformation stresses). The test piece is a simple bolted-up plate assembly of the steel type and thickness under investigation, as shown in

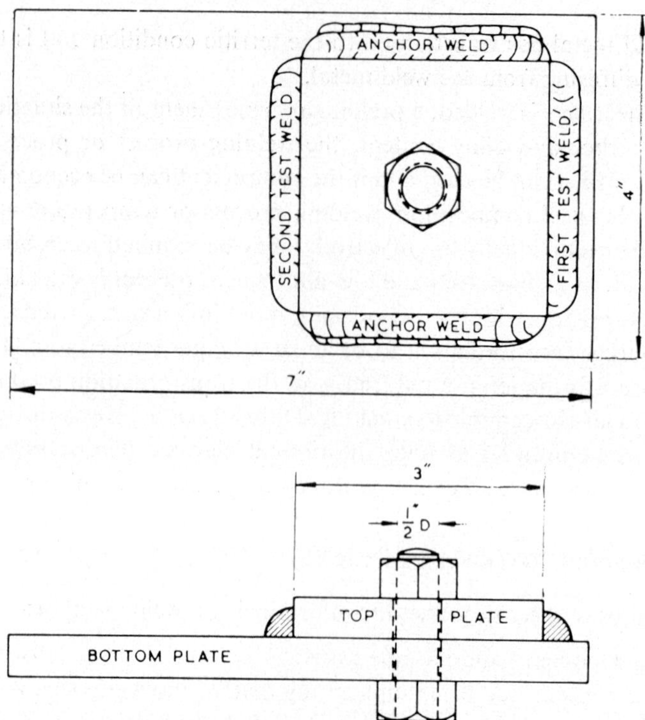

FIG. 5.4.3.5. Controlled thermal severity test for weldability—assembly (Reproduced by kind permission of the Welding Institute)

Figure 5.4.3.5.The assembly is further secured by anchor welds, following which the test welds using the proposed welding process and consumables are carried out either at ambient temperature or at the proposed preheat temperature. The result is based on the presence or absence of cracking when the joint is cut up after cooling and allowing an "incubation" period of 24 h. Sections are then examined under the microscope and results expressed as a weldability index. A weldability index may also be calculated from a formula which gives the "carbon equivalent" of the steel using the steel composition (values in weight per cent), as follows:

$$\text{carbon equivalent} = \text{C}+\frac{\text{Mn}}{20}+\frac{\text{Ni}}{15}+\frac{(\text{Cr}+\text{Mo}+\text{V})}{10}$$

Procedure. The thermal severity number (T.S.N.) of the joint is estimated first; this is a measure of the thickness of plate being welded and is numerically four times the total thickness of plate (in inches) from which the heat may flow from the weld. Then the weldability index (Table 5) of the steel/electrode combination is determined either from C.T.S. tests or by use of the carbon equivalent formula. The size of the weld run and the minimum level of preheat (if any) required to prevent cracking can then be obtained from Table 6.

Summarising, the major factors which influence the incidence of cracking are:

1. Cooling rate.
2. Composition of steel and weld metal.
3. Welding process.

TABLE 5

WELDABILITY INDEX FROM STEEL COMPOSITION

Weldability index letter	Range of carbon equivalent %	
	For class 2—or class 3—electrodes (rutile)	For class 6—electrodes (low hydrogen)
A	Up to 0·20	Up to 0·25
B	0·21–0·23	0·26–0·30
C	0·24–0·27	0·31–0·35
D	0·28–0·32	0·36–0·40
E	0·33–0·38	0·41–0·45
F	0·39–0·45	0·45–0·50
G	More than 0·45	More than 0·50

Reproduced by kind permission of the Welding Institute.

5.4.3.6. STAINLESS STEELS

Stainless steels may be divided into two classes both of which are relatively resistant to corrosion and scaling, due essentially to the chromium content which is 12% minimum (see pp. 445–447). Although they have many similarities to the low alloy steels previously described, with certain susceptibilities to both hot and cold cracking, it is convenient to consider them separately as follows. The "Schaeffler" diagram, Figure 5.4.3.6, illustrates the phases present in stainless steel weld metals of various compositions, expressed as total equivalent chromium and nickel contents, and provides a useful reference to the following notes.

1. *Magnetic*. Martensitic, for example turbine blades and cutlery. Ferritic, for example corrosion-resistant decoration (motor car trim, etc.) and high-temperature furnace parts. The magnetic types contain essentially chromium as the major alloying element; they are heat-treatable if the carbon content exceeds approximately 0·12% and may be welded, but the more hardenable types require pre- and post-weld heat treatments as outlined for the low alloy weld and parent metals. (Martensite areas on diagram.) At high chromium and low nickel equivalent values the structure is always fully ferritic and generally coarse grained and brittle at ambient temperatures. It cannot be refined by post-weld heat treatment.
2. *Non-magnetic*. Austenitic, for example superheater tubes and hangers, steam pipes, and soot-blower nozzles. The non-magnetic types contain sufficient nickel or other austenitiser in addition to chromium to retain the austenite structure at room temperature. Therefore, no hardening occurs due to fast rates of cooling, and welding does not generally require any pre-or post-heating. (Austenite with up to 20% ferrite areas on diagram.)

Possible disadvantages with these steels when subject to welding processes and/or corrosive environments or high-temperature conditions, are:

(i) Hot cracking of weld metal—austenitic weld metal has a tendency to hot cracking unless a small amount of ferrite is present. Normally this is permitted but is often a

TABLE 6

WELDING CONDITIONS FOR THE PREVENTION OF CRACKING IN THE HEAT-AFFECTED ZONE

Thermal severity number (T.S.N.)	Weld-ability index of steel	Minimum plate temperatures (°C)* for welds made using $1\frac{1}{2}$ in. of the following size of electrode per inch of weld					
		10 S.W.G.	8 S.W.G.	6 S.W.G.	4 S.W.G.	$\frac{1}{4}$ in.	5/16 in.
T.S.N. 2 (Two $\frac{1}{4}$-in. plates, etc.)	A	W	W	—	—	—	—
	B	W	W	—	—	—	—
	C	W	W	—	—	—	—
	D	0	W	—	—	—	—
	E	50	W	—	—	—	—
	F	125	25	—	—	—	—
T.S.N. 3 (Two $\frac{3}{8}$-in. plates, or three $\frac{1}{4}$-in. plates, or $\frac{1}{2}$ in. on $\frac{1}{4}$-in. plate, etc.)	A	W	W	W	W	—	—
	B	W	W	W	W	—	—
	C	0	W	W	W	—	—
	D	75	—25	W	W	—	—
	E	100	25	W	W	—	—
	F	150	100	25	W	—	—
T.S.N. 4 (Two $\frac{1}{2}$-in. plates, or three $\frac{1}{3}$-in. plates, or $\frac{3}{4}$ in. on $\frac{1}{4}$-in. plate, etc.)	A	W	W	W	W	W	W
	B	—25	W	W	W	W	W
	C	50	W	W	W	W	W
	D	100	25	W	W	W	W
	E	125	75	0	W	W	W
	F	175	125	75	25	0	W
T.S.N. 6 (Two $\frac{3}{4}$-in. plates, or three $\frac{1}{2}$-in, plates, or 1 in. on $\frac{1}{2}$-in. plate, etc.)	A	—25	W	W	W	W	W
	B	50	W	W	W	W	W
	C	100	25	W	W	W	W
	D	150	100	25	W	W	W
	E	175	125	75	25	—25	W
	F	225	175	125	100	75	0
T.S.N. 8 (Two 1-in. plates, or three $\frac{2}{3}$-in. plates, or $1\frac{1}{2}$ in. on $\frac{1}{2}$-in. plate, etc.)	A	25	W	W	W	W	W
	B	75	0	W	W	W	W
	C	125	75	25	W	W	W
	D	175	125	75	25	0	W
	E	200	150	125	75	50	—25
	F	225	200	175	125	125	50
T.S.N. 12 (Two $1\frac{1}{2}$-in. plates, or three 1-in. plates, or 2 in. on 1-in. plate, etc.)	A	75	25	W	W	W	W
	B	125	75	25	W	W	W
	C	150	125	75	25	0	W
	D	200	175	125	100	75	0
	E	225	200	175	125	100	50
	F	250	225	200	175	150	125
T.S.N. 16 (Two 2-in. plates, or three $1\frac{1}{3}$-in. plates, or 3 in. on 1-in. plate, etc.)	A	75	25	—25	W	W	W
	B	125	75	50	0	—25	W
	C	175	150	125	75	50	0
	D	200	175	175	125	125	50
	E	225	200	200	150	150	100
	F	250	250	225	200	200	150
T.S.N. 24 (Two 3-in. plates, or three 2-in. plates, or larger joints)	A	75	25	—25	W	W	W
	B	125	75	50	0	—25	W
	C	175	150	125	75	75	25
	D	200	175	175	150	125	100
	E	225	200	200	175	175	150
	F	250	250	225	225	200	200

* Temperatures quoted are in 25° steps. W denotes a temperature below −50°C. Reproduced by kind permission of the Welding Institute.

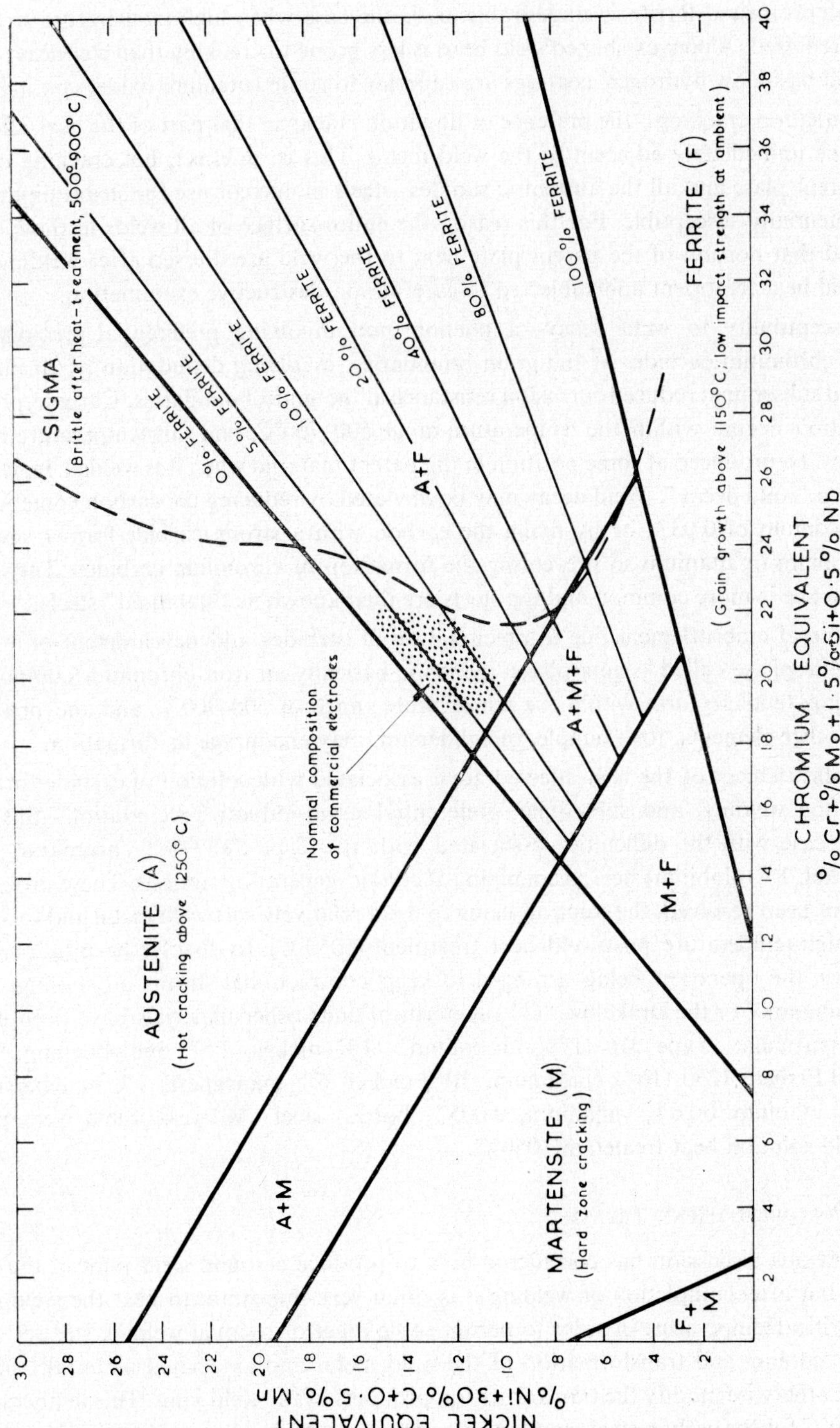

FIG. 5.4.3.6. Constitutional diagram for stainless steel weld metal (Schaeffler)

compromise if ferrite is undesirable, as is the case when high temperature strength is required. A convex-shaped weld bead is less prone to cracking than a concave bead and basic low hydrogen coatings are superior to rutile (titanium oxide) coatings.

(ii) Liquation cracking—the presence of liquation cracks in that part of the heat-affected zone immediately adjacent to the weld metal. This is, in effect, hot cracking of the parent plate and all the austenitic stainless steels in current use for steam piping are potentially susceptible. For this reason the entire surface of all welds in these steels and that portion of the parent plate next to the weld are dressed after welding and final heat treatment and subjected to careful non-destructive examination.

(iii) Susceptibility to weld decay—a phenomenon involving preferential precipitation of chromium carbides at the grain boundaries, involving denudation of chromium and subsequent reduced corrosion resistance at the grain boundaries. Carbide precipitation occurs within the temperature range 500–700°C, and this temperature range must be produced at some position in the parent material when it is welded, hence the term "weld decay". Weld decay may be obviated by reducing the carbon content to a maximum of 0·03% or by fixing the carbon with a strong carbide-former such as niobium or titanium so preventing the formation of chromium carbides. The latter practice is more common and the steels are then known as "stabilised" steels.

(iv) General embrittlement due to precipitation of carbides and development of a hard brittle phase called "sigma-phase" which is basically an iron chromium compound. Sigma tends to form within the temperature range of 500–900°C and the presence of other elements, for example, molybdenum, may encourage its formation.

(v) Embrittlement of the heat-affected zone associated with solution of carbide-formers during welding, and subsequent preferential strain-induced precipitation—this was the case with the difficulties associated with the Type 347 (18% chromium, 12% nickel, 1% niobium) steel steam piping at certain generating stations. These problems have been resolved, the solution being to use a relatively soft weld metal and to apply a high temperature post-weld heat treatment (1050°C), to dissolve harmful precipitates, the operation being arranged to keep contractional strains on cooling, to a minimum. For the Drakelow "C" supercritical units other materials have been used, in particular, Type 316 (17% chromium, 11% nickel, 2·5% molybdenum steel, and Esshete 1250 (16% chromium, 10% nickel, 6% manganese, 1% molybdenum, 1% niobium, 0·26% vanadium, 0·005% boron) steel. All weld joints were post-weld solution heat treated at 1050°C.

5.4.3.7. Post-weld Heat Treatment

The previous discussion has considered how to produce a sound weld joint in the first instance, but after completion of welding it is often very important to heat the weldment to a subcritical temperature in order to permit some relief of residual welding stresses arising from cooling and transformation of the weld metal and associated material, and to temper or otherwise modify the transformation products of the weld joint. This temperature is 600–700°C for the carbon steels and low-alloy ferritic steels, and generally 800°C or above for the austenitic grades. A post-weld stress relief and tempering heat treatment are essential when dealing with a hardenable steel, to ensure that the weld will remain free from cracking when subjected to service stresses. This also applies to certain austenitic steels as outlined in (v) above. Residual stresses due to welding can promote cracking in a corrosive environment (stress corrosion) and a stress relief heat treatment is often required to prevent this.

In the particular case of steels selected for high-temperature service, their microstructure in the heat-affected zone may give rise to embrittlement in the temperature range 500°–700°C. If the weldment involved contains a very high level of stress after welding, it may prove difficult to obtain stress relief without initiating cracks in the heat-affected zone. This problem has to be tackled by attention to design detail so as to minimise stress-concentrating factors.

All heavy section pressure-bearing parts are post-weld heat treated; it may be seen that this requirement can be difficult to apply on site to such items and, therefore, as much work as possible is carried out in the shop where the facilities required, for example large furnaces, are readily available.

5.4.4. Desirable Weldment Properties

Austenitic Type 347 steel steam pipe failures at weld joints have shown essentially circumferential cracking, indicating that stresses other than those due to operating pressure were involved. This suggests that in selecting a suitable weld metal for a given material, consideration of weld properties other than strength alone should be given, and a greater knowledge of the optimum relationship between the properties of the parent metal and the weld, is required. This problem is of particular interest regarding the welding of the $\frac{1}{2}$% CrMoV steam pipe steel, which may be satisfactorily welded using a $2\frac{1}{4}$% Cr, 1% Mo steel weld metal which has reasonable strength and good ductility. In this case a weld metal of matching composition has high strength but inferior ductility and is not recommended.

Summarising, it is considered that weld joint ductility is perhaps the most significant property if accompanied by strength approaching that of the parent material. The problem of ideal property requirements and limits of the weld joint in relation to the parent material, has still not been fully resolved. The solution has to be sought by examination of individual joints and components accompanied by a reasonably accurate assessment of all the stresses involved.

5.4.5. Welding Applications in Construction of C.E.G.B. Plant

All major pressure-containing parts are now of welded construction and some typical examples are given below: readers are referred to Tables 1, 2 and 3 for the materials of construction.

5.4.5.1. Boiler Drums

Drums have attained massive proportions, as exemplified in one type of 500 MW boiler in which the drum is 115 in. long, with an inside diameter of 5 ft 6 in., a shell thickness of $5\frac{3}{4}$ in. and a weight of some 240 tons. The drums are fabricated from sections, the longitudinal joint being welded in a vertical position by electroslag welding, the circumferential seams being welded by the submerged arc multi-layer process. Due to the size of present-day drums, these are often brought to site as two or more cylindrical sections which are welded together on site, using circumferential butt welds. Branches and attachments are still welded essentially by the manual metal arc process using stick electrodes.

5.4.5.2. Boiler Tubes

Tube has been of solid drawn mild steel of about 20 tons/in^2 U.T.S. for many years, but due to the higher pressures in modern boilers a stronger carbon steel of about 27 tons/in^2 U.T.S. is now in general use. A recent innovation is the E.R.W. (electric resistance welded) tube; advantages are improved tolerance in dimensions and reduced cost. Although the composition of the steel strip from which the tubes were made caused some welding troubles due to hot cracking in the parent metal during gas welding on site, these problems are being overcome and a steel of modified composition is now available.

Experience has shown that gas-welded tube joints are particularly prone to defects due to incomplete fusion of the base metal, namely lack of side wall fusion. This is probably associated with the relatively low heat intensity of this process compared with say arc welding and the latter is now generally preferred. Water-side corrosion of highly rated furnace tubes has been associated in certain cases with overpenetration during welding (intrusion of metal into the bore area) and welded joints are therefore avoided in this part of the furnace wherever possible. When such welds are unavoidable the extent of bore intrusion is strictly limited and specified. Argon arc welding allows rather better control of penetration than metal arc welding with a coated electrode, and is currently recommended for this particular operation. Flash butt welding tends to leave an internal protrusion in the tube bore and is not permitted for highly rated sections of the boiler.

Recent developments in boiler construction such as the membrane wall illustrated in Figure 5.4.5.2, require a considerable amount of welding for their construction in addition to the basic tube-to-tube butt joints. In the example shown the membrane is formed by fillet welding a carbon steel strip $\frac{1}{2}$ in. wide $\times$ 3/16 in. thick between adjoining tubes. This operation is carried out by machines in the shop using the carbon dioxide gas-shielded process with a consumable electrode mild steel filler wire.

5.4.5.3. Superheater and Reheater Tubes

Most of the above remarks apply also to these items, but due to the higher operating metal temperatures in the superheater and reheater, creep-resisting ferritic and austenitic alloy steels must be used. The weldability of a specific material must therefore be carefully considered prior to its inclusion in a boiler. Conditions in reheaters are theoretically less onerous due to the lower pressure which allows the use of thinner wall tubes, but because of the temperature the steel used must have adequate high temperature properties.

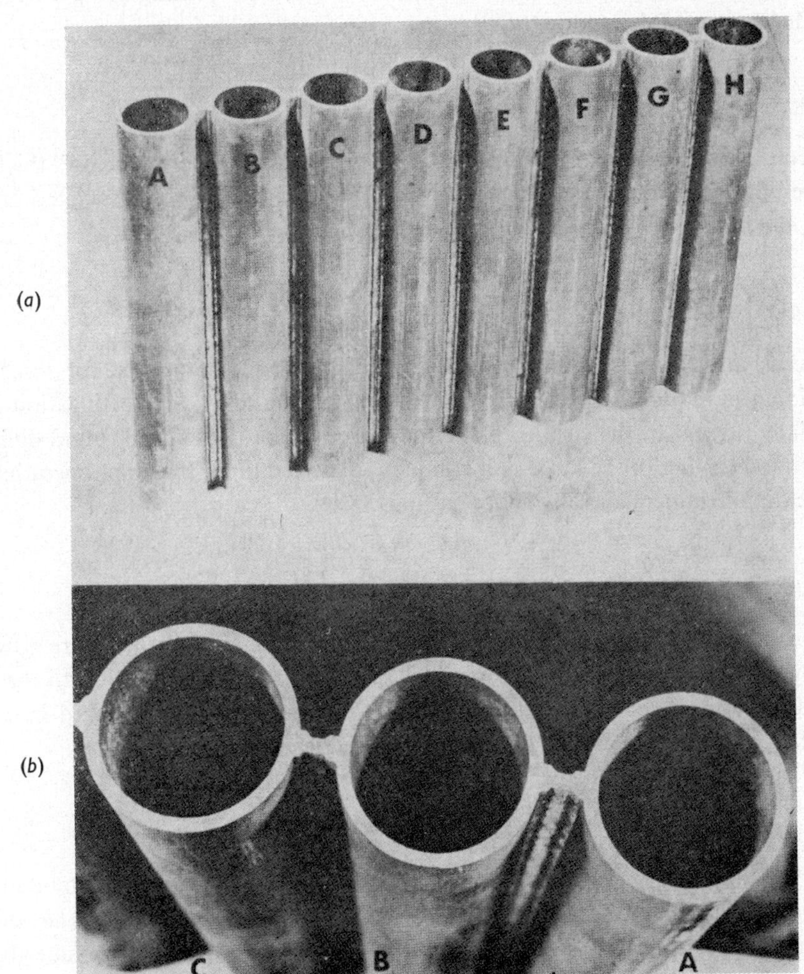

FIG. 5.4.5.2. (a) Membrane panel of boiler tubes. (b) Showing profile of welds between tubes, A, B and C

5.4.5.4. Headers and Manifolds

Expanded tubes are no longer used for tube to header joints and these are now being made by shop welding stubs to the headers. This requires the use of fillet welds which can be subject to bending stresses in operation and their quality must be of the highest order. Element tubes are then butt-welded to these stubs on site, butt-welding being a less critical operation due to its simpler geometry.

5.4.5.5. Steam Pipes and Valves

Welded construction is now used throughout steam pipe lines where pipe wall thicknesses may now be up to 3 in. This includes the joining of wrought pipe with associated branches and valves, the bodies of which are usually castings.

5.4.5.6. Turbine Casings and Steam Chests

These are usually castings, but closed die forgings are now being used for some steam chests. Welding is required for the attachments of pipes and in the rectification of any defects initially present in the casting. In certain cases where it has been found difficult to produce a sound casting in the large and complex shape required, the component has been made by welding together several smaller castings.

5.4.5.7. Turbine Rotors

Forgings are still used almost exclusively for rotors, but certain l.p. rotors are now being produced by welding together hollow forged discs. This permits modification to the design stresses and the use of smaller forgings, which can be more readily produced in a sound condition than the very large forgings required for modern machines.

5.4.5.8. Feed Heaters

The high pressures and temperatures required in the feed water system of modern units have led to the replacement of the tube to tube plate expanded joint in h. p. heaters by a fusion welded joint, and the replacement of copper and copper–nickel alloy tubes by mild steel.

The problem of welding a relatively thin tube to a massive tube plate when many thousands of joints are involved, did not find a ready solution. An early method was to fuse the end of an expanded tube with the tube plate (Fig. 5.4.5.8A); a later development was to machine an annular spigot in the tube plate for each tube to allow the fusion of sections of similar size (Fig. 5.4.5.8B). These methods did not produce welds of high integrity due to the joint design which contained a built-in notch. It was therefore susceptible to fatigue and also not amenable to non-destructive examination. A more satisfactory method is to machine tube stubs in the tube plate which are then butt-welded to the heater tubes by an automatically controlled internally located welding head. The tungsten inert gas (TIG) process is

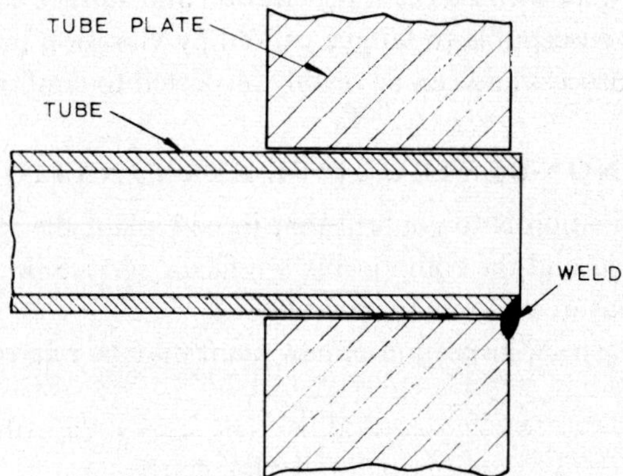

FIG. 5.4.5.8A. Heater tube/tube plate—external plain weld

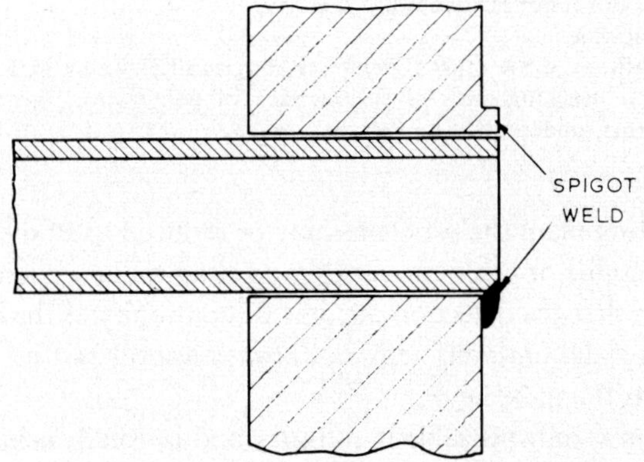

FIG. 5.4.5.8B. Heater tube/tube plate—external spigot weld

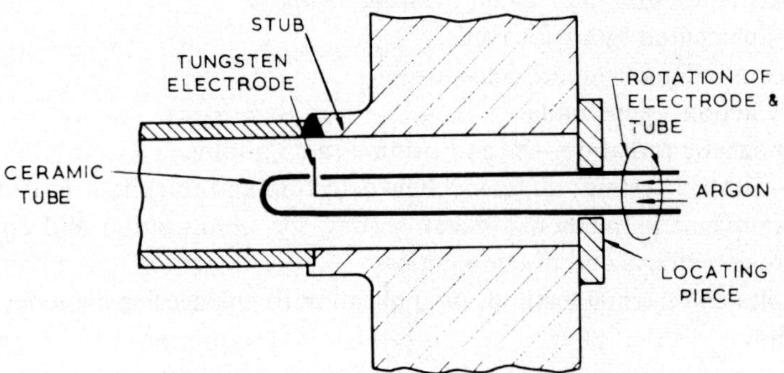

FIG. 5.4.5.8C. Heater tube/tube plate—internal stub weld
(showing Foster–Wheeler internal bore welding head)

employed and the welding head rotates inside the tube, fusing the premachined mating faces to produce a butt weld with excellent penetration and surface contour (Fig. 5.4.5.8c). This type of joint is less susceptible to fatigue caused by vibration in the tube nest, may be made by an automated process and can be readily inspected to confirm weld quality.

5.5. NON-DESTRUCTIVE EXAMINATION

The purpose of examination is to confirm that in new plant the recommended requirements have been observed and the components are fit for service. In the case of operating plant the purpose is to ensure continuing integrity.

From a metallurgical aspect, inspection of new plant may be related to component parts as follows:

Item	Typical defects
Castings	Porosity, internal shrinkage, tears, cracks, scabs.
Forgings	Central unsoundness, cracks.
Plate	Laminations.
Tubes	Laminations, draw marks, defective longitudinal welds in E.R.W. tubes.
Welds	Porosity, cracking, lack of fusion, lack of penetration, overpenetration, slag inclusions, undercut.

In the case of operating plant the problems may be reduced to studies of crack formation (for example, due to fatigue or creep processes) and changes in section due to wear and/or corrosion. Thus we see that the detection of some discontinuity in the structure is an essential requirement. Such a discontinuity may not always appear on the surface, or if it does it may not be visible to the naked eye.

Effective examination is indispensable in industry and is usually achieved by non-destructive testing (N.D.T.) methods, although occasionally quality control procedures demand a destructive test. A number of basic physical phenomena have been exploited for non-destructive techniques and these include the following:

1. Vision—augmented by various aids.
2. Magnetism—magnetic crack detection.
3. Capillary action—penetrants.
4. Electromagnetic radiation—X- and gamma-radiography.
5. Sonics—hammer testing; ultrasonic flaw detection and thickness measurement.
6. Electrical–magnetic methods—metal sorting for composition and condition, wear and corrosion effects and discontinuities.

The principles and normal methods of application to engineering components are briefly described below.

5.5.1. Visual Methods

Any form of inspection must be based upon an initial visual assessment of the geometry of the component and the type and nature of the defect which is likely to be present.

The surface inspection of normally inaccessible locations has been made possible by various illuminated optical devices using a system of lenses and mirrors (for example, introscopes) and these have proved most useful for examining such items as turbine rotor bores and long straight tubes.

The problem of water-side tube corrosion in modern boilers has accented the necessity to inspect the inner surface of long lengths—of the order of 100 ft—of small bore tube down to 2 in. internal diameter, usually containing several bends of varying radii. The lens/mirror system is inadequate for this purpose and two types of camera equipment designed to solve this problem have been developed. These are a miniature conventional camera which can be drawn along the tube, taking photographs by remote control as required, and a miniature television camera which transmits a continuous record of the inner surface of the tube, to be viewed on a monitoring screen by the operator.

This type of equipment makes possible the location of surface scabs due to corrosion during a general survey and it also allows the monitoring of suspect tubes, with the result that it then becomes feasible to replace only proven adversely affected items instead of entire banks of tube in a defective zone.

A considerable hindrance to the application of these methods is the lack of ready access to the interior of boiler tubes, which most modern boiler designs present.

5.5.2. Magnetic Crack Detection

This method is valuable for the detection of cracks which are present at the surface of a component made from a magnetic material. The method is based on the fact that the faces of a crack will tend to form north and south poles if a magnetic flux is established in the component. This flux may be induced by the use of a permanent magnet or an electromagnet, by causing current flow through the component or by wrapping coils around it and then passing a current through the coils. If the system is so arranged that the crack interrupts the flux lines, then it may be revealed by the application of either dry iron powder or iron powder in suspension in a liquid. The particles of iron are attracted to the poles formed at the crack which is then delineated as a black line. This effect may be enhanced by the prior application of a white paint which shows up the line in sharp contrast.

A. C. transformers are often used to supply the necessary flux when searching for surface defects. It has been found possible by the use of heavy d.c. currents—which produce a flux below the surface of the material—to indicate subsurface defects and present developments in this field are most encouraging.

5.5.3. Penetrant Inspection

The method depends upon the ability of a liquid penetrant to fill a crack, is suitable only for the detection of defects which are open at the surface, and is in practice limited essentially to cracks since other surface imperfections do not adequately contain the penetrant. It is necessary for the surface of the component to be in a clean, preferably machined, condition. The method of application is as follows.

The surface is cleaned and degreased and the liquid penetrant applied, being permitted to remain for an appreciable period (preferably several hours), during which it penetrates into any defects which may be open to the surface. The excess penetrant that remains is then

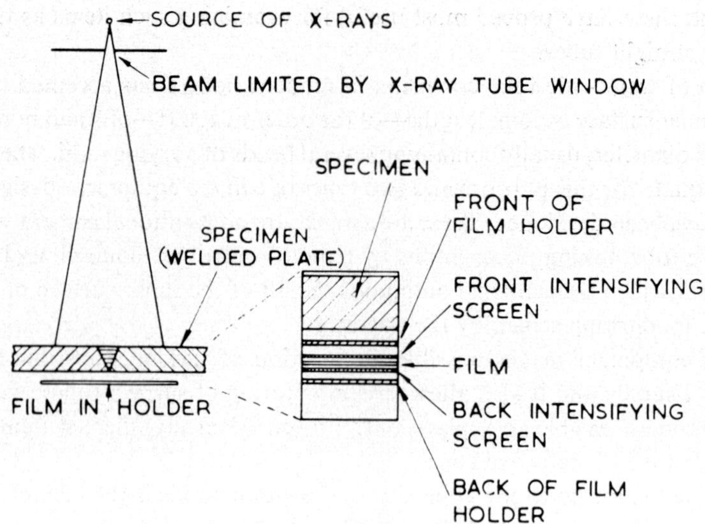

FIG. 5.5.4A. Basic arrangement for radiography (Reproduced by kind permission of the Welding Institute)

removed and an absorbant powder called a developer is sprayed uniformly on to the surface. The developer acts as a blotter and draws out a portion of the penetrant which had previously seeped into the opening. As the penetrant is drawn out it diffuses into the developer forming an indication much wider than the surface opening and, if, as is often the case, the penetrant is coloured with some dye and the developer is white, this forms a readily detectable indication on the surface of the developer.

In both magnetic crack detection and penetrant inspection the use of a fluorescent magnetic ink or penetrant may, in certain cases, simplify inspection if it is possible to view in ultraviolet light.

The penetrant method is most useful in being applicable to virtually all materials, unlike the magnetic methods which are restricted almost entirely to the ferritic steels.

5.5.4. **Radiography**

This method depends on the ability of short-wave radiation (X-rays and gamma-rays) to penetrate materials opaque to visible light, and the fact that this radiation is absorbed according to the density and thickness of the material which may be placed in its path.

Radiography as commonly practised uses a transmission technique by subjecting the test sample to a suitable beam of radiation and noting the various levels of absorption of radiation due to differences of thickness and composition in the sample. These are observed by the use of special photographic film, the emulsion of which is affected in a manner similar to that exposed to ordinary light in a camera. After exposure the film can then be developed and fixed in the normal way, the negative thus providing a permanent record. Figure 5.5.4A shows the basic arrangement.

The sources of radiation originally used were the naturally occurring radioactive elements such as radium, and subsequently X-rays. The scope has now been extended by the more general use of gamma-ray sources in the form of artificially produced radioactive isotopes, for example, cobalt-60, caesium-137, iridium-192 and thulium-170. The radiation from some of these sources has great penetrating power, equivalent to X-rays of up to 2 MeV energy, which makes them particularly valuable when thickness or density is beyond the range of available X-ray equipment. In addition, these sources are often virtually indispensible for site work in locations difficult of access for the relatively bulky X-ray gear. The cost of these sources and their control equipment is usually considerably less than that of an equivalent X-ray machine. If X-rays of sufficient penetration are available, however, their use is in general to be preferred since higher image contrast and shorter exposure times may be achieved. Table 7 lists several radioactive sources at present used in industry, according to their penetrating power in steel and X-ray equivalent given as energising voltage. The recent introduction of high voltage X-rays into the field of radiography has greatly improved penetrating power, sensitivity and definition. For this purpose Betatron equipment in the range 15 to 24 MeV is now being used in industry.

TABLE 7

CHARACTERISTICS OF GAMMA-RAY SOURCES

Source	Half-life	Gamma energies (MeV)	X-ray equivalent (kV)	Range in steel (mm)
Cobalt-60	5·3 years	1·17, 1·33	1250	50–150
Caesium-137	30 years	0·67	670	15–75
Iridium-192	74 days	0·13, 0·29, 0·58 0·60, 0·61	590	10–60
Thulium-170	129 days	0·084	84	In light alloys, < 12

Figure 5.5.4B illustrates the type of defect which can be detected by radiography. The concept of absorption implies the necessity to see significant differences in density in the radiograph, and an accepted standard in present general use is that a difference of 2% should be detectable and confirmed by a device known as a penetrameter or "image quality indicator" (I.Q.I.) which is radiographed along with the subject. Referring to the example in Figure 5.5.4B, this means that in the 1-in. thick welded plate, defects which have a minimum depth of 0·020 in. in line with the incident beam are detectable. The width of the average crack or lamination is usually much less than this and therefore although crack *A* would be detected, crack *B* together with lack of weld fusion at *D*, and plate lamination *C*, could be missed. Heavy porosity *E* and slag lines *F* would almost certainly be detected without difficulty due to their size and geometry. I.Q.I. sensitivity may sometimes be at variance with true defect sensitivity, due to the position of the defect in the sample in addition to its nature and orientation.

Summarising, radiography is excellent for finding porosity, internal shrinkage, slag lines and certain inclusions but severely limited in the detection of planar defects such as cracks

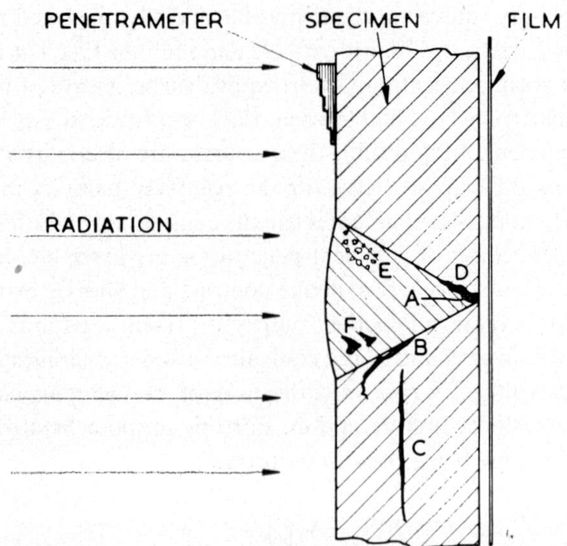

FIG. 5.5.4B. Radiographic detection of defects in 1-in. welded plate

and lack of fusion. This situation can be met by taking further radiographs exactly in line with anticipated planar faults, but it is not an economical procedure and such defects are best sought by another method.

5.5.5. **Ultrasonic Flaw Detection**

Ultrasonic waves are elastic waves of the same nature as sound but whereas the maximum audible frequency is 20,000 c/s, the frequency used for materials inspection is in the range 1–10 Mc/s.

The ultrasonic flaw detection test is based upon the properties exhibited by these waves when transmitted through the material in question. Some of these properties are noted below:

1. Ultrasonic waves can be transmitted in the form of a narrow beam through quite large distances before being adsorbed.
2. Wavelengths of the order of 0·1 in. are readily obtained.
3. At these wavelengths, transmission of ultrasonic waves is affected by the presence of discontinuities in their path. If the discontinuity is larger than the wavelength most of the energy will be reflected in a direction determined by the angle at which the incident beam strikes the defect.

High frequencies are used to obtain the required wavelength; for example, at audible frequencies, e.g. at 0·01 Mc/s, the wavelength in steel is 22·8 in. whilst at 2·5 Mc/s the wavelength is 0·09 in. for compression waves. The latter value is suitable for general application in wrought steel products, the essential requirement being that it must be greater than the anticipated mean grain size (reducing the chance of reflection from grain boundaries) and less than that of the defect it is intended to reveal.

Several types of wave are used. These include compression (longitudinal), shear (transverse), and surface waves. Geometry of the component is of great importance in deciding the type most suitable and reference to Figure 5.5.5A will illustrate the use of (a) compression, and (b) and (c) shear waves, the last two being most commonly used in practice. Shear waves are particularly useful when looking for defects in welds.

The method of application depends upon reflection or attentuation of the waves by discontinuities in the specimen and the reflection (echo sounding type) technique is that most commonly used for the detection of internal defects. The basic ultrasonic flaw detection equipment consists of a piezo-electric transducer (e.g. quartz, barium titanate), mounted in a search unit called a probe. A suitable a.c. voltage is applied to the transducer which is thus caused to vibrate at the required frequency, and a proportion of the energy is passed into the test piece with the aid of a convenient coupling medium such as water or oil. A pulse of energy passes through the sample and it may be picked up by a receiving probe placed either on the opposite face or on the same side as the transmitting probe or by the transmitting probe itself. A suitably oriented planar defect within the specimen will attenuate energy if a transmission technique is used, or the pulse will be reflected back sooner than expected from the far surface if the reflection technique is employed. With the echo reflection method a cathode ray oscilloscope (C.R.O.) is required, as the echoes returning from flaws are accompanied by those from the farther edge of the specimen examined. The C.R.O. separates these echoes on a time base so that, on the display screen, their distances from the origin represent the distances of the echo source from the transmitting/receiving probes (see Fig. 5.5.5B).

It will now be evident that the ultrasonic method is well suited to the location of all types of internal void, particularly planar defects (such as cracks) running normal to the incident beam. Porosity is easy to detect but more difficult to define since the spherical contour, although it always provides one point perfectly oriented for reflection whatever the angle of the incident beam, does cause general scatter of much of the beam energy. Since the method locates defects by reference to a time scale on the cathode ray tube, it may be used for measuring thickness to a high degree of accuracy, with the advantage of requiring access to one side only of the work piece.

Summarising, it may be seen that non-destructive inspection of a component for internal defects may often be most effectively carried out by a combination of radiographic and ultrasonic techniques. Radiography will detect internal voids, tears, porosity, slag, and usually provides a permanent record, whilst ultrasonics properly applied will effectively detect planar defects. Both methods are theoretically applicable to all materials including non-metals, but it is unfortunate that attenuation of ultrasonic energy is so great in coarse-grained austenitic steels that welds cannot be effectively inspected by this method in heavy sections of these materials. For pipework and headers, radiography and penetrant surface inspection of welds must suffice at the present time.

A disadvantage of radiography is that ionising radiations are injurious to health and therefore stringent precautions are required when using an X-ray machine or a gamma-ray source. When working on site this may necessitate the establishment of an extensive prohibited zone to ensure the safety of personnel working in the immediate vicinity, and this may seriously interfere with other operations. Ultrasonic waves of the intensities used for

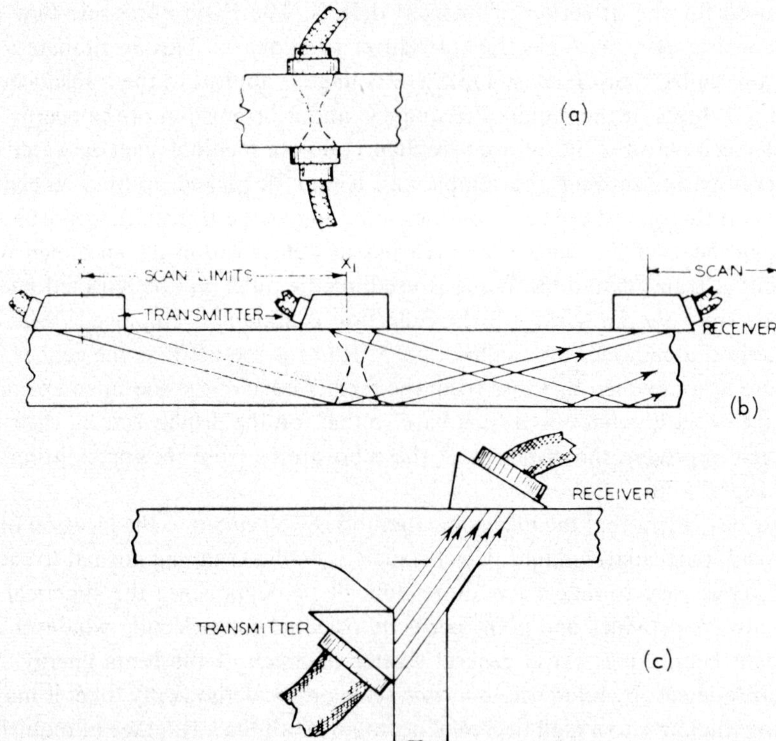

FIG. 5.5.5A. Use of ultrasonic waves in N.D.T. (Reproduced by kind permission of the Welding Institute)

(a) Compression waves
(b) & (c) Shear waves

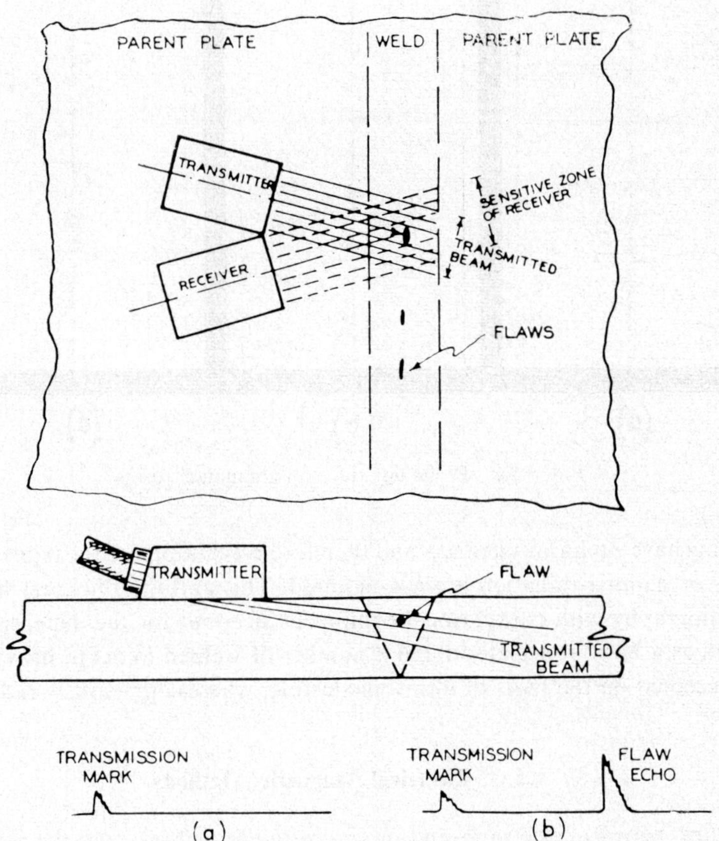

FIG. 5.5.5B. Principles of echo reflection method (Reproduced by kind permission of the Welding Institute)

Cathode ray tube display
(a) No flaws
(b) Single flaw

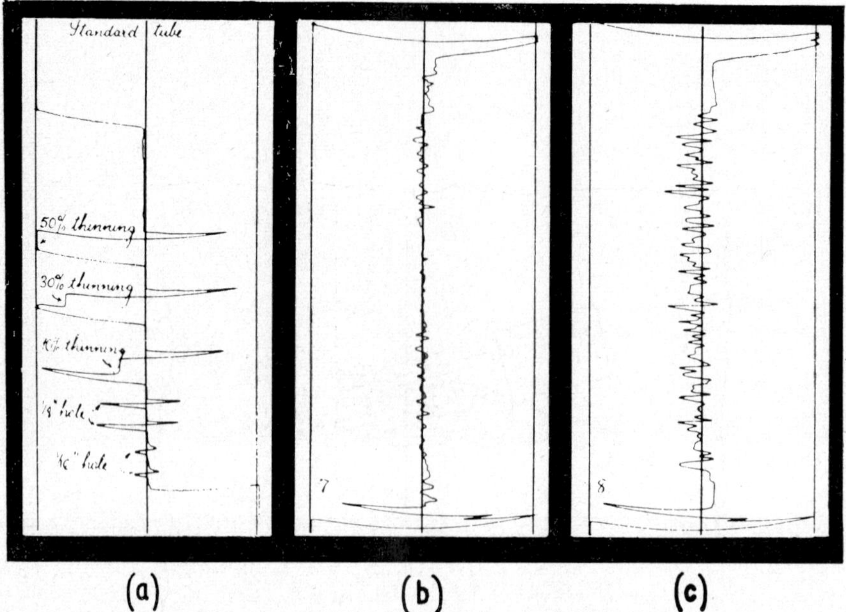

FIG. 5.5.6. "Probolog" traces of aluminium–brass

material testing have no harmful effects and therefore precautions of the type required when using X-rays or gamma-radiation are not required. This and the thickness limitations inherent in radiography with conventional equipment account for the increasing popularity of ultrasonics as a basic test method and a number of welded joints in heavy sections are now being accepted on the basis of ultrasonic testing, whereas previously radiography was specified.

5.5.6. **Electrical/Magnetic Methods**

Relationships between electrical and/or magnetic properties and the constitution of alloys have been known for a considerable period. A detailed study has been made of the characteristics of well-known materials in an attempt to use this information to devise electrical and/or magnetic tests for detecting differences in composition, condition, dimension and the presence of defects. Various instruments have been devised, usually applicable to specific components, on a quality control basis, by comparing a sound item with production through-put.

An instrument of particular interest to the Board is the "Probolog" or "Introview" designed to detect flaws and the effects of corrosion and/or erosion in non-magnetic condenser tubes. The equipment is based upon a probe containing two coils in tandem, which is designed to pass down the bore of the tube to be examined. The primary coil is excited by an alternating current which produces a magnetic flux so inducing a current in the secondary (search) coil. When the probe is placed inside a tube an eddy current is induced in the tube and modifies the current in the secondary coil. The coils form part of a bridge and the electrical system is arranged to permit adjustment so that the output of the coils may be balanced when the probe is placed within a sound tube of the correct dimension and material. Differences in tube properties which are influenced by composition and dimensions—such as incorrect material or loss of material by thinning or pitting—upset the balance of the bridge circuit, producing a voltage which can be made to operate a pen recorder.

The site method of operation is to set up the instrument to give a steady trace whilst the probe is being winched back at a controlled steady rate after having been impelled—usually by compressed air—down a sound tube of the relevant composition. The probe is then "shot" down actual tubes to be tested in the condenser and any change in dimension is recorded on a chart. Comparison with readings taken on sample tubes containing artificial standard defects, permits an assessment of the condition of the tubes under inspection.

Figure 5.5.6 shows three traces obtained by a Probolog instrument whilst examining three tubes of the same aluminium–brass alloy. Trace (a) is of an unused tube containing artificial defects of specific dimensions, whilst traces (b) and (c) are of heat-exchanger tubes *in situ*. Tube (b) was slightly pitted along most of the tube length, due to corrosion, and tube (c) had suffered moderate to severe pitting over much of its surface. The large deflections at each end of the traces are the effect of the tube plate on the instrument.

For inspection of tube in the factory, the probe may be modified to an annular form through which the tubes pass on leaving the production line. This can provide a most useful control—at that stage—on dimensions and composition.

5.6. METALLURGICAL ASPECTS OF POWER PLANT FAILURE

Failure of a component in service may be sudden, the initiation of a crack being immediately followed by its propagation to fracture. This situation, known as brittle fracture, may occur in steel when the temperatures are low, the metal is in a condition of low fracture ductility and the stress system is able to attain a high level momentarily. It has rarely been observed in British power plant experience but remains a matter for research in connection with the design of low-pressure turbine rotors of future large turbines and the need to ensure the integrity of pressure vessels such as boiler drums.

Failure of a component may follow visual evidence of a crack partially penetrating into a metal thickness. The rate of crack propagation and the ultimate possibility of fracture are matters of judgement. To reach such a decision demands knowledge of the stress and temperature conditions to which the metal is subject, its environment and its likely mechanical properties at the time.

Simple reduction of metal sections, for example, by progressive corrosion of a superheater tube wall, may lead to mechanical failure as the applied stress increases.

If material selection has been correctly based upon the pertinent properties, and operating conditions maintained within the limits of a sound design, then the component under consideration will give satisfactory service during the calculated life period. One of the features of modern power plant is the requirement to operate it on an intermittent basis, with perhaps one or two starts a day and weekend shutdown. For those components that have been designed to operate under creep conditions, it is usually assumed that the strain applied to the metal arises principally from the nominal design stress. In practice this strain may be augmented significantly by stress-concentrating features in a component or by strain engendered during plant start-up or shut-down. Stress analysis of these complex situations is a matter of current research activity.

In the following paragraphs the instances of metallurgical failure in power plant will be discussed against particular items. In most cases the examples refer to evidence of cracking rather than complete rupture of the metal. Where it is appropriate the relevance of initial assessment criteria will be mentioned.

5.6.1. **Boiler Plant**

5.6.1.1. Boiler Tubes

These tubes, usually made from carbon steel, are subject to:

(a) High rates of heat transfer.

(b) Bending stresses due to uneven heating, especially at expanded or welded joints into headers or drums.

(c) External erosion from burners and flue gas.

(d) Possible corrosion on the boiler water side.

(e) Occasional manufacturing defects.

(a) High thermal ratings may lead to rapid failure if the internal fluid flow is reduced for any reason. The resultant overheating leads to a failure by creep, characterised by the bulging of the tube with the eventual development of a longitudinal split.

(b) Fatigue cracking due to bending stresses has been observed associated with change of section and/or weld undercut, where tubes have been expanded or welded into headers. The inherent flexibility of a long tube and its relatively simple geometry appear to minimise the incidence of cracking within the tube length due to bending stresses.

(c) Failure may arise due to overstressing of a reduced section of metal.

(d) This has been discussed in a previous chapter (Chapter 2), dealing with "on-load" corrosion, and mentioned in this chapter in Section 5.3.5. It is worth recalling, however, that an unpleasant characteristic of this problem—sudden failure of the boiler tube—arises from embrittlement of the carbon steel due to interaction between atomic hydrogen from the corrosion process and the iron carbide present in the steel. The end product of this reaction is a brittle ferrite of poor mechanical strength. A small addition of a carbide-stabilising element such as chromium to carbon steel may be expected to reduce the risk of embrittlement, and hence violent failure, but not necessarily to prevent the corrosion of the metal. In order to obtain operational experience of the value of using low-alloy steels to combat embrittlement due to corrosion, sample tubes of various alloys containing from 1% to 9% of chromium are to be introduced into some existing boilers. In a recently built boiler, 1% Cr, ½% Mo steel replaces carbon steel in the furnace chamber.

(e) Defects in tube manufacture, although far from being a regular occurrence, can be a cause of serious trouble. Lamination in boiler tubes or score marks arising from the cold drawing of tubes, have given rise to premature failure and may promote corrosion at these regions. This is a quality control problem and it is hoped that advances in the use of non-destructive testing techniques will contribute to its early solution.

5.6.1.2. Boiler Drums

Although some riveted boiler drums are still in existence in older plant, this form of construction has now been entirely supplanted by welding. An early and recurrent problem of riveted drum boilers—so-called "caustic cracking"—deserves mention as one of several examples of stress corrosion cracking in power plant metals.

In this case cracking was initiated at the stress concentration provided by the rivet holes and propagated in an intergranular manner. The concentration of boiler salts, especially caustic soda, due to evaporation at slightly defective seams, provided the corrosive agent. This mechanism led to one boiler explosion and to the setting up of non-destructive testing teams to cope with the detection of cracks in other riveted boiler drums. Ultrasonic methods were developed to detect cracking in the shell below the butt straps, followed by magnetic testing of the plate adjacent to the holes, if removal of the rivets had been judged necessary.

Until recently there had been no example in this country of a potentially catastrophic failure of a welded boiler drum. A recent instance of cracking leading to failure during an over-pressure test prior to operation (see ref. 4) emphasises that the fillet welding of higher strength alloy steel in thick section demands strict attention to design detail and thorough post-fabrication inspection of the weld area, if the risk of fast fracture is to be completely avoided. Some instances of cracking have also been recorded, however, associated with manhole reinforcing rings. A heavy annular ring was welded around the manholes, setting up in the geometries concerned, a system of high residual stresses in the carbon steel that eventually obtained stress relief by cracking. The annular ring was in some cases fitted inside the drum making it very difficult to preheat the weld adequately, thus worsening the situation. An example of the original design of reinforcing ring and a replacement preformed drum end forging are shown in Figures 5.6.1.2(a) and (b).

5.6.1.3. Superheater Tubes

Superheater tubes are subjected to the most severe combination of stress, temperature and corrosive environment of any item of power plant. In modern land boilers their metal temperature, at least 100°F higher than the associated steam temperature, is always high enough to require the use of creep-resistant steels. Reference to Table 1 gives their chemical composition. In addition to high-temperature strength, resistance to corrosion is also important. For example, low-alloy ferritic steel such as $2\frac{1}{4}\%$ Cr, 1% Mo would not be used at metal temperatures above 580°C (1080°F) because of inadequate resistance to corrosion and oxidation over a full service life of 100,000/150,000 h.

Failures in superheater tubes may arise from:

(a) Prior fabrication history.
(b) Faulty heat treatment.
(c) Consequences of welding.
(d) Overheating of the tube metal.
(e) Gas-side corrosion.
(f) Stress corrosion (austenitic steels).

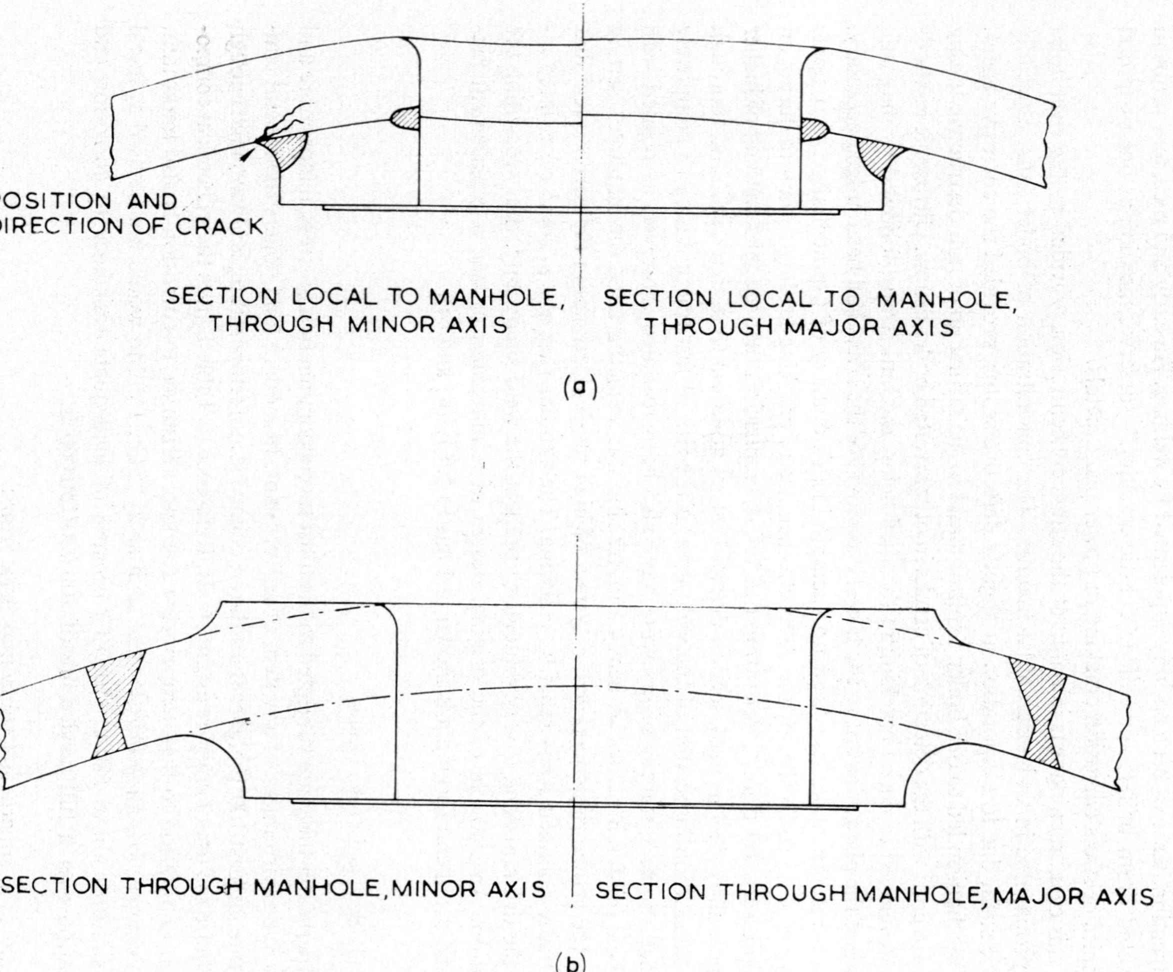

FIG. 5.6.1.2. Boiler drum reinforcement at manhole. (a) Original method of manhole reinforcement. (b) Modified (forged) section with integral stiffening

FIG. 5.6.1.3A. Rupture of cold bent 18 Cr, 12 Ni, 1 Nb austenitic superheater tube

(a) It is common practice to cold bend superheater tube elements to their final shape. The quantitative effect of this process upon the long-life performance of creep-resisting steels is still unknown. Figure 5.6.1.3A shows the appearance of a sudden failure at a cold-bent 18% Cr, 12% Ni, 1% Nb austenitic superheater tube after 43,500 h service.

In this example it is probable that cold forming during fabrication, by reducing the high temperature ductility of the steel, had reduced its creep rupture life.

(b) If heat treatment schedules are not adhered to fully it is possible that the high-temperature strength will be reduced. This problem has been met in American power plant practice with both a 1% Cr, $\frac{1}{2}$% Mo ferritic and an 18% Cr, 8% NiTi austenitic superheater tube steel. Similar cases are rare in British practice, and in this type of failure there is usually some warning of trouble due to the deformation that occurs prior to rupture.

(c) The consequences of welding upon metal properties have been discussed in regard to fabrication. Butt welds in superheater tubes are not usually a location of failure. Fillet welds, either in attaching superheater tubes to headers or attaching spacers and hangers to superheater tubes, are more susceptible locations. Many examples of cracks initiating at the toe of an attachment weld and propagating through the wall of a tube have been reported. Figure 5.6.1.3B is an example of a reheater tube failure where the tube itself is particularly vulnerable because of the relatively thin-walled tube used. The design of attachments of all types is presently under critical examination.

(d) Local overheating due to steam starvation may, as in boiler tubing, give rise to a premature creep rupture failure.

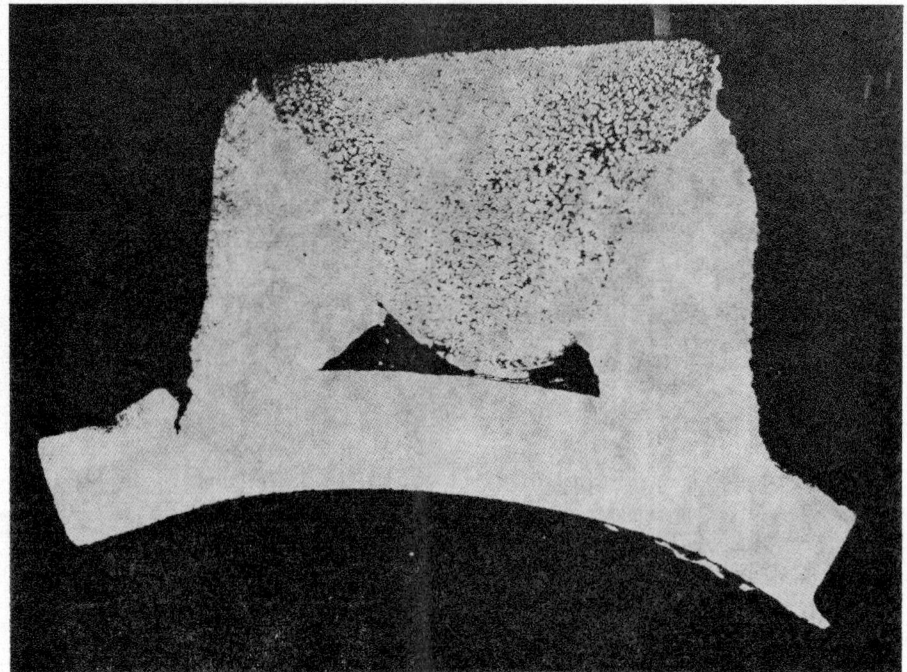

FIG. 5.6.1.3B. Reheater tube attachment weld, showing crack propagating into tube (×5)

(e) Gas-side corrosion due to attack of the metal by the products of coal or oil combustion, has been discussed in a previous chapter. Apart from the relationship between the chromium content of steel and its resistance to corrosive environments at various temperatures, as outlined earlier in this chapter, there is no uniquely metallurgical contribution to be made to this problem at the present time.

A solution to the problem is being sought by the use of clad tubes of which the outer portion is of higher chromium content, and hence more corrosion-resistance, than the steel of the inner portion which is conventional superheater material, chosen primarily for strength at high temperature. Research and site trials are being made.

(f) Austenitic steels are very susceptible to a form of stress corrosion cracking in the presence of water containing chloride or hydroxide. This has very rarely been met in superheater tubing due to the physical distance of the austenitic steel from the evaporation zone and the high degree of superheat (about 180°C) before the austenitic steel tubing is reached. The few examples encountered have been attributed to chloride contamination of the metal surface prior to commissioning of new plant, or to a detail of mal-operation in the early life of the plant.

5.6.2. Steam Pipework and Valves

The incidence of cracking in these components may be considered under the headings of:

(a) Corrosion fatigue.

(b) Thermal fatigue.

(c) Creep rupture arising from fabrication procedures.

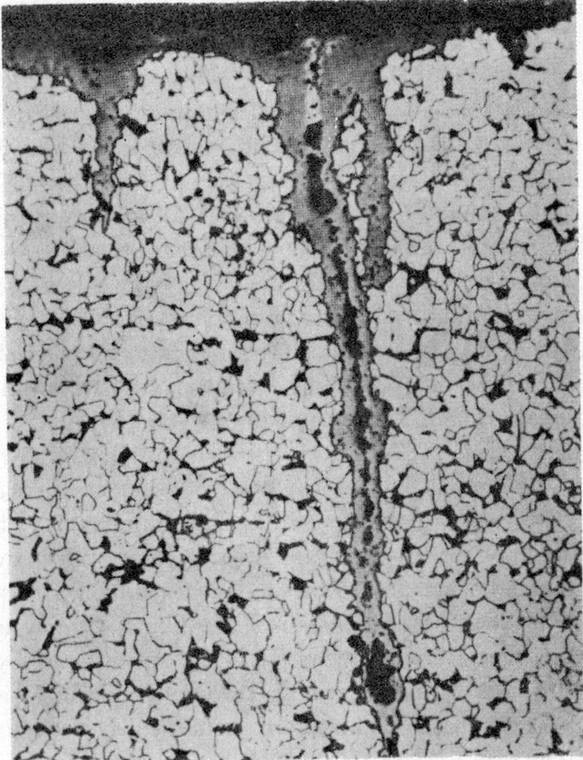

FIG. 5.6.2A. Corrosion fatigue cracks in steam receiver (×135)

(a) Corrosion fatigue is yet another example of cracking initiated by a combination of stress and corrosive environment. Of historical interest, it occurred in power plant operating with steam conditions up to 600 lb/in², 454°C. Steam receivers made of carbon steel were a frequent location of trouble since they often acted as a common drainage point. Typically, water at 252°C (485°F) would impinge on metal at 427°C (800°F) causing it to be strained above its yield stress. Frequent repetition of this occurrence led to transgranular cracks of the type shown in Figure 5.6.2A. Readers interested in this problem of older power plant should consult the reference given at the end of this chapter (see ref. 1).

(b) Thermal fatigue as defined earlier in this chapter involves both creep and fatigue processes. The reader should appreciate that such a broad definition acknowledges some uncertainty about the mechanisms involved, so that any classification of failures within this category is tentative. One example has been noted in some designs of governor and emergency stop valves. In these, made of $\frac{1}{2}$% Mo, $\frac{1}{4}$% V ferritic steel, cracks were initiated in the heads between the balance holes. These were attributed to stresses associated with the non-uniform heating and cooling of the thin webs and relatively massive outer rim. The crack path was transgranular.

(c) The discussion of fabrication problems earlier in this chapter has emphasised the close relation between forming and welding processes and the resultant mechanical properties of the metal. An important point to note is that materials embrittled by fabrication procedures may survive fabrication without visual evidence of cracking, only to initiate cracks prematurely during service.

½% Mo ferritic steel and 18% Cr, 12% Ni, 1% Nb austenitic steel are two examples of such metals. The history of cracking in ½% Mo steam piping is well documented, the report from the Electrical Research Association (see ref. 2) listed in the references being required reading for those concerned with this type of problem. Cracks were often circumferential and generally located at changes in section (for example, at Corwel joints) or adjacent to fillet welds. Initiation of cracks can be attributed simply to stress concentration effects arising from particular geometries and/or the thermal effects of welding upon a steel incapable of redistributing strain without first cracking. Some years ago a similar type of cracking took place in 1% Cr, ½% Mo steam pipe subsequent to the welding on of a slab-type flange, using ½% Mo weld metal. In this case simplification of the geometry overcame the problem. Figure 5.6.2B shows the original (a) and replacement (b) designs of weldment.

Cracking in 18% Cr, 12% Ni, 1% Nb steel is also well documented; the paper by C.E.G.B. authors (see ref. 3 and Section 5.4.3.6 (v)) listed in the references gives a concise treatment of the problem and steps taken to overcome it. In this example it is to be noted that crack initiation has always been observed—for main steam piping—in the circumferential direction. This implies that the direction of the initially dominant stress system has been longitudinal and therefore contributed by stressing other than the simply derived design stress.

The underlying cause of cracking—a metal put into a condition of low ductility, unable, at operating temperature, to dissipate strain peaks without cracking—is common to both these steels. It must be remembered that creep-resistant metals are designed to deform very slowly so that their tolerance to the above engineering situations must be limited. All the cracks detected in these ferritic and austenitic steels followed an intergranular path.

5.6.3. **Turbo-generator Plant**

5.6.3.1. TURBINE CYLINDER CASINGS

Casings are usually made from castings. They are thus subject to surface and shrinkage defects, recourse often being made to welding to restore their integrity. The thermal treatment implicit in these repair procedures will tend to reduce the available high-temperature ductility in the cast parent metal. The rate of temperature rise associated with initial steam flow at start-up, along with the stress-concentrating effects of component geometry and surface finish, may be sufficient to induce permanent (plastic) strain in critical areas.

In practice it is found that the regions most prone to crack initiation are inlet pipes and inlet belt areas, particularly in those casings made in cast ½% molybdenum steel.

The reader will appreciate that this is another example of cracking falling within the general classification of thermal fatigue. Current theory indicates that by a combination of creep and fatigue mechanisms, crack initiation is due to cumulative plastic strain. The analysis of the problem as applied to turbine casings requires careful measurement of temperatures, temperature gradients and rate of temperature increase, in order to get some measure of the strains that may be developed. The significance of this data must then be assessed in relation to the properties of the metal being used.

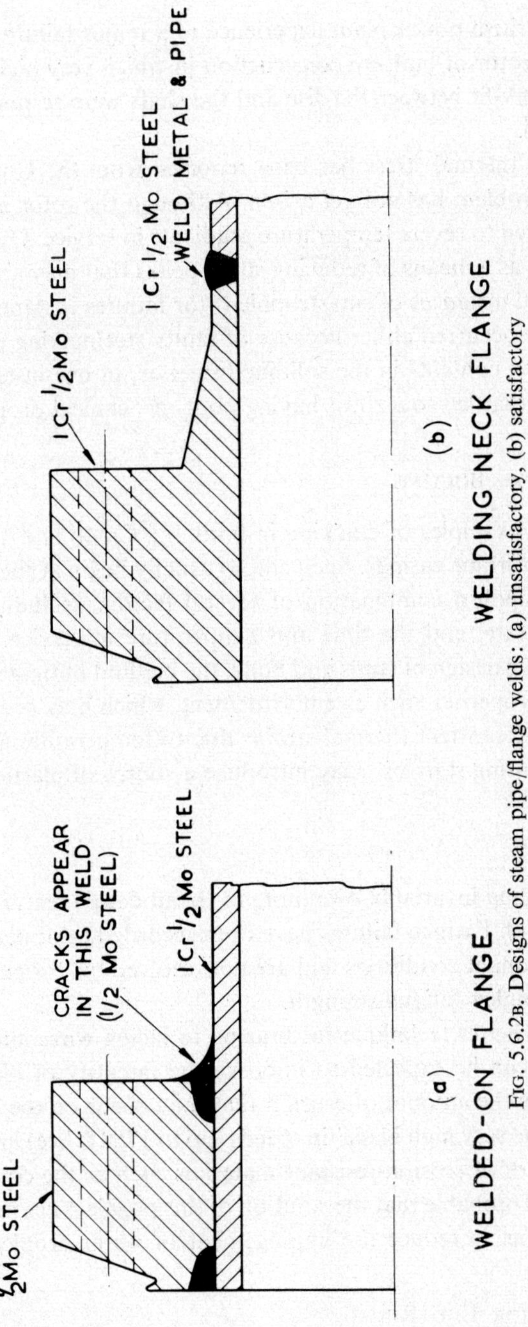

FIG. 5.6.2B. Designs of steam pipe/flange welds: (a) unsatisfactory, (b) satisfactory

5.6.3.2. Turbine Rotors

The only instance in British power plant experience of a major failure of a turbine rotor concerns a low-pressure rotor of built-up construction in which very high stress concentrations at the point of shrink-fit between the disc and the shaft were responsible for a classic fatigue failure (see ref. 5).

A rotor failure due to thermal stress has been reported from the United States, but in British practice such a problem has not yet arisen. Although the rotor is symmetrical and evenly heated it is subjected to severe temperature gradients in service. Over-reliance should not be placed upon creep as a means of reducing stress peaks that may arise in this manner.

There have been several instances of catastrophic rotor failures in American power plant in the past decade. These occurred either because of faulty steelmaking practice which left undesirably high hydrogen contents in the solidified steel or, in one instance, because of a choice of heat treatment that led to a rotor having poor mechanical properties (see ref. 6).

5.6.3.3. High-temperature Bolting

There have been some examples of cracking in both 1% CrMo and 1% CrMoV studs, principally for valve and turbine casings. Such cracks usually begin at the thread root. This type of failure depends upon a combination of several factors, including the amount of plastic strain, the strain rate, and the time and temperature in service. This necessitates careful consideration of the design of studs and bolts, the method of tightening or retightening, and the changes in properties such as embrittlement, which may occur in the material under service conditions. Transient thermal strains due to temperature difference between flange and stud or bolt during start-up, may introduce a source of plastic strain.

5.6.3.4. Turbine Blading

Failures in turbine blading invariably owe more to detail design features than to factors under metallurgical control. Fatigue failures have been recorded at or near the blade roots but are usually due to resonant conditions and are not resolved by, for example, the choice of a metal of somewhat higher fatigue strength.

Indirectly, by applying better techniques of brazing to lacing wires and erosion shields, metallurgical knowledge can be expected to improve the integrity of blading assemblies.

A remaining problem is the amount of erosion that may occur on the l.p. blades of 500 MW turbines in view of the very high blade tip speeds (up to 1780 ft/sec) involved. Although the metallurgical properties of erosion-resistant materials such as the cobalt-based Stellite are being studied, it seems probable that the solution of this problem lies in the development of engineering design so as to reduce the impingement of water droplets on the blades.

5.6.3.5. Alternator Rotor End Rings

The cylinder that encloses the coils of the turbo-alternator is usually made of non-magnetic steel to avoid electrical losses. These are made from one or other of the following iron-based alloys:

C%	Mn %	Ni %	Cr%
0·7	8	8	4
0·5	18	—	4

This component is subject to high stresses (about 20 tons/in^2) both during rotation and when at rest. There has thus been a need for material of high yield strength, values of between 50 and 60 tons/in^2 being obtained with corresponding tensile strengths of about 57 and 67 tons/in^2. To achieve these mechanical properties it is necessary to work-harden the metal: this means that considerable internal stresses may be left in the metal and this has led to two problems.

One of these has been the general susceptibility of this alloy to a usually mild form of stress corrosion cracking illustrated in Figure 5.6.3.5. The stress factor arises mainly from the internal stresses remaining from the original manufacturing method and to a lesser extent from the effects of machining the ring to final size. Corrosive environment, especially due to chloride, may arise in storage conditions in manufacturers works, in power stations prior to commissioning and, in a few cases, by exposure to chloride during operation.

A combination of improved steelmaking techniques and corrosion protection of the end rings prior to their installation into alternators has overcome this difficulty.

The second problem has been the possibility of castastrophic failure of the end ring. One example (see ref. 7) in a Canadian power station was attributed to the stress-raising consequences of ventilation holes drilled in the non-magnetic steel, already heavily cold-worked to achieve the required mechanical properties. Another example, in a British power station, involved an end ring without ventilation holes. In this instancc the stress concentration necessary to initiate and propagate the failure emanated from a region in the end ring itself. The mechanical properties of this end ring showed it to possess a yield stress very close to its ultimate tensile stress, i.e. with little possibility of plastic deformation acting to reduce dangerous peak stresses.

The approach necessary to avoid any repetition of such dangerous failures is to ensure a margin between the stress for yield and that for fracture, adequate to dissipate local stress concentrations and to adjust the metallurgical conditions so that the metal possesses a good level of tensile ductility.

5.7. CONCLUDING REMARKS

This chapter has considered the role of metallurgy in the power industry as it affects the design, construction and operation of plant.

In principle this knowledge should be applied to the design and construction of components so that operational problems caused by some degree of metallurgical inadequacy do not arise. In practice they do and much may be learned from these situations, from which the feedback of information is of vital importance to the designer and constructor of future plant. If this can be achieved the selection criteria for metals will be made more relevant, inspection techniques modernised and the operational history of future power stations less beset by outages stemming from metallurgical factors.

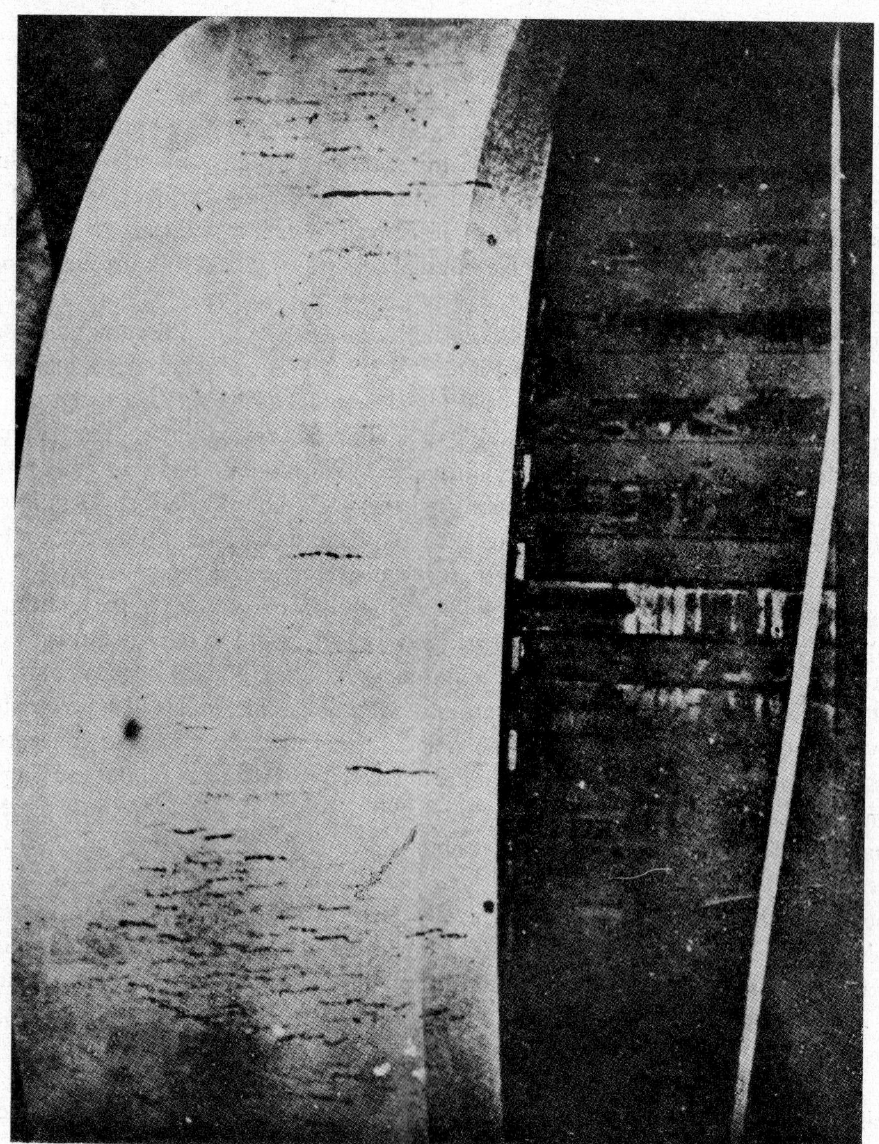

FIG. 5.6.3.5. Rotor end bell showing stress corrosion cracking

TABLE 8

THE SYMBOLS OF THE ELEMENTS OF METALLURGICAL INTEREST

Symbol	Element	Symbol	Element
Ag	Silver	Mg	Magnesium
Al	Aluminium	Mn	Manganese
As	Arsenic	Mo	Molybdenum
Au	Gold	N	Nitrogen
B	Boron	Na	Sodium
Ba	Barium	Nb	Niobium
Be	Beryllium	Ni	Nickel
Bi	Bismuth	O	Oxygen
C	Carbon	P	Phosphorus
Ca	Calcium	Pb	Lead
Cd	Cadmium	Pt	Platinum
Cl	Chlorine	Ra	Radium
Co	Cobalt	Rh	Rhodium
Cr	Chromium	S	Sulphur
Cs	Caesium	Sb	Antimony
Cu	Copper	Si	Silicon
Fe	Iron	Sn	Tin
H	Hydrogen	Ti	Titanium
He	Helium	V	Vanadium
Hg	Mercury	W	Tungsten
Ir	Iridium	Zn	Zinc
K	Potassium	Zr	Zirconium

GLOSSARY OF WELDING TERMS

Crater — in arc welding, a depression at the termination of the bead or in the weld pool beneath the electrode.

Deposited metal — filler metal that has been added during welding operation.

Deposition rate — the weight of metal deposited in unit time.

Downhand welding — the position of welding where it is performed from the upper side of the joint and the face of the weld is approximately horizontal.

Electrode (arc welding) — a "stick" or "wire" from which an arc is struck. It may include bare metal electrode, carbon electrode, composite electrode, covered electrode, lightly coated electrode.

Electrode (resistance welding) — that part or parts of a machine through which the welding current and in most cases, pressure, are applied directly to the work. The electrode may be in the form of a rotating wheel, bar, cylinder, plate, clamp or modification thereof.

Electrode holder — a device used for mechanically controlling the electrode and conducting current to it.

Faying surface — that surface of a member which is in contact with another member to which it is to be joined.

Filler metal — metal to be added in making a weld.

Flash — the molten metal which is expelled or which is squeezed out by the application of pressure, and solidifies around the weld.

Fusion zone — the area of base metal melted, as determined on the cross-section of a weld.

Heat-affected zone — that portion of the base metal which has not been melted but whose mechanical properties or microstructure have been altered by the heat of welding or cutting.

Machine welding — welding with equipment which performs the welding operation under the observation and control of an operator. The equipment may or may not perform the loading and unloading of the work.

Overhead position — the position of welding wherein welding is performed from the underside of the joint.

Parent or base metal — the metal to be welded or cut.

Under-penetration — occurs where the weld metal does not adequately fuse the opposing surface at a weld joint.

Over-penetration — occurs where the weld metal protrudes excessively from the opposing surface of the joint. Known as "icicles" if severe in tube welding.

Porosity — gas pockets or voids in metal.

Preheat temperature — the temperature specified that the base metal must attain in the welding or cutting area, immediately before these operations are performed.

Residual stress — stress remaining in a structure or member as a result of thermal or mechanical treatment or both.

Stress-relief heat treatment — uniform heating of a structure or portion thereof to a sufficient temperature below the critical range, to relieve the major portion of the residual stress, followed by uniform cooling. (*Note.* The terms "normalising", "annealing", etc., are misnomers for this application.)

Vertical position — the position of welding wherein the axis of the weld is approximately vertical.

Weld — a localised coalescence of metal wherein coalescence is produced by heating to suitable temperatures with or without the application of pressure and with or without the use of filler metals. The filler metal either has a melting point approximately the same as the base metals or has a melting point below that of the base metals but above 800°F.

Weldability — the capacity of a metal to be welded under the fabrication conditions imposed, into a specific suitably designed structure, together with the ability to perform satisfactorily in the intended service.

REFERENCES

Corrosion-fatigue Cracking in Steam Pipe Systems at Power Stations. B.E.A. Report, May 1949.

Cracking in Service of 0·5% Mo Steam Pipe. E.R.A. Report, J/T 159.

Cracking in Welded Joints of Austenitic Steel in Power Stations. F. Asbury, B. Mitchell and L. Toft, *British Welding Journal*, November 1960

Report on the Brittle Fracture of a Boiler Drum at Cockenzie Power Station. South of Scotland Electricity Board Report, January 1967

Fatigue Strength of Turbine Shafts with Shrunk-on Discs. M.B. Coyle and S.T. Watson, *J. Inst. Mech. E.*, October 1963.

Large Rotor Forgings for Turbines and Generators. Mochel, Peterson, Conrad and Gunther, *A.S.M.E.*, Vol. 78, No. 7, October 1956.

Report on the Investigation into the Failure of Two 100 MW Turbo-generators, Sir C. Gibb, *J. Inst. Mech. E.*, February 1955.

RECOMMENDED READING

Metallurgy for Engineers, by E. C. Rollason, Arnold.
The Technology and Properties of Ferrous Alloys for High Temperature Use, by M. G. Gemmill, Newnes, 1966.
Metallurgy of Welding, by D. Seferian, Chapman and Hall.
Metallurgy of Welding, Brazing and Soldering, by J. F. Lancaster, George Allen & Unwin Ltd.
Welding Handbook. American Welding Society, Cleaver-Hume Press (5 volumes).
Non-Destructive Testing Handbook. R. C. McMaster, Ed. Ronald Press Co.
A Short Guide to Non-destructive Testing (Compiled from Buxton Lectures). H. L. Carson, Published by C.E.G.B. (Education and Training Department).

EXERCISES

1. Describe the changes which occur in the structure of an alloy of iron and 0·2% carbon, when cooling from the molten state under equilibrium conditions. Sketch the relevant part of the iron–carbon equilibrium diagram to illustrate your answer.

2. Discuss the effect of cooling rate on the solid state transformation characteristics of a 0·2% carbon steel, and indicate the principal effects of alloying elements on this behaviour.

3. Give a simple explanation of "creep" in steel and explain its underlying significance to the successful performance of many power plant components.

4. Describe three different welding techniques and indicate examples of their application to power plant.

5. Show how non-destructive testing techniques can be used to assess the integrity of a high-pressure turbine casing.

6. Select two examples of power-plant component failures to illustrate
(a) an unsuspected design inadequacy,
(b) a difficulty due to the operating regime.
Suggest metallurgical remedies to the problems you have selected.

7. Two possible causes of metal failure are brittle (fast) fracture and cracking due to creep. Discuss these modes of failure and mention one example of each from power plant experience.

8. Outline the principal properties required of a high-temperature steel for:
(a) rotating machinery, and
(b) heat exchangers.

CONTENTS OF VOLUMES 1–8

Volume 1. Planning and Layout

1. Siting and Design of Power Stations
2. Civil Works: Foundations for Building and Plant
3. Civil Works: Superstructure
4. Main Plant Layout
5. Electrical Plant Layout

Volume 2. Mechanical—Boilers; Fuel- and Ash-handling Plant

1. Boiler Unit—Furnaces and Combustion Equipment
2. Boiler Unit—Steam and Water Circuits
3. Boiler Unit—Ancillary Plant and Fittings
4. Boiler Unit—Dust Extraction and Draught Plant
5. Fuel- and Ash-handling Plant

Volume 3. Mechanical—Turbines and Auxiliary Equipment

1. Turbines
2. Turbine Auxiliary Systems
3. Feed Heating Systems
4. Condenser and Cooling Water Plant
5. Miscellaneous Station Services, Pipework and Plant

Volume 4. Electrical—Generator and Electrical Plant

1. The Generator
2. Main Switchgear
3. Electrical Auxiliary System and Equipment
4. Transformers, Reactors and Power Cables
5. Protective Gear and Voltage Regulators

Volume 5. Chemistry and Metallurgy

1. Fuel and Oil
2. Corrosion. Feed and Boiler Water
3. Water Treatment Plant. Cooling Water Systems
4. Plant Cleaning and Inspection
5. Metallurgy and Inspection

Volume 6. *Instrumentation, Controls and Testing*

1. Boiler and Turbine Instrumentation
2. Electrical Instrumentation and Metering
3. Plant Testing
4. Control Systems
5. Automation, Data Loggers and Computers

Volume 7. *Operation and Efficiency*

1. Commissioning and Operation of Boilers and Turbo-generators
2. Electrical Performance and Operation of Turbo-generators and System Operation
3. Safety and Maintenance
4. Practical Aspects of Maintenance
5. Plant Optimisation

Volume 8. *Nuclear Power Generation*

1. Nuclear Physics
2. Nuclear Power Reactors
3. British Nuclear Power Plant
4. Nuclear Power Station Operation
5. Nuclear Techniques